Current Topics in Microbiology 206 and Immunology

Springer-Verlag
Berlin Heidelberg GmbH

Transgenic Models of Human Viral and Immunological Disease

Edited by F.V. Chisari and M.B.A. Oldstone

With 53 Figures

 Springer

Francis V. Chisari, M.D.
Member, Department of Molecular
and Experimental Medicine
Head, Division of Experimental Pathology
The Scripps Research Institute
10666 North Torrey Pines Road
La Jolla, CA 92037, USA

Michael B.A. Oldstone, M.D.
Member, Department of Neuropharmacology
Head, Division of Virology
The Scripps Research Institute
10666 North Torrey Pines Road
La Jolla, CA 92037, USA

Cover illustration: Artists conception of the hepatitis B virus (upper left), and a lymphoid cell (upper right), a fertilized ovum in the process of microinjection (center), and a transagenic mouse (lower).

Cover design: Künkel+Lopka, Ilvesheim

ISBN 978-3-642-85210-7   ISBN 978-3-642-85208-4 (eBook)
DOI 10.1007/978-3-642-85208-4

© Springer-Verlag Berlin Heidelberg 1996
Library of Congress Catalog Card Number 15-12910
Softcover reprint of the hardcover 1st edition 1996

Typesetting: Thomson Press (India) Ltd, Madras
SPIN: 10495396   27/3020/SPS – 5 4 3 2 1 0 – Printed on acid-free paper

Preface

The development of transgenic animal technology was a seminal event which created the opportunity to examine in vivo an extraordinarily wide variety of biological questions that were previously only approachable in vitro. The ability to selectively add or delete defined genes in a living animal represents a quantum advance in the armamentarium of the experimental biologist, the implications of which, notable as they are, are just beginning to be felt.

The application of this technology by a large and growing community of investigators has yielded important insights into fundamental areas of biology as diverse as gene regulation and imprinting, mammalian development, immunological tolerance, and carcinogenesis. It has also made it possible to examine many previously unapproachable questions relevant to the pathogenesis of numerous metabolic, degenerative, neoplastic, viral and autoimmune diseases for which appropriate models did not previously exist. The availability of these systems has also made it possible to examine, for the first time, the complex cascades of events that can be triggered in vivo in response to the novel expression of a single viral or mammalian gene that are not approachable in vitro.

The importance of the collective contributions of the community of scientists who have used this technology to advance the understanding of mammalian biology cannot be overstated, nor can the indebtedness of the scientific community at large to our colleagues who pioneered this technology little more than a decade ago.

The current volume contains a representative sampling of the many ways in which transgenic technology has been applied to investigate the pathogenesis of human viral and immunological diseases. The contributors are leading investigators in the use of transgenic technology for these purposes. The chapters cover a broad assortment of subjects in this area of investigation.

Several chapters describe the use of transgenic mice to examine immunological tolerance and the basis for autoimmunity. For example, the chapter by Peter Doherty reviews what has been learned about T cell effector and memory function in viral immunity from mice in which perforin, class I, and selected cytokine genes have been deleted by gene targeting technology.

Dennis Zaller and Victor Sloan summarize the use of T cell receptor transgenic mice to study mechanisms of central and peripheral tolerance and the use of myelin basic protein T cell receptor transgenic animals to examine the CD4 T cell dependence and immunopathogenesis of experimental autoimmune encephalitis.

Richard Flavell and his colleagues review the consequences of transgenic overexpression of certain viral proteins and cytokines such as tumor necrosis factor alpha and lymphotoxin in the pancreatic beta cells on the development of insulitis and diabetes mellitus.

Robyn Slattery and Jacques Miller continue this story by reviewing evidence that the transgenic expression of I-E and I-A class II genes can modulate the susceptibility of the NOD mouse to diabetes, and they speculate on the mechanisms responsible for protection against diabetes in these animals.

Michael Oldstone and colleagues use targeted expression of lymphocytic choriomeningitis virus proteins in pancreatic beta cells and oligodendrocytes to study the immunological factors and pathways that might be involved in the pathogenesis of diabetes mellitus and multiple sclerosis.

In related studies, Chella David and colleagues review the expression and function of HLA-B27 in transgenic mice, and they describe the characteristics of an animal model of ankylosing spondylitis in HLA-B27/human β-2-microglobulin double transgenic mice.

Another series of chapters describes the cytopathic and antiviral effects of cytokines and the cellular genes they induce. For example, the pathological and protective effects of the overexpression of interferon-γ and several other inflammatory cytokines in a diverse variety of tissue sites are reviewed in the chapter by Kathrin Geiger and Nora Sarvetnick.

In another chapter, Heinz Arnheiter, Otto Haller and their colleagues describe the role of early host defense mechanisms and intracellular immunization pathways in viral pathogenesis using mice transgenic for the interferon inducible antiviral *Mx1* gene.

The final series of chapters examines the creation of transgenic models of viral and virus-like diseases for which traditional models do not exist. For example, Frank Chisari reviews the development of a series of transgenic models of viral hepatitis and hepatocellular carcinoma using subgenomic fragments and full length copies of the hepatitis B virus genome.

Michael Nerenberg and colleagues describe the development of mice that express the HTLV-1 *tax* gene to study multiple aspects of HTLV-1 latency and malignant transformation.

Paul Klotman and Abner Notkins summarize the contributions of transgenic animal models to our understanding of HIV-1 pathogenesis, especially Kaposi's sarcoma and HIV-associated nephropathy, and they provide insight into the basis for tissue restricted expression of CNS strains of HIV-1 and the function of the *nef* gene in vivo.

Stephanie Toggas and Lennart Mucke show that targeted expression of HIV-1 gp120 and interleukin-6 cause histopathological and pathophysiological changes in the brain that resemble those seen in AIDS dementia and suggest the direction for future interventional studies in this disease.

Adriano Aguzzi and colleagues use transgenic technology to demonstrate the pathogenetic potential of human foamy virus, predicting the kind of disease that could be caused by an agent that has not yet been shown to cause disease in humans.

Stanley Prusiner reviews the transgenic approaches he has used to understand the biology and pathogenesis of a family of neurodegenerative diseases caused by an infectious, nonviral, virus-like agent consisting of a modified form of a host-encoded prion protein.

Vincent Racaniello and Ruibao Ren summarize their studies of poliovirus biology and pathogenesis using transgenic mice that express the poliovirus receptor, the first example of the effectiveness of this approach to develop animal models of human viral diseases.

Finally, the volume ends with a review by Y. Ghendon and Paul-Henri Lambert of transgenic models capable of producing complete virus particles that could be infectious for humans, with recommendations for the safe maintenance and distribution of such animals, using the poliovirus receptor transgenic mice as an example.

These chapters convey both the exciting work being done by these investigative teams and the power of the technology they are using to examine many previously unapproachable questions in the areas of viral and immune pathogenesis. As impressive as these advances may be, we expect that they are only the beginning of an avalanche of discovery enabled by transgenic technology that will extend well into the next century and beyond.

La Jolla, California

Francis V. Chisari
Michael B.A. Oldstone

List of Contents

List of Contributors

(Their addresses can be found at the beginning of their respective chapters.)

AGUZZI, A. 243
ARNHEITER, H. 119
BROWN, D.A. 175
CHISARI, F.V. 149
DAVID, C.S. 85
DOHERTY, P.C. 1
EVANS, C.F. 67
FLAVELL, R.A. 33
FRESE, M. 119
GEIGER, K. 101
GHENDON, Y. 327
HALLER, O. 119
HERRATH, M. VON 67
HORWITZ, M.S. 67
KAMBADUR, R. 119
KHARE, S.D. 85
KLOTMAN, P.E. 197
KRATZ, A. 33
LAMBERT, P.-H. 327
LUTHRA, H.S. 85

MARINO, S. 243
MEIER, E. 119
MILLER, J.F.A.P. 51
MUCKE, L. 223
NERENBERG, M. 175
NOTKINS, A.L. 197
OLDSTONE, M.B.A. 67
PRUSINER, S.B. 275
RACANIELLO, V.R. 305
REN, R. 305
RETHWILM, A. 243
RUDDLE, N.H. 33
SARVETNICK, N. 101
SLATTERY, R.M. 51
SLOAN, V.S. 15
TOGGAS, S.M. 223
TSCHOPP, R. 243
XU, X. 175
ZALLER, D.M. 15

Cytotoxic T Cell Effector and Memory Function in Viral Immunity

P.C. Doherty

1 Introduction

The first evidence that virus-specific CD8⁺ cytotoxic T lymphocytes (CTLs) are primary effectors in vivo came shortly after the discovery of major histocompatibility complex (MHC)-restricted CD8⁺ CTL activity (Zinkernagel and Doherty 1974). Adoptive transfer of immune spleen cells from mice of MHC type A into A or (AxB) F1, but not B, virus-infected recipients caused massive inflammatory pathology (Doherty and Zinkernagel 1975; Doherty et al. 1976b) and acute elimination of the pathogen (Blanden et al. 1975a; Zinkernagel and Welsh 1976). This was repeated for situations in which A and B represented different MHC haplotypes or distinct allelic forms of class I MHC glycoproteins (Doherty and Zinkernagel 1975; Blanden et al. 1975b; Doherty et al. 1976a). The analysis was extended to H-2 mutant mice and, again, the spectrum of cross-reactivity for T cell effector function in vivo and CTL activity in vitro seemed to run in parallel (Doherty et al. 1976a; Zinkernagel and Doherty 1979).

The next phase was the development of cloned CD8⁺ T cell lines, which were shown both to have CTL activity and to clear influenza virus from the lung

Department of Immunology, St. Jude Children's Research Hospital, 332 North Lauderdale, Memphis, TN 38105, USA

following adoptive transfer (TAYLOR and ASKONAS 1983). Giving such potent effectors i.v. to carrier mice with disseminated lymphocytic choriomeningitis virus (LCMV) (see Oldstone, this volume) caused acute death from viral pneumonia, reflecting that activated T cells tend to lodge in the infected lung (BYRNE and OLDSTONE 1986). Similarly, injection of small numbers directly into the brains of LCMV-infected mice induced the lethal immunopathology characteristic of this disease, though with little evidence of inflammation (BAENZIGER et al. 1986). Clearly, virus-specific CD8$^+$ T cells were primary mediators of both severe pathology and virus elimination. However, though the results were consistent with a cytotoxic mechanism, the nature of the effector phase was certainly not proven.

The above studies suggested strongly that the virus-specific CD8$^+$ T cells were functioning in vivo (as in vitro) by direct interaction with the virus-infected targets. A further, compelling experiment was the demonstration that i.v. injection of a T cell line specific for one influenza A virus (X) but not another (Y) could clear only X and not Y from the lungs of doubly infected mice (LUKACHER et al. 1984). Other adoptive transfer experiments with LCMV-infected, bone marrow radiation chimeras showed that severe neuropathology was only induced by donor immune T cells of MHC type A, but not B, even though CTLs restricted to both A and B were present in the spleens of the [(A×B)F1→A] chimeric recipients (DOHERTY et al. 1990). Apparently the effector T cells operating in the site of pathology were targeted to virus-infected epithelium (A) rather than to antigen presenting monocyte/macrophages or dendritic cells of bone marrow origin (B), though these stimulator populations were capable of causing CTL development in lymphoid tissue. Still, even if the CD8$^+$ CTLs were interacting directly with virus-infected cells in vivo, not everyone accepted that the final effector mechanism was cytotoxicity rather than (for example) the short-range secretion of potent cytokines (RAMSAY et al. 1993).

The further dissection of the underlying mechanisms has been enormously enhanced by the availability of transgenic mice and mice homozygous (–/–) for the deletion of key genes that impact on T cell differentiation and effector function (KOLLER and SMITHIES 1992). However, it should be kept in mind that such genetically disrupted animals are novel life forms that may be using atypical pathways to achieve an apparently normal end result. This technology shows us the possible limits of a system, but conceptual development needs always to take due account of what is happening in the unmanipulated host (DOHERTY 1993b). The following interfaces both types of analytical approaches.

2 The Proof in the Perforin

The molecular basis of CTL effector function has been a hotly debated topic for many years (PODACK et al. 1991; DOHERTY 1993a; BERKE 1994). Candidate molecules include the serine esterases, or granzymes, events mediated via Fas and its ligand, and perforin. Perforin is a molecule that is secreted by CTLs and operates to induce the assembly of complement-like channels in the plasma membrane of the target cell. Both perforin and the serine esterases are found in the prominent cytoplasmic granules of the CTLs. Following T cell receptor (TCR)-mediated contact between the T cell and target, the granules move along microtubules and then discharge their contents at the cellular interface. Perforin expression has been shown to the prominent in T cells located in sites of virus-induced pathology, though the same is true for the granzymes (MULLER et al. 1989; YOUNG et al. 1989).

The definitive proof that T cell-target contact and CTL effector function are central to virus clearance in vivo has been provided by the analysis of LCMV-infected mice that are homozygous (–/–) for disruption of the perforin gene (KÄGI et al. 1994). Though there is obvious clonal expansion and some inflammatory pathology in these perforin (–/–) mice, they neither develop virus-specific CTL effectors nor eliminate LCMV-infected cells. This does not rule out a role for the serine esterases: perhaps the perforin first makes a hole in the cell membrane which enables the transfer of enzymes that will damage the DNA of the target cell.

More recent evidence (WALSH et al. 1994) indicates that LCMV-specific, perforin (–/–) CD8+ T cells can cause rapid weight loss and death, which could reflect either cytokine mediated effects or the operation of a cytotoxic pathway involving Fas (DOHERTY 1993a). However, it is obvious that secretion of cytokines alone is not sufficient to terminate this infectious process. Even so, the perforin requirement may not necessarily be so absolute for other viruses (GUIDOTTI et al. 1994). Acting in the absence of the CD8+ subset, CD4+ T cells fail to clear LCMV but can terminate the viral pneumonias caused by both influenza virus and Sendai virus (FUNG-LEUNG et al. 1991; EICHELBERGER et al. 1991a; BENDER et al. 1992; HOU et al. 1992; LEHMANN-GRUBE et al. 1993).

3 Significance of CD4+ Cytotoxic T Lymphocytes

There is no consensus on the biological role of CD4+ CTLs (DOHERTY 1993b). In humans, restimulation of measles virus-specific memory T cells has consistently given CD4+ CTL effectors (JACOBSON et al.1988; DHIB-JALBUT and JACOBSON 1994). The same result can be found for Epstein-Barr virus (EBV)-specific CD4+ precursors that are grown out in antigen-driven limiting dilution cultures (SLOBOD et al. 1993). While it is intriguing that human CD4+ CTLs can readily be generated in vitro, this does not establish that cytotoxicity (rather than lymphokine secretion) is a feature of effector mechanisms mediated by this T cell subset in vivo.

More compelling evidence that CD4+ CTLs may be operating in the host, rather than just being an artefact of culture, emerged from experiments with mice homozygous (–/–) for a disruption of the β2-microglobulin (β2-m) gene, the light chain of the class I MHC glycoprotein (KOLLER and SMITHIES 1992; ZIJLSTRA et al.1990; DOHERTY 1993b). These β2-m (–/–) mice express minimal amounts of class I MHC glycoprotein and develop few CD8⁺ T cells. Following infection with LCMV, CD4⁺ CTLs were recovered directly from spleen (MULLER et al.1992). The same was found for lymphocytes isolated by bronchoalveolar lavage (BAL) of the lungs of mice infected intranasally with Sendai virus (HOU et al.1992). In both cases, the kinetics of CD4⁺ CTL development were somewhat slower than that seen normally for the CD8⁺ subset (day 10 rather than day 7) and the level of specific ^{51}Cr release in the standard cytotoxicity assay with virus-infected class II MHC⁺ target cells was generally lower. Also the β2-m (–/–) mice that were infected with the substantially nonlytic LCMV developed a chronic wasting disease (LEHMANN-GRUBE et al.1993; DOHERTY et al.1993), which had been described previously in CD8 (–/–) animals (FUNG-LEUNG et al.1991).

The cytotoxic mechanism used by the virus-specific CD4⁺ CTLs that are generated in vivo in β2-m (–/–) mice is not currently understood (HOU et al. 1993), though a Fas mediated mechanism is likely (STALDER et al.1994) and, by the time this is published, the obvious experiment with CD8-depleted (with monoclonal antibodies) perforin (–/–) mice (KÄGI et al. 1994) should have been done. These activated β2-m (–/–) CD4⁺ T CTLs produce substantial amounts of tumor necrosis factor (TNF), but neutralization of secreted TNF does not block effector function. Experiments with CD8-depleted perforin (–/–) mice may also tell us whether, for example, the clearance of Sendai virus from the respiratory tract reflects that the CD4⁺ effectors are targeted directly to virus-infected epithelial cells that have up-regulated class II MHC glycoprotein expression (WONG et al.1984), perhaps as a consequence of exposure to interferon-γ (IFN-γ). Alternatives are that the CD4⁺ set is operating to provide help for B cells and that the clearance is mediated via virus-specific Ig (SCHERLE et al.1992), or that the production of cytokines is sufficient to terminate the infectious process (RAMSAY et al. 1993). Either of these mechanisms could operate via activated macrophages.

Evidence from Braciale's laboratory indicates that cloned Th1 (IFN-γ), but not Th2 (interleukin-4), cells can protect mice from the consequences of lethal influenza infection (GRAHAM et al. 1994). The CD4⁺ T cells that clear influenza virus from β2-m (–/–) mice are also likely to be Th1 cells and, even following in vivo treatment with monoclonal antibodies (mAbs) to IFN-γ, it is (unlike some of the parasite models) very difficult to switch the response to a Th2 profile (SARAWAR et al.1994). These Th1 effector cells may be operating by any, or all, of the mechanisms discussed above.

4 The CD8⁺ T Cells in β2-m (–/–) Mice

The β2-m (–/–) mice are not completely class I MHC negative (BIX and RAULET 1992a; RAULET 1994). Small amounts of (in particular) H-2D^b can be expressed on the cell surface in the absence of β2-m as isolated heavy chain (FRASER et al. 1987; BIX and RAULET 1992a). This "naked" H-2D^b can act to present sufficient LCMV peptide to target adoptively transferred, virus-specific CD8+ T cells generated in β2-m (+/+) LCMV-infected mice (LEHMANN-GRUBE et al. 1993; DOHERTY et al. 1993). Also, when the β2-m (–/–) mice are injected with allogeneic tumor cells, the very small numbers of CD8+ cytotoxic T lymphocyte precursors (CTLps) that are found in these animals can expand greatly and eventually reject the tumor (APASOV and SITKOVSKY 1994).

The findings from tumor (APASOV and SITKOVSKY 1994) and allograft rejection (ZIJLSTRA et al. 1992), experiments in β2-m (–/–) mice should not, however, be extrapolated directly to the virus models. No evidence for the developement of T cell-mediated effector function was found for either Sendai virus or LCMV infected β2-m (–/–) mice that were depleted of the CD4+ subset (LEHMANN-GRUBE et al. 1993; DOHERTY et al. 1993). Furthermore, radiation chimeras made with β2-m (+/+) bone marrow and irradiated β2-m (–/–) recipients could not clear Sendai virus following CD4– depletion (HOU and DOHERTY 1995). Such (+/+)→(–/–) chimeras develop about 20% of the normal numbers of CD8+ T cells (BIX and RAULET 1992b) and generate potent CTL effector populations that can be isolated from by bronchoalveolar lavage (BAL). In addition, adoptively transferred virus-specific CD8+ memory T cells from β2-m (+/+) mice were unable to terminate the infectious process in respiratory epithelium.

The basic message from the experiments with β2-m (+/+)→(–/–) chimeras is thus that CD8+ effector T cells cannot clear Sendai virus by some "bystander" mechanism involving, for example, secreted cytokines or activated macrophages. The same conclusion is reached from the findings for the perforin (–/–) mice that were infected with LCMV (KÄGI et al. 1994). Virus-specific CD8+ CTLs have to target directly to virus-infected cells. At least for LCMV, virus clearance is also dependent on an intact cytolytic mechanism.

5 The Development of CD4+ and CD8+ Cytotoxic T Lymphocyte Responses

Little has been done to analyze the development of the CD4+ CTLs, as, at least with the mouse models, this effector population only emerges in the absence of the CD8+ subset. The onset of lytic activity for CD4+ T cells from β2-m(–/–) mice is delayed, and the magnitude of lysis is generally lower, when compared with the conventional CD8+ CTL response. A simplistic interpretation of the phenomenon is that the Th1 cells that normally dominate the CD4+ responder profile in virus infections (SARAWAR and DOHERTY 1994; BIRON 1994) are, when driven longer via their TCRs as a consequence of delayed virus clearance, undergoing a further differentiation process that results in the acquisition of lytic activity. Having stated this, it is currently the case that we have little quantitative information on the generation of the Th1 response in virus infections. However, with the development of ELISPOT techniques (FUJIHASHI et al. 1993; SARAWAR et al. 1993;

SARAWAR and DOHERTY 1994) for detecting cytokine-producing lymphocytes and limiting dilution analysis (LDA) for CD4⁺ Th precursors (Thps) (MILLER and REISS 1984), such studies are now in progress for both the Sendai and influenza virus models.

Use of sequential LDA analysis with several different experimental systems is leading to a reasonable definition of the events occurring in a primary, virus-specific CD8⁺ CTL response. The findings reflect that, with the influenza virus and Sendai virus models, productive infection is largely limited to the superficial epithelial cells in the respiratory tract (EICHELBERGER et al. 1991b). The reason for this is that infectious progeny virus is only made by cells expressing a trypsin-like enzyme (WALKER et al. 1992) that cleaves one of the major surface glycoproteins of the virion, the hemagglutinin (HA) of influenza virus and the fusion (F) protein of Sendai virus. As a consequence, there is minimal (if any) evidence for the presence of infectious influenza virus or Sendai virus in lymphoid tissue. The converse is true for LCMV, which (depending on the particular variant) can grow to high titers in lymph nodes, with some of the replication occurring in CD4⁺ T cells (AHMED et al. 1987).

The consequence of these differential patterns of virus growth is that it is generally easy to demonstrate effector CTLs in lymphoid tissue and spleen of mice infected with LCMV 6–9 days previously (ZINKERNAGEL and DOHERTY 1979; LYNCH et al. 1989) while, at least for the less virulent influenza A virus and Sendai virus strains, potent CD8⁺ CTL populations are present only in the respiratory tract (ALLAN et al. 1990; HOU and DOHERTY 1993). Evidence of expanded numbers of CTLp can, however, be found by bulk culture techniques in the regional mediastinal lymph nodes (MLN) as early as 3 days after infection. These CTLp increase in frequency, with a doubling time of about 10–12 h over the subsequent 4–5 days, after which a proportion exit the lymph node (HOU and DOHERTY 1993; DOHERTY 1993c). Some of these blood-bourne CTLp then extravasate into the virus-infected lung and develop further to become the CTLs that terminate the infectious process. The change in functional status from CTLp to effector CTL is presumably a consequence of continuing stimulation via the clonotypic TCR, mediated by direct contact with the virus-infected epithelium. Even so, many of the CTLp that are recovered directly from the pneumonic lung can still be induced to undergo clonal expansion in LDA cultures, indicating that not all the CTLp entering this site of virus-induced pathology become terminally differentiated CTL effectors.

The increase in CTLp frequency is accompanied by changes in patterns of cell surface glycoprotein expression on the responding T cells. The CTLp are L-selectin-low, CD44-high, CD49d (VLA-4)-high (TABI et al. 1988; HOU and DOHERTY 1993; DOHERTY et al. 1994b; TRIPP et al. 1995), a phenotype characteristic of activation that is also found for the effector CTL and persists for months on virus-specific CD8⁺ memory T cells. The assumption from studies with alloreactive CTLp that are present at high frequency prior to any encounter with the inducing antigen is that the naive cells are CD44-low, L-selectin-high (BUDD et al. 1987; TABI et al. 1988; LYNCH et al. 1987; CEREDIG et al. 1987). This has yet, however, to be

proven formally for virus-specific T cells, principally because the precursor frequencies are very low.

6 Cytokines and Cytotoxic T Lymphocyte Effectors

Virus-specific CD8⁺ T cells can, like lymphocytes of this subset primed with other
antigens (KELSO 1990), produce many of the cytokines made by the CD4⁺ "helpers"
(CARDING et al. 1993; SARAWAR and DOHERTY 1994; CARPENTER et al. 1994). The
general impression is, however, that the CD8⁺ set tends to produce smaller
amounts of these biologically active molecules. The CD8⁺ T cells from mice
primed with influenza or Sendai virus have been variously shown to produce TNF,
IFN-γ and interleukin-(IL-2). However, as single cell assays that correlate levels of
(for example) perforin expression with cytokine production have not been done, it
is unclear whether the same T cells function to produce cytokine and to mediate
CTL activity. We have, at this stage, no cell surface marker that allows the
separation (e.g., by flow cytometry) of the cytotoxic effectors from other CD8⁺
T cells that may be responding to the virus. Thus, it is not definitively proven that
cytokine producers and lytic effectors belong to the same, or to different,
lineages, though the former seems more likely. The CD4⁺ CTLs that emerge in the
β2-m (–/–) mice are probably a subset (or a further differentiated state) of the Th1
cells.

The question of the part played by different cytokines in the generation of CTL
effector function has been approached by the in vivo administration of neutralizing
mAbs and by infecting mice that are (–/–) for various cytokine genes (HUANG et al.
1993; KÜNDIG et al. 1993; GRAHAM et al. 1993; KOPF et al. 1994; MÜLLER et al. 1994;
BIRON 1994). Much remains to be done with the cytokine (–/–) mice. The impression at this stage is that the elimination of IL-2, IFN-γ or IL-6 does not greatly
compromise the virus-specific CD8⁺ CTL response, while blocking the function of
IFN-α or IFN-β may have a profound effect. More intensive studies with these
cytokine (–/–) animals are likely to yield a clearer picture of the underlying
compensatory and/or interactive processes.

7 Immune Exhaustion of Virus-Specific CD8⁺ T Cells

Experiments with variants of LCMV that have the capacity to grow in at least a
small proportion of the CD4⁺ T cells indicate that the CD8⁺ CTL response can be
"exhausted" if massive amounts of virus are used in the initial challenge
(MOSKOPHIDIS et al. 1993). A similar effect may be occurring in people infected with
HIV (PANTALEO et al. 1994). Immune exhaustion provides a convenient explanation

for the "high dose immune paralysis" phenomenon that was described many years ago for LCMV by John Hotchin (Suzuki and Hotchin 1971) and shown, in the 1970s, to be reflected in diminished CTL responses and lack of T cell-mediated immuno-pathology (Doherty et al. 1974). The basic idea is that so many cells are infected with virus by the time that the CTL response develops that all the CTLp are driven to become effectors, which eventually undergo apoptosis. The consequence is considered to be selective immunosuppression, leading to persistence of high virus loads.

Such a phenomenon can occur with LCMV because the virus is minimally lytic and can persist in many cell types without obviously compromising normal physiological function. The consequence with a lytic virus, would, however, be death from virus-induced cytopathology (Doherty 1993c). Generalized, acute immune exhaustion is thus probably a fairly unusual occurrence with most pathogens. People infected with HIV may circulate effector CTLs in blood for many years, indicating that there is a process of continuing antigen stimulation in the absence of rapid exhaustion of all the virus-specific CTLp (Riviere et al. 1994). In most infections, CTLs are not found much after the acute (virus clearance) phase of the response is complete, though enhanced numbers of CTLp persist for the life of a laboratory mouse.

A current debate concerns the extent to which transient exhaustion of particular CD8$^+$ CTLp clonotypes may be a feature of acute immune responses (Sprent 1994). It should be possible to resolve this question by comparing the relative prevalence of particular TCR heterodimers expressed on acutely respond-ing and memory T cells. An accessible model may be the H-2D^b-restricted response to LCMV, in which relatively few TCR pairs seem to be used (Pircher et al. 1987; Yanagi et al. 1992). In contrast, the CD8$^+$ CTL clonotypes generated following infection of H-2^b mice with the influenza A viruses and Sendai virus seem to be very diverse (Deckhut et al. 1993; Cole et al. 1994). The potential redundancy is so great that mice that are transgenic for an apparently irrelevant TCRβ chain were able to generate CTL responses specific for influenza virus, Sendai virus and LCMV (Ewing et al. 1994; Doherty et al. 1994a).

A very worthwhile endeavour would be to do longitudinal studies of the maintenance, gain or loss of CTLp clonotypes specific for viruses that may be encountered only once (e.g., vaccinia, Demkowicz and Ennis 1993), cause periodic reinfection (e.g., influenza, Bowness et al. 1993), or persist for life (e.g., EBV, Moss et al. 1992) in humans. People live so much longer than mice! Experimen-tally, it may also be possible to approach the question of clonal dominance using radiation chimeras, made with mixtures of bone marrow from mice that are transgenic for two or more pairs of TCR heterodimers (Kelly et al. 1993; Kearney et al. 1994) specific for a particular complex of viral peptide and class I MHC glycoprotein.

8 T and B Cell Memory and the Recall Response for Cytotoxic T Lymphocyte Precursors

Much of the recent debate about the nature of immunological memory may have been somewhat confused by the perception that the rules governing the long-term persistence of antigen-primed T cells and B cells are likely to be similar(GRAY 1993). There is, however, a fundamental difference (DOHERTY 1995). Protection against secondary infection requires that the effectors of the B cell response, the Ig-producing plasma cells, continue to secrete virus-specific antibody for years after the initial infection. These plasma cells tend to locate in mucosal sites (STAATS et al. 1994) or in bone marrow (HYLAND et al. 1994a,b), microenvironments where there is likely to be substantial constitutive (bone marrow) or antigen-induced (mucosae) cytokine production (CARDING et al. 1993; STAATS et al. 1994). In contrast, the long-term presence of effector T cells is generally undesireable. The consequence of persistent lymphokine production and cytotoxicity is likely to be immunopathology (PANTALEO and FAUCI 1994), particularly where the confrontation between T cell and virus-infected cell is occurring in a key organ, such as the liver (see chapters on hepatitis B virus, HIV and LCMV).

The net result is that, though enhanced numbers of CD8$^+$ CTLp can persist for life (of a laboratory mouse) in the apparent absence of a further encounter with the inducing antigen (MULLBACHER 1994; LAU et al. 1994; HOU et al. 1994; DOHERTY et al. 1994b; MATZINGER 1994), the recall of CTL effector function always takes at least 3–4 days (DOHERTY et al. 1977). This limits the usefulness of established T cell memory, particularly for pathogens that cause most of their pathology at mucosal surfaces.Protection afforded by primed CD8$^+$ CTLp is often (ANDREW et al. 1986; DOHERTY et al. 1989; KLAVINSKIS et al. 1989; CASTRUCCI et al. 1994; LAWSON et al. 1994), though not always (KAST et al.1991), more impressive for viruses that infect solid tissues. Also, the effectiveness of secondary CD8$^+$ T cell responses seems to decline more quickly than might be expected from the analysis of CTLp frequencies, perhaps reflecting the gradual loss of activated phenotype (LIANG et al. 1994; DOHERTY et al. 1994b). Further studies of the long-term persistence and fate of responding T cell clonotypes are clearly needed (see section above).

9 Conclusions

Virus-specific CD8$^+$ T cells act in vivo by a cytotoxic mechanism that requires the expression of perforin, at least in the LCMV model. This and a variety of other studies that have utilized T cell specificity for a particular virus or peptide + class I MHC glycoprotein establish beyond any reasonable doubt that the central event for the CD8$^+$ effector is direct contact with the virus-infected target cell. Whether cytokines secreted by activated CD8$^+$ T cells can also act to promote pathology, or even the clearance of some intracellular pathogens (other than LCMV), has yet to be satisfactorily established. The converse is true for the CD4$^+$ subset, which obviously functions to provide lymphokines: the biological significance (if any) of the CD4$^+$ CTLs that can be generated in β2-m (–/–) mice is currently unclear. Evidence of virus-specific CD8$^+$ T cell memory can be found for years after primary infection. However, (unlike the antibody response) recall to CTL effector function generally takes 3–4 days after in vivo challenge. While protective B cell memory requires the constant presence of effectors (plasma cells) secreting product (antibody), CD8$^+$ T cell memory is balanced to provide primed precursors rather than constitutively activated effectors (CTLs). The consequent delay in the deployment of activated CTLs has to be bourne in mind when designing vaccination strategies, especially for pathogens that cause their primary pathology at mucosal surfaces.

Acknowledgments. Much of this brief review depends on discussion with my colleagues at SJCRH, particularly Sam Hou, Sally Sarawar, Christine Ewing, Ralph Tripp, Lisa Hyland, Mark Sangster and Chris Coleclough. These studies were supported by NIH Grants AI 29579 and AI 31596, and by the American Syrian Lebanese Associated Charities.

References

Ahmed R, King CC, Oldstone MB (1987) Virus-lymphocyte interaction: T cells of the helper subset are infected with lymphocytic choriomeningitis virus during peristent infection in vivo. J Virol 61: 1571–1576

Allan W, Tabi Z, Cleary A, Doherty PC (1990) Cellular events in the lymph node and lung of mice with influenza. Consequences of depleting CD4$^+$ T cells. J Immunol 144: 3980–3986

Andrew ME, Coupar BE, Ada GL, Boyle DB (1986) Cell-mediated immune responses to influenza virus antigens expressed by vaccinia virus recombinants. Microb Pathog 1: 443–452

Apasov SG, Sitkovsky MV (1994) Development and antigen specificity of CD8$^+$ cytotoxic T lymphocytes in β2-microglobulin-negative, MHC class I-deficient mice in response to immunization with tumor cells. J Immunol 152: 2087–2097

Baenziger J, Hengartner H, Zinkernagel RM, Cole GA (1986) Induction or prevention of immunopathological disease by cloned cytotoxic T cell lines specific for lymphocytic choriomeningitis virus. Eur J Immunol 16: 387–393

Bender BS, Croghan T, Zhang L, Small PA Jr (1992) Transgenic mice lacking class I major histocompatibility complex-restricted T cells have delayed viral clearance and increased mortality after influenza virus challenge. J Exp Med 175: 1143–1145

Berke G (1994) The binding and lysis of target cells by cytotoxic lymphocytes: molecular and cellular aspects. Annu Rev Immunol 12: 735–773

Biron CA (1994) Cytokines in immune responses to, and resolution of, virus infection. Curr Opin Immunol 6: 530–538

Bix M, Raulet D (1992a) Functionally conformed free class I heavy chains exist on the surface of β2-microglobulin negative cells. J Exp Med 176: 829–834

Bix M, Raulet D (1992b) Inefficient positive selection of T cells directed by haematopoietic cells. Nature 359: 330–333

Blanden RV, Bowern NA, Pang TE, Gardner ID, Parish CR (1975a) Effects of thymus-independent (B) cells and the H-2 gene complex on antiviral function of immune thymus-derived (T) cells. Aust J Exp Biol Med Sci 53: 187–195

Blanden RV, Doherty PC, Dunlop MB, Gardner ID, Zinkernagel RM, David CS (1975b) Genes required for cytotoxicity against virus-infected target cells in K and D regions of H-2 complex. Nature 254: 269–270

Bowness P, Moss PAH, Rowland-Jones S, Bell JI, McMichael AJ (1993) Conservation of T cell receptor usage by HLA B27-restricted influenza-specific cytotoxic T lymphocytes suggests a general pattern for antigen-specific major histocompatibility complex class I-restricted responses. Eur J Immunol 23: 1417–1421

Budd RC, Cerottini JC, Horvath C, Bron C, Pedrazzini T, Howe RC, MacDonald HR (1987) Distinction of virgin and memory T lymphocytes. Stable acquisition of the Pgp-1 glycoprotein concomitant with antigenic stimulation. J Immunol 138: 3120–3129

Byrne JA, Oldstone MB (1986) Biology of cloned cytotoxic T lymphocytes specific for lymphocytic choriomeningitis virus. VI. Migration and activity in vivo in acute and persistent infection. J Immunol 136: 698–704

Carding SR, Allan W, McMickle A, Doherty PC (1993) Activation of cytokine genes in T cells during primary and secondary murine influenza pneumonia. J Exp Med 177: 475–482

Carpenter EA, Ruby J, Ramshaw IA (1994) IFN-gamma, TNF, and IL-6 production by vaccinia virus immune spleen cells: an in vitro study. J Immunol 152: 2652–2659

Castrucci MR, Hou S, Doherty PC, Kawaoka Y (1994) Protection against lethal lymphocytic chorio-meningitis virus (LCMV) infection by immunization of mice with an influenza virus containing an LCMV epitope recognized by cytotoxic T lymphocytes. J Virol 65: 3486–3490

Ceredig R, Allan, JE, Tabi Z, Lynch F, Doherty PC (1987) Phenotypic analysis of the inflammatory exudate in murine lymphocytic choriomeningitis. J Exp Med 165: 1539–1551

Cole GA, Hogg CG, Woodland DL (1994) The MHC class I restricted T cell response to Sendai virus infection in C57BL/6 mice: a single immunodominant epitope elicits an extremely diverse repertoire of T cells. Int Immunol 6: 1767–1775

Deckhut AM, Allan W, McMickle A, Eichelberger M, Blackman MA, Doherty PC, Woodland DL (1993) Prominent usage of Vβ8.3 T cells in the H-2Dᵇ-restricted response to an influenza A virus nucleoprotein epitope. J Immunol 151: 2658–2666

Demkowicz WE Jr, Ennis FA (1993) Vaccinia virus-specific CD8⁺ cytotoxic T lymphocytes in humans. J Virol 67: 1538–1544

Dhib-Jalbut S, Jacobson S (1994) Cytotoxic T cells in paramyxovirus infection of humans. Curr Top Microbiol Immunol 189: 109–121

Doherty PC (1993a) Cell-mediated cytotoxicity. Cell 75: 607–612

Doherty PC (1993b) Virus infections in mice with targeted gene disruptions. Curr Opin Immunol 5: 479–483

Doherty PC (1993c) Immune exhaustion: driving virus-specific CD8⁺ T cells to death. Trends Microbiol 1: 207–209

Doherty PC (1995) Immune memory to viruses. American Society of Microbiology Lecture. American Society of Virology, Ann Arbor (ASM News, vol 61) pp 68–71

Doherty PC, Zinkernagel RM (1975) Capacity of sensitized thymus-derived lymphocytes to induced fatal lymphocytic choriomeningitis is restricted by the H-2 gene complex. J Immunol 114: 30–33

Doherty PC, Zinkernagel RM, Ramshaw IA (1974) Specificity and development of cytotoxic thymus-derived lymphocytes in lymphocytic choriomeningitis. J Immunol 112: 1548–1552

Doherty PC, Blanden RV, Zinkernagel RM (1976a) Specificity of virus-immune effector T cells for H-2K or H-2D compatible interactions: implications for H-antigen diversity. Transplant Rev 29: 89–124

Doherty PC, Dunlop MB, Parish CR, Zinkernagel RM (1976b) Inflammatory process in murine lymphocytic choriomeningitis is maximal in H-2K or H-2D compatible interactions. J Immunol 117: 187–190

Doherty PC, Efforts RB, Bennink J (1977) Heterogeneity of the cytotoxic response of thymus-derived lymphocytes after immunization with influenza viruses. Proc Natl Acad Sci USA 74: 1209–1213

Doherty PC, Allan W, Boyle DB, Coupar BE, Andrew ME (1989) Recombinant vaccinia viruses and the development of immunization strategies using influenza virus. J Infect Dis 159: 1119–1122

Doherty PC, Allan JE, Lynch F, Ceredig R (1990) Dissection of an inflammatory process induced by CD8⁺ T cells. Immunol Today 11: 55–59

Doherty PC, Hou S, Southern PJ (1993) Lymphocytic choriomeningitis virus induces a chronic wasting disease in mice lacking class I major histocompatibility complex glycoproteins. J Neuroimmunol 46: 11–18

Doherty PC, Hou S, Evans CF, Whitton JL, Oldstone MBA, Blackman MA (1994a) Limiting the available T cell receptor repertoire modifies acute lymphocytic choriomeningitis virus-induced immuno-pathology. J Neuroimmunol 51: 147–152

Doherty PC, Hou S, Tripp RA (1994b) CD8⁺ T cell memory to viruses. Curr Opin Immunol 6: 545–552

Eichelberger MC, Allan W, Zijlstra M, Jaenisch R, Doherty PC (1991a) Clearance of influenza virus respiratory infection in mice lacking class I major histocompatibility complex-restricted CD8⁺ T cells. J Exp Med 174: 875–880

Eichelberger MC, Wand M, Allan W, Webster RG, Doherty PC (1991b) Influenza virus RNA in the lung and lymphoid tissue of immunologically intact and CD4-depleted mice. J Gen Virol 72: 1695–1698

Ewing C, Allan W, Daly K, Hou S, Cole GA, Doherty PC, Blackman MA(1994) Virus-specific CD8+ T-cell responses in mice transgenic for a T- cell receptor beta chain selected at random. J Virol 68: 3065–3070

Fraser JD, Allen H, Flavell RA, Strominger JL (1987) Cell-surface expression of H-2D^b requires N-linked glycans. Immunogenetics 26: 31–35

Fujihashi K, McGhee JR, Beagley KW, McPherson DT, McPherson SA, Huang C-M, Kiyono H (1993) Cytokine-specific ELISPOT assay. Single cell analysis of IL-2, IL-4, and IL-6 producing cells. J Immunol Methods 160: 181–189

Fung-Leung WP, Kundig TM, Zinkernagel RM, Mak TW (1991) Immune response against lymphocytic choriomeningitis virus infection in mice without CD8 expression. J Exp Med 174: 1425–1429

Graham MB, Dalton DK, Giltinan D, Braciale VL, Stewart TA, Braciale TJ (1993) Response to influenza infection in mice with a targeted disruption in the interferon gamma gene. J Exp Med 178: 1725–1732

Graham MB, Braciale VL, Braciale TJ (1994) Influenza virus specific CD4+ Th2 T lymphocytes do not promote recovery from experimental virus infection. J Exp Med 180: 1273–1282

Gray D (1993) Immunological memory. Annu Rev Immunol 11: 49–77

Guidotti LG, Ando K, Hobbs MV, Ishikawa T, Runkel L, Schreiber RD, Chisari FV (1994) Cytotoxic T lymphocytes inhibit hepatiis B virus gene expression by a noncytolytic mechanism in transgenic mice. Proc Natl Acad Sci USA 91: 3764–3768

Hou S, Doherty PC (1993) Partitioning of responder CD8+ T cells in lymph node and lung of mice with Sendai virus pneumonia by LECAM-1 and CD45RB phenotype. J Immunol 150: 5494–5500

Hou S, Doherty PC (1995) Clearance of Sendai virus by CD8+ T cells requires direct targeting to virus-infected epithelium. Eur J Immunol 25: 111–116

Hou S, Doherty PC, Zijlstra M, Jaenisch R, Katz JM (1992) Delayed clearance of Sendai virus in mice lacking class I MHC-restricted CD8+ T cells. J Immunol 149: 1319–1325

Hou S, Fishman M, Gopal Murti K, Doherty PC (1993) Divergence between cytotoxic effector function and tumor necrosis factor alpha production for inflammatory CD4+ cells from mice with Sendai virus pneumonia. J Virol 67: 6299–6302

Hou S, Hyland L, Ryan KW, Portner A, Doherty PC (1994) Virus-specific CD8+ T-cell memory determined by clonal burst size. Nature 369: 652–654

Huang S, Hendriks W, Althage A, Hemmi S, Bluethmann H, Kamijo R, Vilcek J, Zinkernagel RM, Aguet M (1993) Immune response in mice that lack the interferon-gamma receptor. Science 259: 1742–1745

Hyland L, Sangster M, Sealy R, Coleclough C (1994a) Respiratory virus infection provokes a permanent humoral immune response. J Virol 68: 6083–6086

Hyland L, Hou S, Coleclough C, Takimoto T, Doherty PC (1994b) Mice lacking CD8+ T cells develop greater numbers of IgA-producing cells in response to a respiratory virus infection. Virology 204: 234–241

Jacobson S, Sekaly RP, Bellini WJ, Johnson CL, McFarland HF, Long EO (1988) Recognition of intracellular measles virus antigens by HLA class II restricted measles virus-specific cytotoxic T lymphocytes. Ann N Y Acad Sci 540: 352–353

Kägi D, Ledermann B, Burki K, Seiler P, Odermatt B, Olsen KJ, Podack ER, Zinkernagel RM, Hengartner H (1994) Cytotoxicity mediated by T cells and natural killer cells is greatly impaired in perforin-deficient mice. Nature 369: 31–37

Kast WM, Roux L, Curren J, Blom HJ, Voordouw AC, Meloen RH, Kolakofsky D, Melief CJ (1991) Protection against lethal Sendai virus infection by in vivo priming of virus-specific cytotoxic T lymphocytes with a free synthetic peptide. Proc Natl Acad Sci USA 88: 2283–2287

Kearney ER, Pape KA, Loh DY, Jenkins MK (1994) Visualization of peptide-specific T cell immunity and peripheral tolerance induction in vivo. Immunity 1: 327–339

Kelly KA, Pircher H, von Boehmer H, Davis MM, Scollay R (1993) Regulation of T cell production in T cell receptor transgenic mice. Eur J Immunol 23: 1922–1928

Kelso A (1990) Frequency analysis of lymphokine-secreting CD4+ and CD8+ T cells activated in a graft-versus-host reaction. J Immunol 145: 2167–2176

Klavinskis LS, Whitton JL, Oldstone MB (1989) Molecularly engineered vaccine which expresses an immunodominant T-cell epitope induces cytotoxic T lymphocytes that confer protection from lethal virus infection. J Virol 63: 4311–4316

Koller BH, Smithies O (1992) Altering genes in animals by gene targeting. Annu Rev Immunol 10: 705–730

Kopf M, Baumann H, Freer G, Freudenberg M, Lamers M, Kishimoto T, Zinkernagel R, Bluethmann H, Köhler G (1994) Impaired immune and acute-phase responses in interleukin-6-deficient mice. Nature 368: 339–342

Kündig TM, Schorle H, Bachmann MF, Hengartner H, Zinkernagel RM, Horak I (1993) Immune responses in interleukin-2-deficient mice. Science 262: 1059–1061

Lau LL, Jamieson BD, Somasundaram T, Ahmed R (1994) Cytotoxic T-cell memory without antigen. Nature 369: 648–652

Lawson CM, Bennink JR, Restifo NP, Yewdell JW, Murphy BR (1994) Primary pulmonary cytotoxic T lymphocytes induced by immunization with a vaccinia virus recombinant expressing influenza A virus nucleoprotein peptide do not protect mice against challenge. J Virol 68: 3505–3511

Lehmann-Grube F, Lohler J, Utermohlen O, Gegin C (1993) Antiviral immune responses of lymphocytic choriomeningitis virus-infected mice lacking CD8+ T lymphocytes because of disruption of the beta 2-microglobulin gene. J Virol 67: 332–339

Liang S, Mozdzanowska K, Palladino G, Gerhard W (1994) Heterosubtypic immunity to influenza type A virus in mice: effector mechanisms and their longevity. J Immunol 152: 1653–1661

Lukacher AE, Braciale VL, Braciale TJ (1984) In vivo effector function of influenza virus-specific cytotoxic T lymphocyte clones is highly specific. J Exp Med 160: 814–826

Lynch F, Chaudhri G, Allan JE, Doherty PC, Ceredig R (1987) Expression of Pgp-1 (or Ly24) by subpopulations of mouse thymocytes and activated peripheral T lymphocytes. Eur J Immunol 17: 137–140

Lynch F, Doherty PC, Ceredig R (1989) Phenotypic and functional analysis of the cellular response in regional lymphoid tissue during an acute virus infection. J Immunol 142: 3592–3598

Matzinger P (1994) Immunology. Memories are made of this? Nature 369: 605–606

Miller RA, Reiss CS (1984) Limiting dilution cultures reveal latent influenza virus-specific helper T cells in virus-primed mice. J Mol Cell Immunol 1: 357–368

Moskophidis D, Lechner F, Pircher H, Zinkernagel RM (1993) Virus persistence in acutely infected immunocompetent mice by exhaustion of antiviral cytotoxic effector T cells. Nature 362: 758–761

Moss DJ, Burrows SR, Khanna R, Misko IS, Sculley TB (1992) Immune surveillance against Epstein-Barr virus. Semin Immunol 4: 97–104

Mullbacher A (1994) The long-term maintenance of cytotoxic T cell memory does not require persistence of antigen. J Exp Med 179: 317–321

Muller C, Kagi D, Aebischer T, Odermatt B, Held W, Podack ER, Zinkernagel RM, Hengartner H (1989) Detection of perforin and granzyme A mRNA in infiltrating cells during infection of mice with lymphocytic choriomeningitis virus. Eur J Immunol 19: 1253–1259

Muller D, Koller BH, Whitton JL, LaPan KE, Brigman KK, Frelinger JA (1992) LCMV-specific, class II-restricted cytotoxic T cells in beta 2-microglobulin-deficient mice. Science 255: 1576–1578

Müller U, Steinhoff U, Reis LFL, Hemmi S, Pavlovic J, Zinkernagel RM, Aguet M (1994) Functional role of type I and type II interferons in antiviral defense. Science 264: 1918–1921

Pantaleo G, Demarest JF, Soudeyns H, Graziosi C, Denis F, Adelsberger JW, Borrow P, Saag MS, Shaw GM, Sekaly RP, Fauci AS (1994) Major expansion of CD8+ T cells with a predominant Vβ usage during the primary immune response to HIV. Nature 370: 463–467

Pantaleo G, Fauci AS (1994) Tracking HIV during disease progression. Curr Opin Immunol 6: 600–604

Pircher H, Baenziger J, Schilham M, Sado T, Kamisaku H, Hengartner H, Zinkernagel RM (1987) Characterization of virus-specific cytotoxic T cell clones from allogeneic bone marrow chimeras. Eur J Immunol 17: 159–166

Podack ER, Hengartner H, Lichtenheld MG (1991) A central role of perforin in cytolysis? Annu Rev Immunol 9: 129–157

Ramsay AJ, Ruby J, Ramshaw IA (1993) A case for cytokines as effector molecules in the resolution of virus infection. Immunol Today 14: 155–157

Raulet DH (1994) MHC class I-deficient mice. Adv Immunol 55: 381– 421

Riviere Y, Robertson MN, Buseyne F (1994) Cytotoxic T lymphocytes in human immunodeficiency virus infection: regulator genes. Curr Top Microbiol Immunol 189: 65–74

Sarawar SR, Doherty PC (1994) Concurrent production of interleukin-2, interleukin-10, and gamma interferon in the regional lymph nodes of mice with influenza pneumonia. J Virol 68: 3112–3119

Sarawar SR, Carding SR, Allan W, McMickle A, Fujihashi K, Kiyono H, McGhee JR, Doherty PC (1993) Cytokine profiles of bronchoalveolar lavage cells from mice with influenza pneumonia: consequences of CD4+ and CD8+ T cell depletion. Reg immunol 5: 142–150

Sarawar SR, Sangster M, Coffman RL, Doherty PC (1994) Administration of anti-IFN-gamma antibody to β2-microglobulin-deficient mice delays influenza virus clearance but does not switch the response to a T helper cell 2 phenotype. J Immunol 153: 1246–1253

Scherle PA, Palladino G, Gerhard W (1992) Mice can recover from pulmonary influenza virus infection in the absence of class I-restricted cytotoxic T cells. J Immunol 148: 212–217

Slobod KS, Freiberg AS, Allan JE, Rencher SD, Hurwitz JL (1993) T-cell receptor heterogeneity among Epstein-Barr virus-stimulated T-cell populations. Virology 196: 179–189

Sprent J (1994) T and B memory cells. Cell 76: 315–322

Staats HF, Jackson RJ, Marinaro M, Takashi I, Kiyono H, McGhee JR (1994) Mucosal immunity to infection with implications for vaccine development. Curr Opin Immunol 6: 572–583

Stalder T, Hahn S, Erb P (1994) Fas antigen is the mayor target molecule for CD4$^+$ T cell mediated cytotoxicity. J Immunol 152: 1127–1133

Suzuki S, Hotchin J (1971) Initiation of persistent lymphocytic choriomeningitis infection in adult mice. J Infect 123: 603–610

Tabi Z, Lynch F, Ceredig R, Allan JE, Doherty PC (1988) Virus-specific memory T cells are Pgp-1$^+$ and can be selectively activated with phorbol ester and calcium ionophore. Cell immunol 113: 268–277

Taylor PM, Askonas BA (1983) Diversity in the biological properties of anti-influenza cytotoxic T cell clones. Eur J Immunol 13: 707–711

Tripp RA, Hou S, Doherty PC (1995) Temporal loss of the activated L-selectin-low phenotype for virus-specific CD8$^+$ memory T cells. J Immunol 154: 5870–5875

Walker JA, Sakaguchi T, Matsuda Y, Yoshida T, Kawaoka Y (1992) Location and character of the cellular enzyme that cleaves the hemagglutinin of a virulent avian influenza virus. Virology 190: 278–287

Walsh CM, Matloubian M, Liu C-C, Veda R, Kurahara CG, Christensen JL, Huang MTF, Young JD-E, Ahmed R, Clark WR (1994) Immune function in mice lacking the perforin gene. Proc Natl Acad Sci USA 91: 10854–10858

Wong GH, Clark-Lewis I, Harris AW, Schrader JW (1984) Effect of cloned interferon-gamma on expression of H-2 and Ia antigens on cell lines of hemopoietic, lymphoid, epithelial, fibroblastic and neuronal origin. Eur J Immunol 14: 52–56

Yanagi Y, Tishon A, Lewicki H, Cubitt BA, Oldstone MB (1992) Diversity of T-cell receptors in virus-specific cytotoxic T lymphocytes recognizing three distinct viral epitopes restricted by a single major histocompatibility complex molecule. J Virol 66: 2527–2531

Young LH, Klavinskis LS, Oldstone MB, Young JD (1989) in vivo expression of perforin by CD8$^+$ lymphocytes during an acute viral infection. J Exp Med 169: 2159–2171

Zijlstra M, Bix M, Simister NE, Loring JM, Raulet DH, Jaenisch R (1990) β2-microglobulin deficient mice lack CD4$^-$ 8$^+$ cytolytic T cells. Nature 344: 742–746

Zijlstra M, Auchincloss H Jr, Loring JM, Chase CM, Russell PS, Jaenisch R (1992) Skin graft rejection by β$_2$-microglobulin-deficient mice. J Exp Med 175: 885–893

Zinkernagel RM, Doherty PC (1974) Immunological surveillance against altered self components by sensitised T lymphocytes in lymphocytic choriomeningitis. Nature 251: 547–548

Zinkernagel RM, Doherty PC (1979) MHC-restricted cytotoxic T cells: studies on the biological role of polymorphic major transplantation antigens determining T-cell restriction-specificity, function, and responsiveness. Adv Immunol 27: 51–177

Zinkernagel RM, Welsh RM (1976) H-2 compatibility requirement for virus-specific T cell-mediated effector functions in vivo.I. Specificity of T cells conferring antiviral protection against lymphocytic choriomeningitis virus is associated with H-2K and H-2D. J Immunol 117: 1495–1502

Transgenic Mouse Models
of Experimental Autoimmune Encephalomyelitis

D.M. Zaller and V.S. Sloan

1 Demyelinating Autoimmune Diseases

Immunization with early, impure rabies vaccine developed from infected central nervous system (CNS) tissue produced encephalomyelitis in some humans (reviewed in HEMACHUDHA et al. 1987). Based on this observation, several investigators later showed that immunization of animals with heterologous extracts of brain tissue could produce inflammatory lesions in the CNS white matter accompanied by demyelination (reviewed in WEIGLE 1980). In 1947, four investigators independently reported the induction of encephalomyelitis in guinea pigs, monkeys and rabbits using one injection of CNS tissue emulsified in complete Freund's adjuvant (CFA). However, no one was able to induce the disease in mice until Bernard and Carnegie immunized SJL/J mice with pertussis vaccine in addition to spinal cord homogenate or myelin basic protein (MBP) emulsified in CFA (BERNARD and CARNEGIE 1975). Another protein component of myelin, myelin proteolipid apoprotein (PLP) (SATOH et al. 1987; TUOHY et al. 1988) has since been shown to be encephalitogenic.

The clinical signs of experimental autoimmune encephalomyelitis (EAE) include weakness, varying degrees of paralysis up to and including quadriplegia, and eventually death. Pathologically, there are inflammatory infiltrates of both mononuclear and polymorphonuclear cells in the white matter, meninges and

Department of Molecular Immunology, Merck Research Laboratories, RY 80M-124, PO Box 2000 Rahway, NJ 07065, USA

perivascular areas. The lymphocytic infiltrates have been shown to consist of T cells, B cells and macrophages. Additionally, MHC class II, immunoglobulin and MBP have all been found on endothelial cells at the site of CNS lesions (TRAUGOTT et al. 1985).

In the mouse model of EAE, induction of disease is dependent on the genetic background of the host. Mice with the $H-2^u$ or $H-2^k$ haplotype mediate EAE with T cells recognizing an MBP NH_2-terminal epitope. Mice bearing the $H-2^s$ or $H-2^q$ haplotype respond best to a more COOH-terminal epitope of MBP (FRITZ et al. 1985; KONO et al. 1988). However, MHC haplotype is not sufficient to predispose mice towards susceptibility to EAE. While EAE is easily induced in SJL mice, which are $H-2^s$, B10.S mice, which are also $H-2^s$, are resistant. This suggests that genetic background outside of the H-2 locus is also important in determining susceptibility to EAE. Determination of the T cell receptor (TCR) and C5 haplotypes revealed no contribution to susceptibility to induction of EAE (JANSSON et al. 1991).

The pathophysiology of EAE in animals resembles that of the human disease multiple sclerosis (MS). Multiple sclerosis is a CNS demyelinating disease characterized by CNS white matter lesions separated in time and space. The lesions are areas of decreased or absent myelin, with preservation of the underlying axon, and are found throughout the brain, spinal cord and optic nerves. Initially, oligodendrocytes disappear from the plaques. There are attempts at remyelination by immature oligodendrocytes, but eventually, astrocytes proliferate and there is scar formation. During the early phases, there is a perivascular inflammatory infiltrate of lymphocytes, plasma cells and macrophages as well as immunoglobulin deposition within the plaques (RODRIGUEZ 1989). The clinical manifestations of MS are protean and can include weakness of extremities, incoordination, unilateral vision loss (secondary to optic neuritis) and paresthesias. The course is classically relapsing and remitting, but 30% of patients develop a chronic progressive course. Twenty years after onset, only 25% of patients can carry out household or employment duties (SILBERBERG 1992).

Treatment options for patients with MS are limited. Acute exacerbations can be treated with glucocorticoids, but there is no evidence that such treatment affects the ultimate prognosis. Additionally, the side effects of glucocorticoids are well-described and often severe (ANSELL 1991; BOUMPAS 1993). Treatment with β-interferon has recently been shown to reduce the exacerbation rate in patients with relapsing-remitting MS, although the effects on the ultimate progression of the disease are uncertain (reviewed in CONNELLY 1994).

There is strong evidence for an inherited susceptibility to MS. There is an increased concordance in monozygotic twins as well as familial aggregation. A study of 40 sibling pairs afflicted with MS revealed TCRα chain gene haplotypes that are shared in greater frequency than would be expected with random segregation, suggesting that a gene within or closely linked to the TCRα complex influences susceptibility to MS (SEBOUN et al. 1989). However, there is also good evidence for an infectious etiology, especially in the outbreak of MS occurring in the Faroe islands after occupation by British troops during World War II. There were no cases of MS in the islands before the occupation, and sudden development of the disease several years after the occupation began. Additionally, the prevalence of MS increases with distance from the equator (SILBERBERG 1992).

1.1 The Role of T Cells in Experimental Autoimmune Encephalomyelitis

Several lines of evidence demonstrate the essential role of T cells in EAE. GONATAS and HOWARD (1974) showed that T cells were required for the induction of EAE. Lewis rats depleted of T cells were unable to develop EAE or antibodies against MBP. When the rats were reconstituted with thymocytes, EAE developed as it did in the control animals.

Adoptive transfer of EAE in the mouse was first demonstrated in 1976 (BERNARD et al. 1976). The transfer required a large number of lymphocytes that had been activated with CNS homogenate. In 1981, PETTINELLI and McFARLIN transferred the disease in SJL/J mice using lymphocytes that had been activated with purified MBP. The first reported adoptive transfers produced a monophasic illness with minimal demyelination. Other investigators (MOKHTARIAN et al. 1984), using transfer of MBP-sensitized lymph node cells or T cells, were able to produce both acute EAE with significant demyelination, as well as a chronic, relapsing-remitting disease lasting several months. BEN-NUN and LANDO (1983) were able to create T cell lines that mediated EAE in normal mice using relatively small numbers of cells (1×10^6). In 1989, VAN DER VEEN et al. showed that immunization with PLP could also produce T cells capable of adoptively transferring EAE. There was no cross-reactivity between PLP and MBP.

Treatment with monoclonal antibodies against T cells led to similar conclusions. Treatment with anti-CD4 was shown to block the development of EAE (SRIRAM et al. 1988), and several other groups demonstrated improvement in established EAE after administration of anti-CD4 (SRIRAM and ROBERTS 1986; WALDOR et al. 1985). The improvement was manifested by fewer relapses and less severe histopathology.

Antibodies against TCRs specific for MBP were also shown to block induction of or abrogate active EAE. OWHASHI and HEBER-KATZ (1988) developed MBP-specific T cell lines which were capable of inducing severe EAE in the rat. These T cell lines expressed a specific idiotype that was detectable with a monoclonal antibody. Pretreatment with this antibody prevented induction of disease. Analysis of TCR genes of MBP-specific, $H-2^u$-restricted T-cell clones specific for an encephalitogenic epitope of MBP revealed a highly restricted use of TCR V-region segments (ACHA-ORBEA et al. 1988; URBAN et al. 1988). Approximately 80% of the cells used the Vβ8.2 gene segment. A monoclonal antibody against the Vβ8 chain was able to partially inhibit the induction of EAE. Treatment with a combination of anti-Vβ8 and Vβ13 antibodies produced an even more significant reduction in MBP responsiveness and a decrease in escape from antibody protection against disease (ZALLER et al. 1990). These results suggest that, in B10.PL mice, although

80% of MBP peptide-specific T cells use Vβ8, antibody treatment escape is due to the presence of fewer, but still important, MBP-reactive cells using Vβ13.

As early as 1983, it was demonstrated that administration of anti-MHC class II antibodies could alter the course of EAE in mice (SRIRAM and STEINMAN 1983). Later investigations demonstrated that this effect was haplotype-specific (SRIRAM et al. 1987). In F1 offspring of crosses between the EAE-susceptible SJL strain (H-2^s) and the EAE-resistant BALB/c strain (H-2^d), anti-A^d was successful in preventing disease. This effect occurred whether treatment was begun at the time of immunization or following passive transfer of T cells. AHARONI et al. (1991) reported that an antibody specific for the A^s/MBP(89–101) complex inhibited both the in vitro response of T cells to MBP and the induction of EAE.

These lines of evidence demonstrate clearly that EAE is mediated by the interactions of autoreactive, myelin-specific CD4+ T cells with antigen-presenting cells (APCs) expressing MHC class II.

2 Lack of Tolerance to Myelin Basic Protein

The immune response among different mouse strains after challenge with MBP is variable. After immunization with MBP, high responder strains such as SJL and B10.PL are susceptible to EAE, while low responder strains such as BALB/c and C57BL/6 are resistant. However, EAE can consistently be induced in resistant strains by a procedure that involves MBP immunization, restimulation of T cells with MBP in vitro, adoptive transfer of these T cells into naive recipients, followed by in vivo boosting with additional MBP (SHAW et al. 1992). By this protocol, EAE was induced in eleven different EAE-resistant strains representing a wide range of genetic backgrounds and MHC haplotypes. Thus it appears that many, if not all, strains of mice are not completely tolerant to self-MBP.

MBP is specifically expressed in nervous system tissue, where it comprises 30%–40% of CNS myelin and 5%–15% of peripheral nervous system myelin (LEES and BROSTOFF 1984). The nervous system has been considered to be immunologically privileged (BARKER and BILLINGHAM 1977). Under normal circumstances, the blood-brain barrier of tight endothelial junctions prevents access of lymphocytes into the brain. Also, the brain lacks conventional lymphatic drainage, the usual means for trafficking of antigens to secondary lymphoid organs. The anatomic barrier and the lack of efferent lymphatics may account for the prolonged survival of allografts in the brain. Thus the brain and its component tissues may be sequestered from routine lymphoid surveillance. It is possible that MBP might represent a sequestered self-antigen that is never seen by the developing immune system and thus never induces tolerance.

To experimentally assess the degree of tolerance to self-MBP, immune responses to MBP were compared in wild-type and shiverer mice. Shiverer mice are homozygous for a mutation in which the last five exons of the MBP gene are deleted (DUPOUEY et al. 1979; KIRSCHNER and GANSER 1980; READHEAD and HOOD 1990). The first two exons encoding MBP are still present in the shiverer mutation and these exons are transcribed into a short nonpolyadenylated messenger RNA (WIKTOROWICZ and ROACH 1991). However, truncated MBP proteins cannot be detected in shiverer mice (READHEAD and HOOD 1990). Shiverer mice are therefore considered to have an MBP-null phenotype. These mice experience motor tremors and tonic seizures and have a dramatically shortened life span. To determine if the expression of endogenous MBP leads to self-tolerance, the lymph node proliferative responses of wild-type and shiverer mice that had been immunized with MBP were compared. The results are shown in Fig. 1. In mice of two different MHC haplotypes, H-2^d and H-2^u, the responses in the shiverer mice were significantly greater than the responses in the wild-type mice. Thus MBP cannot be considered a sequestered self-antigen, since its expression leads to at least some degree of tolerance. Partial tolerance to MBP implies that some mechanism must exist for presentation of self-MBP to the immune system. Soluble protein exchange between cerebrospinal fluid and lymph, as had been observed for albumin (HARLING-BERG et al. 1989) or incomplete integrity of the blood-peripheral nerve barrier may provide opportunities for APCs to encounter self-MBP.

To shed additional light on the basis for the enhanced immune response to MBP in shiverer mice, responses to individual epitopes of MBP were character-ized. A set of 16 overlapping peptides covering the entire 18.5 KDa form of MBP was used to recall immune responses from mice that were immunized with whole MBP. The results are shown in Fig. 2. In both MHC haplotypes tested, the

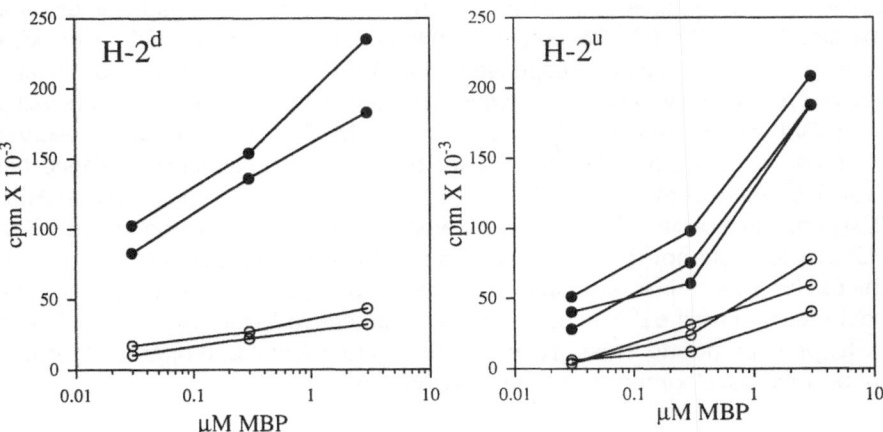

Fig. 1. Myelin basic protein (*MBP*)-deficient shiverer mice are hyper-responsive to MBP. H-2^d or H-2^u, wildtype or shiverer mice were immunized with 50 µg of mouse MBP emulsified in complete Freund's adjuvant. Draining lymph node cells were collected 10 days later and cultured with various concentra-tions of mouse MBP. A total of 3×10^5 lymph node cells were cultured in triplicate wells of 96-well microtiter plates. The plates were incubated at 37°C for 5 days. During the last 16 hours of culture, 1 µCi of [^{3}H] thymidine was added. Cells were then harvested and cpm were determined by liquid scintillation counting. Data points reflect the mean of triplicate samples. Each curve represents the dose response of an individual mouse. *Open circles*, MBP wild-type mice; *filled circles*, MBP deficient mice

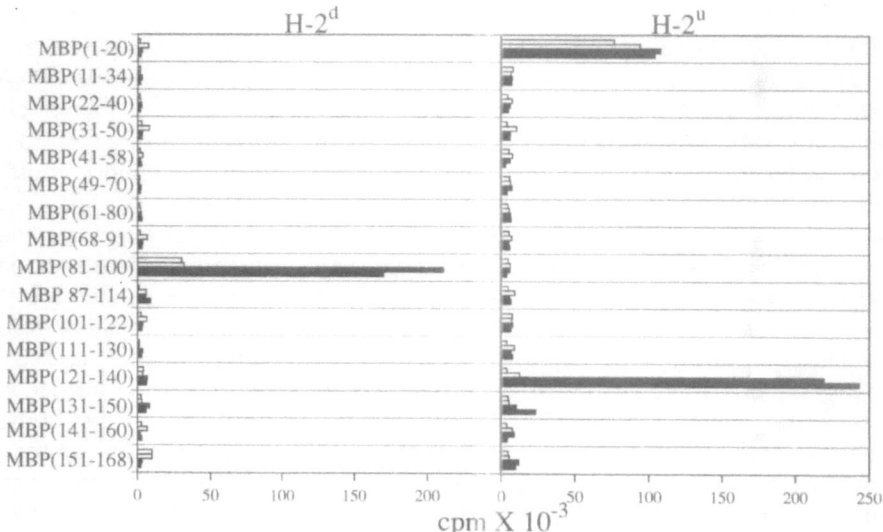

Fig. 2. Epitope mapping of the response to myelin basic protein (*MBP*). Draining lympth node cells from H-2^d or H-2^u, wild-type or shiverer mice that had been immunized with mouse MBP were cultured with 30 µM of the indicated peptides. The assays and analyses were otherwise performed as described in the legend to Fig. 1. Open rectangles, MBP wild-type mice; filled rectangles, MBP deficient mice

hyperresponsiveness to MBP observed in shiverer mice was directed towards single distinct epitopes. In H-2^d shiverer mice, the T cell response was directed towards MBP (81–100). Responses to this epitope in wild-type H-2^d mice were very poor. In H-2^u mice, the response to the MBP(1–20) epitope was comparable in wild-type and shiverer mice; however, only the shiverer mice responded well to the MBP(121–141) epitope. Thus, tolerance to MBP in H-2^u mice was observed for some epitopes of MBP but not for others. Epitope-specific tolerance has previously been demonstrated in the murine response to hen egg lysozyme (HEL) (GAMMON and SERCARZ 1989). After intravenous injection of tolerogenic doses of HEL, mice could not respond to dominant HEL epitopes, but could respond to minor HEL determinants. A lack of tolerance to minor epitopes of self-proteins could account for the presence of mature autoreactive T cells. Activation of these cells, perhaps due to cross-reaction with environmental pathogens, could lead then to the development of autoimmune disease.

3 T Cell Receptor Transgenic Mice

T cell receptor transgenic mice offer a powerful model system for studying tolerance to MBP and the role of MBP-specific T cells in the EAE disease process.

Given the vast diversity of the TCRs expressed in any individual, it was a formidable task to study the development of specific T cells before the development of transgenic technology. The normal mouse has approximately 10^8 different TCRs (VON BOEHMER 1990). The transgenic animal, by contrast, has essentially only one receptor with a single defined specificity. Since the problem of the low frequency of T cells specific for a single antigen is eliminated in TCR transgenic mice, they are ideal for studying mechanisms of tolerance.

In transgenic mice expressing a rearranged TCR-β chain, endogenous TCR β genes do not rearrange (BORGULYA et al. 1992; RIMM et al. 1989; UEMATSU et al. 1988). Thus, allelic exclusion of the β-chain locus is due to negative feedback on VβDβJβ recombination. In contrast, in α-chain transgenics, endogenous α chains can rearrange (HEATH and MILLER 1993). It was subsequently demonstrated that the degree of expression of endogenous α-chain is determined by the presence or absence of positive selection (PETRIE et al. 1993). A model for T cell development was proposed in which cells that fail positive selection continue to rearrange α chains until positive selection takes place (PETRIE et al. 1993). Nevertheless, there is always some "leakiness" in endogenous α-chain expression, which varies from transgenic to transgenic.

The first TCR transgenic mice were created in 1988 by von Boehmer's group, using a TCR specific for the HY antigen (TEH et al. 1988) and by Loh's group, using a TCR specific for the L^d class I•MHC antigen (SHA et al. 1988a,b). The TCRs used in these transgenic mice were derived from CD8-positive T cells clones. Both groups found that positive selection of T cells expressing the transgenic TCR occurred only in mice with the appropriate MHC class I haplotype and that transgenic T cells were remarkably skewed towards the expression of CD8. Both groups also showed that when the transgenic T cells developed in mice expressing the relevant self-antigen, tolerance occurred via intrathymic clonal deletion. The clonal deletion occurred at or before the double-positive stage. These results suggested that the interactions among the TCR, the CD8 and the self-MHC determined the developmental fate of MHC class I-restricted T cells. These observations were soon extended to MHC class II-restricted TCR transgenic mice (BERG et al. 1989; KAYE et al. 1989). In these models, a dramatic skewing towards CD4+ T cells was observed when the transgenic T cells developed in the appropriate MHC haplotype.

3.1 T Cell Receptor Transgenic Models of Peripheral Tolerance

Not all antigens can be expressed intrathymically, so that some tolerance must be induced in the periphery. Transgenic mouse technology has brought forth valuable model systems for elucidating the mechanisms of peripheral tolerance. For example, MORAHAN et al. (1991) created transgenic mice that have rearranged TCR genes encoding an anti-H-2K^b TCR and mice expressing H-2K^b in pancreatic β cells. They then used monoclonal antibodies (mAbs) to follow anti-H-2K^b T cells in double transgenic mice expressing H-2K^b. Most of the T cells present were specific for H-2K^b. However, the double transgenic mice were tolerant of H-2K^b, as demonstrated by their inability to reject H-2K^b+ skin grafts. As the pancreatic β cells died off, gradually eliminating the tolerogen, tolerance decreased. Significantly, there was neither deletion of the potentially autoreactive T cells nor down-regulation of either CD8 or the TCR, supporting the concept of peripheral tolerance by T cell anergy.

OHASHI et al. (1991) created transgenic mice that coexpressed the lymphocytic choriomeningitis virus (LCMV) glycoprotein (GP) in pancreatic β-islet cells and a rearranged TCR specific for the LCMV-GP antigen. Transgenic T cells were not tolerized but nevertheless remained quiescent in vivo due to a lack of appropriate T cell activation. Activation by LCMV infection abolished peripheral tolerance resulting in T cell-mediated diabetes. These experiments establish T cell clonal ignorance as an additional mechanism for maintaining functional tolerance in vivo.

In another model, NOD transgenic mice were created using TCR genes derived from a CD4+ clone that was islet antigen-specific, H-2g^7-restricted and diabetogenic (KATZ et al. 1993). T cells from these mice were not clonally deleted in the thymus or in the periphery. In addition, T cells from the TCR transgenic mice were not anergic, in that they could respond quite well to islet cells in vitro, and aggressively infiltrated the pancreatic islets in vivo. The onset of disease in the transgenic animals was only slightly earlier that in the NOD controls animals, although the disease penetrance was somewhat higher in the transgenic group. Thus there was a significant delay (at least 17 weeks) before the animals developed overt autoimmune disease despite the presence of large numbers of functionally autoreactive T cells early in development.

In a series of elegant experiments, SCHONRICH et al. (1991, 1992) created transgenic mice expressing K^b in specific tissues: hepatocytes, keratinocytes or neuroectodermal cells. They then crossed these mice with anti-K^b-specific TCR transgenics and followed the fate of the anti-K^b-specific T cells. Using skin grafts, they demonstrated tolerance to the K^b antigen in the double transgenic mice. However the basis for the observed tolerance differed between the various animals. In the hepatocyte-K^b F1 mice and the neuroectodermal-K^b F1 mice, tolerance was achieved principally by down-modulation of TCR and CD8 molecules. This down-modulation could be reversed by in vitro stimulation with antigen in the neuroectodermal-K^b F1 mice but not in the hepatocyte-K^b F1 mice. In the keratinocyte-K^b F1 mice tolerance was achieved by clonal anergy.

The studies summaried above, and similar studies, demonstrate the impossibility of defining a single mechanism of peripheral tolerence. In various transgenic models, it can be demonstrated that tolerance can be achieved by either intra-thymic or extrathymic clonal delection, TCR or coreceptor down-modulation, T cell anergy, or T cell clonal ignorance. Models of split tolerance have also been described in which tolerance is maintained in vivo (skin grafts are not rejected) but not in vitro (T cell activation can be observed in mixed lymphocyte cultures). In addition, active suppression is another potential mechanism for achieving tolerance to peripheral antigens. Tolerance must thus be viewed as a multifaceted phenomenon.

3.2 Myelin Basic Protein-Specific T Cell Receptor Transgenic Mice

MBP-specific TCR transgenic mice can be used to follow the developmental fate of autoreactive T cells and the ability of these T cells to mediate EAE. TCR genes derived from an A^u-restricted, MBP(1–11)-specific T cell hybridoma, 172.10 (URBAN et al. 1988) were used to generate the first such MBP-specific TCR transgenic mice (GOVERMAN et al. 1993). Separate transgenic lines expressing rearranged TCRα and TCRβ genes were established by injection of C57BL/6 × DBA/2 F$_2$ oocytes. Virtually 100% of the T cells in the TCRβ line expressed the transgenic β-chain, while about 70% of the T cells in the TCRα line expressed the transgenic α-chain (Fig. 3). After intercrossing and backcrossing to B10.PL, the two lines were bred together to generate mice expressing the complete MBP-specific TCR. In these double transgenic mice, there was a marked skewing towards the CD4

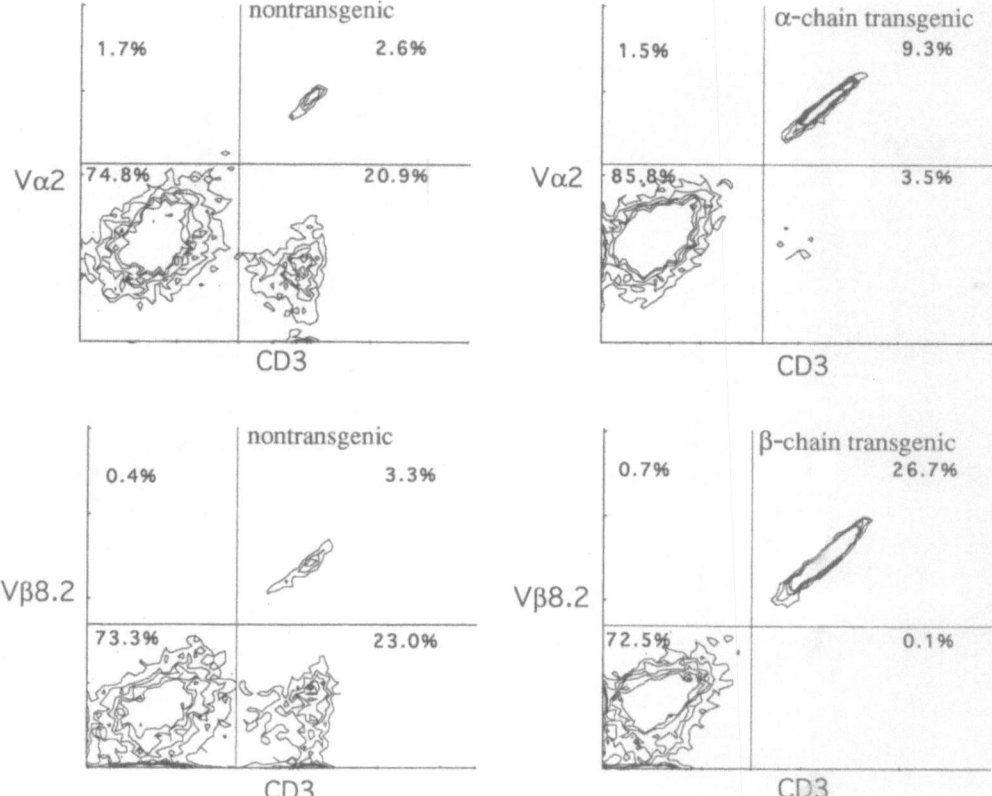

Fig. 3. Expression of a myelin basic protein (MBP)-specific T cell receptor (TCR) in transgenic mice. Spleen cells from nontransgenic, TCR α-chain transgenic, orTCR β-chain transgenic mice were double stained with a FITC-conjugated antibody specific for CD3, followed by a phycoerythrin (PE)-conjugated antibody specific for either Vβ8.2 or Vα2. Percentages of singly and doubly staining cells are indicated

compartment in both the thymic and peripheral T cell populations (Fig. 4). This was expected since the TCR transgenes were obtained from an MHC class II-restricted T cell hybridoma. Very few mature T cells developed when the TCR transgenes were crossed onto mice with H-2^b or an H-2^d haplotype, reflecting the requirement of the restricting H-2^u haplotype for positive selection. The fact that large numbers of mature CD4+ T cells expressing the transgenic TCR developed in mice of the H-2^u haplotype indicates that MBP(1–11)-specific T cells are not negatively selected despite the fact that MBP is a self-antigen.

Although mature T cells expressing the transgenic receptor could be found in the transgenic mice, it was possible that they might have been rendered anergic. To test this, the ability of transgenic T cells to proliferate and secrete interleukin-2 (IL-2) in response to exogenously added MBP peptide was measured. T cells from transgenic mice, but not from nontransgenic littermates, responded vigorously in

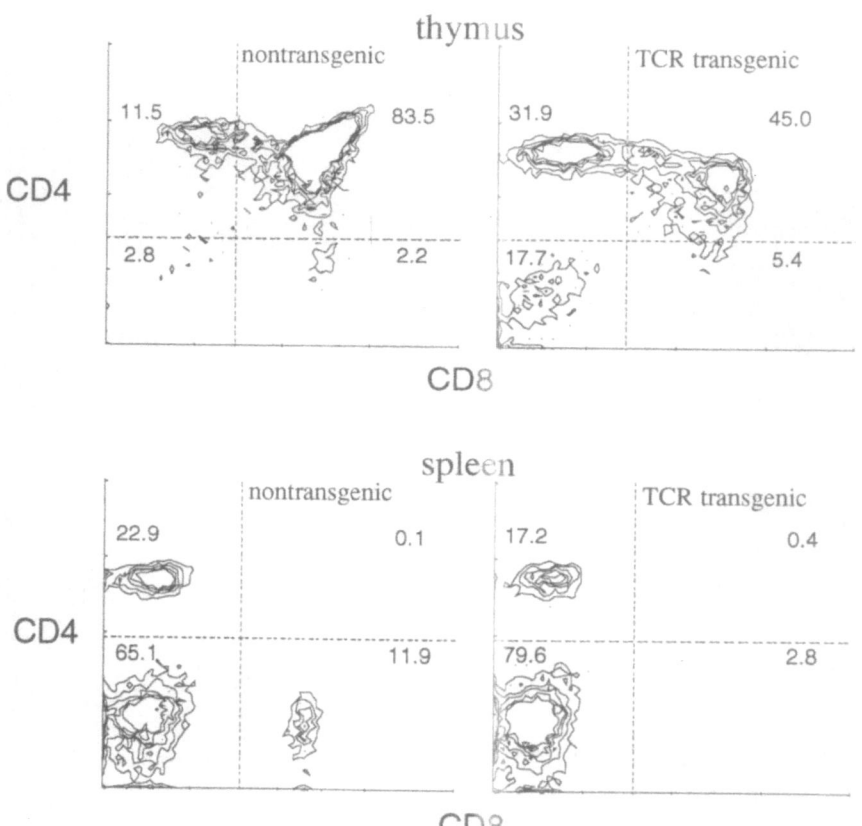

Fig. 4. Skewing towards the CD4 compartment in myelin basic protein (MBP)-specific T cell receptor (TCR) transgenic mice. Thymocytes or splenocytes were double stained with a FITC-conjugated antibody specific for CD8, followed by a PE-conjugated antibody specific for CD4. Percentages of singly and doubly staining cells are indicated

these assays (Fig. 5). Thus the MBP-specific transgenic T cells were functionally competent. Spontaneous MBP-specific autoantibodies could not be detected in unimmunized transgenic mice. However, after immunization with MBP, very high titers of serum anti-MBP antibodies could be measured (Fig. 6). This demonstrated that the MBP-specific T cells were capable of functioning as helper cells in vivo. Despite the presence of large numbers of functionally competent MBP-specific T cells, the transgenic mice in a B10.PL background rarely developed spontaneous EAE when housed in a specific pathogen-free facility. The incidence

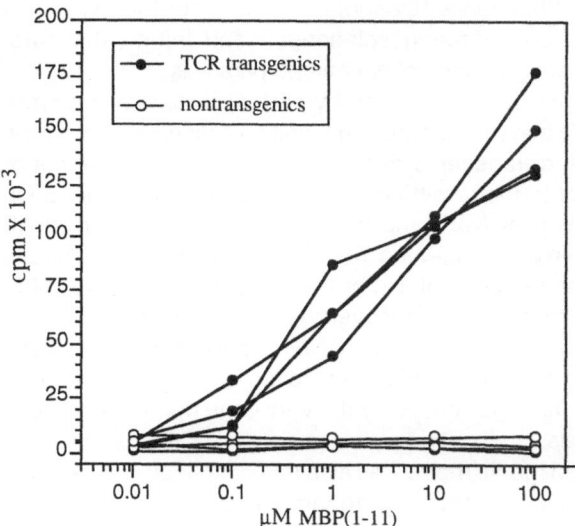

Fig. 5. T-cells from myelin basic protein (MBP)-specific T cell receptor (TCR) transgenic mice can respond to MBP. Spleen cells from unimmunized nontransgenic or TCR transgenic mice were cultured with various concentrations of MBP peptide at 5×10^4 CD3+ cells per well. ³H-thymidine incorporation data were collected and analyzed as described in the legend to Fig. 1

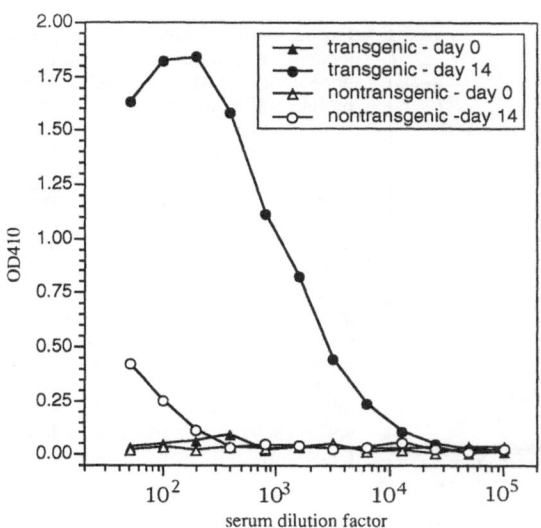

Fig. 6. T cells from myelin basic protein (MBP)-specific T cell receptor (TCR) transgenic mice can function as helper cells. Transgenic or nontransgenic mice were immunized intraperitoneally with MBP(1–20) peptide emulsified in complete Freund's adjuvant. Serum samples were collected immediately before immunization and 14 days later. Specific antibodies in the sera were measured using an anti-MBP-specific ELISA

of spontaneous paralysis was higher in transgenic animals housed in a conventional facility with a high pathogen load, raising the intriguing possibility that an environmental factor or factors might initiate autoimmune responses in these mice.

Severe EAE could be induced in these transgenic animals by the injection of pertussis toxin alone (Fig. 7). Pertussis toxin is known to enhance the induction of EAE in several species of animals and is usually required in murine models of EAE. The precise role of pertussis toxin in inducing EAE is unclear due to its pleiotropic physiological effects. One possibility is that pertussis toxin increases the vascular permeability of the CNS, allowing lymphocytes to pass through the blood-brain barrier (LINTHICUM 1982). Once transgenic T cells enter the CNS, they may become activated after encountering self-antigen and initate the auto-immune disease process resulting in demyelination and paralysis.

The paralysis induced in the transgenic mice by pertussis toxin was severe (grade 3 or grade 4, see Fig. 7) and permanent. Remission of disease symptoms was not observed in any of the transgenic mice, even after several months of observation. This is in contrast to the remitting paralysis most often observed in nontransgenic mice by injection of MBP and pertussis toxin. Studies on EAE induction in CD8-deficient animals suggested that CD8+ T cells could play an important role in mediating the recovery phase of the disease (JIANG et al. 1992) KOH et al. 1992). It is possible that the lack of recovery in the transgenic mice could be related to the fact that there are relatively few CD8+ T cells in these animals. Alternatively, the failure of the transgenic mice to recover from severe paralysis could be due extensive tissue damage caused by the very large numbers of MBP-specific T cells in these animals.

This MBP-specific TCR transgenic model has also revealed additional genetic factors that influence the development of autoimmune disease. Transgenic

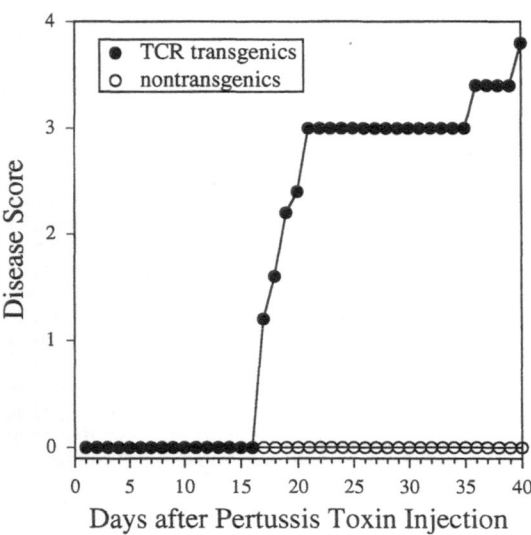

Fig. 7. Pertussis toxin, by itself, can induce experimental autoimmune encephalitis (EAE) in myelin basic protein (MBP)-specific T cell receptor (TCR) transgenic mice. Transgenic or nontransgenic mice were immunized intravenously with 400 ng of pertussis toxin on day 0 and day 2. Animals were scored daily for disease severity using the following grading scale: grade 1, loss of tail tone; grade 2, hind limb weakness and difficulty turning over; grade 3, hind limb paralysis; grade 4, hind limb and fore limb paralysis; grade 5, death. Data was collected for 19 nontransgenic mice and 16 transgenic mice. The average disease score within each group is plotted as a function of time

animals that had been back-crossed for ten generations with B10.PL mice were mated with several different inbred mouse strains. F1 mice from these crosses were evaluated for the development of MBP-specific T cells and the incidence of spontaneous EAE. These experiments have established that both MHC-linked and MHC-unlinked genes can contribute to the maturation of MBP-specific T cells and to the development of autoimmunity in mice.

These MBP-specific TCR transgenic mice have been used to study the phenomenon of high zone tolerance (CRITCHFIELD et al. 1994). MBP-specific transgenic T cells were shown to undergo apoptosis when stimulated with high doses of antigen. Apoptosis of these T cells was shown to depend on a sequence of events that invloved T cell activation and progression through the cell cycle followed by the reengagement of the TCR by antigen. It was speculated that this apoptotic program might provide a feedback mechanism that can down-modulate the immune response when it becomes too intense.

A similar MBP-specific TCR transgenic mouse line was recently established from another MBP(1–11)-specific, A^u-restricted T cell clone (LAFAILLE et al. 1994). Similar to the previously described transgenic mice, these animals contained large numbers of CD4+, functionally competent T cells, but exhibited only a low frequency of spontaneous EAE. The transgenic mice were then crossed with RAG-1 deficient mice to create animals that contained T cells expressing the transgenic TCR but no other B or T cells. Interestingly, 100% of the RAG-1 deficient TCR transgenic animals developed spontaneous EAE. This indicates that CD4+, MBP-specific T cells can cause EAE in the absence of any other lymphocytes. The complete absence of functional B cells in these animals clearly demonstrates that autoantibodies are not required for disease induction in this model. The data also suggest that some as yet uncharacterized population of lymphocytes can function to prevent EAE. These protective lymphocytes might derive from the relatively few CD4 or CD8 T cells in the transgenic mice that express endogenously rearranged TCR genes. Regulatory T cells that can modulate EAE have been described in several different experimental systems (ELLERMAN et al. 1988; GAUR et al. 1993; JIANG et al. 1992; KOH et al. 1992; MILLER et al. 1992; SUN et al. 1988; ZHANG et al. 1993). Alternatively, the MBP-specific transgenic T cells may themselves acquire a protective function after stimulation by antigen presenting B cells that are absent in RAG-1 deficient mice.

4 Future Directions

Myelin basic protein-specific TCR transgenic mice have provided powerful model systems for studying EAE. These models can be further exploited to yield new insights into this autoimmune disease process. It is clear that both genetic and environmental factors can affect the incidence of spontaneous EAE in the transgenic animals. Controlled breeding experiments can be used to enumerate the genes that influence the frequency of spontaneous EAE and eventually can be used to pinpoint the genetic location and identity of these genes. Environmental factors such as diet can be systematically altered in order to evaluate their effect(s) on the rate of occurrence of spontaneous EAE. The influence of environmental pathogens can be evaluated by introducing specific infectious agents into the animals. Adoptive transfer of T cells derived from nontransgenic mice into MBP-specific TCR transgenic mice may aid in the identification of T cell populations that modulate the remission phase of the disease. In addition, immunologically relevant molecules invloved in the induction of EAE can be identified and characterized by perturbing the expression of these molecules in the transgenic mice and then monitoring the outcome. These and related transgenic animal models have enormous potential as tools to enhance our understanding of the auto-immune disease process.

Acknowledgments. The authors thank Hans Zweerink and Ed O'Neill for critical reading of the manuscript.

References

Acha-Orbea H, Mitchell DJ, Timmermann L, Wraith DC, Tausch GS, Waldor MK, Zamvil SS, McDevitt HO, Steinman L (1988) Limited heterogeneity of T cell receptors from lymphocytes mediating autoimmune encephalomyelitis allows specific immune intervention. Cell 54: 263–273

Aharoni R, Teitelbaum D, Arnon R, Puri J (1991) Immunomodulation of experimental allergic encephalomyelitis by antibodies to the antigen-Ia complex. Nature 351: 147–150

Ansell BM (1991) Overview of the side effects of corticosteroid therapy. Clin Exp Rheumatol 9 Suppl 6: 19–20

Barker CF, Billingham RE (1977) Immunologically privileged sites. Adv Immunol 25: 1–28

Ben-Nun A, Lando Z (1983) Detection of autoimmune cells proliferating to myelin basic protein and selection of T cell lines that mediate experimental autoimmune encephalomyelitis (EAE) in mice. J Immunol 130: 1205–1209

Berg LJ, Pullen AM, Fazekas DSGB, Mathis D, Benoist C, Davis MM (1989) Antigen/MHC-specific T cells are preferentially exported from the thymus in the presence of their MHC ligand. Cell 58: 1035–1046

Bernard CCA, Carnegie PR (1975) Experimental autoimmune encephalomyelitis in mice: immunologic response to mouse spinal cord and myelin basic proteins. J Immunol 114: 1537–1540

Bernard CCA, Leydon J, Mackay IR (1979) T cell necessity in the pathogenesis of experimental autoimmune encephalomyelitis in mice. Eur J Immunol 6: 665–660

Borgulya P, Kishi H, Uematsu Y, von Boehmer H (1992) Exclusion and inclusion of alpha and beta T cell receptor alleles. Cell 69: 529–537

Boumpas DT (1993) Glucocorticoid therapy for immune mediated diseases: basic and clinical correlates. Ann Intern Med 119: 1198–1208

Connelly JF (1994) Interferon-beta for multiple sclerosis. Ann Pharmacother 28: 620–616

Critchfield JM, Racke MK, Zuniga PJ, Cannella B, Raine CS, Goverman J, Lenardo MJ (1994) T cell deletion in high antigen dose therapy of autoimmune encephalomyelitis. Science 263: 1139–1143

Dupouey P, Jacques C, Bourre JM, Cesselin F, Privat A, Baumann N (1979) Immunochemical studies of myelin basic protein in shiverer mouse devoid of the major dense line of myelin. Neurosci Lett 12: 113–119

Ellerman KE, Powers JM, Brostoff SW (1988) A suppressor T-lymphocyte cell line for autoimmune encephalomyelitis. Nature 331: 265–267

Fritz RB, Skeen MJ, Chou C-HJ, Garcia M, Egoov IK (1985) Major histocompatibility complex-linked control of the murine response to myelin basic protein. J Immunol 134: 2328–2332

Gammon G, Sercarz E (1989) How some T cells escape tolerance induction. Nature 342: 183–185

Gaur A, Ruberti G, Haspel R, Mayer JP, Fathman CG (1993) Requirement for CD8+ cells in T cell receptor peptide-induced clonal unresponsiveness. Science 259: 91–94 (published erratum appears in Science (1993) 260 (5106): preceding 281

Gonatas NK, Howard JC (1974) Inhibition of experimental autoimmune encephalomyelitis in rats severely depleted of T cells. Science 186: 839–841

Goverman J, Woods A, Larson L, Weiner LP, Hood L, Zaller DM (1993) Transgenic mice that express a myelin basic protein-specific T cell receptor develop spontaneous autoimmunity. Cell 72: 551–560

Harling-Berg C, Knopf PM, Merriam J, Cserr HF (1989) Role of cervical lymph nodes in the systemic humoral immune response to human serum albumin microinfused into rat cerebrospinal fluid. J Neuroimmuol 25: 185–193

Heath WR, Miller JF (1993) Expression of two alpha chains on the surface of T cells in T cell receptor transgenic mice. J Exp Med 178: 1807–1811

Hemachudha T, Griffin DE, Giffels JJ, Johnson RT, Moser AB, Phanuphak P (1987) Myelin basic protein as an encephalitogen in encephalomyelitis and polyneuritis following rabies vaccination. N Engl J Med 316: 369–374

Jansson L, Olsson T, Hojeberg B, Holmdahl R (1991) Chronic experimental autoimmune encephalo-myelitis induced by the 89–101 myelin basic protein peptide in B10RIII (H-2r) mice. Eur J Immunol 21: 693–699

Jiang H, Zhang SI, Pernis B (1992) Role of CD8+ T cells in murine experimental allergic encephalomyelitis. Science 256: 1213–1215

Katz JD, Wang B, Haskins K, Benoist C, Mathis D (1993) Following a diabetogenic T cell from genesis through pathogenesis. Cell 74: 1089–1100

Kaye J, Hsu ML, Sauron ME, Jameson SC, Gascoigne NR, Hedrick SM (1989) Selective development of CD4+ T cells in transgenic mice expressing a class II MHC-restricted antigen receptor. Nature 341: 746–749

Kirschner DA, Ganser AL (1980) Compact myelin exists in the absence of myelin basic protein in the shiverer mutant. Nature 283: 207–210

Koh DR, Fung LW, Ho A, Gray D, Acha OH, Mak TW (1992) Less mortality but more relapses in experimental allergic encephalomyelitis in CD8−/− mice. Science 256: 1210–1213

Kono DH, Urban JL, Horvath SJ, Ando DG, Saavedra RA, Hood L (1988) Two minor determinants of myelin basic protein induce experimental allergic encephalomyelitis in SJL/J mice. J Exp Med 168: 213–227

Lafaille JJ, Nagashima K, Katsuki M, Tonegawa S (1994) High incidence of spontaneous autoimmune encephalomyelitis in immunodeficient anti-myelin basic protein T cell receptor mice. Cell 78: 399–408

Lees MB, Brostoff SW (1984) Proteins in myelin. In: Morell P (ed) Myelin. Plenum, New York, pp 197–224

Linthicum DS (1982) Development of acute autoimmune encephalomyelitis in mice: factors regulating the effector phase of the disease. Immunobiology 162: 211–220

Miller A, Lider O, Roberts AB, Sporn MB, Weiner HL (1992) Suppressor T cells generated by oral tolerization to myelin basic protein suppress both in vitro and in vivo immune responses by the release of transforming growth factor beta after antigen-specific triggering. Proc Natl Acad Sci USA 89: 421–425

Mokhtarian F, McFarlin DE, Raine CS (1984) Adoptive transfer of myelin basic protein-sensitized T cells produces chronic relapsing demyelinating disease in mice. Nature 309: 356–358

Morahan G, Hoffmann MW, Miller JF (1991) A nondeletional mechanism of peripheral tolerance in T-cell receptor transgenic mice. Proc Natl Acad Sci USA 88: 11421–11425

Ohashi PS, Oehen S, Buerki K, Pircher H, Ohashi CT, Odermatt B, Malissen B, Zinkernagel RM, Hengartner H (1991) Ablation of "tolerance" and induction of diabetes by virus infection in viral antigen transgenic mice. Cell 65: 305–317

Owhashi M, Heber-Katz E (1988) Protection from experimental allergic encephalomyelitis conferred by a monoclonal antibody directed against a shared idiotype on rate T cell receptors specific for myelin basic protein. J Exp Med 168: 2153–2164

Petrie HT, Livak F, Schatz DG, Strasser A, Crispe IN, Shortman K (1993) Multiple rearrangements in T cell receptor alpha chain genes maximize the production of useful thymocytes. J Exp Med 178: 615–622

Pettinelli CD, McFarlin DE (1981) Adoptive transfer of experimental allergic encephalomyelitis in SJL/J mice after in vitro activation of lymph node cells by myelin basic protein: requirement for Lyt 1+ 2-lymphocytes. J Immunol 127: 1420–1423

Readhead C, Hood L (1990) The dysmyelinating mouse mutations shiverer (shi) and myelin deficient (shimld). Behav Genet 20: 213–234

Rimm IJ, Bloch DB, Seidman JG (1989) Allelic exclusion and lymphocyte development. Lessons from transgenic mice. Mol Biol Med 6: 355–364

Rodriguez M (1989) Multiple sclerosis: basic concepts and hypothesis. Mayo Clin Proc 64: 570–576

Satoh J, Sakai K, Endoh M, Koike F, Kunishita T, Namikawa T, Yamamura T, Tabira T (1987) Experimental allergic encephalomyelitis mediated by murine encephalitogenic T cell lines specific for myelin proteolipid apoprotein. J Immunol 138: 179–184

Schonrich G, Kalinke U, Momburg F, Malissen M, Schmitt VA, Malissen B, Hammerling GJ, Arnold B (1991) Down-regulation of T cell receptors on self-reactive T cells as a novel mechanism for extrathymic tolerance induction. Cell 65: 293–304

Schonrich G, Momburg F, Malissen M, Schmitt VA, Malissen B, Hammerling GJ, Arnold B (1992) Distinct mechanisms of extrathymic T cell tolerance due to differential expression of self antigen. Int Immunol 4: 581–590

Seboun E, Robinson MA, Doolittle TH, Ciulla TA, Kindt TJ, Hauser SL (1989) A susceptibility locus for multiple sclerosis is linked to the T cell receptor beta chain complex. Cell 57: 1095–1100

Sha WC, Nelson CA, Newberry RD, Kranz DM, Russell JH, Loh DY (1988a) Positive and negative selection of an antigen receptor on T cells in transgenic mice. Nature 336: 73–76

Sha WC, Nelson CA, Newberry RD, Kranz DM, Russel JH, Loh DY (1988b) Selective expression of an antigen receptor on CD8-bearing T lymphocytes in transgenic mice. Nature 335: 271–274

Shaw MK, Kim C, Ho KL, Lisak RP, Tse HY (1992) A combination of adoptive transfer and antigenic challenge induces consistent murine experimental autoimmune encephalomyelitis in C57BL/6 mice and other reputed resistant strains. J Neuroimmunol 39: 139–149

Silberberg DH (1992) The demyelinating diseases. In: Cecil RL, Wyngaarden JB, Bennett JC (eds) Cecil textbook of medicine. Saunders, Philadelphia, pp 2196–2200

Sriram S, Roberts CA (1986) Treatment of established chronic relapsing experimental allergic encephalomyelitis with anti-L3T4 antibodies. J Immunol 136: 4464–4469

Sriram S, Steinman L (1983) Anti I-A antibody suppresses active encephalomyelitis. Treatment model for diseases linked to IR genes. J Exp Med 158: 1362–1367

Sriram S, Topham DJ, Carroll L (1987) Haplotype-specific suppression of experimental allergic encephalomyelitis with anti I-A antibodies. J Immunol 139: 1485–1489

Sriram S, Carroll L, Fortin S, Cooper S, Ranges G (1988) In vivo immunomodulation by monochonal anti-CD4 antibody. II. Effect on T cell response to myelin basic protein and experimental allergic encephalomyelitis. J Immunol 141: 464–468

Sun D, Ben NA, Wekerle H (1988) Regulatory circuits in autoimmunity: recruitment of counter-regulatory CD8+ T cells by encephalitogenic CD4+ T line cells. Eur J Immunol 18: 1993–1999

Teh HS, Kisielow P, Scott B, Kishi H, Uematsu Y, Bluthmann H, von Boehmer H (1988) Thymic major histocompatibility complex antigens and the alpha beta T-cell receptor determine the CD4/CD8 phenotype of T cells. Nature 335: 229–233

Traugott U, Raine CS, McFarlin DE (1985) Acute experimental allergic encephalomyelitis in the mouse: immunopathology of the developing lesion. Cell Immunol 91: 240–254

Tuohy VK, Lu Z, Sobel RA, Laursen RA, Lees MB (1988) A synthetic peptide from myelin proteolipid protein induces experimental allergic encephalomyelitis. J Immunol 141: 1126–1130

Uematsu Y, Ryser S, Dembic Z, Borgulya P, Krimpenfort P, Berns A, von Boehmer H, Steinmetz M (1988) In transgenic mice the introduced functional T cell receptor beta gene prevents expression of endogenous beta genes. Cell 52: 831–841

Urban JL, Kumar V, Kono DH, Gomez C, Horvath SJ, Clayton J, Ando DG, Sercarz EE, Hood L (1988) Restricted use of T cell receptor V genes in murine autoimmune encephalomyelitis raises possibilities for antibody therapy. Cell 54: 577–592

Van der Veen RC, Trotter JL, Clark HB, Kapp JA (1989) The adoptive transfer of chronic relapsing experimental allergic encephalomyelitis with lymph node cells sensitized to myelin proteolipid protein. J Neuroimmunol 21: 183–191

Von Boehmer H (1990) Self recognition by the immune system. Eur J Biochem 194: 693–698

Waldor MK, Sriram S, Hardy R, Herzenberg LA, Herzenberg LA, Lanier L, Lim M, Steinman L (1985) Reversal of experimental allergic encephalomyelitis with monoclonal antibody to a T-cell subset marker. Science 227: 415–417

Weigle WO (1980) Analysis of autoimmunity through experimental models of thyroiditis and allergic encephalomyelitis. Adv Immunol 30: 159–273

Wiktorowicz M, Roach A (1991) Regulation of myelin basic protein gene transcription in normal and shiverer mutant mice. Dev Neurosci 13: 143–150

Zaller DM, Osman G, Kanagawa O, Hood L (1990) Prevention and treatment of murine experimental allergic encephalomyelitis with T cell receptor Vbeta-specific antibodies. J Exp Med 171: 1943–1955

Zhang J, Medaer R, Stinissen P, Hafler D, Raus J (1993) MHC-restricted depletion of human myelin basic protein-reactive T cellsby T cell vaccination. Science 261: 1451–1454

The Contribution of Insulitis
to Diabetes Development
in Tumor Necrosis Factor Transgenic Mice

R.A. Flavell[1,2], A. Kratz[3], and N.H. Ruddle[3]

[1] Section of Immunobiology, 310 Cedar Street, FMB 412 Yale University School of Medicine, New Haven, CT 06520-8011, USA
[2] Howard Hughes Medical Institute, Yale University School of Medicine, New Haven, CT 06520-8077, USA
[3] Department of Epidemiology and Public Health, Yale University School of Medicine, New Haven, CT 06520-8034, USA

1 Introduction

1.1 Characteristics of Insulin-Dependent Diabetes Mellitus

The hallmark of insulin-dependent diabetes mellitus (IDDM) is infiltration in (insulitis) and around (peri-insulitis) the islets of Langerhans, the site of synthesis of insulin (EISENBARTH 1986; GEPTS 1965). Destruction of these islets occurs and, as a consequence, the ability of the patient to produce insulin is eliminated, resulting in frank diabetes. Evidence accumulated over a number of years has shown that IDDM is determined by both genetic and environmental factors. A breakthrough in the genetics was made by the observation of a strong association between IDDM and HLA-DQβ, and later DQα alleles (TODD 1990). These early studies, however, showed clearly that this disease is polygenic in nature, since possession of these HLA susceptibility alleles only predisposed one to, but did not confer a high probability of, IDDM occurrence. In addition, multiple families, including some with identical twins, carrying susceptibility to IDDM have been studied. Strikingly, the results showed that, even when one twin develops IDDM, the probability of the second twin developing disease is only 20%–30%. This has led to the conclusion that environmental effects also play a key role in the development of diabetes (EISENBARTH 1986; TODD 1990).

The study of IDDM has been greatly facilitated by the availability of good animal models. In both biobreeding (BB) rat and the nonobese diabetic (NOD) mouse a form of IDDM similar to that found in humans is observed. The NOD mouse (KIKUTANI and MAKINO 1992) has proved to be a particularly suitable model due to the reproducibility and high frequency of disease and the availability of transgenic technology to genetically manipulate disease in these animals. The general features of NOD IDDM parallel those of human disease. Islets of Langerhans become infiltrated with peri-insulitis occurring around 3–4 weeks of age and gradually developing in intensity, ultimately converting to insulitis. β-cell destruction ensues, as evidenced by the loss of detectable insulin. Diabetes frequencies depend on the particular colony and may reach of up 90% in females, and somewhat less in males, by approximately 6 months of age. As in human IDDM, genetic analysis of NOD diabetes shows that this disease is polygenic. The MHC (Idd-1) genes have been well characterized: The NOD mouse has an unusual MHC class II I-A allele which, like its human homologue HLA-DQ, carries the same small amino acid residue at position 57; this striking parallel in structure between these two rare alleles further supports the relevance of the NOD model to human disease (TODD 1990). The role of a number of other recessive genes predisposing to IDDM has been demonstrated over the last few years, in particular using genome mapping techniques (HATTORI et al. 1986; TODD et al. 1991; WICKER et al. 1987, 1989). This has led to the identification of genes on chromosomes 1, 3, 6, 7, 11 and 14, all of which play a role in increasing the frequency of IDDM (GHOSH et al. 1983). No individual gene is absolutely required for disease, with the exception of Idd-1. Rather, these genes seem to play a cumulative role in increasing the incidence of IDDM.

1.2 Diabetes is a Multistep Process

The relationship between the inflammatory response and development of diabetes is an intriguing one. The role of environmental factors, discussed above, in the development of IDDM coupled with the correlation of a variety of different infectious agents with development of disease suggested that infection might play a role as a key environmental factor. A favored hypothesis (antigen mimicry) has been that infectious agents might encode gene products which are sufficiently similar in sequence to islet-specific proteins, such that an immune response primed against an infectious agent would then precipitate a response against self-antigen. Alternative hypotheses to be considered are that infection or other insults can lead to the establishment of a chronic inflammatory state, which itself can play a role in the development of autoimmune disease by the enhanced presentation of self-antigen to infiltrating lymphocytes (to be discussed below). Finally, the recent observations that some T lymphocytes contain two T cell receptors (TCRs), with two α chains combined with a common β, has raised the possibility that initial priming of a T cell by an infectious agent could lead to the recognition of self-antigens through the second receptor. While this seems to be a low probability event, such a model has neither been excluded nor appropriately tested.

Is insulitis either a necessary or a sufficient condition for the development of IDDM? As discussed above, one potential hypothesis is that an inflammatory response is sufficient to precipitate IDDM. Alternatively, the development of IDDM might require the prior infiltration of tissue provoked by environmental factors, and thus insulitis might play a predisposing role in disease. The alternative hypothesis to explain the association between insulitis and diabetes is that insulitis is an effect rather than a cause of diabetes and reflects the consequences of activation of a few self-reactive lymphocytes followed by the attraction of other inflammatory cells to the tissue by the mediators released by these activated cells. Even under this hypothesis, however, it is possible that bystander cells drawn to a tissue by the autoimmune response might play role in exacerbation of IDDM and possible tissue destruction. The analysis of genes associated with IDDM has suggested that certain genes predispose to insulitis and diabetes (notably Idd-3 and Idd-10 on chromosome 3), whereas two genes on chromosome 1 seem to predispose predominantly to peri-insulitis (GHOSH et al. 1993). Likewise, certain strains of NOD mice, such as NOD/WEHI, seem to show insulitis but little diabetes (CHARLTON et al. 1989). These results suggest again that insulitis may be a necessary, but possibly not a sufficient, condition for diabetes. However, the frequency of diabetes in NOD mice is notoriously variable from colony to colony and is strongly influenced by a variety of factors, including diet.

1.2.1 The Role of Inflammation

Vascular addressins for example, have been implicated in the inflammation associated with diabetes. 1. Inflammation is a multistep process involving multiple pairs of interacting ligand-receptor molecules. 2. A role for some of these has been suggested in diabetes and provides some insight into the mechanism of insulitis and sialitis in the NOD mouse. Inflamed islets express peripheral lymph node addressin (PNAd) and mucosal addressin cell adhesion molecule (MadCAM- 1), two antigens characteristic of high endothelial venules (FAVEEUW et al. 1994; HANNINEN et al. 1993) normally expressed in peripheral lymph nodes and Peyer's patches, respectively. Inflamed salivary glands of NOD mice express PNAd but not MadCAM-1. When compared to peripheral lympho- cytes, the lymphoid cells infiltrating these organs express lower levels of L-selectin, a receptor for these ligands. This may be because T cells previously exposed to antigen express reduced levels of L-selectin, and these cells, rather than naive T cells, preferentially enter tissue. It has not been determined whether expression of these ligands contributes to or is a consequence of in- flammation, since their expression is only seen in inflamed islets. A role for one of these ligands in insulitis is suggested from studies demonstrating that treat- ment of NOD mice with Mel-14 antibody inhibits the transfer of diabetes (YANG et al. 1993). Sialitis was not studied in these mice. There is also convincing evidence for a role of the VLA-4-VCAM-1 system in the pathogenesis of NOD diabetes. VCAM-1 expression is increased on blood vessels in NOD islets (BARON et al. 1993), and treatment of NOD mice with anti-VLA-4 antibody (YANG et al. 1993) or NOD recipients of NOD spleen cells with anti-VCAM-1 antibody (BARON et al. 1993) markedly decreases insulitis and inhibits disease development. The evi- dence is less compelling for a role for the LFA-1/ICAM-1 system in IDDM pathogenesis. Though LFA-1 and ICAM-1 are expressed in salivary glands and islets of NOD mice (FAVEEUW et al. 1994), anti-ICAM-1 treatment does not inhibit the transfer of diabetes in NOD mice (BARON et al. 1993). These studies suggest that the increased expression of ICAM-1 could be a consequence, rather than a contributing cause of inflammation.

1.2.2 The Role of the Immediate Tumor Necrosis Factor Ligand-Receptor Family in Diabetes

Though a number of cytokines have been implicated in IDDM, we have chosen to concentrate on tumor necrosis factors (TNFs). As potent mediators of inflam- mation, TNF-α and lymphotoxin (LT; also known as LT-α and TNF-β) may play important roles in the initiation of inflammatory responses and the resultant pathology. CD4[+] T lymphocytes that infiltrate the islets of Langerhans in the pancreas of the NOD mouse (HELD et al. 1990) and BB rat (JIANG and WODA 1991), express a variety of cytokine genes, including TNF-α. TNF-α, and to a lesser extent LT, synergize with interleukin-1 (IL-1) in the cytolysis of β-cells in vitro (MANDRUP-POULSEN et al. 1987). Both cytokines synergize in vitro with interferon-γ (IFN-γ) to induce class II expression in the islets of Langerhans (PUJOL-BORRELL et al. 1987), and they are potent inducers of other cytokines such as IL-1 or IL-6 that can contribute to inflammation (PAUL and RUDDLE 1988). TNF-α also induces expression of intercrines or chemokines (OPPENHEIM et al. 1991), a class of mole- cules that has been associated with inflammatory pathology in several human disease states and could be important direct contributors to autoimmune disease. The loci encoding human and murine TNF-α and LT map within the MHC (MULLER et al. 1987; Nedwin et al. 1985), which is linked to IDDM susceptibility in both the human population (POCIOT et al. 1991) and the NOD mouse. A more tenuous genetic involvement is suggested by the linkage of (though so far not identity of) Idd-6 to the p55-TNF receptor. However, the observation that systemic administration of high doses of TNF-α or LT to adult animals inhibits diabetes in the BB rat (SATOH et al. 1990) and NOD mouse (JACOB et al. 1990; SATOH et al. 1989; SEINO et al. 1993) suggests that these cytokines could actually prevent diabetes. Another possibility is that they contribute to pathogenesis at one stage of diabetes and to protection at another stage. This interesting suggestion is given credence by the recent observations concerning administration of anti-LT/ TNF antibody or cytokine to NOD mice up to 3 weeks of age (YANG et al. 1994). In these studies, administration of TNF-α to newborn NOD mice increased the incidence of diabetes to 100% and markedly decreased the age of onset. Treatment with a monoclonal antibody (TN3. 19. 12) that neutralizes both LT and TNF-α (SHEEHAN et al. 1989) prevented the development of diabetes and led to a reduction in insulitis. The fact that there was no effect on sialitis suggests either that the inflammation in salivary glands is not influenced by LT and/or TNF-α or that the effect is through a membrane form of one or both of these molecules, since TN3. 19.12 only reacts with their soluble forms. T cells from NOD mice treated with TN3.19.12 also exhibited a decreased proliferative response to glutamic acid decarboxylase (GAD). The fact that responses to ovalbumin were normal suggests that this is not a generalized immunosuppression. Taken together, these studies suggest that LT and or/TNF-α play a crucial role in early stages of NOD disease. They suggest that one or both of the cytokines contribute to the inflammatory process in islets and then, perhaps through the recruitment of antigen-presenting cells (APCs), to the development of an immunologically specific response to islet antigens.

2 The Tumor Necrosis Factor Ligand-Receptor Family in Pancreatic Inflammation

2.1 Characteristics of the Tumor Necrosis Factor Ligand-Receptor Family

The three members of the immediate TNF family whose genes are linked in the MHC are: tumor necrosis factor (TNF; TNF-α), lymphotoxin (LT; LT-α; TNF-β) and lymphotoxin-β (LT-β). LT was the first member to be described. It was identified as a cytotoxic product of antigen activated lymph node cells in delayed type hypersensitivity in the rat (RUDDLE and WAKSMAN 1967, 1968) and as a product of activated human peripheral blood cells (GRANGER and WILLIAMS 1968). LT (and

TNF-α) can kill target cells by a process called programmed cell death or apoptosis (SCHMID et al. 1986, 1987). LT is secreted from activated CD4 or CD8 cells and has also been detected in autoimmune infiltrates. It forms a homo-trimeric complex and interacts with either the p55 (CD120a) or p75 (CD120b) TNF receptor (TNF-R). The crystal structure of the LT-p55 TNF-R complex reveals an interaction of three receptor moleculels with the LT trimer (BANNER et al. 1993). LT-β is a recently described member of the TNF gene complex (BROWNING et al. 1993; MILLET and RUDDLE 1994). The type II transmembrane protein encoded by this gene (WARE et al. 1992) is highly homologous to LT and TNF-α. It appears to require LT for transport to the cell surface (ANDROLEWICZ et al. 1992) and forms a heterotrimeric complex with LT on the surface of activated lymphocytes. The function of the LT-LT-β complex is unknown, though it may serve to deliver a membrane-associated form of LT by cell-contact. It is not known whether LT-β can form a homotrimeric complex in the absence of LT and whether it can be released in a manner similar to TNF-α. A receptor which binds the LT-LT-β complex has recently been described and is a member of the larger TNF-R family (CROWE et al. 1994).

TNF-α, originally described as a product of activated macrophages, is also produced by activated CD4 and CD8 T cells and astrocytes and microglia (LIEBERMAN et al. 1989; SAWADA et al. 1989). TNF-α has also been demonstrated in lesions of numerous autoimmune diseases including the insulitis of the NOD mouse. TNF-α can be produced in two different forms: secreted and active as a trimer or as a type II transmembrane molecule. It appears to require a metallo-proteinase in order to be secreted, as recent reports indicate that its production in vitro and biologic activity is inhibited by metalloproteinase inhibitors (McGEEHAN et al. 1994; MOHLER et al. 1994). Some reports suggest that membrane associated TNF-α is biologically active (AVERSA et al. 1993; KRIEGLER et al. 1988). However, the nature of that interaction with the known receptors is not defined and actually difficult to envision given the available information concerning the necessity for simultaneous receptor interaction with the groove formed between two mono-meric ligands (BANNER et al. 1993).

The members of the immediate MHC-linked TNF family are related to a larger molecular family consisting of a group of structurally related receptors and a group of highly homologous ligands. The members of the receptor family are characterized by repeats of 12 cysteines and signal many of the biologic activities ascribed to p55 and p75 TNF-R including apoptosis and stimulation of cell growth. The receptor family includes the two TNF-R, low affinity nerve growth factor (NGF) receptor, fas CD40, CD30, CD27, OX40, and several viral open reading frames (LOETSCHER et al. 1992). Ligands for several of these recep-tors have been identified and, with the exception of NGF, are highly homologous to TNF-α, LT-β and LT-α. In addition, most are type II transmembrane proteins, e.g., TNF-α and LT-β.

2.2 The Role of Tumor Necrosis Factor Family Members in Lymphocyte Homing and Inflammation

Numerous biologic activities are carried out by LT and TNF-α, which appear to play key roles in inflammatory processes. Both LT and TNF-α induce expression of adhesion molecules on human umbilical vein cells in vivo alone or in synergy with IFN-γ (JOHNSON and POBER 1990). TNF-α injection in vivo results in inflammation and expression of ELAM-1 by 2 h and ICAM-1 by 9 h (MUNRO et al. 1989). In vitro, TNF-α induces E-selectin mRNA on human umbilical vein endothelial cells (BEVILACQUA et al. 1989) and ICAM-1 on isolated mouse pancreatic β-cells (PRIETO et al. 1992). Both LT and TNF-α are potent inducers of inflammation when expressed inappropriately in vivo as demonstrated in mice transgenic for LT or TNF-α (PICARELLA et al. 1992, 1993). These effects may be due in part to up-regulation of adhesion molecules including VCAM-1 and ICAM-1 (PICARELLA et al. 1993; Kratz and RUDDLE 1994; see below). TNF and/or LT up-regulation of adhesion molecules is also demonstrated in experimental allergic encephalomyelitis (EAE). We have shown in a mouse model of this inflammatory autoimmune disease that anti-LT/TNF-α antibody treatment of mice that received encephalitogenic T cells inhibits transfer of clinical signs of disease and results in a marked reduction in spinal cord inflammation (RUDDLE et al. 1990). This effect is due in large part to inhibition of VCAM-1 expression (BARTEN and RUDDLE 1994) indicating a crucial role of LT and/or TNF-α in the pathogenesis of EAE through an effect on adhesion molecule expression.

Though variations in their molecular regulation account for some differences in the biologic roles of TNF family members (MILLET and RUDDLE 1994), evidence suggests that individual members of the family play unique roles. We have recently reported that mice engineered so as to lack expression of LT (LT–/– mice) express normal levels of TNF-α but have no lymph nodes or Peyer's patches and exhibit marked splenic disorganization (DE TOGNI et al. 1994). This indicates that LT does possess unique properties with regard to lymphoid organogenesis, perhaps through an effect on expression of homing receptors or ligands.

The precise role of the individual TNF-Rs in biologic activities, particularly inflammation, has not been resolved, nor have any of the known receptors been clearly implicated in lymphocyte homing. However, the fact that LT–/– mice do not develop lymph nodes and exhibit splenic disorganization and deranged lympho-cyte homing to the liver and lung suggest that LT and an LT receptor could very well be involved in this process through induction of homing ligands or receptors. Since the intracellular domains of the p55 and p75 TNF-R are distinct, different biologic activities would be intuited. Furthermore, they are independently regu-lated (HOHMANN et al. 1991); in general the p55 TNF-R is expressed constitutively on most cells, and the p75 receptor is less widespread and can be induced in lymphocytes (WARE et al. 1991) and endothelial cells (SLOWIK et al. 1993). Never-theless, the two receptors appear to mediate many of the same activities. Most in vitro studies suggest a crucial role for p55 TNF-R in the induction of at least some adhesion molecules (MACKAY et al. 1993). MACKAY et al. (1993) have demonstrated in in vitro studies that p55 TNF-R regulates expression of ICAM-1, E-selectin, VCAM-1, and CD44 in human umbilical vein endothelial cells, while α-2 integrin expression is mediated through both p55 and p75 TNF-R. Despite the suggestion that p55 TNF-R regulates adhesion molecule expression in vitro, p75-/- mice do not develop an inflammatory lesion after subcutaneous injection of TNF, nor do they exhibit the usual liver inflammatory process after infection with *Schistosoma mansoni* (Moore, personal communication), suggesting a crucial role for that receptor in some types of inflammation. Neither the p55 nor p75 TNF-R is involved with the process of lymphoid organogenesis, as mice deficient for either or both receptors exhibit normal lymph nodes and Peyer's patches and splenic organization (Smith, personal communication). This suggests that a different receptor system is involved in LT-mediated homing, perhaps a new receptor or the recently described TNF-RrP that binds the LT-α LT-β complex (CROWE et al. 1994).

3 Rat Insulin Promoter Transgenic Mice and Islet Inflammation

3.1 Effect of Expression of Transgene-Derived Tumor Necrosis Factor or Lymphotoxin

In order to evaluate the role of TNF family members in inflammation, we and others (HIGUCHI et al. 1992; PICARELLA et al. 1992, 1993) have developed mice transgenic for the rat insulin promoter (RIP), driving expression of individual TNF family members, to target cytokine expression to the β-cells in the islets of Langerhans in the pancreas. In RIP-TNF-α mice, transgene-derived TNF-α mRNA and protein are produced in the islets (HIGUCHI et al. 1992; PICARELLA et al. 1993). RIP-LT mice express the transgene in pancreas, kidney, and skin (PICARELLA et al. 1992). Expression of TNF-α or LT in islets results in a marked cellular infiltrate.

In RIP-LT mice, the pattern of individual islets is heterogeneous with some islets uninvolved, some with interstitial accumulation (peri-insulitis) and some partially or completely infiltrated by mononuclear cells (insulitis). In general, more than 40% of the islets exhibit some form of inflammation even at one month of age and this proportion remains fairly constant over time. The histologic appearance of RIP TNF-α pancreata is slightly different, and suggests a pattern of migration from the capillary network within the islet and from the vasculature outside the islet capsule. It may be that the difference in histologic pattern in the islets in the two types of mice is due to a difference in pattern of expression of the two cytokines, i.e., that TNF-α is both secreted and membrane bound (thus in proximity to the islet capillary bed), whereas LT in this presumably LT-β-negative tissue is only secreted, influencing predominantly the vasculature outside the capsule. RIP-LT mice also exhibit inflammatory exudate in the kidney and ruffled fur consistent with transgene mRNA expression. In situ hybridization shows that the RIP-LT and

RIP-TNF-α transgenes are expressed predominantly by the islets, by β-cells and not by infiltrating cells. Immunocytochemistry of frozen pancreatic sections and FACS analysis of leukocytes from infiltrated islets obtained from either RIP-TNF-α or RIP-LT mice indicate that 50% of the infiltrating cells are either CD4+ or CD8+ T cells. Double negative cells are predominately B220+ IgM+ B cells. Some F4/80+ macrophages are also seen.

TNF-α and LT are not unique in their ability to induce inflammation when expressed in pancreatic islet cells. β-Cell directed IL-2 expression also results in islet inflammation (ALLISON et al. 1992; ELLIOTT and FLAVELL 1994), and the phenotype of the infiltrating cells in islets of mice with low level RIP-IL-2 expression is very similar to that of RIP-TNF-α and RIP-LT mice. That is, there are CD4+ and CD8+ T cells and macrophages in peri-islet and intra-islet locations. Other founder mice which appeared to have a higher expression of the transgene develop a widespread pancreatitis. None of these infiltrates appears to be autoimmune or antigen-specific, nor do they lead without further activation to specific β-cell destruction and diabetes. Mice transgenic for IL-10 driven by the human insulin promoter (WOGENSEN et al. 1993) also develop a pancreatic infiltrate consisting of CD4+ and CD8+ T cells, B cells and macrophages. This infiltrate is not limited to the islets, as is the case with RIP-TNF-α, RIP-LT, and low level RIP-IL-2 mice; rather it is widespread throughout the pancreas. An inflammatory exudate is also seen in the case of insulin promoter driven IFN-γ expression, though in this case evidence of antigen-specific activation and β-cell destruction is seen (SARVETNICK et al. 1988, 1990).

The observation that several different islet directed cytokines induce pancreatic inflammation with some characteristics similar to those observed in RIP-LT or RIP-TNF-α mice suggests several possible nonmutually exclusive mechanisms to explain this. It may be that expression of each individual cytokine is sufficient to set into motion a cascade of events such as changes in adhesion molecules that leads to inflammation. This would suggest several redundant systems. Another possibility is that individual cytokines induce each other and set the process in motion.

3.2 Changes in Antigen Expression Are Seen in Pancreata of Transgenic Mice

Islets of Langerhans normally show low levels of MHC class I and no expression of MHC class II. TNFs are known to increase MHC expression, which could result in increased immunogenicity and play a role in the induction of autoimmunity. Significant up-regulation of MHC class I (H-2Kb) is most evident on the β-cells of islets in LT mice, where fewer islets are obscured by infiltrating lymphocytes as is the case in TNF-α mice. MHC class II expression is apparent on some infiltrating cells in both TNF-α and LT transgenic mice and presumably reflects the fact that B cells are strongly class II-positive. Faint expression of class II is suggested on RIP-LT islet tissue but has not been properly quantified and was difficult to evaluate because of the presence of infiltrating cells in the area.

3.3 Islet Expression of the Transgenes Is Associated with Endothelial Cell Activation and Expression of Adhesion Molecules

In our original studies with RIP-LT and RIP-TNF-α mice, we reported increased staining for VCAM-1 and ICAM-1 on frozen sections of islets of RIP-LT- or RIP-TNF-α-positive mice (PICARELLA et al. 1993). However, it was difficult to interpret these observations because these adhesion molecules were expressed at high levels particularly in the areas of infiltrating leukocytes. In fact, ICAM-1 was expressed on the infiltrating cells, consistent with the fact that B and T cells can be ICAM-1-positive. Thus it was not clear whether the adhesion molecule expression was a consequence of, a contributor to, or unrelated to inflammation in this setting. To address this problem, HIGUCHI et al. (1992) crossed RIP-TNF-α transgenic mice onto a SCID background and we have crossed RIP-LT transgenic mice with mice carrying the RAG-2 null mutation (KRATZ and RUDDLE 1994). These systems make it possible to study the effects of TNF-α or LT in an in vivo setting in the absence of other cytokines derived from mature lymphocytes.

HIGUCHI et al. (1992) have crossed RIP-TNF-α transgenic mice with animals homozygous for the SCID mutation, which have no T or B cells. The animals develop a fibrotic reaction in their pancreatic islets, which is as severe in the transgenic SCID mice as in the transgenic animals with a normal immune system. In addition, a strong infiltration of F4/80-positive cells is present, which is more marked than in non-SCID transgenic mice. As all the islet cells, except β-cells, stain with an antibody specific for ICAM, it is not possible in this system to explore up-regulation of this adhesion molecule in the continuous presence of the cytokine. By electron microscopy, the changes in the endothelial cells of transgenic SCID mice were found to be similar to those in the immuno-competent animals, namely, a thickening of the cytoplasm of the endothelial cells.

In similar studies, we have crossed RIP-LT transgenic mice with animals homozygous for the RAG-2 null mutation (KRATZ and RUDDLE 1994). The RAG-2 protein is necessary for rearrangement of both the TCR and the immunoglobulin genes, and RAG-2 deficient mice have been shown to lack mature T and B cells (SHINKAI et al. 1992). We found a marked up-regulation of ICAM-1 and VCAM-1 on endothelial cells within the islets of RIP-LT transgenic RAG-2 null mice. No other cells in the pancreas expressed these adhesion molecules. Electron microscopic studies revealed that the endothelial cells showed the typical morphology of an inflammatory site, with an increased thickness and irregular protrusions of their cytoplasm, as well as enlarged, plump nuclei. These findings confirm that LT alone, in the absence of other cytokines derived from mature lymphocytes, is sufficient to activate endothelial cells in vivo.

Changes which can be found in the endothelium of RIP-LT transgenic RAG-2 null mice are also be seen in high endothelial venules, vessels characteristic of lymph nodes and Peyer's patches. This is of particular interest in light of a recent report noted above, that mice deficient in LT expression lack peripheral lymph nodes and Peyer's patches, demonstrating that LT is necessary for the normal development of these organs (DE TOGNI et al. 1994). The results obtained in RIP-LT transgenic RAG-2 null mice suggest that LT is not only absolutely necessary but also sufficient for certain aspects for lymph node development. This raises the tantalizing possibility that the same mechanisms which are of importance for the formation of the islet infiltrate in the RIP-LT transgenic animals are also involved in the formation of lymphoid organs in organogenesis and that RIP-LT transgenic mice will allow elucidation of at least some of these mechanisms.

3.4 The Transgenic Mice Do Not Develop Diabetes and Their Infiltrating Cells Do Not Appear to Be Activated

FACS analysis shows a very low level of IL-2 receptor expression in the pancreatic infiltrate of the transgenic mice. There is also very little TNF expression in the infiltrating cells in RIP-TNF mice. This pattern is not very different from that of prediabetic NOD mice, in which very low numbers of cells expressing IL-2 receptor or TNF-α or LT were apparent in the infiltrates. Despite the fact that there is remarkable mononuclear infiltration in the islets of the transgenic mice, none of them develop diabetes even after 12 months of observation. Founders and progeny of crosses with C57BL/6 show insulin staining even in infiltrated islets, normal total pancreatic insulin levels, normal urine glucose and normal blood glucose. F1 progeny of RIP-TNF x NOD mice showed essentially the same picture as those animals crossed to B6, that is, inflammation but no β-cell destruction, reduction in insulin content, or diabetes in mice observed up to 2 years of age.

4 Cytokine Transgenics and Antigen Presentation and Recognition

4.1 Islet-Specific Costimulation Converts Tumor Necrosis Factor-Induced Insulitis to Diabetes

The results described above establish clearly that expression of TNF-α or LT in the islets is sufficient to cause insulitis or peri-insulitis, respectively. However, neither RIP-TNF-α nor LT transgenic mice develop any signs of diabetes, even up to 1 year of age. An inflammatory infiltrate is therefore not a sufficient condition to cause disease. The key issue becomes whether this can be a predisposing condition for autoimmunity. To test this we have analyzed transgenic mice expressing the human costimulatory molecule hB7-1, also directed by the insulin promoter (GUERDER et al. 1994). T cells require two signals to become activated. The first signal, mediated through the TCR, constitutes antigen recognition, but activation is only obtained if there is a second concomitant signal delivered through a variety of costimulatory ligands on APCs interacting with cognate receptors on T cells. Of these ligands, the B7 family of molecules, B7-1 and B7-2, are considered to be some of the most important. These are expressed on all functional APCs and have two receptors on T cells, CD28 and CTLA-4. Strong evidence shows that the CD28 receptor provides a profound costimulatory signal which is sufficient for the activation of T cells in conjunction with stimulation through the TCR. Accordingly, we generated transgenic mice expressing hB7-1 on the islets. The first question that we wished to address was whether the presence of both signals on a tissue such as the islets of Langerhans would convert these cells into bonafide APCs, and secondly, whether the ability of such cells to present antigen and therefore activate T cells would be sufficient to lead to autoimmunity. Islets expressing hB7 are good APCs in vitro (GUERDER et al. 1994). However, hB7-1 transgenic mice only develop IDDM in approximately 2% of the cases. The critical question from the focus of this review is whether TNF-mediated inflammation can influence the development of autoimmunity in these mice. The potential role of TNF-mediated insulitis in predisposing hB7 mice to autoimmunity was therefore tested by crossing transgenic mice expressing either TNF-α (GUERDER et al. 1994) or LT (unpublished observations) with hB7. Strikingly, while TNF mice never develop IDDM and hB7 mice rarely exhibit clinical signs of this disease, 100% of double transgenic TNF-α x B7 mice become diabetic within 3–5 weeks of birth, and most B7 x LT mice also become diabetic at a very high frequency. Thus, the combination of an activating environment (presence of hB7) and a strong inflammatory response (TNF) is sufficient to provoke rapid autoimmunity. TNF-α promotes a strong lymphocytic infiltrate and up-regulates MHC class I levels. We cannot at this stage distinguish which of these effects is important; clearly both may play a role.

The IDDM seen in hB7 x TNF double transgenic mice is genuinely auto-immune. Strong evidence for antigen specificity in this system is obtained first by the demonstration that destruction is accompanied by the eliciting of islet-specific T cells which proliferate in vitro (GUERDER et al. 1994). Second, engrafting under the kidney capsule of B7-expressing islets into TNF x B7 double transgenic mice causes rapid destruction of such grafts. The insulitis provoked by TNF therefore likely predisposes towards genuine autoimmunity in this system.

4.2 Tumor Necrosis Factor Potentiates Autoimmune Destruction of Viral Glycoprotein-Expressing Islets upon Viral Infection

A second interesting model in which an inflammatory response mediated by TNF can influence autoimmune processes comes from the studies of OHASHI et al. (1997), utilizing a lymphocytic choriomeningitis virus (LCMV) glycoprotein (GP) transgene directed to the islets of Langerhans by the RIP. These transgenic mice do not develop immune tolerance to this antigen, presumably because of its failure to express in the thymus. Instead, a state of immune ignorance develops in which circulating T cells capable of recognizing the LCMV GP product exist, but normally do not become activated. Infection of the mice with LCMV, however, primes these cells and leads to autoimmune destruction of the islets of Langerhans. We presume that these cells are primed in lymphoid organs, and activated cells subsequently migrate to the tissue and cause autoimmune destruction. Interestingly, infection of these transgenic mice with vaccinia constructs expressing LCMV GP does not lead to diabetes. This presumably results from the priming of insufficient numbers of GP-reactive cells, since this failure can be overcome by crossing the GP mice onto mice transgenic for a TCR which recognizes this antigen. In this case, the majority of T cells are LCMV-reactive and therefore presumably a critical number of T cells are readily activated by the infection with vaccinia (OHASHI et al. 1993). It is not really clear why vaccinia does not activate sufficient T cells in mice with a normal T cell repertoire, but this could have to do with the biodistribution of the virus or any number of other variables.

Interestingly, vaccinia-GP virus can induce autoimmunity directed towards GP in mice that are double transgenic for GP and TNF-α on the islets of Langerhans. Thus, presumably here, as in the case described above for B7, local infiltration in the islets of Langerhans promoted by secretion of TNF-α is sufficient to promote recruitment of the low frequency of activated cells; these cells then mediate destruction of the islets of Langerhans.

4.3 Islet-Specific Interleukin-2 Potentiates Autoimmune Destruction of Islets by Low Affinity T Cells

Expression of a variety of other cytokines in the islets of Langerhans also appears to lead to islet inflammation. Thus, as described above, IL-2 directed by the RIP leads to islet infiltration (ALLISON et al. 1992; ELLIOTT and FLAVELL 1994). In one case (ALLISON et al. 1992), this infiltration did not lead to autoimmune destruction, consistent with the TNF results described above. In another study, this infiltrate led to destruction of islets and concomitant diabetes at approximately 4–6 months of age (ELLIOTT and FLAVELL 1994). However, in this case, destruction of the islets was caused by apparently nonautoimmune mechanisms, since no evidence for specific autoimmunity could be detected. Despite the fact that the destruction was nonautoimmune, it did appear to be mediated by the cellular infiltrate. Irradiation of recipients eliminated the inflammatory infiltrate and retarded development of diabetes until infiltration was reestablished. It therefore seems likely that the major effect of IL-2 is to provoke infiltration and activation of natural killer (NK)-like cells, possibly lymphokine-activated killer (LAK) cells.

Effects of IL-2 transgenes, however, can also be demonstrated on antigen-specific processes. Thus, the original rationale for the expression of IL-2 was to provide help for those T cells unable to produce their own IL-2. Specifically, anergic CD4 T cell clones cannot produce IL-2, but can be activated by IL-2. In addition, many CD8 T cells are unable to produce IL-2 but require this cytokine for maturation. In mice transgenic for an H-2K^b antigen directed by the RIP, moderate levels of H-2K^b are expressed on the β-cells of the islets, but very low levels appear to be expressed in the thymus and are responsible for the deletion of high affinity T cells. Thus, when these transgenic mice, which carry a transgene-encoded TCR specific for H-2K^b, are analyzed, low affinity cells are found in the periphery. Interestingly, in these transgenic mice autoimmune diabetes normally does not occur, presumably because the transgene-encoded H-2K^b is not seen in an immunogenic context due to the absence of both costimulators on the islet cells and costimulatory cytokines in the local environment. This activation does, however, occur if a RIP-IL-2 transgene is crossed into this background to generate triple transgenic mice (HEATH et al. 1992). In this case, there is autoimmune destruction of the islets of Langerhans. It appears likely that the mechanism underlying this destruction is the provision of help, in the form of IL-2, to these T cells, thereby permitting inactivation of CTLs and concomitant destruction of islets. It is not clear what role the inflammation mediated by IL-2 plays in this process, nor is it clear to which extent nonspecific destruction by LAK-like cells could play a role in this diabetes. Nonetheless, this is the second case in which a cytokine with inflammatory properties can precipitate autoimmunity in a system where this normally does not occur. Interestingly, a parallel result can be obtained by the triple transgene combination of RIP-K^b and K^b-specific TCR transgenes together with islet-encoded B7 (Guerder and Flavell, unpublished). Therefore, it also appears that activating circumstances are provided in this case and presumably render the T cells independent of exogenous IL-2 by the provision of costimulation directly from the islet cells which are subsequently destroyed.

5 Summary and Conclusion: Insulitis Predisposes to Diabetes

In summary, the inflammatory response mediated by cytokines such as TNF can promote recruitment of lymphocytes to a tissue. Moreover, if other conditions are met, this can provide a predisposing role to autoimmune disease. TNFs induce the appearance of adhesion molecules (and presumably, therefore, extravasation of lymphocytes into tissue from the vasculature) and increase the levels of MHC class I on tissue. However, it is not clear which of these effects plays the key role in induction of disease. This should be the subject of further study. The data substantiate the hypothesis that chronic inflammation might play a precipitating role in autoimmunity and could be one of the environmental factors of importance in the development of so many autoimmune diseases.

References

Allison J, Malcolm L, Chosich N, Miller JFAP (1992) Inflammation but not autoimmunity occurs in transgenic mice expressing constitutive levels of interleukin-2 in islet β cells. Eur J Immunol 22: 1115–1121

Androlewicz MJ, Browning JL, Ware CF (1992) Lymphotoxin is expressed as a heteromeric complex with a distinct 33-kDa glycoprotein on the surface of an activated human T cell hybridoma. J Biol Chem 267: 2542–2547

Aversa G, Punnonen J, deVries JE (1993) The 26-kD transmembrane form of tumor necrosis factor-α on activated CD4+ T cell clones provides a costimulatory signal for B cell activation. J Exp Med 177: 1575–1585

Banner DW, D'Arcy A, Jaanes W, Gentz R, Schoenfeld H-J, Broger C, Loetscher H, Lesslauer W (1993) Crystal structure of the soluble human 55 kd TNF receptor-human TNF-β complex: implications for TNF receptor activation. Cell 73: 431–445

Baron JL, Madri JA, Ruddle NH, Hashim G, Janeway CA (1993) Surface expression of α4 integrin by T cells is required for their entry into brain parenchyma. J Exp Med 177: 57–68

Barten D, Ruddle NH (1994) Vascular cell adhesion molecule-1 modulation by TNF in experimental allergic encephalomyeleits. J Neuroimmunol 51: 123–133

Bevilacqua MP, Stengelin S, Gimbrone MA, Seed B (1989) Endothelial leukocyte adhesion molecule 1: an inducible receptor for neutrophils related to complement regulatory proteins and lectins. Science 243: 1160–1165

Browning JL, Ngam-ek A, Lawton P, DeMarinis J, Tizard R, Chow EP, Hesson C, O'Brine-greco B, Foley SF, Ware CF (1993) Lymphotoxin-β, a novel member of the TNF family that forms a heteromeric complex with lymphotoxin on the cell surface. Cell 72: 847–856

Charlton B, Bacelj A, Slattery RM, Mandel TE (1989) Cyclophosphamide-induced diabetes in NOD/WEHI mice. Diabetes 38: 441–447

Crowe PD, VanArsdale TL, Walter BN, Ware CF, Hession C, Ehrenfels B, Browning JL, Din WS, Goodwin RG, Smith CA (1994) A lymphotoxin-β-specific receptor. Science 264: 707–710

De Togni PD, Goellner J, Ruddle NH, Streeter PR, Fick A, Mariathasan S, Smith SC, Carlson R, Shornick LP, Strauss-Schoenberger J, Russell JH, Karr R, Chaplin DD (1994) Abnormal development of peripheral lymphoid organs in mice deficient in lymphotoxin. Science 264: 703–707

Eisenbarth GS (1986) Type I diabetic mellitus. A chronic autoimmune disease. Engl J Med 314: 1360–1368

Elliott EA, Flavell RA (1994) Transgenic mice expressing constitutive levels of IL2 in islet β cells develop diabetes. Int Immunol 6: 1629–1637

Faveeuw C, Gagnerault M-C, Lepault F (1994) Expression of homing and adhesion molecules in infiltrated islets of Langerhans and salivary glands of nonobese diabetic mice. J Immunol 152: 5969–5978

Gepts W (1965) Pathologic anatomy of the pancreas in juvenile diabetes mellitus. Diabetes 14: 619–633

Ghosh S, Palmer SM, Rodrigues NR, Cordell HJ, Hearne CM, Cornall RJ, Prins J-B, McShane P, Lathrop GM, Peterson LB, Wicker LS, Todd JA (1993) Polygenic control of autoimmune diabetes in nonobese diabetic mice. Nature Genet 4: 404–409

Granger GA, Williams TW (1968) Lymphocyte cytotoxicity in vitro: activation and release of a cytotoxic factor. Nature 218: 1253–1254

Guerder S, Picarella D, Linsley PS, Flavell RA (1994) Costimulator B7-1 confers antigen-presenting cell function to parenchymal tissue and in conjunction with tumor necrosis factor a leads to autoimmunity in transgenic mice. Proc Natl Acad Sci USA 91: 5138–5142

Hanninen A, Taylor C, Streeter PR, Stark LS, Sarte JM, Shizuru JA, Simell O, Michie SA (1993) Vascular addressins are induced on islet vessels during insulitis in nonobese diabetic mice and are involved in lymphoid cell binding to islet endothelium. J Clin Invest 92: 2509–2515

Hattori M, Buse JB, Jackson RA, Glimcher L, Dorf ME, Minami M, Makino S, Moriwaki K, Kuzuya H, Imura H, Strauss WM, Seidman JG, Eisenbarth GS (1986) The NOD mouse: recessive diabetogenic gene in the major histocompatibility complex. Science 231: 733–735

Heath WR, Allison J, Hoffmann MW, Schonrich G, Hammerling G, Arnold B, Miller JFAP (1992) Autoimmune diabetes as a consequence of locally produced interleukin-2. Nature 359: 547–549

Held W, MacDonald HR, Weissman IL, Hess MW, Mueller C (1990) Genes encoding tumor necrosis factor α and granzyme A are expressed during development of autoimmune diabetes. Proc Natl Acad Sci USA 87: 2239–2243

Higuchi Y, Herrera P, Muniesa P, Huarte J, Belin D, Ohashi P, Aichele P, Orci L, Vassalli J-D, Vassalli P (1992) Expression of a tumor necrosis factor α transgene in murine pancreatic β cells results in severe and permanent insulitis without evolution towards diabetes. J Exp Med 176: 1719–1731

Hohmann H-P, Brockhaus M, Baeuerle PA, Remy R, Kolbeck R, Vanloon APGM (1991) Expression of the types A and B tumor necrosis factor (TNF) receptors is independently regulated, and both receptors mediate activation of the transcription factor, NF-κB. J Biol Chem 265: 22409–22417

Jacob CO, Aiso S, Michi SA, McDevitt HO, Acha-Orbea H (1990) Prevention of diabetes in nonobese diabetic mice by tumor necrosis factor (TNF): similarities between TNF-α and interleukin 1. Proc Natl Acad Sci 87: 968–972

Jiang Z, Woda BA (1991) Cytokine gene expression in the islets of the diabetic biobreeding/Worcester rat. J Immunol 146: 2990–2994

Johnson DR, Pober JS (1990) Tumor necrosis factor and immune interferon synergistically increase transcription of HLA class I heavy- and light chain genes in vascular endothelium. Proc Natl Acad Sci USA 87: 5183–5187

Kikutani H, Makino S (1992) The murine autoimmune diabetes model: NOD and related strains. Adv Immunol 51: 285–322

Kratz A, Ruddle NH (1994) Lymphotoxin alone, in the absence of other lymphocyte derived cytokines, activates endothelial cells in vivo and interferes with normal islet development. Eur Cytokine Netw 5: 203

Kriegler M, Perez C, DeFay K, Albert I, Lu SD (1988) A novel form of TNF/cachectin is a cell surface cytotoxic transmembrane protein: ramifications for the complex physiology of TNF. Cell 53: 45

Lieberman AP, Pitha P, Shin HS, Shin ML (1989) Production of tumor necrosis factor and other cytokines by astrocytes stimulated with lipopolysaccharide or a neurotropic virus. Proc Natl Acad Sci USA 86: 6348–6352

Loetscher H, Brockhaus M, Dembic Z, Gallati H, Gentz R, Gubler U, Lahm H-W, Lustig A, Pan Y-CE, Schlaeger E-J, Tabuchi H, Zulauf M, Lesslauer W (1992) Two distinct human TNF receptors: purification, molecular cloning, and expression. In: Osawa T, Bonavida B (eds) Tumor necrosis factor: structure-function relationship and clinical application. Karger, Basel, pp 34–47

Mackay F, Loetscher H, Stueber D, Gehr G, Lesslauer W (1993) Tumor necrosis factor-α (TNF-α) induced cell adhesion to human endothelial cells is under dominant control of one TNF receptor type, TNF-R55. J Exp Med 177: 1277–1286

Mandrup-Poulsen T, Bendtzen K, Dinarello CA, Nerup J (1987) Human tumor necrosis factor potentiates human interleukin 1-mediated rat pancreatic β-cell cytotoxicity. J Immuol 139: 4077–4082

McGeehan GM, Becherer JD, Bast RC Jr, Boyer CM, Champion B, Connolly KM, Conway JG, Fordun PO, Karp S, Kidao S, McElroy AB, Nichols J, Pryzwansky K, Sonoenen F, Sekut L, Truesdale A, Verghese M, Warner J, Ways JP (1994) Regulation of tumour necrosis factor-α processing by a metalloproteinase inhibitor. Nature 370: 558–561

Millet I, Ruddle NH (1994) Differential regulation of lymphotoxin (LT), lymphotoxin-β (LT-β), and TNF-α in murine T cell clones activated through the TCR. J Immunol 152: 4336–4346

Mohler KM, Sleath PR, Fitzner JN, Cerretti DP, Alderson M, Kerwar SS, Torrance DS, Otten-Evans C, Greenstreet T, Weerawarna K, Kronheim SR, Petersen M, Gerhart M, Kozlosky CJ, Marck CJ, Black RA (1994) Protection against a lethal dose of endotoxin by an inhibitor of tumour necrosis factor processing. Nature 370: 218–220

Muller U, Jongeneel V, Nedospasov SA, Lindahl KF, Steinmetz M (1987) Tumor necrosis factor and lymphotoxin genes map close to H-2D in the mouse major histocompatibility complex. Nature 325: 265–267

Munro JM, Pober JS, Cotran RS (1989) Tumor necrosis factor and interferon-γ induce distinct patterns of endothelial activation and associated leukocyte accumulation in skin of Papio anubis. Am J Pathol 135: 121–133

Nedwin GE, Naylor SL, Sakaguchi AY, Smith D, Jarret NJ, Pennica D, Goeddel DV, Gray PW (1985) Human lymphotoxin and tumor necrosis factor genes: structure, homology and chromosomal localization. Nucleic Acids Res 13: 6361–6373

Ohashi PL, Oehen S, Buerki K, Pircher H, Ohashi CT, Odermatt B, Malissen B, Zinkernagel RM, Hengartner H (1991) Ablation of "tolerance" and induction of diabetes by virus infection in viral antigen transgenic mice. Cell 65: 305–317

Ohashi PL, Oehen S, Aichele P, Pircher H, Odermatt B, Herrera P, Higuchi Y, Buerki K, Hengartner H, Zinkernagel RM (1993) Induction of diabetes is influenced by the infectious virus and local expression of Class I and TNF-alpha. J Immunol 11: 5185–5194

Oppenheim JJ, Zachariae COC, Mukaida N, Matsushima K (1991) Properties of the novel pro-inflammatory supergene "intercrine" family. Annu Rev Immunol 9: 617–648

Paul NL, Ruddle NH (1988) Lymphotoxin. Annu Rev Immunol 6: 407–438

Picarella DE, Kratz A, Li C-B, Ruddle NH, Flavell RA (1992) Insulitis in transgenic mice expressing TNF-β (lymphotoxin) in the pancreas. Proc Natl Acad Sci USA 89: 10036–10040

Picarella DE, Kratz A, Li C-B, Ruddle NH, Flavell RA (1993) Transgenic TNF$_f$α production in islets leads to insulitis, not diabetes: distinct patterns of inflammation in TNF-α and TNF-β transgenic mice.J Immunol 149: 4136–4150

Pociot F, Molvig J, Wogensen L, Worsaae H, Dalboge H, Baek L, Nerup J (1991) A tumor necrosis factor beta gene polymorphism in relation to monokine secretion and insulin-dependent diabetes mellitus. Scand J Immunol 33: 37–49

Prieto J, Kaaya EE, Juntti-Berggren L, Berggren P-O, Sandler S, Biberfeld P, Patarroyo M (1992) Induction of intercellular adhesion molecule-1 (CD54) on isolated mouse pancreatic β cells by inflammatory cytokines. Clin Immunol Imunopathol 65: 247–253

Pujol-Borrell R, Todd I, Doshi M, Bottazzo GF, Sutton R, Gray D, Adolf GR, Feldmann M (1987) HLA class II induction in human isleT cells by interferon-γ plus tumour necrosis factor or lymphotoxin. Nature 326: 304–306

Ruddle NH, Waksman BH (1967) Cytotoxic effect of lymphocyte-antigen interaction in delayed hypersensitivity. Science 157: 1060–1062

Ruddle NH, Waksman BH (1968) Cytotoxicity mediated by soluble antigen and lymphocytes in delayed hypersensitivity. III. Analysis of mechanism. J Exp Med 128: 1267–1279

Ruddle NH, Bergman C, McGrath KM, Lingenheld EG, Grunnet ML, Padula SJ, Clark RB (1990) An antibody to lymphotoxin and tumor necrosis factor prevents transfer of experimental allergic encephalomyelitis. J Exp Med 172: 1193–1200

Sarvetnick N, Liggitt D, Pitts SL, Hansen SE, Stewart TA (1988) Insulin-dependent diabetes mellitus induced in transgenic mice by ectopic expression of class II MHC and interferon-gamma. Cell 52: 773–782

Sarvetnick N, Shizuru J, Liggitt D, Martin L, McIntyre B, Gregory A, Parslow T, Stewart T (1990) Loss of pancreatic islet tolerance induced by β-cell expression of interferon-γ. Nature 346: 844–847

Satoh J, Seino H, Abo T, Tanaka S-I, Shintani S, Ohta S, Tamura K, Sawai T, Nobunaga T, Oteki T, Kumagai K, Toyota T (1989) Recombinant human tumor necrosis factor α suppresses autoimmune diabetes in nonobese diabetic mice. J Clin Invest 84: 1345–1348

Satoh J, Seino H, Shintani S, Tanaka S-I, Ohteki T, Masuda T, Nobunaga T, Toyota T (1990) Inhibition of type 1 diabetes in BB rats with recombinant human tumor necrosis factor-α. J Immunol 145: 1395–1399

Sawada M, Kondo N, Suzumura A, Marunouchi T (1989) Production of tumor necrosis factor-alpha by microglia and astrocytes in culture. Brain Res 491: 394–397

Schmid DS, Tite JP, Ruddle NH (1986) DNA fragmentation: manifestation of targeT cell destruction mediated by cytotoxic T cell lines, lymphotoxin-secreting helper T cell clones and cell-free lymphotoxin-containing supernatant. Proc Natl Acad Sci USA 83: 1881–1886

Schmid DS, Hornung R, McGrath KM, Ruddle NH (1987) TargeT cell DNA fragmentation is mediated by lymphotoxin and tumor necrosis factor. Lymphokine Res 61: 195–202

Seino H, Takahashi K, Satoh J, Zhu XP, Sagara M, Masuida T, Nobunaga T, Funahashi I, Kajikawa T, Toyota T (1993) Prevention of autoimmune diabetes with lymphotoxin in NOD mice. Diabetes 42: 398–404

Sheehan KCF, Ruddle NH, Schreiber RD (1989) Generation and characterization of hamster monoclonal antibodies that neutralize murine tumor necrosis factors. J Immunol 142: 3884–3893

Shinkai Y, Rathbun G, Lam K-P, Oltz EM, Stewart V, Mendelsohn M, Charron J, Datta M, Young F, Stall AM, Alt FW (1992) RAG-2-deficient mice lack mature lymphocytes owing to inability to initiate V(D)J rearrangement. Cell 68: 855–867

Slowik MR, DeLuca LG, Flers W, Pober JS (1993) Tumor necrosis factor (TNF) activates human endothelial cells through the p55 TNF receptor but the p75 receptor contributes to activation at low TNF concentration. Am J Pathol 143: 1724–1730

Todd JA (1990) Genetic control of autoimmunity in type 1 diabetes. Immunol Today 11: 122–129

Todd JA, Aitman TJ, Cornall RJ, Ghosh S, Hall JRS, Hearne CM, Knight AM, Love JM, McAleer Ma, Prins J-B, Rodrigues N, Lathrop M, Pressey A, DeLarato NH, Peterson LB, Wicker LS (1991) Genetic analysis of autoimmune type 1 diabetes mellitus in mice. Nature 351: 542–547

Ware CF, Crowe PD, Vanarsdale TL, Andrews JL, Grayson MH, Jerzy R, Smith CA, Goodwin RG (1991) Tumor necrosis factor (TNF) receptor expression in T lymphocytes: differential regulation of the type 1 TNF recepto during activation of resting and effector cells. J Immunol 147: 4229–4238

Ware CF, Crowe PD, Grayson MH, Androlewicz MJ, Browning JL (1992) Expression of surface lymphotoxin and tumor necrosis factor on activated T, B, and natural killer cells. J Immunol 149: 3881–3888

Wicker LS, Miller BJ, Coker LZ, McNally SE, Scott S, Mullen Y, Appel MC (1987) Genetic control of diabetes and insulitis in the nonobese diabetic (NOD) mouse. J Exp Med 165: 1639–1655

Wicker LS, Miller BJ, Fischer PA, Pressey A, Peterson LB (1989) Genetic control of diabetes and insulitis in the nonobese diabetic mouse: pedigree analysis of a diabetic $H-2^{nod/b}$ heterozygote. J Immunol 142: 781–784

Wogensen L, Huang X, Sarvetnick N (1993) Leukocyte extravasation into the pancreatic tissue in transgenic mice expressing interleukin-10 in the islets of Langerhans. J Exp Med 178: 175–185

Yang X-D, Karin N, Tisch R, Steinman L, McDevitt HO (1993) Inhibition of insulitis and prevention of diabetes in nonobese diabetic mice by blocking L-selectin and very late antigen 4 adhesion receptors. Proc Natl Acad Sci USA 90: 10494–10498

Yang X-D, Tisch R, Singer SM, Liblau L, Cao Z, Schreiber R, Nagata S, McDevitt H (1994) The role of TNF in lymphocyte development and in the immunopathogenesis of insulin-dependent diabetes. Eur Cytokine Netw 5: 184

Influence of T Lymphocytes and Major Histocompatibility Complex Class II Genes on Diabetes Susceptibility in the NOD Mouse

R.M. Slattery[1] and J.F.A.P. Miller[2]

1 Introduction

NOD mice spontaneously develop an insulin-dependent diabetes mellitus (IDDM) which shares many features with the human disease. In the NOD mouse, insulitis begins at about 30–40 days of age in 100% of cases, but hyperglycemia develops by 80 days in only 40%–90% of females and by 250 days in 5%–20% of males, depending on the colony (Baxter et al. 1989).

[1] John Curtin School of Medical Research, Australian National University, Canberra, ACT 2605, Australia
[2] The Walter and Eliza Hall Institute of Medical Research, Post Office Royal Melbourne Hospital, Victoria 3050, Australia

At least nine genes contribute to IDDM susceptibility (GARCHON 1992), the major component of this susceptibility being the inheritance of certain class II genes encoded by the major histocompatibility complex (MHC). Thus, the MHC of NOD mice lacks I-E gene expression and expresses a unique I-A molecule (I-A^{NOD}) (ACHA-ORBEA and McDEVITT 1987). The gene encoding the I-A α chain is identical to the I-Aα^d allele, but the I-Aβ^{g7} allele is unique.

Several β-cell autoantigens have been described as having a role in the disease process, although it is not clear where each contributes in the cascade of pathogenetic events. Most recently, tolerance has been induced to glutamic acid decarboxylase (GAD) and this has been shown to prevent NOD mice from developing IDDM (TISCH et al. 1993; KAUFMAN et al. 1993). It is not known, however, whether GAD is the relevant autoantigen responsible for initiating the disease or whether it exerts its effect downstream of that event. Nor is it known whether the immune response to GAD is initiated by class I-restricted CD8$^+$ T cells or class II-restricted CD4$^+$ T cells.

2 Cellular Components of the Disease Process

2.1 Role of T Lymphocytes

Considerable evidence shows that the disease process in NOD mice is T-cell dependent. Neonatally thymectomized NOD mice and athymic nude NOD mice do not develop IDDM (OGAWA et al. 1985; MAKINO et al. 1986). Immunosuppressive regimens, such as administration of cyclosporin A, anti-thymocyte serum and monoclonal anti-Thy-1 antibody, all have an inhibitory effect on the development of IDDM (MORI et al. 1986; KIDA et al. 1985; HARADA and MAKINO 1986). Furthermore, IDDM can be adoptively transferred from hyperglycemic NOD donors to young or irradiated syngeneic recipients by purified T cells (BENDELAC et al. 1987; MILLER et al. 1988)

2.1.1 CD4$^+$ T Lymphocytes

To define which T cells were required for the adoptive transfer of IDDM, unfractionated T cells or cells depleted of CD4$^+$ T cells from diabetic donors were transferred to young irradiated or to neonatal NOD recipients (MILLER et al. 1988; BENDELAC et al. 1987). Although unfractionated T cells did transfer the disease, the removal of CD4$^+$ T cells from these prevented the adoptive transfer of IDDM, indicating a crucial role for CD4$^+$ T cells in the pathogenesis of IDDM.

Studies involving the adoptive transfer of spleen and lymph node cells from cyclophosphamide(CYP)-treated NOD mice into thymectomized, T cell-depleted recipients also indicated an essential role for CD4$^+$ T cells in the transfer of insulitis.While unfractionated cells or cells depleted of CD8$^+$ T cells transferred insulitis, cells depleted of CD4⁺ T cells were unable to do so (HANAFUSA et al. 1988).

Further evidence for the pivotal role of CD4⁺ T cells in both insulitis and IDDM comes from studies with monoclonal anti-CD4⁺ antibody. Mice treated continuously with this were protected from insulitis and IDDM, provided treatment was begun by 2–3 weeks of age, before the onset of insulitis (KOIKE et al. 1987). In mice with established insulitis, sustained anti-CD4 antibody treatment removed the infiltrating cells (SHIZURU et al. 1988), but insulitis recurred when treatment ceased (CHARLTON and MANDEL 1989). Anti-CD4 antibody also protected NOD mice from the accelerated form of diabetes induced by CYP treatment (CHARLTON and MANDEL 1988). In the isograft model of disease recurrence, neither islet isografts nor allografts (cultured in high oxygen) were destroyed in normoglycemic NOD recipients, but both were destroyed in diabetic NOD mice. This disease recurrence was abrogated by monoclonal anti-CD4 antibody given in the peritransplant period, indicating an essential role for CD4⁺ T lymphocytes in the initiation of disease in transplanted tissue (WANG et al. 1988, 1991).

2.1.2 CD8⁺ T Lymphocytes

Not only CD4⁺ but also CD8⁺ T cells seem essential for the autoimmune destruction of β cells in NOD mice. CYP-induced diabetes was prevented when the mice were treated with anti-CD8 monoclonal antibody (CHARLTON et al. 1988). Adoptive transfer experiments indicated an essential role for CD8⁺ T cells, since their removal from the donor population prevented the transfer of IDDM to neonatal syngeneic recipients and to almost all irradiated recipients (BENDELAC et al. 1987; MILLER et al. 1988). Support for the essential role of CD8⁺ T cells is also evident in experiments involving the transfer of spleen cells from diabetic mice into young NOD mice treated with anti-CD8 antibody.

Recently, it has been shown that NOD mice which lack the expression of β-2 microglobulin as a result of gene targeting, and therefore lack CD8⁺ T cells, do not develop insulitis (KATZ et al. 1993; SERREZE et al. 1994; WICKER et al. 1994). This finding appears to be inconsistent with previous studies showing lack of CD8⁺ T cells involvement in insulitis. Hence it is still not clear whether CD8⁺ T cells are important during the *initiation* state of the disease to "help" CD4⁺ T cells cause early damage to β cells. Furthermore, the role of CD8⁺ T cells in the final *effector* stage of the disease remains controversial. Although there is in vitro evidence for CD8⁺ T cell-mediated cytotoxic destruction of NOD islets, which is unaffected by the removal of CD4⁺ T cells (NAGATA et al. 1989), there is also evidence to suggest that CD8⁺ T cells are not the final effector cells. This is because 3 weeks after adoptive transfer of spleen cells, anti-CD8 antibody no longer prevented transfer whereas anti-CD4 antibody did, and in the isograft model, anti-CD8 antibody failed to prevent disease recurrence, whereas anti-CD4 antibody was effective (WANG et al. 1991).

2.2 Role of B Lymphocytes

As stated above, cell transfer experiments showed that IDDM can be transferred by purified T cells. These findings do not, however, exclude the possibility that recipient B cells may have been recruited by the donor T cells to participate in the disease process. Alternatively, B cells may have been essential in the activation of donor T cells prior to transfer. Using neonatal recipitents, BENDELAC et al. (1989) showed that anti-μ antibody treatment of newborn recipients did not affect the adoptive transfer of IDDM by T cells from diabetic donors. It seems therefore that effective transfer does not require recruitment of B lymphocytes, although the possibility that T cells from the diabetic donors may have been initially activated by interaction with B cells is not excluded.

Autoantibodies have been detected in the sera of NOD mice, but their significance is not clear. Insulin autoantibodies were found in diabetic and nondiabetic NOD mice (PONTESSILLI et al. 1987; ZIEGLER et al. 1989), and the incidence in nondiabetic females was higher than in nondiabetic males (MICHEL et al. 1989). Anti-64kDa antibodies were detectable in NOD mice at the time of weaning and disappeared within weeks of the onset of IDDM (ATKINSON and MACLAREN 1988). Islet cell antibodies were present by day 15 in female NOD mice (REDDY et al. 1990).

Since neither B cells nor sera from diabetic NOD mice could transfer disease, the role of antibodies in the autoimmune process is thought to be secondary to β cell damage (COOKE 1990).

2.3 Role of Macrophages

Macrophages have been implicated in autoimmune β cell destruction because they are evident in the islet infiltrates and because silica (which is selectively toxic to macrophages) has prevented the disease (CHARLTON et al. 1988; LEE et al. 1988). More recently, it has been shown that inhibiting the ability of recipient macrophages to migrate to sites of inflammation by treatment with specific monoclonal antibodies prevented the transfer of IDDM by diabetic spleen cells (HUTCHINGS et al. 1990). It seems therefore that macrophages and both CD4[+] and CD8[+] T cells have an essential role in the pathogenesis of the disease.

2.4 Evidence for a Bone Marrow Defect

NOD bone marrow (BM) cells have been shown to transfer insulitis and diabetes to (NOD × NON) F1 recipients (SERREZE et al. 1988), and to irradiated mice of the following strains: (NOD × C57BL/10) F1 (WICKER et al. 1988), C57BL and B10.BR (LA FACE and PECK 1989), and C3H/HeN (IKEHARA et al. 1990). Based on these findings, a possible explanation for the defect in NOD mice may be the failure of NOD BM-derived MHC class II[+] cells resident in the thymus to negatively select potentially autoreactive T cells during T cell differentiation.

In the converse system, BM cells from nondiabetes-prone BALB/c nude mice (IKEHARA et al. 1985) and B10.BR mice (LA FACE and PECK 1989) transferred protection to irradiated recipients. This suggests that disease may be prevented by replacing NOD BM-derived cells in the thymus with BM-derived cells from strains of mice not prone to develop IDDM. Allogeneic responses in mixed lymphocyte cultures and T cell independent B cell mitogen responses were reported to be normal by IKEHARA et al. (1985), but no evidence was given in either report for MHC-restricted T cell responsiveness. Since the mice were irradiated before reconstitution with allogeneic BM, they would not be expected to have many remaining peripheral antigen-presenting cells (APCs) of NOD genotype. These would be replaced by allogeneic BM-derived cells. The epithelium of the thymus in these mice (which is essential to impart MHC restriction on T cells) would still be of NOD genotype. Hence BALB/c-derived or B10.BR-derived thymocytes maturing through the recipient NOD thymus would be expected to be restricted by NOD MHC only. Peripheral T cells would thus be expected to be "blind" to any antigen (foreign or self) presented by the donor-type BM-derived APCs. The lack of insulitis in these mice could therefore have been due to an inability of the reconstituted T cells to perceive the appropriate autoantigens.

Alternatively, if MHC restriction in the thymus can be imparted by both thymic epithelium and BM-derived cells, immunocompetent T cells should be generated in these mice. The lack of β cell destruction would in this case be explained on the basis of negative selection imparted by the protective MHC class II elements of the BM-derived cells in the thymus.

An experimental protocol to overcome the possible "blindness" of the T cells in the above model involved the use of diabetes-resistant (NOD × NON) F1 BM cells as donors for irradiated NOD recipients. In this combination, the APCs were of F1 origin and hence able to present antigen to NOD-restricted T cells. IDDM was prevented but no data were given on T cell numbers and T cell responsiveness (SERREZE and LEITER 1991). In subsequent experiments, SERREZE and LEITER (1991) transferred to NOD mice a mixed population of NOD.H-2nb1 and NOD BM cells and reported that the majority were protected from disease although recipients of NOD BM were not. In a similar experimental system, BENDELAC et al. (1989) showed that T cell-depleted F1 BM cells from (NOD × BALB/c) F1, (NOD × C57BL/6)F1 or (NOD × CBA/Ca)F1 mice failed to prevent IDDM, although the incidence was reduced. The results are, however, inconclusive since all the recipient mice had lost their state of chimerism by 12 weeks, and all but one mouse had lost their state of tolerence to the BM donor-type parental skin grafts by 9 weeks.

Taken together, the chimeric data suggest that the BM defect in NOD mice may manifest itself within the thymus by the failure of BM-derived cells expressing NOD MHC class I or II molecules to negatively select potentially autoreactive T cells. Introducing BM-derived cells with non-NOD genotypes would allow such cells in the thymus to delete autoreactive NOD CD8$^+$ and CD4$^+$ T cells through interactions with the non-NOD MHC-encoded class I and class II molecules.

3 MHC Class II and Disease in NOD Mice

3.1 Protection from Disease by I-E

The lack of I-E expression is an obvious potential defect in NOD mice and came under investigation initially. Almost complete prevention of autoimmune insulitis was obtained by back-crossing the I-E gene from mice transgenic for I-Eα^d or I-Eα^k into NOD mice (NISHIMOTO et al. 1987; BÖHME et al. 1990). Subsequently, the same result was achieved in a genetically cleaner model by microinjecting the I-Eα^d gene directly into ferilized NOD eggs (UEHIRA et al. 1989; LUND et al. 1990). However in both NOD-I-E lines studied by LUND et al. (1990), there was peri-islet infiltration in 16%–46% of islets, whereas in mice from the NOD-I-E line of UEHIRA et al. (1989), there was complete protection from insulitis. Such differences in the degree of protection may be related to the age of the mice examined for insulitis or to the level or pattern of I-E expression in individual transgenic lines.

The role of the NOD MHC and of I-E was studied using H-2 congenic NOD mice by PODOLIN et al. (1993). Mild insulitis but no IDDM was found in mice that had a NOD background but a non-NOD, I-E$^+$ MHC. F1 mice congenic for I-E$^+$ MHC, which also carried one copy of the NOD MHC, showed mild insulitis and a small percentage (3%–5%) of the older mice (9–17 months) developed diabetes as well. Although this was the first report of spontaneous diabetes in NOD mice bearing an I-E gene, the data from previous reports are not likely to be significantly different when the numbers and age of the mice analyzed are taken into account. UEHIRA et al. (1989) did not report any insulitis at 4.5 months but did not follow their mice for diabetes incidence beyond that age. LUND et al. (1990) reported complete protection from insulitis and diabetes in their mice at 12 months and PODOLIN et al. (1993) observed insulitis in three of 108 mice and diabetes in three out of 62 mice for an age range up to 17 months.

CYP was reported to induce diabetes in a small proportion of I-E$^+$ F1 MHC congenic NOD mice, but it was unable to do so in I-E^d transgenic NOD males or females (LUND et al. 1990; UNO et al. 1992).

BÖHME et al. (1990) investigated the mechanism of I-E protection by back-crossing wild-type and promoter-mutated I-Eα^k transgenes onto the NOD genetic background. Mice were created so as to have I-E present on different cell subsets. Almost complete protection from insulitis was observed when I-E was expressed on all class II expressing compartments (thymic epithelium, thymic medulla, B cells and macrophages), but when expression was lacking from any one of these compartments, no protection was afforded. Unfortunately, complementary "compartment" I-E transgenic mice were not crossed to demonstrate whether a combination of all "expressing" compartments did confer protection. The wild type I-E transgenic mouse used did not control for possible perturbations in cell to cell signaling that may have resulted from deletions of I-E regulatory elements. Nevertheless, ignoring this issue, it seems likely that protection requires I-E to be expressed in both thymic and peripheral cell compartments.

Using BM chimeras, PARISH et al. (1993) also investigated the mechanism of I-E protection. Control NOD-I-E irradiated recipients of NOD-I-E BM showed no disease, and NOD-I-E BM cells protected irradiated NOD mice from diabetes whereas NOD BM cells did not. It should be noted that, although protection of NOD mice from disease was afforded by NOD-I-E BM cells, 16% of the islets in these chimeras did have infiltrates (compared to 72% in controls). In a second experiment, the BM chimeras were grafted with a thymus. In this case, NOD mice reconstituted with NOD-I-E BM were protected from diabetes regardless of whether they had been grafted with a thymus from a NOD or NOD-I-E donor. Furthermore NOD mice given NOD BM showed a high incidence of disease regardless of the source of the grafted thymus. Thus, as suggested in the earlier studies of BM chimeras, the NOD-I-E thymus is not capable of affording protection in the presence of NOD BM, and NOD-I-E BM can transfer protection even in the presence of a NOD thymus. These results indicate that the radioresistant, non-BM-derived cells of the thymus are not involved in conferring diabetes susceptibility or resistance in chimeric mice.

The data from congenic and transgenic NOD mice generally indicate that I-E reduces the severity of insulitis and protects from diabetes for up to 9–17 months. Diabetes is inducible by CYP in I-E-protected congenic NOD female mice but not in the smaller number of transgenic I-E protected male mice analyzed. The protection by I-E requires both thymic and extrathymic expression and can be transferred by BM cells from NOD-I-E to NOD recipients.

3.2 Protection from Disease by I-A

The first external domain of the NOD I-A β chain has five consecutive nucleotide substitutions from 248–252 which leads to two radical amino acid changes at residue 56 and 57. These unique substitutions are in a region that is usually conserved between human and murine species (ACHA-ORBEA and MCDEVITT 1987). In Caucasoids MHC DQβ chains having aspartic acid (asp) at residue 57 have been associated with a dominant nonsusceptibility to IDDM, whereas DQβ chains with noncharged amino acids have been associated with susceptibility (TODD et al. 1987). In the homologous region of the murine I-A β chain, most mice have a proline at residue 56 and asp at 57, but the NOD mouse has histidine at 56 and serine at 57.

In 1990, three groups reported that like I-E, I-A was also able to protect NOD mice from diabetes. Although no protection was afforded by incorporation of the I-A β gene alone, transgenic expression of both I-Aα^k and I-Aβ^k dramatically reduced the severity of insulitis and diabetes (SLATTERY et al. 1990; MIYAZAKI et al. 1990). LUND et al. (1990) transgenically introduced a His-56 to Pro-56 mutant of the Aβ^{g7} allele and found that this modified β chain also conferred protection from IDDM.

To examine the importance of residue 57 of the NOD I-A β chain, MIYAZAKA et al. (1990) produced I-A^k transgenic mice which lacked Asp 57 (IAα^k I-Aβ^{k57ser}).

Despite this lack of Asp at residue 57, the mice were protected from IDDM. This finding argues against residue 57 as being the single crucial diabetogenic element of the I-A molecule, although clearly this region of the class II molecule has an important influence in diabetogenesis.

Subsequent to the original description of non-NOD I-A conferring protection, there has been only two reports investigating the mechanism involved. SLATTERY et al. (1993) transferred BM cells depleted of T cells from NOD I-A^k transgenic mice to neonatal NOD mice and these were protected from IDDM for more than 7 months. In order to ascertain at what level protection operated, whether in the thymus or in the periphery, 30 million T cells were transferred from NOD-I-A^k mice to either thymectomized irradiated NOD or to NOD-severe combined immuno-deficiency disease (scid) mice. In both these recipients, a high incidence of dia-betes developed. These results indicate that protection from IDDM in NOD-I-A^k mice can be mediated by BM cells alone and is not due to the deletion or permanent silencing of potentially autoreactive T cells.

Consistent with this and in contrast to I-E$^+$ transgenic NOD mice, the administration of CYP to NOD-I-A^k transgenic mice induced diabetes in 33% of female mice and 14% of male mice (SLATTERY et al. 1993).

SINGER et al. (1993) also demonstrated that NOD mice transgenic for a non-NOD I-A molecule, I-A^d, were protected from IDDM (diabetes developed in 10% of transgenic females compared to 45% of nontransgenic females by 30 weeks of age). Transfer of 5–10 million spleen cells from diabetic NOD mice to irradiated NOD recipients induced diabetes, but to achieve this, using cells from diabetic NOD-I-A^d transgenic mice, required many more spleen cells. Thus, increasing the number of spleen cells from diabetic transgenic mice to as many as 20–50 million per recipient transferred disease to only two of seven recipient irradiated NOD mice and to only two of six irradiated NOD-I-A^d mice. Furthermore, 4–10 million purified T cells from transgenic mice injected together with total spleen cells from diabetic NOD mice into irradiated NOD recipients gave an incidence of diabetes from 25% to 73% compared with a higher incidence from 86%–91% when purified spleen cells from nontransgenic NOD had been cotransferred with the diabetic spleen cells. Since no I-A^{d+} cells were transferred with the purified T cells, the results imply either that the I-A^d-restricted transgenic T cells protected in a non-MHC restricted manner, or cross-reacted with the NOD I-A^{NOD} to confer protection, or that I-A^{NOD}-restricted T cells were themselves committed to a protective phenotype. It was suggested that protection in this model may have been due to clonal diversion, e.g., the differentiation of T cells to nonpathogenic effector cells of the Th2 type, but no direct evidence was given for this possibility.

3.3 The Class II Defect in NOD Mice

The defect that leads to autoimmune insulitis in NOD mice may be the failure of I-A^{NOD} to present peptides which can readily be bound and presented by I-E, I-A^k or IAα^d-I-Aβ^{g57pro}. Such a defect may be important intrathymically during the development of the T cell repertoire, or extrathymically when peptide affinity for MHC is essential for the initiation of T cell responses. Presentation of such a peptide intrathymically may be necessary either for the negative selection of T cells reactive against the putative islet cell antigen or for the positive selection of T cells able to exert a protective effect. If negative selection were inefficient (because of the low affinity of binding of a particular peptide to I-A^{NOD}), potentially auto-aggressive T cells would migrate to the periphery. They may remain quiescent until triggered by an up-regulated or neoantigen presented in association with I-A^{NOD} molecules (GAMMON and SERCARZ 1989). Alternatively, low affinity of a particular peptide for I-A^{NOD} may lead to a defect in positive selection of T cells with suppressive properties essential to prevent the responses of activated auto-aggressive T cells. Defects in suppression have indeed been proposed for the pathogenesis of diabetes in both humans (CHANDY et al. 1984) and NOD mice (YASUNAMI and BACH 1988; CHARLTON et al. 1989; SEMPE et al. 1994).

4 Possible Mechanisms of Protection by Non-NOD Class II MHC Molecules

4.1 Intrathymic Mechanisms

4.1.1 Deletion or Permanent Silencing

When T cells were transferred from protected NOD-I-A^k mice into NOD-scid recipients or into lethally irradiated and thymectomized NOD mice, they were able to cause IDDM in a large percentage of the recipients (SLATTERY et al. 1993). This indicates that autoreactive T cells have neither been deleted nor made anergic by transgenic incorporation of I-A^k. Since autoreactive T cells persist in NOD-I-A^k mice without causing IDDM in most mice, class II-mediated protection cannot be explained by negative selection of potentially autoreactive T cells and must involve some form of immunoregulation. There are indeed many examples of the persistence of potentially autoreactive T cells circulating in healthy individuals and experimental animals (MILLER and FLAVELL 1994).

In I-E transgenic mice, there is no direct evidence to show that I-E protects from IDDM by negative selection. On the contrary, BÖHME et al. (1990) showed that although transgenic I-E expression in selective compartments of the thymus facilitated negative selection of T cells, I-E expression in these situations did not protect from diabetes.

4.1.2 Immunoregulation

Autoreactive T cells may be prevented from attacking β cells by immunoregulatory T cells that have been selected by protective class II molecules during thymic development. Positive selection of T cells by class II is, however, normally imparted by the thymic epithelium and yet protection could be transferred by BM-derived cells given to NOD mice lacking the protective I-A^k (SLATTERY et al. 1993) or I-E (PARISH et al. 1993) on thymic epithelium. Hence any mechanism invoking immunoregulatory T cells selected by class II molecules requires that such selection takes place on BM-derived cells.

An immunoregulatory T cell may act in one of several ways. It may directly suppress the autoreactive cell, although it is difficult to envisage how a class II-restricted T cell might recognize another T cell. The regulatory T cells could recognize, in association with protective class II molecules, peptide ligands of autoreactive T cells and then provide help for class I-restricted T cells that could suppress the autoreactive T cells. If the protective T cell was class I-restricted, it is difficult to understand why protection was not transferred to NOD-scid mice that have the appropriate class I restricting elements. Alternatively, an immuno-regulatory T cell may indirectly suppress autoreactive T cell via a third cell type which might interact with both T cell types. The APCs which express the restriction element for both T cells is a likely candidate. Such an explanation for class II-mediated protection from IDDM in NOD mice requires that the protective class II molecules be expressed not only on positive selection elements within the thymus but also on APCs in the periphery.

4.2 Extrathymic Mechanisms

4.2.1 Epitope Capture

I-E or I-A on peripheral APCs could protect transgenic NOD mice from IDDM by competing for autoantigenic determinants involved in the cascade of events leading to autoimmunity (DENG et al. 1993). If the affinity of the diabetogenic peptide is greater for I-E or I-A^k than for I-A^{NOD}, the relevant peptides may be sequestered away from I-A^{NOD} thus preventing the activation of the autoreactive T cells. Since the processing and presentation of self-antigens is thought to be an intracellular event, protection by antigen capture would require that the protective class II molecules and the I-A^{NOD} molecules be expressed on the same cell. SLATTERY et al. (1993), however, showed that NOD mice were protected from IDDM by the neonatal injection of NOD-I-A^k BM cells. Although the donor NOD-I-A^k BM-derived cells in the recipient mice expressed both I-A^k and I-A^{NOD}, they constituted <10% of the recipient's total lymphoid compartment. Almost all the APCs expressed only I-A^{NOD} and hence would be able to process and present peptide products in association with I-A^{NOD}. There would thus be no competition from I-A^k molecules. Despite this, the autoreactive T cells were not activated to autoaggression in these chimeras.

Consistent with this observation in transgenic NOD mice are the findings of SERREZE and LEITER (1991), who showed that while NOD BM cells transferred diabetes to 75% of reconstituted irradiated NOD (MHC F1 congenic) recipients, a

1:1 mixture of NOD BM and NOD.H-2nb1 BM cells transferred diabetes to only 7% such recipients. In the situation of the mixed BM cells, approximately 50% of the hematopoietic cells expressed only I-A^{NOD}. Despite this, there was a significant reduction in the extent of activation of the autoreactive T cells.

4.2.2 Antigen Clearance

If, as implied by the work of SLATTERY et al. (1993) and BÖHME et al. (1990), the APC plays a crucial role in class II mediated protection, it seems likely that this is achieved through some immunoregulatory T cells. This may occur as a result of the expansion of a population of I-E-or I-A^K-restricted T cells which would compete for the autoantigen thereby circumventing the initiation of a significant I-A^{NOD}-restricted autoimmune cascade (Fig. 1a,b).

4.2.3 Antigen-Presenting Cell Paralysis

Alternatively, the I-E-or I-A-restricted T cells may paralyze the APCs presenting autoantigenic peptides so that interaction with autoimmune T cells no longer leads to activation.

The human MHC class II molecules, the HLA-DR1 heterodimer, occurs as a dimer of dimers with an orientation that would allow simultaneous interactions with two T cell receptor (TCR) complexes, which have been proposed to act as "molecular tweezers" on the surface of APCs (PLOEGH and BENAROCH 1993). The bringing together of two class II molecules occupied by the same peptide would allow a higher avidity interaction between class II molecules, TCRs and associated CD3, CD4 coreceptors and associated kinases, and hence the signal transduction cascade that leads to T cell activation. The class II defect in NOD mice, responsible for initiating autoimmunity, may simply be that a lower antigenic threshold allows APCs to induce T cell activation. In the transgenic NOD mice, the increased diversity of class II molecules on APCs may limit superdimer formation and thereby impede proper signal transduction (Fig. 1b), or alternatively result in the delivery of a qualitatively different signal, perhaps one that results in clonal diversion (Fig. 1c).

Although the evidence suggests that the APC is central to the regulation of the autoreactive T cells, the delivery of a diversion signal may not necessarily go directly via the APC. Irrespective of whether the immunoregulatory process results from a diversion signal, or the absence of a signal, the effect is unlikely to be permanent. This would explain why the transfer of T cells from protected NOD-I-A^k mice into nontransgenic NOD mice ultimately results in autoimmune diabetes (SLATTERY et al. 1993).

4.2.4 Immune Deviation

Why self-reactive T cells are not normally activated is not fully understood. The level of antigen presented or the mode of presentation may determine which

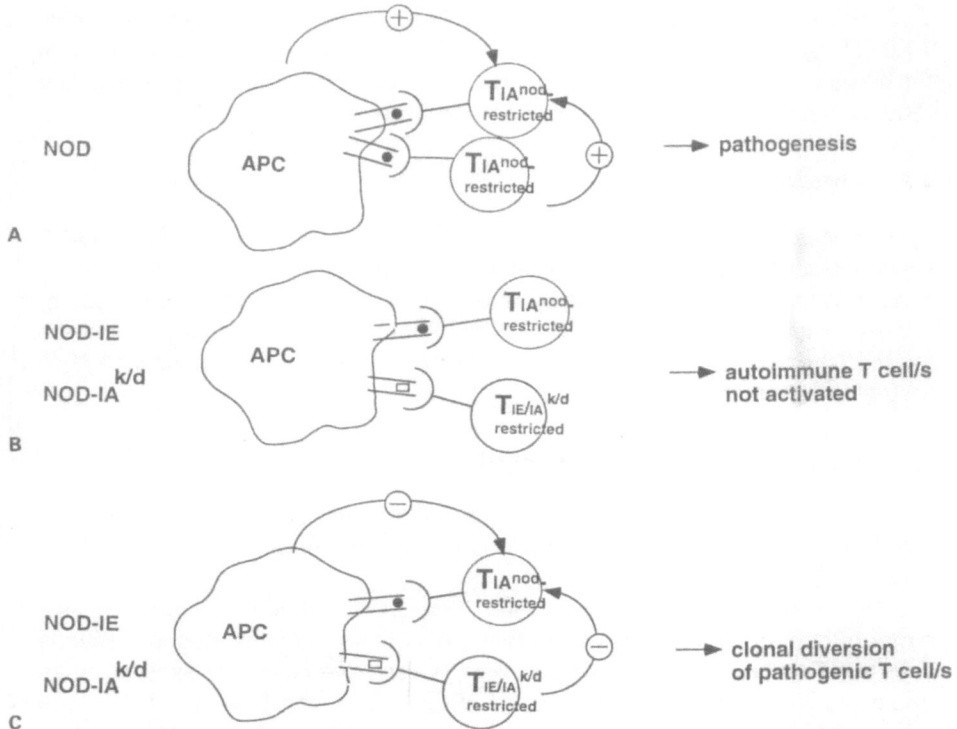

Fig. 1A-C. Proposed mechanism of class II-mediated protection in NOD mice. **A** At low antigen threshold, the antigen-presenting cell (APC) in NOD mice is competent to bring about an activation signal to the engaged T cells. The T cells are activated by virtue of their shared T cell receptor (TCR) specificity and ability to dimerize class II molecules of shared isotype and allotype which are presenting the same peptide antigen. It is possible that the activation signal is one that results from direct T cell-T cell interaction brought about by the proximity imposed by class II dimers on the APC or indirect signaling via the APC itself. **B** At the same antigen concentration, the APC from NOD class II transgenic mice is incompetent to bring about an activation signal to the engaged T cells. This is because the engaged T cells are restricted to different class II allotypes or isotypes and are therefore unable to cause dimerization. **C** Alternatively, APCs from NOD class II transgenic mice may bring about a qualitatively different signal as a result of the engagement by T cells which are restricted to different class II allotypes or isotypes. The different signal may lead to clonal diversion of pathogenic T cells

T cell type is activated. It has recently been proposed that protection mediated by class II genes in transgenic NOD mice may be related to the development of Th2-type T cells (SINGER et al. 1993). The evidence for a skewing of the T cell subsets towards the Th1 type in the pathogenesis of IDDM in NOD mice, however, remains controversial. Although interleukin (IL)-10 given systemically to NOD mice reduced the incidence of IDDM (PENNLINE et al. 1994), the disease was accelerated in 33% of transgenic NOD mice expressing, in their β cells, IL-10, which was stated, though not shown, to promote a Th2-like lymphokine pattern (WOGENSEN et al. 1994). In these mice, IL-10 may simply have exerted a chemotactic and local inflammatory response, as was demonstrated for IL-2 (ALLISON et al. 1992).

Clonal diversion has also been implicated in another model of autoimmune diabetes. Double transgenic mice expressing the influenza hemagglutinin on islet β cells and a TCR transgene specific for a class II-restricted hemagglutinin peptide were either resistant or susceptible to spontaneous IDDM depending on their genetic background. The resistant strains characteristically showed a non-MHC-encoded predisposition towards a Th2 response, since their CD4+ T cells produced high levels of IL-4 (Scott et al. 1994). The T cells producing the IL-4 were not, however, directly linked to the autoimmune response.

5 Summary

Central to the autoimmune pathogenesis of IDDM in NOD mice is the MHC class II region. In all models studied to date, expression of NOD MHC class II genes is essential for disease development suggesting a crucial role for I-A^{NOD}-restricted presentation of autoantigen. Protection has been afforded by transgene incorporation of other non-NOD class II genes and many models have been proposed to account for this effect. It is now clear that protection is not achieved by deletion or permanent silencing of all autoreactive T cell clones. It also appears that expression of these genes is required both intra- and extrathymically. It still remains to be determined what role these genes may have in the various compartments and how the autoreactive cells are held in check in protected NOD transgenic mice. Currently, the most likely explanation is that intrathymic expression of non-NOD class II genes is required for the positive selection of class II-restricted immunoregulatory T cells, while peripheral expression is necessary to bring about the interaction of these cells in a tricellular complex with NOD autoantigen-specific T cells and APCs, so that the response can be deviated to a nonpathogenetic one. Whether this process is active or passive is not known.

Acknowledgments. We thank Dr. Brett Charlton for helpful suggestions and critical reading of the manuscript. Work from Prof. Miller's laboratory was supported by the National Health and Medical Research Council of Australia and by NIH grant AI-29385.

References

Acha-Orbea H, McDevitt HO (1987) The first external domain of the nonobese diabetic mouse class II I-A beta chain is unique. Proc Natl Acad Sci USA 84: 2435–2439

Allison J, Malcolm L, Chosich N, Miller JFAP (1992) Inflammation but not autoimmunity occurs in transgenic mice expressing constitutive levels of interleukin-2 in islet B cells. Eur J Immunol 22: 1115–1121

Atkinson MA, Maclaren NK (1988) Autoantibodies in nonobese diabetic mice immunoprecipitate 64, 000-Mr islet antigen. Diabetes 37: 1587–1590 (published erratum appears in Diabetes (1989) 38(5): inside front cover)

Baxter AG, Adams MA, Mandel TE (1989) Comparison of high- and low-diabetes-incidence NOD mouse strains. Diabetes 38: 1296–1300

Bendelac A, Boitard C, Bach JF, Carnaud C (1989) Neonatal induction of allogeneic tolerance prevents T cell-mediated autoimmunity in NOD mice. Eur J Immunol 19: 611–616

Bendelac A, Boitard C, Bedossa P, Bazin H, Bach JF, Carnaud C (1988) Adoptive T cell transfer of autoimmune nonobese diabetic mouse diabetes does not require recruitment of host B lymphocytes. J Immunol 141: 2625–2628

Bendelac A, Carnaud C, Boitard C, Bach JF (1987) Syngenic transfer of autoimmune diabetes from diabetic NOD mice to healthy neonates. Requirement for both L3T4+ and Lyt-2+ T cells. J Exp Med 166: 823–832

Böhme J, Schuhbaur B, Kanagawa O, Benoist C, Mathis D (1990) MHC-linked protection from diabetes dissociated from clonal deletion of T cells. Science 249: 293–295

Chandy KG, Charles AM, Kershnar A, Buckingham B, Waldeck N, Gupta S (1984) Autologous mixed lymphocyte reaction in man. XV. Cellular and molecular basis of deficient autologous mixed lymphocyte response in insulin-dependent diabetes mellitus. J Clin Immunol 4: 424–428

Charlton B, Mandel TE (1988) Progression from insulitis to beta-cell destruction in NOD mouse requires L3T4+ T-lymphocytes. Diabetes 37: 1108–1112

Charlton B, Mandel TE (1989) Recurrence of insulitis in the NOD mouse after early prolonged anti-CD4 monoclonal antibody treatment. Autoimmunity 4: 1–7

Charlton B, Bacelj A, Mandel TE (1988) Administration of silica particles or anti-Lyt2 antibody prevents beta-cell destruction in NOD mice given cyclophosphamide. Diabetes 37: 930–935

Charlton B, Bacelj A, Slattery RM, Mandel TE (1989) Cyclophosphamide-induced diabetes in NOD/WEHI mice. Evidence for suppression in spontaneous autoimmune diabetes mellitus. Diabetes 38: 441–447

Cooke A (1990) An overview on possible mechanisms of destruction of the insulin-producing beta cell. Curr Top Microbiol Immunol 164: 125–142

Deng H, Apple R, Clare-Salzler M, Trembleau S, Mathis D, Adorini L, Sercarz E (1993) Determinant capture as a possible mechanism of protection afforded by major histocompatibility complex class II molecules in autoimmune disease. J Exp Med 178: 1675–1680

Gammon G, Sercarz E (1989) How some T cells escape tolerance induction. Nature 342: 183–185

Garchon HJ (1992) Non-MHC-linked genes in autoimmune diseases. Curr Opin Immunol 4: 716–722

Hanafusa T, Sugihara S, Fujino KH, Miyagawa J, Miyazaki A, Yoshioka T, Yamada K et al. (1988) Induction of insulitis by adoptive transfer with L3T4+Lyt2- T-lymphocytes in T-lymphocyte-depleted NOD mice. Diabetes 37: 204–208

Harada M, Makino S (1986) Suppression of overt diabetes in NOD mice by anti-thymocyte serum or anti-Thy 1, 2 antibody. Jikken Dobustu 35: 501–504

Hutchings P, Rosen H, O'Reilly L, Simpson E, Gordon S, Cooke A (1990) Transfer of diabetes in mice prevented by blockade of adhesion-promoting receptor on macrophages. Nature 348: 639–642

Ikehara S, Ohtsuki H, Good RA, Asamoto H, Nakamura T, Sekita K, Muso E et al. (1985) Prevention of type I diabetes in nonobese diabetic mice by allogenic bone marrow transplantation. Proc Natl Acad Sci USA 82: 7743–7747

Ikehara S, Kawamura M, Takao F, Inaba M, Yasumizu R, Than S, Hisha H et al. (1990) Organ-specific and systemic autoimmune diseases originate from defects in hematopoietic stem cells. Proc Natl Acad Sci USA 87: 8341–8344

Katz J, Benoist C, Mathis D (1993) Major histocompatibility complex class I molecules are required for the development of insulitis in non-obese diabetic mice. Eur J Immunol 23: 3358–3360

Kaufman DL, Clare-Salzler M, Tian J, Forsthuber T, Ting GS, Robinson P, Atkinson MA et al. (1993) Spontaneous loss of T-cell tolerance to glutamic acid decarboxylase in murine insulin-dependent diabetes. Nature 366: 69–72

Kida K, Kaino Y, Miyagawa T, Gotoh Y, Matsuda H, Kono T (1985) Effect of cyclosporin on insulitis and islet cell surface antibody in non-obese diabetic mice. Congress of the International Diabetes Federation, Madrid

Koike T, Itoh Y, Ishii T, Ito I, Takabayashi K, Maruyama N, Tomioka H et al. (1987) Preventive effect of monoclonal anti-L3T4 antibody on development of diabetes in NOD mice. Diabetes 36: 539–541

La Face DM, Peck AB (1989) Reciprocal allogeneic bone marrow transplantation between NOD mice and diabetes-nonsusceptible mice associated with transfer and prevention of autoimmune diabetes. Diabetes 38: 894–901

Lee KU, Amano K, Yoon JW (1988) Evidence for initial involvement of macrophage in development of insulitis in NOD mice. Diabetes 37: 989–991

Lund T, O'Reilly L, Hutchings P, Kanagawa O, Simpson E, Gravely R, Chandler P et al. (1990) Prevention of insulin-dependent diabetes mellitus in non-obese diabetic mice by transgenes encoding modified I-A beta-chain or normal I-E alpha-chain. Nature 345: 727–729

Makino S, Harada M, Kishimoto Y, Hayashi Y (1986) Absence of insulitis and overt diabetes in athymic nude mice with NOD genetic background. Jikken Dobutsu 35: 495–498

Michel C, Boitard C, Bach JF (1989) Insulin autoantibodies in non-obese diabetic (NOD) mice. Clin Exp Immunol 75: 457–460

Miller BJ, Appel MC, O'Neil JJ, Wicker LS (1988) Both the Lyt-2+ and L3T4+ T cell subsets are required for the transfer of diabetes in nonobese diabetic mice. J Immunol 140: 52–58

Miller JFAP, Flavell R (1994) T cell tolerance and autoimmunity in transgenic models of central and peripheral tolerance. Curr Opin Immunol 6: 892–899

Miyazaki T, Uno M, Uehira M, Kikutani H, Kishimoto T, Kimoto M, Nishimoto H et al. (1990) Direct evidence for the contribution of the unique I-ANOD to the development of insulitis in non-obese diabetic mice. Nature 345: 722–724

Mori Y, Suko M, Okudaira H, Matsuba I, Tsuruoka A, Sasaki A, Yokoyama H et al. (1986) Preventive effects of cyclosporin on diabetes in NOD mice. Diabetologia 29: 244–247

Nagata M, Yokono K, Hayakawa M, Kawase Y, Hatamori N, Ogawa W, Yonezawa K et al. (1989) Destruction of pancreatic islet cells by cytotoxic T lymphocytes in nonobese diabetic mice. J Immunol 143: 1155–1162

Nishimoto H, Kikutani H, Yamamura K, Kishimoto T (1987) Prevention of autoimmune insulitis by expression of I-E molecules in NOD mice. Nature 328: 432–434

Ogawa M, Maruyama T, Hasegawa T, Kanaya T, Kobayashi F, Tochino Y, Uda H (1985) The inhibitor effect of neonatal thymectomy on the incidence of insulitis in non-obese diabetic mice. Biomed Res 6: 103–106

Parish NM, Chandler P, Quartey-Papafio R, Simpson E, Cooke A (1993) The effect of bone marrow and thymus chimerism between non-obese diabetic (NOD) and NOD-E transgenic mice, on the expression and prevention of diabetes. Eur J Immunol 23: 2667–2675

Pennline KJ, Roque GE, Monahan M (1994) Recombinant human IL-10 prevents the onset of diabetes in the nonobese diabetic mouse. Clin Immunol Immunopathol 71: 169–175

Ploegh H, Benaroch P (1993) MHC class II dimer of dimers. Nature 364: 16–17

Podolin PL, Pressey A, DeLarato NH, Fischer PA, Peterson LB, Wicker LS (1993) I-E+ nonobese diabetic mice develop insulitis and diabetes. J Exp Med 178: 793–803

Pontessilli O, Carotenuto P, Gazda LS, Pratt PF, Prowse SJ (1987) Circulating lymphocyte populations and autoantibodies in non-obese diabetic (NOD) mice: a longitudinal study. Clin Exp Immunol 70: 84–93

Reddy S, Biddy N, Elliott RB (1990) Longitudinal study of islet cell antibodies and insulin autoantibodies and development of diabetes in non-obese diabetic (NOD) mice. Clin Exp Immunol 81: 400–405

Scott B, Liblau R, Degermann S, Marconi LA, Ogata L, Caton AJ, McDevitt HO et al. (1994) A role for non-MHC genetic polymorphism in susceptibility to spontaneous autoimmunity. Immunity 1: 73–82

Sempe P, Richard M-F, Bach J-F, Boitard C (1994) Evidence of CD4+ regulatory T cells in the non-obese diabetic male mouse. Diabetologia 37: 337–343

Serreze DV, Leiter EH (1991) Development of diabetogenic T cells from NOD/Lt marrow is blocked when an allo-H-2 haplotype is expressed on cells of hemopoietic origin, but not on thymic epithelium. J Immunol 147: 1222–1229

Serreze DV, Leiter EH, Christianson GJ, Greiner D, Roopenian DC (1994) Major histocompatibility complex class I-deficient NOD-B2mnull mice are diabetes and insulitis resistant. Diabetes 43: 505–509

Serreze DV, Leiter EH, Worthen SM, Shultz LD (1988) NOD marrow stem cells adoptively transfer diabetes to resistant (NOD x NON)F1 mice. Diabetes 37: 252–255

Shizuru JA, Taylor EC, Banks BA, Gregory AK, Fathman CG (1988) Immunotherapy of the nonobese diabetic mouse: treatment with an antibody to T-helper lymphocytes. Science 240: 659–662

Singer SM, Tisch R, Yang XD, McDevitt HO (1993) An Abd transgene prevents diabetes in nonobese diabetic mice by inducing regulatory T cells. Proc Natl Acad Sci USA 90: 9566–9570

Slattery RM, Kjer NL, Allison J, Charlton B, Mandel TE, Miller JFAP (1990) Prevention of diabetes in non-obese diabetic I-Ak transgenic mice. Nature 345: 724–743

Slattery RM, Miller JFAP, Heath WR, Charlton B (1993) Failure of a protective major histocompatibility complex class II molecule to delete autoreactive T cells in autoimmune diabetes. Proc Natl Acad Sci USA 90: 10808–10810

Tisch R, Yang XD, Singer SM, Liblau RS, Fugger L, McDevitt HO (1993) Immune response to glutamic acid decarboxylase correlates with insulitis in non-obese diabetic mice. Nature 366: 72–75

Todd JA, Bell JI, McDevitt HO (1987) HLA-DQ beta gene contributes to susceptibility and resistance to insulin-dependent diabetes mellitus. Nature 329: 599–604

Uehira M, Uno M, Kurner T, Kikutani H, Mori K, Inomoto T, Uede T et al. (1989) Development of autoimmune insulitis is prevented in E alpha d but not in A beta k NOD trasgenic mice. Int Immunol 1: 209–213

Uno M, Miyazaki T, Uehira M, Nishimoto H, Kimoto M, Miyazaki J, Yamamura K (1992) Complete prevention of diabetes in transgenic NOD mice expressing I-E molecules. Immunol Lett 31: 47–52

Wang Y, McDuffie M, Nomikos IN, Hao L, Lafferty KJ (1988) Effect of cyclosporine on immunologically mediated diabetes in nonobese diabetic mice. Transplantation 46 Suppl: 106S

Wang Y, Pontesilli O, Gill RG, La RF, Lafferty KJ (1991) The role of CD4+ and CD8+ T cells in the destruction of islet grafts by spontaneously diabetic mice. Proc Natl Acad Sci USA 88: 527–531

Wicker LS, Miller BJ, Chai A, Terada M, Mullen Y (1988) Expression of genetically determined diabetes and insulitis in the nonobese diabetic (NOD) mouse at the level of bone marrow-derived cells. Transfer of diabetes and insulitis to nondiabetic (NOD X B10) F1 mice with bone marrow cells from NOD mice. J Exp Med 167: 1801–1810

Wicker LS, Leiter EH, Todd JA, Renjilian RJ, Peterson E, Fischer PA, Podolin PL et al. (1994) Beta 2-microglobulin-deficient NOD mice do not develop insulitis or diabetes. Diabetes 43: 500–504

Wogensen L, Lee MS, Sarvetnick N (1994) Production of interleukin 10 by islet cells accelerates immune-mediated destruction of beta cells in nonobese diabetic mice. J Exp Med 179: 1379–1384

Yasunami R, Bach JF (1988) Anti-suppressor effect of cyclophosphamide on the develop ment of spontaneous diabetes in NOD mice. Eur J Immunol 18: 481–484

Ziegler AG, Vardi P, Ricker AT, Hattori M, Soeldner JS, Eisenbarth GS (1989) Radioassay determination of insulin autoantibodies in NOD mice. Correlation with increased risk of progression to overt diabetes. Diabetes 38: 358–363

Virus-Induced Autoimmune Disease: Transgenic Approach to Mimic Insulin-Dependent Diabetes Mellitus and Multiple Sclerosis

M.B.A. Oldstone, M. von Herrath, C.F. Evans, and M.S. Horwitz

1 Introduction

Transgenic technology has been used by virologists for two main purposes. One has been to evaluate cell-specific expression and function in vivo of specific viral proteins. Of the many such studies performed, one prominent example is the work of Levine and colleagues (MARKS et al. 1989) on the expression of SV40 T antigen that resulted in choroid plexus papillomas; these experiments led to identification of p53 bound to T antigen. As another example, NERENBERG et al. (1987) used transgenic expression of the TAT protein of HTLV-1 to establish the oncogenic potential of TAT. Lastly, NELSON and colleagues (unpublished data) probed expression of the immediate early 72 KDa protein of human cytomegalovirus and located this early regulatory protein in the islets of Langerhans, smooth muscle wall of large arteries, retina and salivary gland. All of these studies were similar in that the authentic promoter of the particular viral gene being expressed was employed. The second major avenue followed by virologists utilizing transgenic technology has been to create animal models of viral pathogenesis. Examples of this application are described in separate chapters of this volume and include, first, the isolation and expression of the gene encoding the human poliovirus receptor in mice by REN and RACANIELLO (1992), REN et al. (1990), and

Division of Virology, Department of Neuropharmacology, The Scripps Research Institute, 10666 N. Torrey Pines Road, La Jolla, CA 92037, USA

independently by HORIE et al. (1994) and KOIKE et al. (1991), allowing study of the virus' tropism, pathogenesis of the disease it causes, and design of new protective vaccines. Second, CHISARI et al. (1987, 1989) utilized the albumin promoter to express hepatitis B virus surface antigen in the liver to study virus-induced chronic liver disease, hepatocellular carcinoma, and the role of the immune system in either clearing virus of causing immunopathology. The third example is that of TOGGAS et al. (1994) who used an astrocyte-specific promoter to express HIV-1 gp120 in transgenic mice. The brain lesions of these mice closely mimic those of AIDS neuropathology in humans including neuronal injury, microglia activation and gliosis.

We have also adapted the transgenic approach to study virus-induced autoimmune disease. As reviewed previously in this series (OLDSTONE 1989a), infectious agents, particularly viruses, are implicated in autoimmunity and can induce autoimmune responses by a variety of unique mechanisms. Autoimmune diseases are associated with many factors that play a role in the initiation and continuance of disease. Among the factors identified, host genes, immune responses, and infecting viruses are most often implicated (SINHA et al. 1990; YOON 1990). To better understand this process, we began studies on two important autoimmune diseases of humans, insulin-dependent diabetes mellitus (IDDM) and multiple sclerosis (MS). Both are T lymphocyte mediated autoimmune diseases; in IDDM, insulin-producing β cells located in the pancreatic islets of Langerhans are destroyed and, in MS, myelin sheaths that insulate axons and are formed by oligodendrocytes are destroyed. Our plan was to express a viral gene either in β cells located in the pancreatic islets of Langerhans or in oligodendrocytes by using cell-specific promoters. The viral gene, for all practical purposes, becomes a "self" gene in the transgenic mice that is passed on to their progeny. Selection of the appropriate viral gene requires that the protein it encodes elicits a vigorous cytotoxic T lymphocyte (CTL) response, the kinetics and chemistry of which are known and manipulatable.

Although the symptoms and signs can be rapid in onset, a common clinical feature of autoimmune disease is the latent or lag period from the suspected onset until later in life when full-blown disease occurs. Epidemiological studies (reviewed in SADOVNIK and EBERS 1993) of persons who develop MS have illuminated this concept. On the basis of these investigations, it is clear that patients born in a geographical area where the incidence of MS is high (e.g., Canada) who have resettled where the incidence is lower (e.g., the tropics, Africa) still retain heightened susceptibility to this disease. The converse is also true in that those born in geographic areas of lower incidence and move to locations with a heightened incidence retain the low incidence for developing MS. Similar scenarios accompany other autoimmune disorders including IDDM (GREEN 1990; LA PORTE et al. 1985). Additionally, genetics play a role in host susceptibility to these disorders. HLA linkage for both IDDM (MICHELSEN et al. 1990) and MS (reviewed in OKSENBERG et al. 1993) is well established. Yet, environmental agents or factors also play a role, as the study of monozygotic twins has determined (GREEN 1990; EBERS et al. 1986; KOTZIN 1993).

To integrate these various components of genetics, environmental factors (i.e., microbe or virus) and T lymphocyte reactivity into one theme, we hypothesize that two interrelated but separate events occur in both IDDM and MS and perhaps in other autoimmune diseases. First, an initial exposure to a virus early in life results in restricted and low level expression of a viral gene in the appropriate target cell, i.e., β cells of the islets of Langerhans for IDDM, oligodendrocytes for MS. These events by themselves need not cause disease, especially if the host is hyporesponsive or tolerant to the viral product. Indeed, this set of circumstances may occur if the infectious agent also infects cells of the immune system (OLDSTONE 1989b, 1991) or mimics a host self-protein (OLDSTONE 1987). Alternatively, the first event could be the expression of a self-protein that shares antigenic determinants with a virus. Later in life, the second event occurs: infection with the same infectious agent or one with cross-reacting antigenic determinants. The result is an immune response to the infecting agent; this response, designed to clear the new infection, also generates reactive lymphocytes or antibodies that eventually localize to the target cell (i.e., β cells, oligodendrocytes) with subsequent progression to a specific autoimmune disease. Figure 1 summarizes these ideas.

The scenario pictured in Fig. 1 also allows for multiple MHC and background genes to control the susceptibility of a host to various infectious agents and to control the varying strengths of immune responses generated by infectious agents. Although not discussed here, the model illustrated also allows for an initial microbial infection early in life to cause an acute onset autoimmune disease due not to hyporesponsiveness or tolerance but to the induction of cytokines.

This chapter brings together and discusses the experimental evidence to test the hypothesis depicted in Fig. 1.

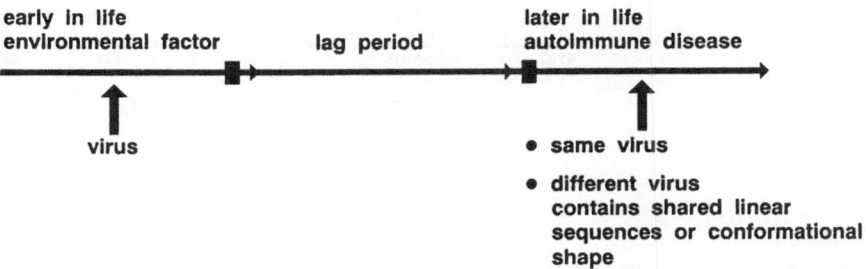

Fig. 1. How viruses may induce autoimmune diseases like diabetes or multiple sclerosis. Infection by a virus or another microbe does not induce disease by itself, but viral antigen persists in a certain cell type (insulin producing beta cells of the islets of Langerhans in the pancreas, thyroid hormone producing cells, oligodendrocytes, etc.). Later in life, when the organism encounters the same or a related microbe, a response is induced against the virus which cross-reacts with self (viral) antigen and leads to progressive tissue damage and disease

2 A Transgenic Model of Virus-Induced Insulin-Dependent Diabetes Mellitus

To test this hypothesis, the insulin promoter (RIP) was used to create separate lines of transgenic (tg) mice whose pancreatic β cells expressed either the nucleoprotein (NP) or glycoprotein (GP) gene of lymphocytic choriomeningitis virus (LCMV). Tg mice were formed on the H-2^b or H-2^d background, and the viral GP or NP genes were selected because they elicit strong T lymphocyte responses that have been carefully studied and dissected in both murine MHC haplotypes (OLDSTONE et al. 1988; WHITTON et al. 1989; WHITTON 1990). The construct used to generate these tg mice is shown in Fig. 2.

Several tg lines were created, and three questions were asked. First, does a viral gene expressed in a mouse as a transgene cause disease? Second, can humoral and cellular immune responses to this viral gene product (anti-self) be generated later in life? Third, can the generated anti-viral (anti-self) immune response lead to IDDM? The results of experiments performed to answer these three questions appear in the text.

The viral gene, introduced as early as the egg stage of the animal, became incorporated into the germline and was expressed in islet cells, thereby becoming a self-gene. To answer the first question, the viral transgene rarely produced IDDM at any time in the lives of these animals. Tolerance induced was primarily

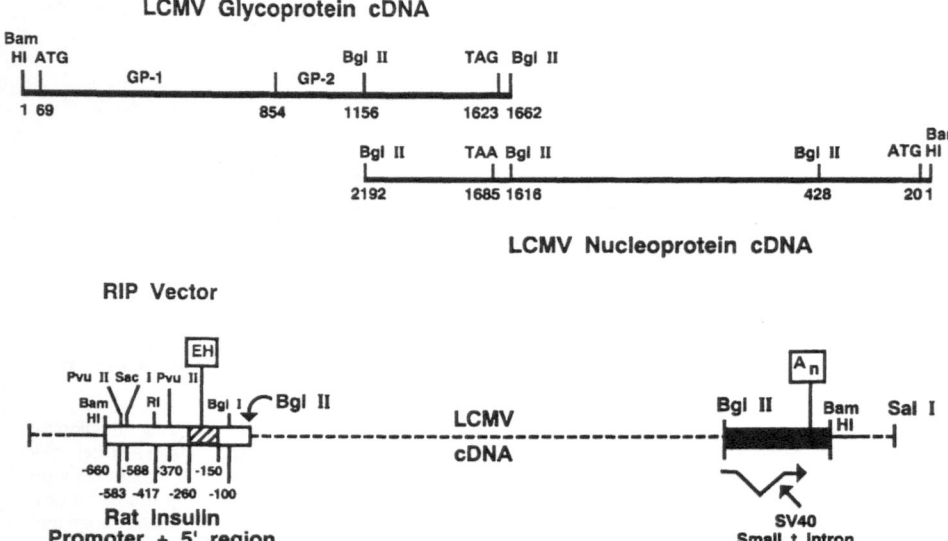

Fig. 2. The lymphocytic choriomeningitis virus (LCMV) cDNA clones and rat insulin promoter (RIP) vector product used to establish transgenic mice

peripheral because: (1) challenge with the same virus in later life caused IDDM (OLDSTONE et al. 1991) in >90% of these tg mice (Fig. 3), and (2) lymphocytes obtained from the spleens of such tolerant mice could be primed in vitro to generate anti-viral (self) CTLs when incubated with *Drosophila* cells transfected with and expressing the correct MHC class I molecule and the appropriate viral NP or GP peptide (Lewicki and Oldstone, unpublished data). Further, the lymphocytes recovered from the islets of Langerhans from tg mice, when tested by in vitro assay, were virus (self)-specific and MHC restricted and, by in vivo assay, homed to islets of Langerhans of LCMV tg mice expressing the correct viral gene but not to those expressing an incorrect gene. That is, lymphocytes from RIP LCMV-NP tg mice, upon adoptive transfer, homed only to the islets of RIP LCMV-NP tg but not of LCMV-GP tg mice. Conversely, lymphocytes isolated from RIP LCMV-GP tgs homed only to islets in RIP LCMV-GP tg mice but not LCMV-NP tg mice. Thus, these results showed that immune response to the viral transgene could be generated and cause disease. Further, the data obtained in our studies (OLDSTONE et al. 1991) were also independently and concurrently by OHASHI et al. (1991).

In the Ohashi report (OHASHI et al. 1991), the RIP was used to express the GP gene of LCMV WE strain, and upon challenge with LCMV, IDDM occurred rapidly within 7–14 days after inoculation. The Ohashi model was skewed for rapid onset disease by utilizing double transgenics made by crossing RIP LCMV-GP with mice expressing the T cell receptor for one of the two peptide regions on the viral GP molecules recognized by CTLs (OHASHI et al. 1991). By contrast, our group expressed the NP and GP of LCMV Armstrong strain in β cells of the islets of Langerhans and noted both rapid onset and slow onset models of IDDM (OLDSTONE et al. 1991; VON HERRATH et al. 1994a). Agreeing with the report by OHASHI et al. (1991), the rapid onset IDDM occurred within 7–21 days; in contrast to their observations, the slow onset IDDM was delayed in most tg lines until 2–6 months after viral challenge (OLDSTONE et al. 1991; VON HERRATH et al. 1994a). Table 1 lists the various tg lines generated and their phenotypes.

3 Molecular Dissection of Slow and Rapid Onset of Virus-Induced Insulin-Dependent Diabetes Mellitus

To determine the factors that distinguish rapid onset from slow onset IDDM, a tg line called GP 34–20 that developed IDDM within 14 days after viral challenge and a slow onset line named NP 25–3 that developed IDDM 2–6 months after viral challenge were studied (see Table 1). As illustrated in Fig. 4, hypoglycemia developed within 7–14 days after viral challenge in all GP 34–20 mice with a mean glucose level of 442 mg/dl $\pm$ 33 (standard error of the mean). By contrast, hyperglycemia did not occur in either uninfected (blood glucose mean 167 $\pm$ 4) or virally infected 25–3 mice (mean blood glucose 189 $\pm$ 13) at that time point.

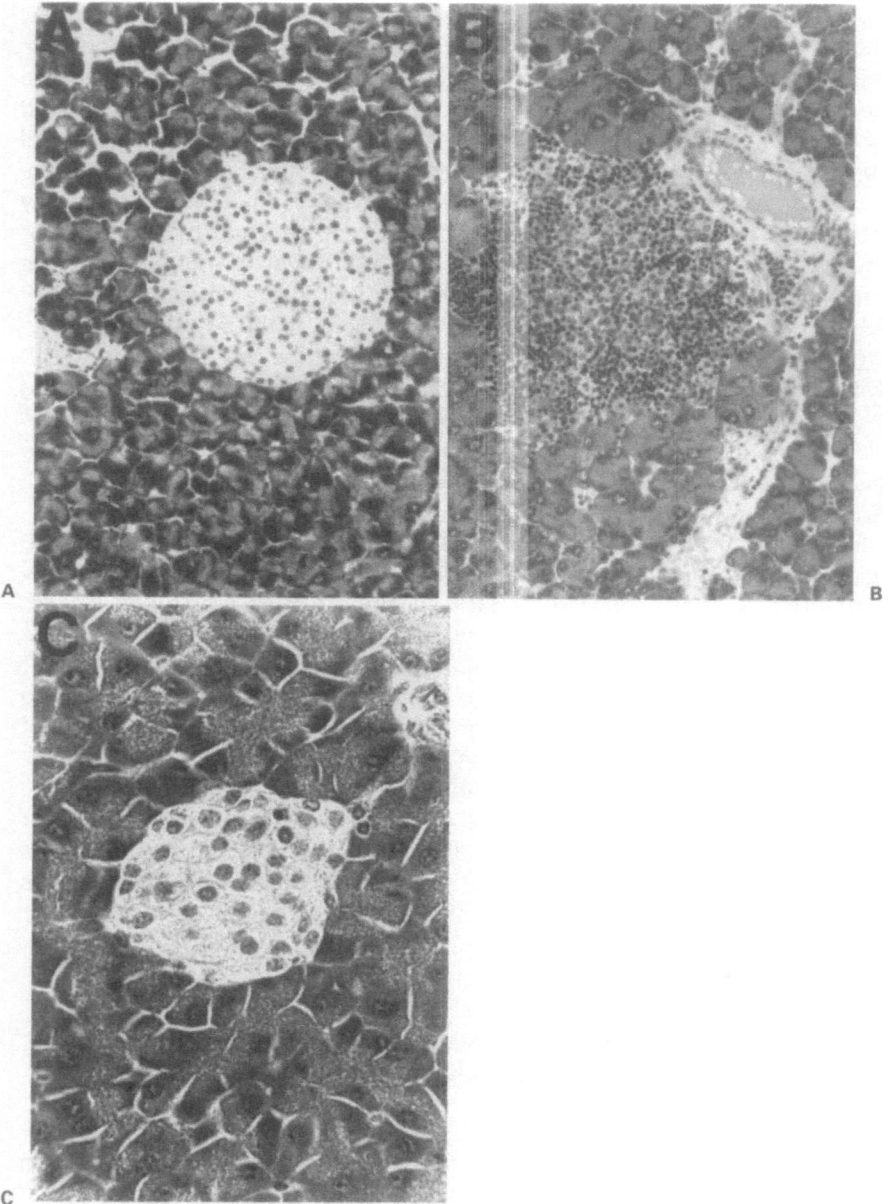

Fig. 3A–C. Islets of Langerhans from RIP LCMV-NP transgenic mice (see Fig. 2). Greater than 95% of mice expressing the transgene in their islets throughout their lives show normal islet morphology and absence of infiltrating lymphocytes **A** unless challenged with a virus that shares antigen determinants with the transgene. In this case **B** destruction of the islets with an accompanying inflammatory response occurs. Less than 5% of mice expressing the transgene in their islets spontaneously develop elevated blood glucose. The islets of these mice **C** show swollen cells, ground glass appearing cytoplasm and no inflammatory infiltrate. The elevated blood sugar for these <5% transgenic mice is likely due to overexpression of the transgene and a storage-like disorder of the β-cells

Table 1. Fate of the RIP- LCMV transgenic lines in the presence or absence of a viral challenge in later life[a]

| Line designation | Transgene only | | | | Transgene + LCMV[b] | | | | Onset IDDM |
| | Blood glucose mg/dl | | Pancreatic Insulin | | Blood glucose mg/dl | | Pancreatic Insulin | | |
	Number >250	Mean ± 1 SE	µg/ml	insulitis	Number >250	Mean ±1 SE	µg/ml	insulitis	
RIP NP 25-3	0/50	155±5	55±10	Nil	44/50	461±34	3±1	+++	2–5m
RIP NP 25-5	1/25	186±20	78±6	Nil	18/20	413±26	8±3	+++	2–5m
RIP NP 25-7	2/30	210±33	48±17	Nil	17/2	429±30	10±9	+++	2–5m
RIP NP 25-19	0/25	170±7	67±5	Nil	22/25	472±15	2±15	+++	2–5m
RIP NP 33-27	0/40	186±8	ND	Nil	19/20	404±26	4±1	+++	<1m
RIP NP 34-20	1/26	189±13	60±9	Nil	23/28	436±22	2±1	+++	<1m
RIP GP 12-5	0/25	170±14	ND	Nil	18/21	390±36	ND	+++	2–5m
RIP GP 34-20	0/50	203±7	54±8	Nil	18/21	470±21	2±1	+++	<1m
RIP GP 34-21	0/25	184±6	61±4	Nil	22/25	439±19	3±1	+++	2–5m
RIP GP 64-6	2/30	178±8	ND	Nil	12/20	318±32	16±8	+++	2–5m
SV40 GP31	0/25	129±7	46±9	Nil	0/16	146±11	51±8	Nil	Nil
bxd (non-transgenics)	0/25	172±12	58±10	Nil	0/19	179±10	62±12	Nil	Nil

LCMV, lymphocytic choriomeningitis virus; RIP, rat insulin promoter.
[a]see Figs. 1 and 3.
[b]1×10^5 PFU at 6–8 weeks of age.

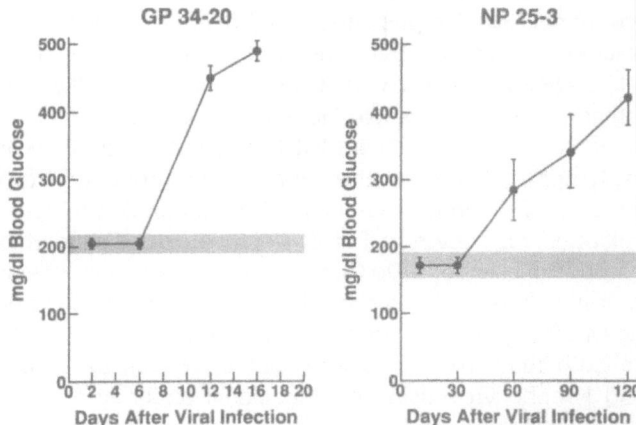

Fig. 4. Blood glucose levels in RIP LCMV transgenic mice (see Fig. 2) after LCMV infection. Two month old RIP GP 34–20 (no thymic expression of transgenic) or RIP NP 25–3 (thymic expression of transgene) were inoculated with 2×10^5 pfu LCMV (ARM) ip (12–20 mice per group). Blood glucose was measured weekly for 4 weeks, thereafter monthly. Mean values ± 1 SEM are displayed. Littermates that did not receive LCMV showed no elevation of blood glucose during the displayed test period and, the mean ± 4 SEM for their blood glucose levels is displayed as the *shaded area*

However, 2–6 months after viral inoculation, NP 25–3 mice developed IDDM as judged from their mean blood glucose levels of 422 ± 46 mg/dl.

Both the rapid and slow onset groups of tg mice had mononuclear cells infiltrating their islets and these were identified mainly as CD8[+] and CD4[+]

lymphocytes. Macrophages were also noted, not in the infiltrate but along the borders of the islets of Langerhans. Paralleling the different kinetics for the two forms of IDDM, these infiltrates appeared significantly earlier in GP 34–20 mice than in NP 25–3 mice.

A search for tissues expressing the viral transgenes indicated that the rapid onset IDDM mice expressed their transgene only in the pancreas; in contrast, tg mice with the slow onset IDDM expressed the transgene in the thymus as well as the pancreas (Fig. 5). Multiple other tissues including spleen, kidney, heart, muscle, lung, brain, and liver did not express the transgene in either the slow or rapid onset IDDM.

Other studies were done to evaluate the role and activity of lymphocytes participating in the IDDM. Lymphocytes harvested from the islets of Langerhans were able to recognize MHC-restricted LCMV-specific targets, to traffic specifically to islets bearing the appropriate transgene following adoptive transfer and, on occasion, to cause insulitis and IDDM in some of the tg recipients. Functional analysis of the lymphocytes according to phenotype in adoptive transfer experiments indicated that adoptively transferred CD8-bearing lymphocytes caused insulitis. CD4-bearing lymphocytes failed to transfer disease but responded in a MHC-restricted viral antigen-specific proliferation assay (von Herrath, unpublished data). Additionally, LCMV-specific CD8$^+$ lymphocytes from rapid onset (GP 34–20) mice (H-2^b) had equivalent affinities and avidities to CTLs generated in non-tg H-2^b mice. To the contrary, slow onset CD8$^+$ lymphocytes from (NP 25–3) tg mice (H-2^b) failed to generate a primary CTL response but generated secondary responses (Table 2). CTLs made by H-2^d NP 25–3 tg mice were of less affinity and avidity than CTLs generated in non-tg control littermates (VON HERRATH 1994a).

Because CD8$^+$ lymphocytes from slow onset IDDM tg mice were of lower affinity and avidity than lymphocytes from the rapid onset group or from controls, the possible role of CD4$^+$ lymphocyte help in the generation of virus-induced IDDM was evaluated. As shown in Fig. 6, depletion of CD4-bearing lymphocytes by using a monoclonal antibody to CD4 did not alter the incidence or kinetics of IDDM occurring in the rapid onset model (GP 34–20). By contrast, after depletion of CD4-bearing lymphocytes from the slow onset tg model (NP 25–3), no IDDM developed. In both the rapid and slow onset IDDM, as expected, depletion of CD8-bearing lymphocytes using a monoclonal antibody to CD8 aborted the development of IDDM (Fig. 6).

4 Major Histocompatability Complex Genes and Thymic Selection

Mice with slow onset IDDM, as exhibited in tg line 25–3, expressed the viral transgene in both the β cells of the islets of Langerhans and in the thymus. To better understand the role of MHC and other non-MHC linked background genes,

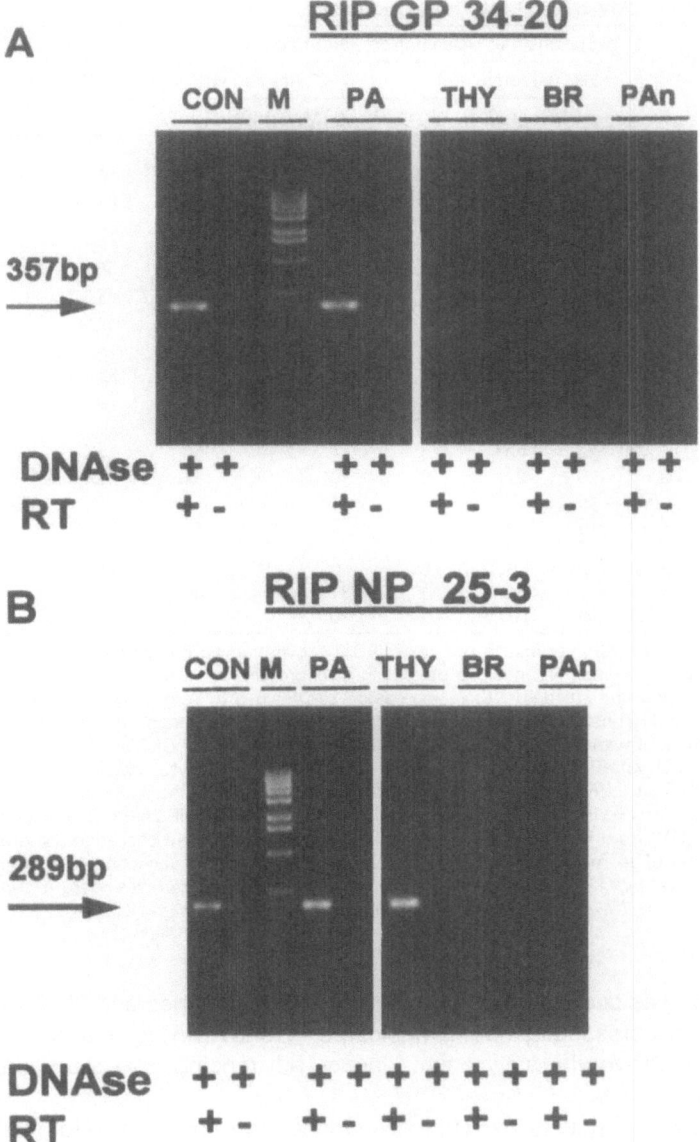

Fig. 5A,B. Expression of LCMV GP or NP RNA in RIP LCMV transgenic mice (see Fig. 2) as determined by reverse transcriptase PCR (RT-PCR). **A** GP 34–20 mice expressed LCMV GP only in their pancreas (*PA*) but not in the brain (*BR*), or thymus (*THY*). All samples were treated with RNAse-free RQ1 DNAse prior to RT-PCR to eliminate DNA contamination. Control samples without RT were run simultaneously with each sample. Pancreatic samples from nontransgenic littermates (*PAn*) did not show any transgene product. Control RNA was obtained from spleens of mice that were persistently infected with LCMV. **B** LCMV-NP was expressed in pancreas (*PA*) and thymus (*THY*) of RIP-NP transgenic mice, but not in brain (*BR*) or pancreas (*PAn*) of nontransgenic littermates. *M*, markers of RNA size

Table 2. CTL Tolerance to LCMV in RIP-LCMV Transgenic Mice

CTL activities found in spleens at various times post infection	E:T	Specific ^{51}Cr Release from target cells (%)						
		H-2^b infected with			H-2^d infected with			H-2^q infected with ARM
		ARM	VV/GP	VV/NP	ARM	VV/GP	VV/NP	
splenic								
H-2^d d7	50:1	3	0	0	70	1	62	0
	25:1	2	0	0	63	0	50	ND
H-2^b d7	50:1	62	51	33	0	0	1	2
	25:1	53	48	19	0	0	0	ND
H-2^q d7	50:1	3	ND	ND	4	ND	ND	56
F1/BALB x C57BL6 d7	50:1	66	32	26	59	1	39	4
	25:1	51	17	11	50	2	29	2
RIP GP34-20 H-2^b d7	50:1	40	27	25	3	2	1	0
	25:1	27	17	14	1	0	2	ND
RIP NP25-3 H-2^b d7	50:1	42	38	5	3	1	2	1
	25:1	39	28	2	1	0	0	ND
d60 S°	5:1	34	26	1	1	0	2	2
	2.5:1	28	22	0	0	0	0	ND
d120 S°	5:1	26	8	13	1	5	6	0
	2.5:1	14	5	8	1	0	0	ND
CD4 depleted d120 S°	5:1	11	0	0	1	4	2	1
	2.5:1	2	0	0	0	1	0	ND
RIP NP25-3 H-2^d d7	50:1	1	ND	1	31	1	22	
	25:1	0	ND	1	20	0	12	

CTL, cytotoxic T lymphocyte; RIP, rat insulin promoter; LCMV, lymphocytic chorimeningitis virus; ARM, Armstrong; VV, vaccinia virus; NP, nucleoprotein; GP, glycoprotein.
CTL assays were performed in a standard ^{51}Cr release assay run for 5–6 h. Target cells were H-2^b (MC57), H-2^d (BalbC17) or H-2^q fibroblasts uninfected or infected with LCMV (strain Armstrong) 48h prior to the assay (MOI=1), or with VV recombinants expressing the GP or NP of LCMV (vv/NP, vv/GP) 8 h prior to the assay (MOI=3). Effector cells were either primary CTL splenocytes obtained 7 days after infection with 1x10^5 pfu LCMV of nontransgenic H-2^b (C57BL6), H-2^d (Balbc/ByJ), H-2^b x H-2^d (Balb/c x C57BL6, F1) or H-2^q (SWR/J) mice, from transgenic RIP GP 34-20, RIP NP 25-3 H-2^b and H-2^d mice or secondary (S°) CTL from RIP NP-H-2^b mice 60 and 120 days after LCMV challenge. Where indicated, CD4 depletion of tg mice was carried out prior to LCMV infection (injecting 0.1mg/ml monoclonal antibody to CD4 i.v.). E : T=effector to target cell ratio used. All samples were run in triplicates and SE was <12%. (See VON HERRATH et al. 1994a for details).

the 25–3 line (bxd) was back-crossed to the H-2^d, H-2^b haplotypes and 25–3 tg mice (k x b) were back-crossed to the H-2^b haplotype. As shown in Fig. 7, all these tg lines developed IDDM with slow onset kinetics but ranging across a wide window of time.

Hence, whereas H-2^b mice (C57BL6) (back-crossed from b x d mice for five generations) developed IDDM by 2–6 months after viral challenge, those of the H-2^d (Balb) background (back-crossed from bxd mice for five generations) required only 1–2 months to manifest IDDM and those with H-2^b background (back-crossed from H-2^k (CBA) x H-2$^{b/bm1}$ B1O.BR mice for five generations) genes took the longest and had the lowest incidence of diabetes. The specific component of the NP molecule recognized by H-2^b mice is located at the carboxy end from amino acid sequence 396–404, whereas the component restricted by H-2^d is at the amino portion of the molecule, positions 118–127 (WHITTON et al. 1989).

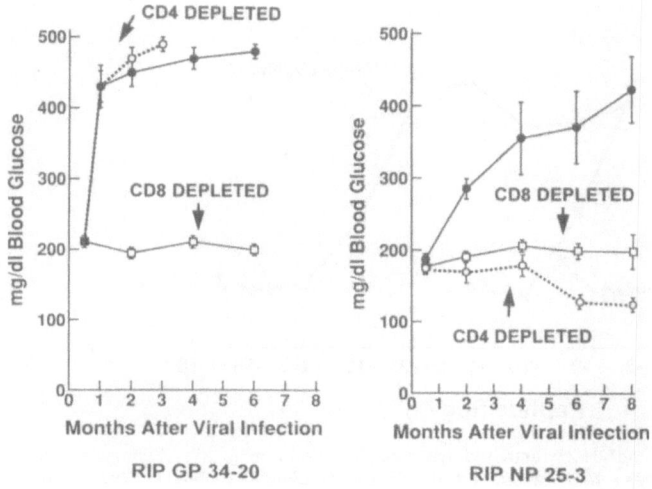

Fig. 6. Blood glucose levels in RIP-LCMV transgenic mice (see Fig. 2) depleted of CD4 and CD8 lymphocytes before receiving 1×10⁵ pfu LCMV ip. *Left panel*, results for GP 34–20 mice, *right panel*, RIP NP 25-3 H-2ᵇ mice. Group held ten or more mice and the *bar* indicates 1 SEM. *Closed circles* in both panels show results from transgenic mice that had normal CD4 or CD8 lymphocytes (not depleted) and received 1×10⁵ pfu LCMV-ARM ip. *Open symbols* show results from immunodepleted mice. FACS analysis demonstrated that depletion of CD4 or CD8 lymphocytes was >96%. Results displayed here for CD4 and CD8 depletion of H-2ᵇ RIP-NP transgenic mice were also found with H-2ᵈ RIP-NP transgenic mice

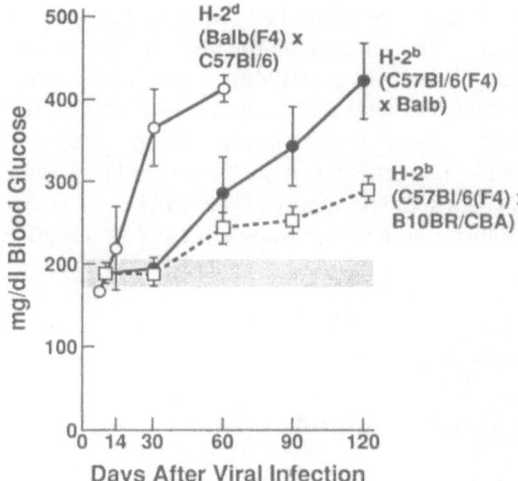

Fig. 7. Effects of different genetic backgrounds on virus induced IDDM in RIP-LCMV transgenic mice (see Fig. 2). Two month old RIP NP 25-3 mice were bred from b×d to Balb/c (H-2ᵈ), C57BL6 or B10BR/CBA (H-2ᵇ) backgrounds and were inoculated with 1 × 10⁵ pfu LCMV i.p. Blood glucose was measured weekly for 4 weeks and thereafter monthly. Mean value ± 1 SEM is displayed. Littermates that did not receive viral inoculation did not show elevated blood glucose values during the observation period. For these controls the mean ± 4 SEM is displayed as the *shaded area*

Although H-2ᵇ NP 25–3 tg mice failed to generate a primary LCMV CTL response after viral challenge, their H-2ᵈ tg counterparts did so but with markedly lower activities/affinities (Table 2). To account for these differences in affinities, dilutions of the H-2ᵇ or H-2ᵈ NP peptide were made over a peptide dose range of

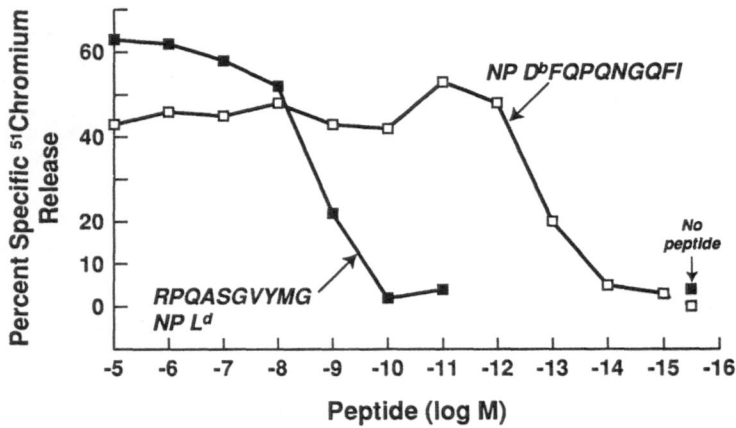

Fig. 8. Lytic activity of H-2^b of H-2^d restricted lymphocytic choriomeningitis virus-nucleoprotein (LCMV-NP) specific CTL clones when reacted with uninfected syngenic target cells located with the appropriate virally restricted peptide. Peptides from LCMV-NP recognized by H-2^d of H-2^b-restricted LCMV-specific CTLs were used at log dilutions to coat MHC matched H-2^b (MC57) or H-2^d (BalbCL7) target cells. Specific killing is expressed in % ^51Cr release after 5 h in a standard ^51Cr release (CTL) assay. SEM was <10% for all values displayed and data were confirmed in three independent experiments. These NP peptides are the only NP peptide recognized by LCMV Armstrong-specific CTL. (From WHITTON et al. 1989; OLDSTONE et al. 1991)

10^{-5} to 10^{-15} molar. As shown in Fig. 8 at 10^{-13} M, peptide FQPQNGQFI still sensitized H-2^b targets for lysis by D^b-restricted anti-LCMV CTLs. In contrast, $10^{-8.5}$ M, or 4 logs more peptide, was required to sensitize H-2^d targets for lysis by LCMV CTLs. This and other data obtained by expressing the viral NP gene under control of the Thy1.2 promoter (VON HERRATH et al. 1994b) denoted an inverse association between the affinity for MHC-viral (NP) peptide and CTLs. Ten thousand-fold fewer H-2^b NP peptide molecules were needed to sensitize H-2^b targets for CTL killing than H-2^d NP. This difference in binding affinity likely explains the presence of low affinity CTLs in the periphery of H-2^d tg mice following primary CTL response, compared to the absence of primary CTLs in H-2^b NP25–3 tg mice.

5 A Transgenic Model for Virus-Induced Multiple Sclerosis

To determine the generality of the concept cartooned in Fig. 1, the LCMV Armstrong NP or GP were expressed specifically in oligodendrocytes using the myelin basic protein (MBP) promoter (Gow et al. 1992). Utilizing the construct shown in Fig. 9, several tg lines were established (Table 3.)

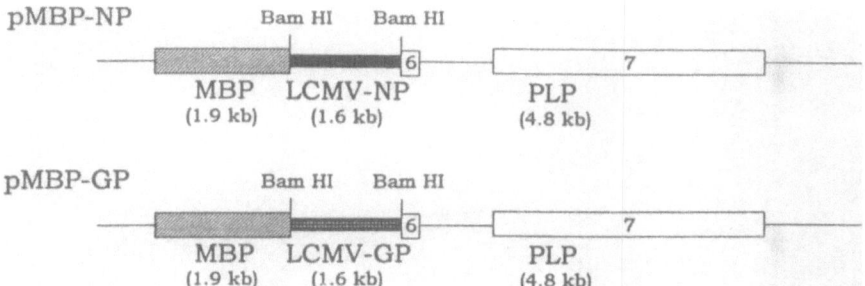

Fig. 9. DNA constructs used to express the lymphocytic choriomeningitis virus nucleoprotein (LCMV NP) or glycoprotein (GP) in oligodendrocytes. The LCMV NP or GP genes were cloned in between the myelin basic protein *(MBP)* promoter, and exons 6 and 7 and intron 6 of the proteolipid protein *(PLP)* gene which provide polyadenylation and splicing signals

Table 3. Summary of the MBP-LCMV transgenic lines

Line	Trans-genic		Uninfected			Infected (LCMV)[a]		
			Lymphocytes infiltrating the brain[b]		Clinical symptoms[c]	Lymphocytes infiltrating the brain[b]		Clinical symptoms[c]
		MHC	CD4	CD8		CD4	CD8	
MBP-NP-H5	NP	H-2^d	0/6	0/6	None	2/10	8/10	Yes
MBP-NP-H28	NP	H-2^d	0/6	0/6	None	2/11	7/11	Yes
MBP-NP-H60	NP	H-2^d	0/6	0/6	None	3/7	5/7	Yes
MBP-NP-M40	NP	H-2^s	0/6	0/6	None	1/8	5/8	Yes
MBP-NP-M69	NP	H-2^s	0/6	0/6	None	2/7	4/7	Yes
MBP-GP-S27	GP	H-2^b	0/6	0/6	None	1/3	3/3	Yes
MBP-GP-S46	GP	H-2^b	0/6	0/6	None	2/3	3/3	Yes
Nontransgenic	–	H-2^d	0/6	0/6	None	0/17	4/17	None
Nontransgenic	–	H-2^s	0/6	0/6	None	0/9	2/9	None
Nontransgenic	–	H-2^b	0/6	0/6	None	0/6	0/6	None

MBP, myelin basic protein; LCMV, lymphocytic choriomeningitis virus; NP, nucleoprotein; GP, glycoprotein.
[a] Mice were infected intravenously with LCMV at 2 x 10^6 plaque forming units.
[b] Mice were sacrified at 3 weeks postinfection. CD4 and CD8-positive cells were identified by immuno histo chemical staining of sagittal brains sections. The total number of lymphocytes infiltrating the brain parenchyma across equivalent fields was counted. Greater than 20 positive CD4 cells or 50   positive CD8 cells per section was considered a positive result.
[c] Mice were observed clinically for up to 1 year. Transgenic mice displayed symptoms ranging from loss of coordination to seizures.

Despite expression of the transgene, none of the mice showed CNS disease during their lifespan. However, when such tg mice were challenged with LCMV, a proportion developed neurologic signs seizures and incoordination. Examinations of CNS tissues showed the infiltration of primarily CD8⁺, but also in some instances CD4⁺, lymphocytes in the brain parenchyma (Fig. 10). Interestingly, these infiltrating lymphocytes persisted as long as 6 months after viral challenge and were often associated with the activation of microglia.

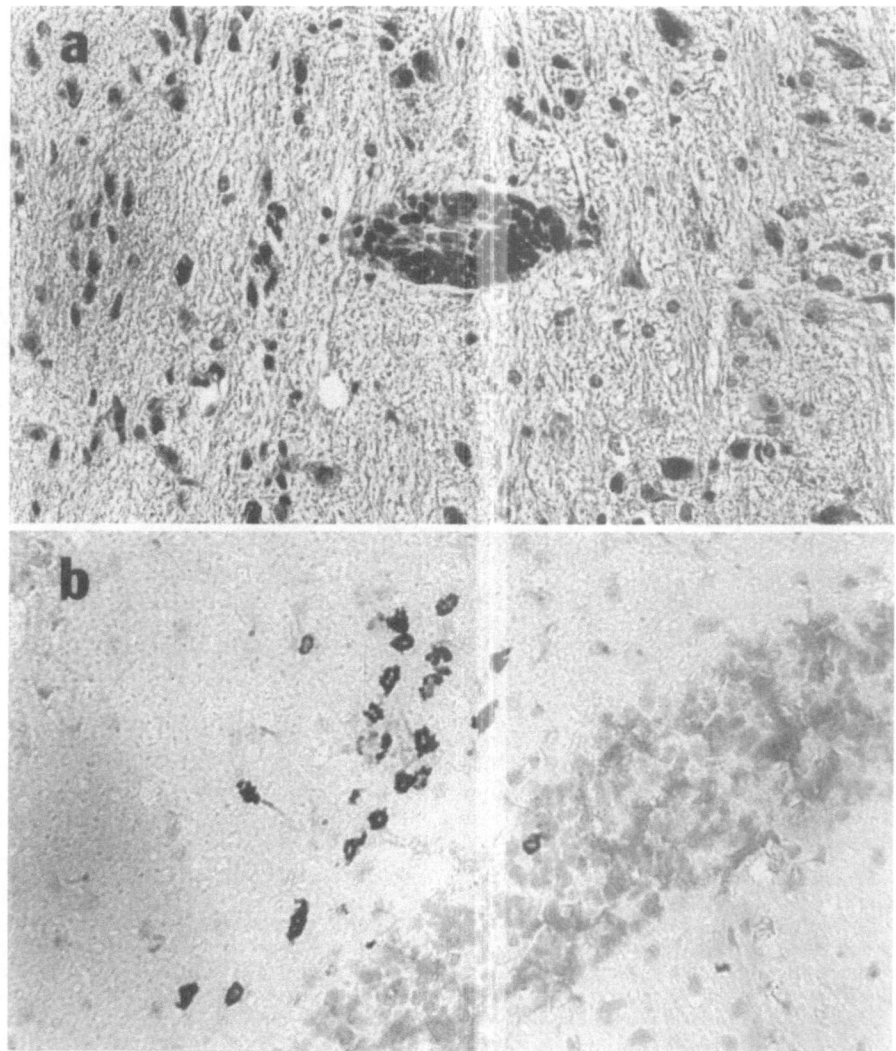

Fig. 10a,b. Mononuclear cell infiltration in the brains of lymphocytic choriomeningitis virus nucleo-protein (LCMV NP) transgenic mice following infection with LCMV. Mice transgenic for the LCMV NP gene were sacrificed 3 weeks postinfection with LCMV. Brain sections were stained with hematoxylin and eosin **a** to show perivascular mononuclear cell infiltration, or with monoclonal antibodies to CD8 **b** to identify the primary infiltrating cell type, in this example in the white matter of the cerebellum. Similar findings occurred in mice studied 6 months after viral challenge

6 Conclusion

Microbial agents, particularly viruses, are implicated in autoimmune diseases, according to multiple findings. Through mechanisms of replication in lymphocyte subsets, activation of lymphocytes, release of cytokines, and regulation of the expression of MHC class I and class II molecules, viruses can prime an immune response against self. Additionally, viruses can infect and selectively replicate in antigen presenting cells and unique lymphocyte subsets and disorder their functions. Finally, viruses can contain chemical structures that mimic normal host self-proteins, an event termed "molecular mimicry", and, by eliciting an immune response against themselves, can also generate a cross-reactive immune response against self. We have begun to explore the role of specific viral gene products transplanted in unique differentiated cells as participants in the complex etiology of autoimmune disease. The technology of cloning viral genes and expressing them in vivo under cell-specific promoters allows this experimental approach. As shown in this review, expression of a viral transgene in β cells of the islets of Langerhans or in oligodendrocytes, per se, need not cause disease by itself, although there is evidence in other systems that such expression of viral or host genes can produce disease (Lo et al. 1988; ALLISON et al. 1988; ROMAN et al. 1990), presumably in part due to over-expression of the tg and a storage-like disorder. In our models, the "silent" expressions of a viral gene is not associated with disease over the lifetime of the animal. However, when the host becomes infected with a virus encoding the same gene as the transgene or one closely related to it, a resultant immune response directed against the virus may also recognize the transgene leading to progressive T cell-mediated disorder and destruction of the tissue expressing the viral (self) gene with subsequent autoimmune disease. The evidence for this is firmly established in the IDDM model (OLDSTONE et al. 1991; VON HERRATH et al. 1994a) and still under exploration for the MS model (HORWITZ et al. 1994). This multifactorial process is further dependent on whether the viral transgene is expressed in both the thymus and the disease-related cell or target tissue. Thymic expression influences negative selection of responder lymphocytes and thus delays the onset of an autoimmune disorder (VON HERRATH et al. 1994a). An interesting notion is the possibility that a search for potential etiologic agents in autoimmune disorders involving a lag or latent period should include analysis of the thymus as well as the tissue targeted for autoimmune attack. Further, the MHC haplotype or other background genes of an individual undergoing autoimmune dysfunction play a role in the affinity of binding of the transgene products to the MHC molecule and thus influence the degree of negative selection that occurs or influences the vigor of the resulting immune response. Thus, current technical ability to express host or viral genes in unique cell populations and to make double or triple tg mice, in which various cytokine genes or lymphocyte activation genes can be expressed along with the viral gene, offer a unique possibility for molecular dissection of autoimmunity. The end results of these kinds of studies should provide leads to the prevention and treatment of human autoimmune disease.

Acknowledgments. This is publication number 8791-NP from the Department of Neuropharmacology. The Scripps Research Institute, La Jolla, CA 92037. This work was supported in part by USPHS grants NS12428 and AI09484, NIH Training Grant AG00080 (CFE), MH19185 (MSH), National Multiple Sclerosis Society Fellowships FA 1041-A-1 (CFE) and FA 1111-A-1 (MSH) and a Juvenile Diabetes Foundation Fellowship #393357 (MvH).

References

Allison J, Campbell I, Morahau G, Mandil E, Harrison C, Miller JFAP (1988) IDDM in transgenic mice resulting from over-expression of class I molecules in pancreatic β-cells. Nature 333: 529–533

Chisari F, Filippi P, Burns J, McLachlan A, Popper H, Pinkert C, Palmiter P, Brinster R (1987) Structural and pathological effects of synthesis of hepatitis B virus large envelope polypeptide in transgenic mice. Proc Natl Acad Sci USA 84: 6909–6913

Chisari F, Klopchin K, Moriyama T, Pasquinelli C, Dunsford H, Sell S, Pinkert C (1989) Molecular pathogenesis of hepatocellular carcinoma in hepatitis B virus transgenic mice. Cell 58: 1145–1156

Ebers GC, Bulman DE, Sadovnick AD, Paty DW, Warren S, Hader W, Murray TJ, Seleand TP, Duquette P, Grey T et al. (1986) A population-based study of multiple sclerosis in twins. N Engl J Med 315: 1638–1642

Gow A, Friedrich VL Jr, Lazzarini RA (1992) Myelin basic protein gene contains separate enhancers for oligodendrocyte and Schwann cell expression. J Cell Biol 119: 605–616

Green A (1990) The role of genetic factors in development of insulin dependent diabetes mellitus. Curr Top Microbiol Immunol 164: 3–17

Horie H, Koike S, Kurata T, Sato-Yoshida Y et al. (1994) Transgenic mice carrying the human poliovirus receptor: new animal model for study of poliovirus neurovirulence. J Virol 68: 681–688

Horwitz MS, Evans CF, Lazzarini RA, Oldstone MBA (1994) A transgenic model for immune mediated demyelination. FASEB J 8: 1142

Koike A, Taya C, Kurata T, Abe S et al. (1991) Transgenic mice susceptible to poliovirus. Proc Natl Acad Sci USA 88: 951–955

Kotzin BL (1993) Twins and T-cell responses. 364: 187–188

La Porte RE, Tajima N, Akerblom HK, Berlin MD, Brosseau J, Chrisley M, Drash AI, Fishbein H, Green A, Hamman R, Harris M, King H, Zaron Z, Neil A (1985) Geographic difference in the risk of IDDM: the importance of registries. Diabetes Care 8 Suppl: 101–107

Lo D, Burkly L, Widera G, Cowing L, Flavell RA, Palmiter R, Brinster R (1988) Diabetes and tolerance in transgenic mice expression class II MHC molecules in pancreatic β-cells. Cell 53: 159–168

Marks J, Lin J, Hinds P, Miller D, Levine A (1989) Cellular gene expression in papilloma of the choroid plexus from transgenic mice that express the simian 40 large T antigen. J Virol 63: 790–797

Michelsen B, Dyberg T, Visssing H, Serup P, Lernmark A (1990) HLA-DQ and -DX genes in IDDM. Curr Top Microbiol Immunol 164: 57–70

Nerenberg M, Henricks J, Reinolds R, Khoury G, Jay G (1987) The tat-gene of HTLV-1 induces mesenchymal tumors in transgenic mice. Science 237: 1324–1329

Ohashi P, Oeheu S, Buerski K, Pircher H, Ohsashi C, Odermatt B, Malisseu B, Zinkernagel RM, Hengartner H (1991) Ablation of tolerance and induction of diabetes by virus in viral antigen transgenic mice. Cell 65: 305–317

Oksenberg JR, Begovich AB, Erlich HA, Steinman L (1993) Genetic factors in multiple sclerosis. JAMA 270 (19): 2362–2369

Oldstone MBA (1987) Molecular mimicry and autoimmune disease. Cell 50: 819–820

Oldstone MBA (ed) (1989a) Molecular mimicry. Cross-reactivity between microbes and host proteins as a cause of autoimmunity. Curr Top Microbiol Immunol 145

Oldstone MBA (1989b) Viral persistence. Cell 56: 517–520

Oldstone MBA (1991) Molecular anatomy of viral persistence. J Viral 65: 6381–6386

Oldstone MBA, Whitton JL, Lewicki H, Tishon A (1988) Fine dissection of a nine aminoacid glycoprotein epitope, a major determinant recognized by LCMV-specific class-I restricted H-2^b CTL. J Exp M 168: 559–570

Oldstone MBA, Nerenberg M, Southern P, Price J, Lewicki H (1991) Virus infection triggers IDDM in a transgenic model: role of the anti-self (virus) immune response. Cell 65: 319–331

Ren R, Racaniello V (1992) Human poliovirus receptor gene expression and poliovirus tissue tropism in transgenic mice. J Virol 66: 296–304

Ren R, Constantini F, Gorgacz E, Lee J, Racaniello V (1990) Transgenic mice expressing a human poliovirus receptor: a new model for poliomyelitis. Cell 63: 353–362

Roman L, Simons LF, Hammer RE, Sambrook JF, Getlinig MJH (1990) The expression of influenza virus hemagglutinin in the pancreatic β-cells of transgenic mice results in autoimmune diabetes. Cell 61: 383–396

Sadovnick AD, Ebers GC (1993) Epidemiology of multiple sclerosis: a critical overview. Can J Neurol Sci 20: 17–29

Sinha A, Lopez T, McDevitt H (1990) Autoimmune Disease: the failure of self tolerance. Science 248: 1380–1387

Toggas S, Masliah E, Rockenstein EH, Rall GF, Abraham CR, Mucke L (1994) Central nervous system damage produced by expression of the HIV-1 coat protein gp120 in transgenic mice. Nature 367: 188–193

Von Herrath MG, Dockter J, Oldstone MBA (1994a) How virus induces a rapid or slow onset IDDM in a transgenic model. Immunity 1: 231–242

Von Herrath MG, Dockter J, Nereberg M, Gairin JE, Oldstone MBA (1994b) Thymic selection and adaptability of CTL responses in transgenic mice expressing a viral protein in the thymus. J Exp Med 180: 1901–1910

Whitton JL, Tishon A, Lewicki H, Gebhard J, Cook T, Salvato M, Joly E, Oldstone MBA (1989) Molecular analysis of a five amino-acid cytotoxic T-lymphocyte epitope: an immuno-dominant region, which induces nonreciprocal CTL cross-reactivity. J Virol 63(10): 4303–4310

Whitton JI (1990) Lymphocytic choriomeningitis virus CTL. Semin Virol 1: 257–261

Yoon JW (1990) The role of viruses and environmental factors in induction of diabetes. Curr Top Microbiol Immunol 164: 95–124

Role of HLA-B27 in Spondyloarthropathies

S.D. Khare[1], H.S. Luthra[2], and C.S. David[1]

1 Introduction

HLA-B27 associated spondyloarthropathies are a group of inflammatory diseases primarily affecting the spine and peripheral joints. Ankylosing spondylitis (AS) is a prototypic disorder of spondyloarthropathies. Approximately 0.1% of the Caucasian population suffers from AS in North America. Some 90% of patients with this disorder carry the HLA-B27 gene, which is present in 8% of the Caucasian population (Gran and Husby 1994). There are nine subtypes of HLA-B27 (B*2701 to B*2709), differing from each other by a few amino acid residues (Khan 1994). Sequence analysis of α1, α2 and α3 domains of HLA-B27 from a patient and a normal individual did not show any differences (Coppin and McDevitt 1986). This

[1] Department of Immunology, Mayo Clinic and Medical School, 200 First St. SW, Rochester, MN 55905, USA
[2] Department of Rheumatology, Mayo Clinic and Medical School, Rochester, MN 55905, USA

suggests that other genetic and/or environmental factors besides HLA-B27 are required for disease susceptibility/pathogenesis. Enterobacteria including *Klebsiella pneumoniae, Shigella flexneri, Salmonella typhimurium, Yersinia enterocolita* and *Chlamydia trachomatis* have been isolated from patients with spondyloarthropathies and are potential triggers because of the temporal relationship between infection and disease onset or reactivation. Some bacterial proteins (from *Klebsiella, Yersinia*, etc.) have amino acid sequences that mimic HLA-B27, which may initiate breakdown of tolerance to self-antigens or allow autoimmune responses to develop in these patients. In order to understand the mechanism(s) involved in pathogenesis and the role of HLA-B27 in this disease, several laboratories generated HLA-B27 transgenic mice and rats. HLA-B27 transgenic mice were either asymptomatic or developed mild disease (IVANYI et al. 1991). Our studies showed increased pathogenicity of *Yersinia enterocolitica* 0:8 WA in HLA-B27 transgenic mice (NICKERSON et al. 1990a). HAMMER et al. (1990), using HLA-B27 transgenic rats, observed spontaneous inflammatory disease. Recent studies have given us new insights into the structure and function of HLA-B27 molecule and its subtypes. These observations should set the stage for understanding the role of this gene and that of bacteria in the pathogenesis of these diseases.

2 Structure of HLA-B27

Of all MHC class I molecules, HLA-B27 has received the most attention because of its strong association with spondyloarthropathies. The crystal structures of HLA-A2 and HLA-B27 have provided new insights into our understanding of the structure and function of class I molecules (BJORKMAN et al. 1987; MADDEN et al. 1992). The basic structure of HLA-B27 is very similar to those of HLA-A2 and HLA-Aw68 and shares ~90% similarity with most of the HLA class I molecules (Fig. 1). MHC class I molecules are expressed on all lymphoid cells and present antigenic peptides to CD8-positive T cells. Class I molecules contain three extracellular domains ($\alpha1$, $\alpha2$, $\alpha3$) and are expressed on the cell surface associated with $\beta2$-microglobulin ($\beta2m$) and antigenic peptides in their peptide-binding cleft. Conserved or nearly conserved side chains are found at both ends of the peptide-binding cleft formed by 3-dimensional fold of the $\alpha1\alpha2$ domain of MHC class I (BJORKMAN et al. 1987; SAPER et al. 1991). Even though $\alpha3$ domain and $\beta2m$ do not have direct contact with the antigenic peptide in the peptide-binding groove, protein–protein ($\alpha3$ and $\beta2m$ with $\alpha1\alpha2$) interaction may influence the conformation and thus function of class I molecules. MHC class I genes show distinct polymorphism, with amino acid differences primarily in the first domain (PARHAM et al. 1988). The structural differences between HLA-B27 and other HLA-A molecules that have been studied involve minor packing rearrangements of the $\alpha3$ domain and $\beta2m$ relative to $\alpha1\alpha2$, and shifts in the conformations of loops. The shifts in domain packing in the HLA-A2 molecule appear to reflect the relatively

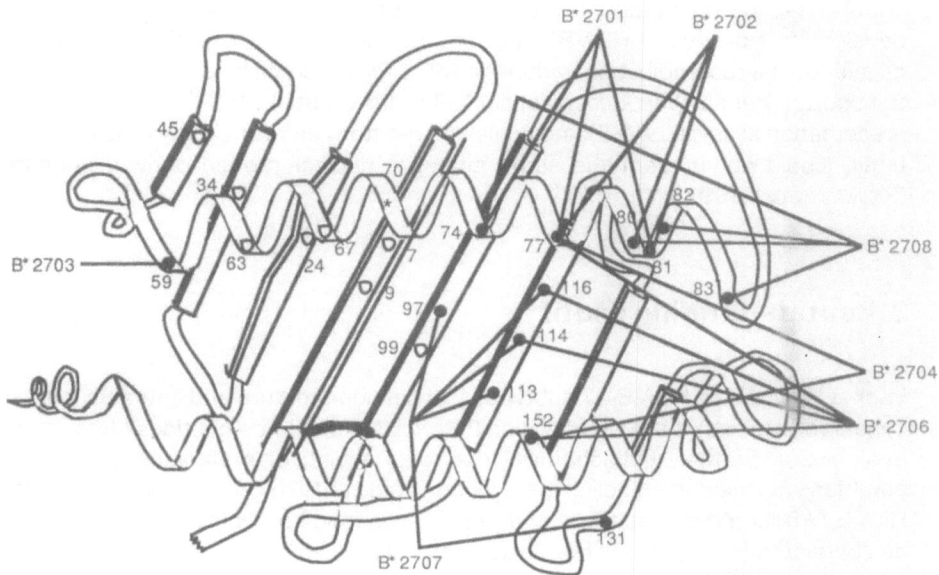

Fig. 1. Positions of amino acid residues of HLA-B27 subtypes on a schematic ribbon diagram of the antigen-binding cleft of B*2705. (Modified from MACLEAN 1992)

flexible association of α3 with α1α2, which involves only a small number of interdomain contacts (SAPER et al. 1991). In the HLA-B27 molecule, an additional α3-α1α2 contact involving a hydrogen bond from the HLA-B locus residue Arg-239 (on α3) to the main chain of carbonyl oxygen of Arg-48 (on α1) is seen, though it does not require a major adjustment of the main chain at either residue. The α3 and β2m packing under the β sheet of α1α2 domains does not completely cover the sheet, causing the molecule to form a concave surface beneath the peptide-binding domain (MADDEN et al. 1992). This surface may also have some role in the formation of multimeric forms of class I molecules (KRISHNA et al. 1992), which possibly are empty and fail to bind and present exogenously supplied peptides. Dimer formation of class I molecules has also been reported in in vivo (CAPPS et al. 1993). HLA-B27 molecules are more stable than other HLA-A or H-2 molecules and appear to bind exogenously supplied peptides (BENJAMIN et al. 1991). Dimer formation of these class I molecules in vivo has been shown in the absence of β2m.

The HLA-B locus is closely linked to several other genes which may influence assembly, transport and expression of class I molecules. TAP and Lmp genes are polymorphic and map to the MHC class II region (ZHOU et al. 1993). Proteosomes, Lmp 2 and Lmp 7 are probably involved in the assembly of class I molecules which the transporter genes TAP-1 and TAP-2 function in transporting the assembled class I molecules to the cell surface. A preliminary study of polymorphism in HLA-linked transporter genes in individuals with AS has revealed no disease suscepti-bility alleles in 22 individuals tested (COLONNA et al. 1992). MAKSYMOWYCK et al.

(1994) have demonstrated the association of Lmp-2 polymorphism with the development of iritis in HLA-B27-positive individuals with AS.. Recent studies have shown a recombination within the MHC region containing proteosome and transporter genes (VAN ENDERT et al. 1992; POWIS et al. 1993). In an antigen presentation assay, HLA-B27 molecule derived from different individuals within a family could not present the same influenza nuclear protein derived peptide (PAZMANY et al. 1992).

3 Peptide-Binding Motif

Pocket B (Fig. 1) of HLA-B-27 subtypes has a unique feature and interacts with a conserved arginine residue at position 2 from B*2705 eluted peptides (JARDETZSKY et al. 1991). Further, radiochemical pool sequencing of metabolically labelled peptides was used to establish the anchoring side chain motif of the endogenous HLA-B27-bound peptides (ROJO et al. 1993). Sequence of arginine-labelled pepti- des extracted from a panel of B27 subtypes indicated the requirement of arginine at position 2 (P2). Peptide from mutants which showed alteration in the structure of pocket B and also at the pocket B rim could not change this effect of the residue 2 (R2) motif. Mutation at position 45 is important as it affects the binding motif at P2. Even though many peptides bind to multiple subtypes of HLA-B27, some peptides bind with greater altered efficiency. This suggests that polymorphism outside pocket B may also be important in peptide binding. Further analysis of B*2705 motif using T2 (transporter and proteosome deficient) cells and ME-1 antibody showed Arg at P9 as a secondary anchor and not P1 (WEN et al. 1994). However Arg at P1 affected the binding reactivity by ME-1 antibody. Also a minimum ME-1 reactivity was seen in the presence of either of Asn, Asp, Glu or Pro at P1. A resistant HLA-B27 subtype, B*2703, binds to a subset of disease susceptible B*2705 bound peptides especially when Arg or Lys are at P1 (COLBART et al. 1994). Recently, peptide-binding specificity of five HLA-B27 subtypes (B*2701, B*2703, B*2704, B*2705 and B*2706) has been described by using synthetic peptides carrying the primary anchor residue motifs by combination of amino acid residues at P2 and P9 (TANIGAKI et al. 1994). The motif R-K and R-R bind to B*2705. In addition to R-K and R-R, a new motif H-R is accepted by B*2703, B*2704 and B*2706 but not by B*2701. Other B*2702 and B*2705 motifs, R-H, R-L, R-A, R-F and R-G were accepted by all B27 subtypes.

4 Peptides Bound to HLA-B27

The 3-dimensional structure of B*2705 (a subtype of HLA-B27 which is associ- ated with the majority of AS patients) suggests that peptides, mostly nanomers, are bound in extended conformation and anchored at both ends through both peptide terminals and R2 (MADDEN et al. 1991, 1992). Conservation of arginine as an "anchor" side chain at the second peptide position, which is bound in a potentially HLA-B27-specific pocket, may have some role in the association of HLA-B27 with AS and other diseases. Recently using an unusual monoclonal antibody (Mab) directed against HLA-B27 (MARB4), peptides much longer than nine amino acids were eluted. The HLA-B27 eluted peptides were ~8–33 amino acids long. The major population of bound peptides had substantial amounts of arginine at P1–9 (40%–50%) and as expected arginine at P2 (URBAN et al. 1994). These long peptides were speculated to represent intermediates in the formation of nonamers or adventitiously bound peptides.

Immunoprecipitation of MHC molecule and acid elution of MHC-bound peptide is the key to addressing the issues related to anchor residues, peptide motif and sequence similarity with known amino acid sequences. Currently, several anti-HLA-B27 monoclonal antibodies are available. They recognize free heavy chain, conformational molecule and, more recently, peptide+HLA-B27 (WEISS et al. 1990a; WANG et al. 1994). As discussed earlier peptides acid-eluted from HLA-B27 showed arginine as an anchor residue at P2. These peptides from ME-1 immunoprecipitated HLA-B27 molecule were ~9–13 amino acids long. Sequence analysis of these peptides exhibited homology with human histone H3, heat shock proteins (human Hsp89a and Hsp89b), ribosomal proteins from yeast and rats, HIV gag p24 and gp120. Surprisingly, two peptides showed sequence similarity with the HLA-B27 molecule itself (JARDETZKY et al. 1991). It is possible that the two HLA-B27 peptides may have been generated at the time of papain cleavage or acid treatment and are not specifically bound in the antigen-binding groove. Alternatively, they could have been generated by intracellular alternate splicing, as described in both nonclassical and classical MHC class I molecules (ULKER et al. 1990; ABU-HADID et al. 1994). Recently much longer peptides from MARB4 immunoprecipitated HLA-B27 molecules have been described (URBAN et al. 1994). A maximum of 20% of these peptides could bind with arginine at P2 in the B pocket and the NH2-terminal deep in the A pocket, where a pentagonal hydrogen-binding network (formed by Y7, Y59, Y171 and a water molecule) accomodates the charged group. The remaining 80% of long peptides are present with no arginine at P2 in the B pocket but rather at P3, P4 or P5 in the B pocket and the NH_2-terminal extending beyond the A pocket. Y59 may be critically involved in the binding of long peptides of MARB4-reactive HLA-B27 molecule. It is important to note that MARB4 Mab does not react to the HLA-B27 subtype B*2703. Interestingly, this is the only subtype which is not associated with AS. Similarly other Mabs directed against HLA-B27 molecules might precipitate B27 with other peptides in the peptide-binding groove. Recent studies have described that conformation of H-chain of class I molecule is influenced intracellularly by β2m and some mutations at the contact sites in β2m may influence the peptide binding by HLA-B27 molecule (DANLICZYK and DELOVITCH 1994; FUKAZAWA et al. 1994). However, role of α3 domain is still undefined.

5 HLA-B27 Subtypes in Ankylosing Spondylitis

Of the nine subtypes of HLA-B27, eight (B*2701–08) are defined by DNA sequence. Seven of these subtypes differ by charged amino acid on isoelectric focusing. About 80%–90% of the B27-positive Caucasian population are B*2705. Possibly other subtypes have arisen by point mutation at sites shown in Fig. 1. The second most common subtype is B*2702, which differs from B*2705 at positions 77, 80 and 81 (Table 1). B*2704 and B*2706 are limited to Asians. B*2703 subtype is very common among African blacks, as 61% of the HLA-B27-positive individuals carry this subtype. B*2701, B*2707 and B*2708 are limited to a few families. All of these subtypes differ from each other on the basis of between one and eight amino acid substitutions.

The question is whether all or some subtypes of B27 are involved in disease susceptibility. It is difficult to analyze the association of HLA-B27 subtypes with AS since the HLA-B27 molecule is present in low frequency among the population. A large scale study from Europe showed association of B*2705 and B*2702 subtypes with AS among Caucasians, native Americans and Siberians. B*2704 and B*2706 were associated with As in Asians. Amino acids at position 59 and 114 are different in B*2703 and B*2706, respectively. Since these two subtypes are not associated with AS, residues at position 59 and 114 may be involved in disease resistance. These data also suggest association of a common B27 epitope in AS and do not support the arthritogenic peptide theory. In a recent review, five different models for HLA-B27 association with AS were described (MACLEAN 1992).

A few recent studies have described presentation of peptides by subtypes of HLA-B27 to restricted cytotoxic T lymphocytes (CTLs). CTLs restricted to B*2702, B*2705 and B*2704 subtypes could present the same viral nuclear peptide derived from Epstein-Barr virus. However clonal analysis showed that B*2704 restricted response is not shared by other B27 subtypes. This study suggests functional overlap in the selection and presentation of CTL epitopes

Table 1. Amino acid sequence variation in HLA-B27 subtypes[a]

	Amino acid residue										
	59	74	77	80	81	97	113	114	116	131	152
B*2705	Tyr	Asp	Asp	Thr	Leu	Asn	Tyr	His	Asp	Ser	Val
B*2701	–	Tyr	Asn	–	Ala	–	–	–	–	–	–
B*2702	–	–	Asn	Ile	Ala	–	–	–	–	–	–
B*2703	His	–	–	–	–	–	–	–	–	–	–
B*2704	–	–	Ser	–	–	–	–	–	–	–	Glu
B*2706	–	–	Ser	–	–	–	–	Asp	Tyr	–	Glu
B*2707	–	–	–	–	–	Ser	His	Asn	Tyr	Arg	–
B*2708[b]	–	–	Ser	Asn	–	–	–	–	–	–	–

Dash indicates amino acid identity with B*2705 at that position.
[a] Sequence of B*2709 is not included.
[b] Subtype B* 2708 also differs at position 82 and 83 with other B27 subtypes with Arg and Gly, respectively.

(BROOKS et al. 1993). The difference in peptide presentation between B*2705 and B*2703 showed the importance of the P1 side chain in maintaining high affinity peptide binding to B*2703. B*2705-binding peptides with Arg or Lys at P1 bind with greater affinity to B*2703 (COLBERT et al. 1994). Thus, disease resistant B*2703 may not be able to bind and present arthritogenic peptide.

6 Bacteria as Triggering Agents

As described earlier, only a small percentage of HLA-B27-positive individuals suffers from AS; obviously, other factors are involved. Isolation of *Yersinia enterocolitica* (MERILAHTI-PALO et al. 1991), *Chlamydia trachomatis* (HUGHES et al. 1991), and bacterial lipopolysaccharide (LPS) from *Salmonella* and *Shigella* (GRANFORS et al. 1990, 1991) from the involved joints further point to a trigger in the causation of disease process. Indirect involvement of pathogens has been demonstrated by the presence of antibodies to microbes (MAKI-IKOLA et al. 1991) including *Salmonella* (MAKI-IKOLA and GRANFORS 1992) and *Klebsiella* (SHODJAI-MORADI et al. 1992). *Yersinia*-specific T cell clones (VINER et al. 1991) and immune complexes (LAHESMAA-RANTALA et al. 1987) have been demonstrated in the synovial fluid in patients with AS. Recent studies have demonstrated the presence of bacterial antigen-reactive T cells from synovial fluid of AS patients. Recently, HERMANN and YU (1993) have described *Yersinia*- and *Salmonella*-specific HLA-B27-restricted CD8+ T cells derived from synovial fluids of patients with reactive arthritis and AS. In this study CTLs from five of the six patients showed autoreactivity by killing uninfected B27 cell lines.

7 Collagen as an Autoantigen in Spondyloarthropathies

In a recent study, 19 nonamer peptides bearing the B27-binding motif from type I, II and XI collagens were identified and tested for binding affinity to HLA-B27 (GAO et al. 1994). Four of these peptides bound to HLA-B27 with high affinity. One of these peptides also showed CTL reactivity in a HLA-B27-restricted manner with cells from a reactive arthritis patients. This study demonstrate that collagen molecule may also be a target autoantigen in some AS patients.

8 Possible Molecular Mimicry

The exact mechanism by which microbes/microbial products may interact with the HLA-B27 molecule in the development of AS is unknown, but one of the hypotheses invokes molecular mimicry. This explanation is based on the reactivity of anti-B27 Mab with certain arthritis-causing bacteria. Several mechanisms for the possibility of antigenic mimicry of the host immune system have been proposed. Autoantibodies to HLA-B27 seen in AS patients support such possibilities (SCHWIMMBECK et al. 1987). In this report, a six amino acid sequence, QTDRED, is shared by the hypervariable domain of HLA-B27 and a *Klebsiella* protein. Other studies have also shown such mimicry by using HLA-B27 reactive Mabs with bacterial antigens (Table 2). Even though the region between amino acids 70 and 78 on the α1 domain of HLA-B27 shares similarity with *Yersinia*, *Shigella*, *Salmonella* and *Klebsiella* (LAHESMAA et al. 1991), no definite cross-reactivity could be observed by T cell proliferation and EIA. Table 2 lists some of these conflicting reports regarding antigenic mimicry with HLA-B27.

9 Expression and Function of HLA-B27 in Transgenic Animals

In late 1980's, several centers generated B27 transgenic animals (SAVARIRAYAN et al. 1988; TAUROG et al. 1988; WEISS et al. 1990a). With the exception of KRIMPENFORT et al. (1987), who used a B*2702 construct, a B*2705 construct was used for the generation of B27 transgenic mice. Transgenic rats with a varying number of copies of B27 and human β2m (hβ2m) were produced by HAMMER et al. (1990). In our laboratory (SAVARIRAYAN et al. 1988), transgenic mice were generated by microinjecting a 6.5 kb *Eco*RI fragment containing the entire gene into (B6XSJL)F1 mouse embryo. Embryos were implanted in pseudopregnant mice and offspring were screened by Southern blotting for the presence of the B27 gene. By back-crossing, several lines of transgenic mice were obtained on different H-2 backgrounds. The HLA-B27 gene product was expressed on mouse lymphoid cells in the absence of hβ2m (TAUROG et al. 1988; SAVARIRAYAN et al. 1988). High expression of HLA-B27 was observed in mice homozygous for H-2b, H-2f, H-2s, H-2p, H-2r and H-2k haplotypes (NICKERSON et al. 1990b). Mice with a H-2v haplotype expressed HLA-B27 at an intermediate level; however, minimal expression was observed in H-2q and H-2d haplotypes. Using recombinant mouse strains, a low level of HLA-B27 expression was mapped to H-2D or a closely linked gene. Except for H-2D or a closely linked gene, high expression of HLA-B27 was also dependent on the presence of the hβ2m transgene (WEISS et al. 1990b). In the absence of hβ2m, cell surface expression of HLA-B27 in transgenic mice was recognized as alloantigenic by CTLs (TAUROG et al. 1988). T cells from transgenic mice were also functional, as they responded to the same HLA-B27 restricted influenza peptide which was recognized by human HLA-B27 restricted cytotoxic T cell lines (WEISS et al. 1990b). REBOUL et al. (1991) characterized T cells from transgenic mice which were also involved in the xeno-recognition of HLA. A pronounced xenogenic HLA response by mouse T cells

Table 2. Molecular mimicry between HLA-B27 and microbes in ankylosing spondylitis

Microbes	Mimicry defined by	Comments	Reference
Yersinia enterocolitica Chlamydia tracheomatis	Serology	No mimicry with HLA-B27	ROUDIER et al. (1985)
Klebsiella pneumoniae	Serology??		OGASAWARA et al. (1986)
Klebsiella	Serology	Lymphocyte reactivity using anti-klebsiella sera	SINGH et al. (1986)
Klebsiella pneumoniae	Serology	Failure of anti-K.pneumoniae antibody with PBMC of AS patients	CAMERON et al. (1987)
Yersinia Shigella Klebsiella	Serology	Mab B27.M1 and B27.M2 cross-reactivity	VAN BOHEMEN et al. (1984)
??Microbes	Serology	Mab Ye-2 cross-reactivity with bacteria	KONO et al. (1985)
Shigella envelope protein	Serology	Anti-B Mab(B27M1 and B27M reactivity with 36 and 23 Da molecule of Shigella	RAYBOURNE et al. (1988)
Yersinia pseudutuberculosi	Serology	A 19-kDa molecule of Yersinia cross-reacts with HLA-B27	CHEN et al. (1987)
Klebsiella pneumoniae nitrogenase	Serology and T cell	Using peptides shared by HLA-B27.1 and K. pneumoniae	TSUCHIYA et al. (1989)
Klebsiella pneumoniae	Serology??	In articular tissue of AS patient	HUSBY et al. (1989)
Yersinia	Serology	Tetrapeptide shared by Yop1 (outer memb protein of Yersinia) and HLA-B27	LAHESMAA et al. (1990)
Yersinia, Salmonella	Serology	Molecular mimicry (of micrcobes) in 33% of patients located in the same place of B27 molecule (amino acids 70–78 in the variable region of $\alpha1$ helix	LAHESMAA et al. (1991)
Yad1 of Yersinia	T cells and antibody	No lymphoproliferation using tetrapeptide of Yad1 which mimics B27	LAHESMAA et al. (1992)
Klebsiella pneumoniae	Serology	Autoantibody to HLA-B27 in the sera of AS patients mimics Klebsiella	SCHWIMMBECK et al. (1987)
Mycobacterium tuberculosis hsp60	Serology	Serum reactivity from AS patients with HLA-B27 cells transfected with M.tuberculosis hsp60 gene	KELLNER et al. (1994)

PBMC, peripheral blood mononuclear cells; Mab, monoclonal antibody; As, ankylosing spondylitis.

was observed in a transgenic mouse with a hybrid (B27$\alpha1\alpha2$/Kd$\alpha3$) HLA-B27 (KALINKE et al. 1990). Some disparity on the interaction of class I+ peptide with mouse CD8+ T cells could be overcome by introducing this hybrid transgene into mice.

10 Expression of HLA-B27 in B27/Human β2-Microglobulin Double Transgenic Animals

The double transgenic B27/hβ2m animals were produced by mating single transgenic mice (Ivanyi et al. 1991; Pedrinasi et al. 1994) or by microinjecting ova with both B27 and hβ2m genes (Hammer et al. 1990). Though B27 paired with endogenous β2m in rats and mice, expression was increased several fold when the hβ2m gene was introduced into the mice (unpublished observations). Such increased expression of HLA-B27 is observed in mice with low copy numbers (<10 copies) of hβ2m transgene. In rats, expression of HLA-B27 was not dependent on a high copy number of either B27 or hβ2m transgenes (Hammer et al. 1990).

11 Animal Models of Ankylosing Spondylitis

As yet, no animal model of disease with similarities to human AS is available. Rats with HLA-B27 and hβ2m transgene share some similarities with human AS (Hammer et al. 1990). Susceptibility of arthritis in these rats is dependent on the copy number of transgene products of both HLA-B27 and hβ2m. Rats with <50 copies of the transgene were resistant to arthritis. In the study of Hammer et al. (1990), two lines of rats which develop spontaneous arthritis had >50 copies of both the transgenes. These rats developed disease at the age of 10 weeks which was characterized by arthritis involving peripheral and vertebral joints, diarrhea, nail changes, skin lesions and inflammation of the male genital tract and heart. Spleen cells from nontransgenic rats receiving skin graft from transgene positive donors showed cytolysis when tested against L cells transfected with B27, suggesting that the transgene product is recognized in a conventional manner by allogenically primed T cells. Further, a third healthy line of transgenic rats with ~20 copy number of HLA-B27 and hβ2m developed arthritis when homozygous for the transgene locus (Taurog et al. 1993). The increase in the number of B27 molecules was not simply a consequence of the inflammatory process. The distal colon was the early site of inflammation and showed a high level of expression of B27 in cells of the lamina propria, but not in all colonic epithelial cells. Interestingly, all disease-prone lines showed early high mRNA expression of B27 in the thymus and spleen, including in utero.

The cellular basis of multisystem inflammatory disease has been investigated by transferring cells from disease-prone transgenic rats to irradiated nontransgenic animals (Breban et al. 1993). Successful transfer of disease required engraftment of bone marrow cells, suggesting that precursor bone marrow cells are required for disease induction and not the mature cells. A more recent study (Breban et al. 1993) from the same group showed that the disease in arthritis susceptible transgenic rats is T cell-dependent. Transfer of bone marrow cells from HLA-B27 and hβ2m double transgenic animals into euthymic animals transferred the disease; however, no arthritis was seen when cells were transferred into athymic rats. Thus the data indicate that the disease is T cell-dependent but thymus-independent and that HLA-B27 is definitely involved in the pathogenesis of spondyloarthritis-like disease in rats. Recently, the same group has also shown that disease-susceptible rats produced and kept in a germ-free environment developed milder disease with no gut inflammation. This suggests a role for environmental factors in the aggressive form of disease.

A milder form of joint disease has also been described in HLA-B27/hβ2m double transgenic mice (IVANYI et al. 1991). These mice develop ankylosing enthesopathy (ANKENT) of the ankle and tarsel joints. ANKENT in mice is greatly dependent on H-2 haplotype, sex, age and environment. H-2^k mice have relatively more risk of disease susceptibility than mice with other H-2 haplotypes. The investigators used the B*2702 gene, which is the second most common B27 subtype among AS patients. In a large group of animals, about 30% B27+hβ2m+, H-2k (B10.BR) mice developed ANKENT compared to ~20% single-positive (B27-negative, hβ2m-positive) mice. The disease was present in most of the males. A similar observation with a lower percentage of arthritic mice was observed with the H-2b haplotype. Thus, it is possible that some H-2 or H-2-linked genes also influence the development of arthritis in these mice. The influence of environmental factors was very low, since B27+hβ2m+ mice developed arthritis in a specific pathogen-free colony. Further, a specific pathogen causing the disease could not be identified, as administration of different strains of *Yersinia* with various doses and routes did not influence the development of arthritis. B27-negative mice also developed arthritis, although disease was very mild and with low frequency. This is not a good mouse model for B27-linked spondyloarthropathy.

12 Bacteria as an Arthritis-Inducing Agent in Mice

A very small proportion (up to 5%) of B27-positive individuals develop AS. The development of arthritis in these patients often occurs after infection with enteric bacteria. This suggests a role for enteric bacteria in the development or perpetuation of arthritis. We have previously reported a significantly higher rate of hind limb paralysis in HLA-B27 transgene-positive mice than in nontransgene littermates following exposure to *Y.enterocolitica* 08WA. Paraspinal abscesses were observed in the paralyzed animals (NICKERSON et al. 1990a). In this study, transgene-positive animals intravenously immunized with 10^4 *Y. enterocolitica* 08WA also showed increased mortality compared to nontransgenic littermates. We have also demonstrated a role for H-2d and H-2q genes in the susceptibility to *Yersinia*-induced arthritis (NICKERSON et al. 1990c) in the absence of HLA-B27. Only H-2q and H-2d mice with clonal deletion of certain Vβ populations were susceptible to *Yersinia*-induced arthritis (YIA). Since I-E, D and Mls are known to delete certain Vβ populations, these genes might also be important in the susceptibility to YIA in mice. Studies on YIA have also been done in Balb/C and C3H/HeJ mice (DE LOS TOYOS et al. 1992). Oral challenge with *Y. enterocolitica* 0:3 with a pretreatment of desferrioxamine led to development of arthritis in joints. *Yersinia* were isolated from the involved joints at the time of acute inflammation in mice. A definite mixed infiltration of synovia, joint spaces and soft tissue was observed in arthritic mice. In addition to isolation of live bacteria, complete loss of articular cartilage leading to an anomalous ossification the filling joint space was observed even 2–6 months after the arthritic ankles/paws were examined. A long-term anti-lipopolysac-charide response was also seen in both arthritic and nonarthritic mice after oral infection with *Yersinia* (DE LOS TOYOS et al. 1993). Similarly, *Y. enterocolitica* infection and YIA has also been reported in rats (MERTZ et al. 1991; KOOL et al. 1991; MERILAHTI-PALO et al. 1992).

The possibility of cross-reactivity between bacteria and HLA-B27, was also examined in B27/hβ2m double transgenic mice (SINGH et al. 1994). The cross-reactive QTDRED sequence in B*2705 and *Klebsiella* showed T cell tolerance in B27/hβ2m double transgenic mice. The authors suggested the pathogenesis of AS may be related to B27-bound self-peptide and not the cross-reactivity.

13 Is There a Role for HLA-B27 in Spondyloarthropathy?

This question remains unanswered. One approach to addressing this question would be to have an ideal animal model of spondyloarthropathies. The most studied double transgenic rat model has >100 copies of both the transgenes and thus the changes seen in these rats could be the result of graft vs host disease. This model has many features similar to human spondyloarthropathies, with inflammation involving the gastrointestinal tract, peripheral and vertebral joints, male genital tract, skin, nail and heart. Since a high copy number is essential for this to occur, it is possible that a high copy number of any MHC gene could cause similar disease. Antigen-specific HLA-B27-restricted immune responses have been shown only in a few patients with AS. This raises the question of the role of B27 in disease. Is it an antigen presenting molecule? Is it involved in molecular mimicry with bacterial antigens? Or is HLA-B27 an autoantigen? To address these questions we need an animal model which is transgenic for HLA-B27 and human β2m, with a low copy number of both transgenes. Such an animal will allow study of other MHC region genes (for example, the role of MHC class II, Tap and Lmp genes) and non-MHC genes (cytokine genes). Using knockout mice, the role of endogenous β2m could be tested using the human β2m transgene in β2m-deficient animals. In such animals, the influence of infectious agents and environmental factors can be determined. Our laboratory is currently trying to generate such a mouse model for spondyloarthropathy.

Acknowledgments. These studies were supported by NIH grant AR-39875 and by funds from the Minnesota Arthritis Foundation.

References

Abu-Hadid MM, Fuji H, Sood AK (1994) Alternatively spliced MHC class I mRNAs show specific deletion of sequences encoding the extracellular polymorphic domain. Int Immunol 6: 323–337

Benjamin RJ, Madrigal JA, Parham P (1991) Peptide binding to empty HLA-B27 molecules of viable human cells. Nature 351: 74–77

Bjorkman PJ, Saper MA, Samraoui B, Bennet WS, Strominger JL, Wiley DC (1987) Structure of the human class I histocompatibility antigen HLA-A2. Nature 329: 506–512

Breban M, Hammer RE, Richardson JA, Taurog JD (1993) Transfer of the inflammatory disease of HLA-B27 transgenic rats by bone marrow engraftment. J Exp Med 178: 1607–1616

Brooks JM, Murrey RJ, Thomas WA, Kurilla MG, Rickinson AB (1993) Different HLA-B27 subtypes present the same immunodominant Epstein-Barr virus peptide. J Exp Med 178: 879–887

Cameron FH, Russell PJ, Easter JF, Wakefield D, March L (1987) Failure of anti-*Klebsiella pneumoniae* antibodies to cross react with peripheral blood mononuclear cells from patients with ankylosing spondylitis. Arthritis Rheum 30: 300–305

Capps GG, Robinson BE, Lewis KD, Zuniga MC (1993) In vivo dimeric association of class I MHC heavy chains. J Immunol 151: 159–169

Chen J-H, Kono DH, Yong Z, Park MS, Oldstone MB, Yu DTU (1987) A *Yersinia pseudotuberculosis* protein which cross reacts with HLA-B27. J Immunol 139: 3003–3011

Colbert RA, Rowland-Jones SL, McMichal AJ, Frelinger JA (1994) Differences in peptide presentation between B27 subtypes: the importance of the P1 side chain in maintaining high affinity peptide to B*2703. Immunity 1: 121–130

Collona M, Bresnahan M, Bahram S, Strominger JL, Spices T (1992) Allelic variations of the human putative transporter involved in antigen processing. Proc Natl Acad Sci USA 89: 3932–3936

Coppin HL, McDevitt HO (1986) Absence of polymorphism between HLA-B27 genomic exon sequences isolated from normal donors and ankylosing spondylitis patients. J Immunol 137: 2168–2172

Danliczyk UG, Delovitch TI (1994) β2-microglobulin induces a conformational change in an MHC class I H chain that occurs intracellularly and is maintained at the cell surface. J Immunol 153: 3353–3541

De los Toyos JR, Menendez P, Sampedro A, Hardisson C (1992) *Yersinia enterocolitica* 0:3-induced arthritis in mice: microbiological and histopathological information. APMIS 100: 455–464

De los Toyos JR, Diaz R, Hardisson C (1993) Protection against the lethal effect and arthritogenic capacity of *Yersinia enterocolitica* serovar 0:3 for mice. FEMS Microbiol Immunol 76: 289–298

Fukazawa T, Hermann E, Edidin M, Juan W, Huang F, Kellner H, Floege J, Farahmandian D, Williams KM, Yu DTU (1994) The effect of β2-microglobulins on the conformation of HLA-B27 detected by antibody and by CTL. J immunol 153: 3543–3550

Gao X-M, Woodsworth P, McMichal A (1994) Collagen-specific cytotoxic T lymphocyte responses in patients with ankylosing spondylitis and reactive arthritis. Eur J Immunol 24: 1665–1670

Gran JT, Husby G (1994) Ankylosing spondylitis: prevalence and demography. In: Klippel JH, Dieppe PA (eds) Rheumatology. Mosby, London, pp 1–6

Granfors K, Jalkanen S, Lindberg AA, Maki-Ikola D, van Essen R, Lahesmaa-Rantala R, Isomaki H, Saario R, Arnold WJ, Toivanen A (1990) Salmonella polysaccharide in synovial cells from patients with reactive arthritis. Lancet 335: 685–688

Granfors K, Jalkanen S, Toivanen A, Koski J, Lindberg AA (1992) Bacterial polysaccharide in synovial fluid cells in Shigella triggered reactive arthritis. J Rheumatol 19: 500

Hammer RE, Maika SD, Richardson JA, Tang J-P, Taurog JD (1990) Spontaneous inflammatory disease in transgenic mice expressing HLA-B27 and human β2m: an animal model of HLA-B27-associated human disorders. Cell 63: 1099–1112

Hermann E, Yu DTC (1993) HLA-B27-restricted CD8 T cells derived from synovial fluids of patients with reactive arthritis and ankylosing spondylitis. Lancet 342: 646–650

Hughes RA, Hyder E, Treharne JD, Keat AC (1991) Intra-articular chlamydial antigen and inflammatory arthritis. Q J Med 291: 575–588

Husby G, Tsuchiya N, Schwimmbeck PL, Keat A, Pahle JA, Oldstone MB, Williams RC Jr (1989) Crossreactive epitope with *Klebsiella pneumoniae* nitrogenase in articular tissues of HLA-B27+ patients with ankylosing spondylitis. Arthritis Rheum 32: 437–445

Ivanyi P, Eulderink F, Alphan V, Capkova J, Gaeda K, Heesemann J, Hoebe-Hewryk B, Kievits F, Lokhorst W, Pla M, Weinreick S, Zurcher C (1991) Joint diseases in B27 transgenic mice. In: Lipsky PE, Taurog JD (eds) HLA-B27+ spondyloarthropathies. Elsevier, Amsterdam, pp 71–83

Jardetzky TS, Lane WS, Robinson RA, Madden DR, Wiley DC (1991) Identification of self peptides bound to purified HLA-B27. Nature 353: 326–329

Kalinke U, Arnold B, Hammerling GJ (1990) Strong xenogeneic HLA response in transgenic mice after introducing an $\alpha 3$ domain into HLA-B27. Nature 348: 642–644

Kellner H, Wen J, Wang J, Raybourne RB, Williams KM, Yu DTU (1994) Senum antibodies from patients with ankylosing spondylitis and Reiter's syndrome are reactive with HLA-B27 cells transfected with the *Mycobacterium tuberculosis* hsp 40 logene. Infect Immun 62: 284–291

Khan MA (1994) Spondyloarthropathies. Curr Opin Rheumatol 6: 351–353

Kono DH, Ogasawara M, Effors RB, Park MS, Walford RL, Yu DTU (1985) Ye-1 monoclonal antibody that cross reacts with HLA-B27 lymphoblatoid cell lines and an arthritis causing bacteria. Clin Exp Immunol 61: 503–508

Kool J, Ruseler van Embden JGH, van Lieshout LMC, de Visser H, Gerrits-Boeye MY, van den Berg WG, Hazenberg MP (1991) Induction of arthritis in rats by soluble peptidoglycan-polysaccharide complexes produced by human intestinal flora. Arthritis Rheum 34: 1611–1616

Krimpenfort P, Rudenko G, Hochstenbach F, Guessow D, Berns A, Plough H (1987) Crosses of two independently derived transgenic mice demonstrate functional complementation of the genes encoding heavy (HLA-B27) and light (beta 2 microglobulin) chains of HLA class I antigens. EMBO J 6: 1673–1676

Krishna S, Benaroch P, Pillai S (1992) Tetrameric cell-surface MHC class I molecules. Nature 357: 164–167

Lahesmaa R, Skurnik M, Granfors K, Mottonen T, Saario R, Toivanen A, Toivanen P (1990) A tetrapeptide shared by *Yersinia* outer membrane protein Yop 1 and HLA-B27; immune response in patients with *Yersinia*-triggered reactive arthritis. Scand J Rhematol Suppl 87: 70–71

Lahesmaa R, Skurnik M, Vaara M, Leirisalo-Repo M, Nisssila M, Granfors K (1991) Molecular mimickry between HLA-B27 and *Yersinia*, *Salmonella*, *Shigella* and *Klebsiella* within the same region of HLA-$\alpha 1$-helix. Clin Exp Immunol 86: 399–404

Lahesmaa R, Skurnik M, Granfors K, Mottonen T, Saario R, Toivanen A, Toivanen P (1992) Molecular mimicry in the pathogenesis of spondyloarthropathies. A critical appraisal of cross reactivity between microbial antigens and HLA-B27. Br J Rheumatol 31: 221–229

Lahesmaa-Rantala R, Granfors K, Isomaki H, Toivanen A (1987) *Yersinia*-specific immune complexes in synovial fluid of patients with *Yersinia*-triggered reactive arthritis. Ann Rheum Dis 46: 510–514

Maclean L (1992) HLA-B27 subtypes: implications for the spondyloarthropathies. Ann Rheum Dis 51: 929–931

Madden DR, Gorga JC, Strominger JL, Wiley DC (1991) The structure of HLA-B27 reveals nonamer self peptide bound in an extended conformation. Nature 353: 321–325

Madden DR, Gorga JC, Strominger JL, Wiley DC (1992) The three-dimensional structure of HLA-B27 at 2.1 A° resolution suggests a general mechanism for tight binding to MHC. Cell 70: 1035–1048

Maki-Ikola O, Granfors K (1992) Salmonella-triggered reactive arthritis. Lancet 339: 1096–1097

Maki-Ikola O, Lehtinen K, Granfors K, Vanionpaa R, Toivanen P (1991) Bacterial antibodies in ankylosing spondylitis. Clin Exp Immunol 84: 472–475

Maksymowyck WP, Wessler A, Schmitt-Egenolf M, Suarez-Almozer M, Ritzel G, Borstel RCV, Pazderka F, Russel AS (1994) Polymorphism in an HLA linked proteosome gene influences phenotypic expression of disease in HLA-B27 positive individuals. J Rheumatol 21: 665–669

Merilahti-Palo R, Soderstorm K-O, Lehesmma-Rantala R, Granfors K, Toivanen A (1991) *Yersinia enterocolitica* in the synovial membrane in *Yersinia*-triggered reactive arthritis. Ann Rheum Dis 50: 81–90

Merilahti-Palo R, Gripenberg-Lerche C, Soderstorm K-O, Toivanen P (1992) Long term follow up of SHR rats of experimental *yersinia* associated arthritis. Ann Rheum Dis 51: 91–96

Mertz AKH, Batford SR, Curschellas E, Kist MJ, Gondolf KB (1991) Cationic *Yersinia* antigen-induced chronic allergic arthritis in rats: A model for reactive arthritis in humans. J Clin Invest 87: 632–642

Nickerson CL, Luthra HS, Savarirayan S, David CS (1990a) Susceptibility of HLA-B27 transgenic mice to *Yersinia enterocolitica* infection. Hum Immunol 28: 382–396

Nickerson CL, Hanson J, David CS (1990b) Expression of HLA-B27 in transgenic mice is dependent on the mouse class I H2-D genes. J Exp Med 172: 1255–1261

Nickerson CL, Hogen KL, Luthra HS, David CS (1990c) Effect of H-2 genes on expression on HLA-B27 and *Yersinia*-induced arthritis. Scand J Rheumatol Suppl 87: 85–90

Ogasawara M, Kono DH, Yu DTY (1986) Mimicry of human histocompatibility HLA-B27 antigens by *Klebsiella pneumoniae*. Infect Immun 51: 901–908

Parham P, Lomen CE, Lawlor DA, Ways JP, Holmes N, Coppin HL, Salter RD, Wan A-M, Ennis PD (1988) Nature of the polymorphism in HLA-A, -B and -C molecules. Proc Natl Acad Sci USA 85: 4005–4009

Pazmany L, Rowland-Jones S, Huet S, Hill A, Sutton J, Murrey R, Brooks J, McMichael A (1992) Genetic modulation of antigen presentation by HLA-B27 molecules. J Exp Med 175: 361–369

Pedrinaci S, Hanson J, David C (1994) Hierarchy in the assembly of HLA-B27 and HLA-Cw3 molecules in transgenic mice. Immunogenetics 39: 130–137

Powis SH, Tonks S, Mockridge I, Kelly AP, Bodmer JG, Trowsdale J (1993) Alleles and haplotype of the MHC-encoded ABC transporters Tap1 and Tap2. Immunogenetics 37: 373–380

Raybourne RB, Bunning VK, Williams KN (1988) Reaction of anti-HLA-B monoclonal antibodies with envelope proteins of *Shigella* species: evidence for molecular mimicry in the spondyloarthropathies. J Immunol 140: 3489–3495

Reboul M, Frangoulis B, Rocca A, Degos L, Pla M (1991) Recognition of HLA-B27 by mouse cytotoxic T-cell clones: a transgenic mouse model. Immunogenetics 34: 196–200

Rojo S, Garcia F, Villadangos JA, Lopez de Castro JA (1993) Changes in the repertoire of peptide bound to HLA-B27 subtypes and to site-specific mutants inside and outside pocket B.J Exp Med 177: 613–620

Roudier J, De Montclos H, Thouvenot D, Chomel JJ, Guillermet FN, Betuel H (1985) Absence of cross reaction between HLA-B27 and *Yersinia enterocolitica* and *Chlamydia tracheomatis* in reactive arthritis and ankylosing spondylitis. Clin Rheumatol 4: 487

Saper MA, Bjorkman PJ, Wiley DC (1991) Refined structure of the human histocompatibility antigen HLA-A2 at 2.6 A° resolution. J Mol Biol 219: 277–319

Savarirayan S, Prakash S, Banarjee S, Haqqi T, Little R, Hanson J, McCormick J, Nickerson CL, David CS (1988) Expression of HLA-B27 in association with mouse β2-microglobulin in transgenic mice. Fed Proc 2: A889

Schwimmbeck PL, Yu DT, Oldstone MB (1987) Autoantibodies to HLA-B27 in the sera of HLA-B27 patients with ankylosing spondylitis and Reiter's syndrome. Molecular mimicry with *Klebsiella pneumoniae* as potential mechanism of autoimmune disease. J Exp Med 166: 173–181

Singh B, Milton JD, Woodrow JC (1986) Ankylosing spondylitis, HLA-B27 and *Klebsiella*: a study of lymphocyte reactivity of anti-*klebsiella* sera. Ann Rheum Dis 45: 190–197

Singh B, Dillion T, Lauzon J, Fraza E, Russell AS (1994) Tolerance to the HLA-B27 and *Klebsiella pneumoniae* cross reactive epitope in mice transgenic for HLA-B2705 and human β2-microglobulin. J Rheumatol 21: 670–674

Shodjai-Moradi F, Ebringer A, Abul Jadayel I (1992) IgA antibody response to *Klebsiella* in ankylosing spondylitis measured by immunoblotting. Ann Rheum Dis 51: 233–237

Tanigaki N, Fruci D, Vigneti E, Strace G, Rovero P, Londei M, Butler RH, Tosi R (1994) The peptide binding specificity of HLA-B27 subtype. Immunogenetics 40: 192–198

Taurog JD, Lowen L, Forman J, Hammer RE (1988) HLA-B27 in inbred and noninbred transgenic mice: cell surface expression and recognition as an alloantigen in the absence of human β2-microglobulin. J Immunol 141: 4020–4023

Taurog LD, Maika SD, Simmons WA, Breban M, Hammer RE (1993) Susceptibility to inflammatory disease in HL-B27 transgenic rat lines correlates with the level of B27 expression. J Immunol 150: 4168–4178

Tsuchiya N, Husby G, Williams RC Jr (1989) Studies of humoral and cell-mediated immunity to peptide shared by HLA-B27.1 and *Klebsiella pneumoniae* nitrogenase in ankylosing spondylitis. Clin Exp Immunol 76: 354–360

Ulker N, Lewis KD, Hood LE, Stroynowski I (1990) Activated T cells transcribe an alternatively spliced mRNA encoding a soluble form of Qa-2 antigen. EMBO J 9: 3839–3847

Urban RG, Chicz RM, Lane WS, Strominger JL, Rehm A, Kenter MJH, UytdeHaag FGCM, Plough H, Uchanska-Ziegler B, Ziegler A (1994) A subset of HLA-B27 molecules contains peptide much longer that nonamers. Proc Natl Acad Sci USA 91: 1534–1538

Van Bohemen CG, Grumet FC, Zanen HC (1984) Identification of HLA-B27M1 and M2 cross reactive antigens in *Klebsiella*, *Shigella* and *Yersinia*. Immunology 52: 607–610

Van Endert PM, Lopez MR, Patel SD, Monaco JJ, McDevitt HO (1992) Genomic polymorphism, recombination and linkage disequilibrium in human major histocompatibility complex-encoded antigen processing genes. Proc Natl Acad Sci USA 89: 11594–11597

Viner NJ, Bailey LC, Life PF, Bacon PA, Gaston JSH (1991) Isolation of *Yersinia*-specific T-cell clones from the synovial membrane and synovial fluid of a patients with reactive arthritis. Arthritis Rheum 34: 1151–1157

Wang J, Yu DTU, Fukazawa T, Kellner H, Wen J, Cheng X-K, Roth G, Williams KM, Raybourne RB (1994) A monoclonal antibody that recognizes HLA-B27 in context of peptides. J Immunol 152: 1197–1205

Weiss EH, Schliesser H, Rietmullar G, Kievits F, Ivanyi P, Brem G (1990a) The copy number and the presence of human beta 2 microglobulin control cell surface expression of HLA-B27 antigen transgenic mice with a 25 kb B27 gene fragment. In: Egorov I, David CS (eds) Transgenic mice and mutants in MHC research. Springer, Berlin Heidelberg New York, pp 205–213

Weiss EH, Schliesser G, Botteron C, McMichael A, Riethmuller G, Kievits F, Ivanyi P, Brem G (1990b) HLA class I transgenic mice as model system to study MHC-restricted antigen recognition in man. Scand J Rheumatol Suppl 87: 91–96

Wen J, Wang J, Kuipers JG, Guang F, Williams KM, Raybourne RB, Yu DTU (1994) Analysis of HLA-B*2505 peptide motif, using T2 cells and monoclonal antibody ME1. Immunogenetics 39: 444–446

Zhou P, Cao H, Smart M, David C (1993) Molecular basis of genetic polymorphism in major histocompatibility complex-linked proteosome gene (Lmp-2). Proc Natl Acad Sci USA 90: 2681–2684

The Influence of Cytokines on the Central Nervous System of Transgenic Mice

K. Geiger and N. Sarvetnick

1 Introduction

Cytokines are peptide regulatory factors which have been known for more than 40 years. In the past, extensive studies have contributed to a large body of data concerning their structure and their activities. The study of their function included the addition of cytokines to selected cell lines in vitro and the use of inhibitory antibodies and recombinant molecules in vivo. Recently, the production of mice transgenic for cytokine genes, or of mice whose genes encode either cytokines or their receptors, inactivated by homologous recombination (knockout mice), allowed new insight into the function of cytokines in vivo and revealed a whole network of related biological interactions. In contrast to the normal situation, in which the production of cytokines is mostly localized and transient, in the

Department of Neuropharmacology, The Scripps Research Institute, CVN10, 10666 North Torrey Pines Road, La Jolla, CA 92037, USA

transgenic animals the presence of a transgene, or the complete absence of a targeted gene product, throughout the life of an animal often has consequences different from those that follow the repeated injection of cytokines, which have a short half-life in vivo, or the administration of antibodies, which may not reach every site of cytokine production.

Several methods have been used to generate transgenic mice, such as microinjection of DNA into the pronucleus (GORDON 1989; MOUNTZ and ZHOU 1990) or transfer of foreign genes into mouse embryos by either retroviral vectors, or infection with vaccinia virus (RAMSHAW et al. 1992). Procedures for targeted recombination have been reviewed by CAPECCHI (1989).

Transgenic mouse models have been used extensively to analyze the activities of cytokines in several organs (TAVERNE 1993). However, this review concentrates on the influences of cytokines on the central nervous system (CNS), including the retina of the eye, and on their ability to induce in transgenic animal models CNS diseases that simulate those in humans. The cytokines discussed here and the transgenic approaches choses are listed in Table 1.

2 Specific Immunological Properties of the CNS

Generally speaking, the CNS is regarded as an immunologically privileged site. This property is shared by the brain and the eye, whose retina is considered part of the CNS. Structurally, an effective blood-tissue barrier consisting of a specialized vasculature containing endothelial cells with tight junctions protects the CNS from circulating blood and, thereby, from most circulating cytokines, migrating lymphocytes and mononuclear cells. Additionally, the CNS is essentially free from a lymphatic system that could capture potential antigens, and its neurons do not constitutively express MHC class I, which have a pivotal role in the generation of antigen-specific immune responses (BENACERRAF 1981; WONG et al. 1984). Pathological events within the CNS may result in the breakdown of the blood-brain barrier. However, in the eye, repeated insults are necessary to penetrate the blood-retina barrier. Additionally, the eye is protected by immuno-suppressive cytokines present in the intraocular fluids (STREILEIN 1993).

In the CNS, because only a limited number of cell types is present, each can be analyzed for the influence of cytokines. Neurons of different characteristics and functions are influenced in their development by various cytokines (MERRIL 1992; SATOH et al. 1988). However, this review targets the relationship between the induction of disease and host-derived cytokines and, therefore, focuses on glial cells – a major target of cytokine influence.

Table 1. Summary of cytokine genes expressed in the CNS of transgenic mice

Transgene	Promoter or vector	Expression	Pathology	References
Hu IL-1β	Lens α-crystallin	Eye	Microphthalmus, ocular inflammation	ENWUAGU et al. (1994)
Hu IL-2/Hu IL-2RL	MHC class I	MHC class I expressing cells	Pneumonia, loss of Purkinje cells in the cerebellum	ISHIDA et al. (1989)
Hu IL-2	MT-1	Thymocytes serum	None	KROEMER et al. (1991)
Mu IL-3	GFAP	Astrocytes eyes, brain	Polymorphonuclear, inflammation in eyes and brain, cataract	CHIANG et al. (1994)
Mu IL-6	GFAP	Astrocytes eyes, brain	Neuropathy and angiogenesis in cerebellum, ataxia seizures, death	CAMPBELL et al. (1993)
Mu GM-CSF	MLV	Macrophages	Macrophage overgrowth, ocular inflammation, cataract wasting, paralysis, death, retinal degeneration	LANG et al. (1987)
Mu IFN-β	MT-1	Liver, testes, brain	Antiviral activities in serum, male sterility	IWAKURA et al. (1988)
Mu IFN-γ	Lens α-crystallin	Eye	Microphthalmus, ocular inflammation, cataract, retinal degeneration	WAWROUSEK et al. (1994)
Mu IFN-γ	Rhodopsin	Eye	Ocular infiltration, cataract, retinal degeneration, loss of intraocular immune privilege, increased resistance to viral infection of the CNS	GEIGER et al. (1994a)
Mu MHC class I	GFAP	Astrocytes, brain	Induction of antigen-specific lysis of virally infected astrocytes	RALL et al. (1994)
Bov bFGF	RSV-LTR	Muscle, heart, brain	Increased resistance to hypoxemic brain damage	MACMILLAN et al. (1993)

Hu, human; Mu, Murine; Bov, bovine; MT-1, metallothionein promoter; MLV, Moloney leukemia virus promoter; RSV-LTR, Rous sarcoma virus long terminal repeat; GFAP, glial fibrillary acidic protein.

2.1 Glial Cells

As opposed to microglia, macroglia consist of oligodendrocytes, astrocytes, and ependymal cells. Oilgodendrocytes function in the myelination of nerve fibers, allowing efficient nerve impulse conduction through the insulation of axons (MORELL and NORTON 1980), but probably do not respond directly to immunological stimuli (BENVENISTE 1992). These cells are recognizable by their expression of myelin basic protein (MBP), which indicates the extent of their maturation (RANSCHT et al. 1982), or labeling with galactocerebroside (RAFF et al. 1978). Astrocytes, the most numerous of the glial cells, can be identified by the presence of glial fibrillary protein (GFAP) (BIGNAMI et al. 1972). Their functions in the CNS include the regulation of ions and metabolites (JOHNSTON and ROOTS 1976), participation in the neurotransmitter transport system, and induction of the blood-brain

barrier by contact with the vascular endothelium (JANZER and RAFF 1987). Furthermore, astrocytes retain the capacity to proliferate in response to injury, represent
immunocompetent cells within the CNS that can present MHC class I- and class
II-encoded antigens upon stimulation (JOHN et al. 1992; MASSA et al. 1992), and
secrete various cytokines (BENVENISTE 1992). The significance of these features will
be discussed later in the context of inflammation in the CNS. In contrast to the
macroglia, microglia are considered as specialized tissue macrophages in the
brain. Presumably they originate from bone marrow-derived monocytes and
probably populate the CNS after vascularization (PERRY and GORDON 1988). Microglia can be identified by a number of surface antigens they share with other
macrophage-like cells such as immunoglobulin Fc receptors, type 3 complement
receptors and β-2 integrins (PERRY et al. 1985). Their major function is the phago-
cytosis of cell debris, which may be important in the modeling of the developing
CNS (PERRY and GORDON 1988). Also, microglial cells have been demonstrated to
express MHC antigens upon stimulation (WEINSTEIN et al. 1990) and are probably
involved with inflammation and repair processes in the adult brain.

3 Cytokines in Inflammatory Diseases of the CNS

Neurological disorders associated with inflammatory changes and dysfunction of
glial cells include multiple sclerosis (MS), Alzheimer's disease and viral infections.
In particular, infections with HIV and human herpes viruses are linked to neurological disorders such as AIDS dementia complex. Common to all these diseases is
the important role of the hosts' immune system and of multiple cytokines,
including interleukin (IL)-1, IL-6, tumor necrosis factor (TNF)-α, and interferon
(IFN)-γ (BENVENISTE 1992), in the development of CNS damage. The possible
sources of such damage could be either indirect toxicity, such as neural degeneration, or generation of atypical products by damaged cells, such as the neurofibril-
lary tangles in Alzheimer's disease, or demyelination, a hallmark of MS (ADAMS
1977). Other mechanisms may rely on stimulation of immunocompetent cells
within the CNS. In addition, cytokines may have an important role in the reaction
of the CNS to pathogenic stimuli exemplified by trauma and hypoxemic damage.
    Viral infections of the CNS can induce several forms of disease. For example,
herpes viruses infect neurons and glial cells and develop either acute cytolytic
activity, causing severe encephalitis and necrosis within the CNS (NEELEY et al.
1985), or persistent infection, with little or no activity (MITCHELL et al. 1994; ROCK
and FRASER 1983). The frequency of lytic herpes infections in immmunocom-
promised individuals per se demonstrates a connection between the outcome of
viral infection and the status of the host immune system (GUILOFF and TAN 1992).
Numerous cytokines have been found in the CNS of patients with virally induced
encephalitis; among these TNF-α, IL-2, IL-6, IFN-α/β and IFN-γ probably have a
major impact on the outcome of disease (CAMPBELL et al. 1994; PRICE et al. 1988;

SHANKAR et al. 1992). In addition, the interferons are capable of exerting antiviral activity on their own. However, it is currently unclear how cytokines may influence the development of latency or reactivation of viruses.

HIV has a specific role in CNS infections since it's action is linked to secondary infections with other microorganisms (GUILOFF and TAN 1992). HIV can infect macrophages, macrophage-derived multinucleated giant cells, glial cells and vascular endothelial cells (SHARER 1992). Neurons probably remain free from HIV gene products, despite the presence of neurological symptoms (PEPOSE et al. 1985; PRICE et al. 1988, 1990; SCHLOTE 1991). However, HIV virions, themselves, do not exert direct neurotoxic effects (BERNTON et al. 1992). Results concerning the involvement of infected monocytes and microglia on HIV-related damage of the CNS are controversial (BERNTON et al. 1992; GIULIAN et al. 1993; VAZEUX 1990). Apart from the possible interaction with other viruses, several lymphokines and cytokines secreted by infected glia could participate in the development of pathology. Especially IL-1, TNF-α and TGF-β1 are suspected to play an important role in the pathogenesis of HIV encephalitis (VITKOVIC et al. 1993).

4 Transgenic Animals as Models of Inflammatory CNS Diseases

In numerous experiments, proinflammatory cytokines have been expressed ectopically in the CNS to model facets of human disease and to test these cytokines' capacities for inducing infiltration by inflammatory cells. Since massive inflammation of the brain can cause immediate death, several research groups have used the eyes of transgenic mice as a nonvital part of the CNS to target ectopic cytokine expression.

4.1 Interleukin-1

This cytokine is expressed constitutively in the brain during fetal development and occurs in two forms, IL-1α and IL-1β, which are products of two different genes. In the adult animal, predominant sources of IL-1 are activated macrophages. Endothelial cells, keratinocytes, B cells, microglia and astrocytes can secrete this cytokine in response to stimulation. IL-1 participates in the activation of T cells by increasing the expression of both IL-2 and IL-2 receptor and can enhance the differentiation of B cells. IL-1 is probably a major mediator of inflammation in the CNS and can induce proliferation of astrocytes. Additionally, IL-1 induces the production of various cytokines by astrocytes and the expression of several other inflammatory metabolites such as prostaglandins and collagenases. IL-1 also promotes leukocyte adhesion (ARAI et al. 1990). Transgenic expression of human IL-1β was achieved by using the lens α-crystallin promoter yielding ectopic production of this cytokine in the lens of the eye (WAWROUSEK et al. 1994). Resulting animals developed local angiogenesis and inflammation of the eye probably mediated by increased expression of the adhesion molecules LFA-1 and ICAM-1. These manifestations were restricted to the eye and not apparent in the brain. Systemically, this local cytokine expression had the effect of lowering responses to IL-1 and lipopolysaccharide, possibly due to partial tolerance to IL-1 and/or down-regulation of IL-1 receptors. Regardless of the cause, the presence of such systemic effects indicates a breakdown of the blood-retina barrier.

4.2 Interleukin-2

Interleukin-2 is produced mainly by T cells and serves as a growth factor for lymphocytes, natural killer (NK) cells and monocytes (KLIMPEL et al. 1989; SMITH 1992). IL-2 induces the expression of other cytokines, such as IFN-γ, a cytokine that has a pivotal role in Th1-type immune responses (MOSMANN and COFFMAN 1989). In transgenic mice, expression of the human IL-2 and IL-2 receptor light chain gene under control of the MHC class I promoter is accompanied by expression of the transgene in the thymus, spleen, bone marrow, lung, skin and muscles (ISHIDA et al. 1989). Subsequently, NK cells infiltrate the lungs of these animals, suggesting that IL-2 preferentially promotes NK cells under such conditions. The animals have gait disturbances and a loss of Purkinje cells in the cerebellum as well, but show no inflammation within the brain. Since the expression of this cytokine under a different promoter (metallothionein; KROEMER et al. 1991) does not induce neuronal dysfunction, these results may be linked to a change in the regulation of neuronal development within the cerebellum.

4.3 Interleukin-3

A product of activated T cells, IL-3 supports the growth and differentiation of various hematopoietic cell lines and contributes to local inflammation by stimulating the growth of mast cells and the secretion of histamine (ARAI et al. 1990). Mice made transgenic for IL-3 under control of the GFAP promoter (CHIANG et al. 1994) express the cytokine on astrocytes in the brain and in the eye. The resulting mice suffer from inflammatory changes in both organs. In the eye, such changes include polymorphonuclear cell infiltration linked to a complete loss of the photoreceptors and thinning of the inner retinal layers, indicating a loss of the blood-retina barrier.

4.4 Interleukin-6

Interleukin-6 is a pleiotropic cytokine that participates in inflammatory responses. The cytokine is secreted by a number of cell types including fibroblasts, B cells, monocytes, endothelial cells, microglia and astrocytes. Several other cytokine, including IL-1, TNF-α and IFN-γ, are capable of enhancing the expression of IL-6, which in turn can stimulate the secretion of acute phase proteins like fibrinogen and C-reactive protein (VAN SNICK 1990). Overexpression of IL-6 in astrocytes of the CNS was induced in transgenic mice under the control of GFAP promoter, resulting in severe tremors, hind limb weakness, ataxia and seizures (CAMPBELL et al. 1993). Mice with abundant IL-6 expression were smaller than their nontransgenic litter mates and died at the age of 3–10 weeks. Neuropathology included nerve degeneration, astrocytosis, angiogenesis and induction of factor VIII-related antigens in the brain. Changes were most pronounced in the cerebellum and included vacuolation of neurons and loss of dendritic ramifications. Similar neuronal pathology is found in a variety of neurologic diseases as different as scrapie (HOGAN et al. 1987) and HIV encephalitis (VITKOVIC et al. 1993). These findings suggest that IL-6 may participate in the induction of generalized neuronal toxicity mechanisms.

4.5 Granulocyte/Macrophage Colony-Stimulating Factor

The granulocyte/macrophage colony-stimulating factor (GM-CSF) is produced by bone marrow stromal cells and participates in the regulation of hematopoiesis inducible by activated T cells in inflammatory responses. GM-CSF interacts with other proinflammatory cytokines such as IL-3 and IL-6 (MARCINKIEWICZ 1990). In mice, GM-CSF has been hyperexpressed on hematopoietic cells under the control of the Moloney murine leukemia virus long terminal repeat (LTR) (LANG et al. 1987). These animals exhibited elevated levels of GM-CSF in the serum, urine, peritoneal cavity and eye, resulting in an increase of presumably autoactivated macrophages. In addition to inflammatory destruction and increased fibrosis of the striated muscles, macrophages accumulated in the eyes and the retina degenerated, but interestingly, the brain underwent no damage. Transgenic animals suffered from a wasting syndrome and died by 2–4 months of age. There is evidence that the cytokines IL-1α and b-fibroblast growth factor (b-FGF), which are found at increased levels in the eyes of these mice, may contribute to the development of such pathology (CUTHBERTSON et al. 1990). However, transgenic mice which express b-FGF under the Rous sarcoma virus (RSV-LTR) promoter in the brain do not develop overt pathology (MACMILLAN et al. 1993). GM-CSF-expressing transgenic mice could present a useful model of fibrotic diseases, which apparently can be caused by inadequate proliferation of hematopoietic cells. Also illustrated here are the potentially damaging influences of macrophages on neuronal tissue.

4.6 Interferons

As mentioned above, interferons exist in three forms, α, β, and γ. IFN-α and β are secreted by various leukocytes and participate in the immune response. IFN-γ derives mainly from activated T cells and NK cells. All three forms of IFN are secreted by virus infected cells and have antiviral properties. IFN-γ additionally enhances the activity of macrophages and NK cells and induces MHC class I and class II molecule presentation on a variety of cells, including astrocytes (SUZUMURA et al. 1986). This cytokine is one of the main players in Th1-type immune responses (MOSMANN and COFFMAN 1989). When IFN-β was systemically hyperexpressed under the metallothionein (MT)-1 promoter, transgenic mice stimulated with cadmium expressed the cytokine in the testes, the liver and the brain (IWAKURA et al. 1988). The animals had no inflammatory changes but were infertile due to degeneration of the seminiferous tubules and showed antiviral activity caused by a synergistic effect between IFN-α and β, demonstrating mutual regulation by these cytokines.

IFN-γ seems not to be essential for development of the immune system, since IFN-γ knockout mice show no major abnormalities of their lymphoid tissues (DALTON et al. 1993). However, these animals die from a normally sublethal virus infection, underlining the importance of the cytokine's antimicrobial effects. Other mice lacking the IFN-γ receptor show a similar phenotype, combined with an abnormality in the concentration of IgM (HUANG et al. 1993).

IFN-γ has been produced in the eyes of transgenic mice by using either of two regulating elements. The lens α-crystallin promoter or the rhodopsin promoter. The use of the α-crystallin promoter yields overexpression of IFN-γ in the lens (EGWUAGU et al. 1994), resulting in microphthalmus and massive inflammatory destruction of the whole eye, including the lens and retina. Under the rhodopsis promoter, IFN-γ expressed in the photoreceptors of the retina induces inflammatory changes of the eye, cataracts and loss of photoreceptors, without influencing the development of the eye (GEIGER et al. 1994a). Since α-crystallin is known to be expressed before birth, the microphthalmus observed in this model may result from developmental dysregulation caused by IFN-γ expression during fetal life. Regardless of the promoter used, the cytokine's expression and inflammatory changes were restricted to the eye, indicating the interference of an efficient barrier between the eye and the brain.

4.6.1 Expression of Interferon-γ in the Retinas of Transgenic Mice

Transgenic mice with IFN-γ expressed ectopically in the eye were created to test the influence of IFN-γ on the morphology of ocular neurons and on the immunological properties of the intraocular environment. The rhodopsin promoter (LEM et al. 1991; ZACK et al. 1991) was used to direct expression of the cytokine specifically to the eye's photoreceptors. The intraocular cellular infiltration, corneal clouding, cataract and loss of photoreceptors that resulted resembled the retinal degeneration in the GM-CSF mice described above. Additionally, in the transgenic (rhoγ) mice, antigenic changes, including an increase of GFAP, neurofilaments, the adhesion molecules LFA-1 and ICAM-1, and factor VIII-related antigen and the expression of MHC class I and class II in the retina occurred (GEIGER et al. 1994a).

Inflammatory changes appeared in the rhoγ mice at 2 weeks of age, progressing to a maximum at 4–6 weeks, and receding thereafter. Cellular infiltrates found in the cornea, the iris, the ciliary body, the vitreous and the retina consisted of lymphocytes and polymorphonuclear cells. Inflammation was accompanied by the outer retina's loss of structure. These changes became evident at 1–2 weeks of age when inflammation was still minimal. By 12–14 weeks, the outer nuclear layer had disappeared. Of the cell types identified in the inflammatory infiltrates, NK cells and cytotoxic T cells were situated in the inner retinal layers, which remained virtually undamaged. Introduction of the transgene into SCID mice, which have no mature T or B cells, did not influence the amount of retinal damage, excluding T and B cells as the source of photoreceptor loss. Similarly, lack of MHC class I or class II expression had no influence on the development of retinal pathology (GEIGER and SARVETNICK 1994b).

IFN-γ enhanced the expression of ICAM-1, and other molecules that modify the migration of leukocytes, on retinal pigment epithelium cells (ELNER et al. 1992), thus influencing the function of the blood-retina barrier in transgenic mice (AROCKER-METTINGER et al. 1992; SLIGH et al. 1993). The unusual expression of MHC class II antigens on retinal pigment epithelial cells (BIGNAMI and DAHL 1979; HISCOTT et al. 1984) could indicate that the latter assume the function of antigen presenting cells (LIVERSIDGE and FORRESTER 1992) in this transgenic system. However, the observed photoreceptor degeneration was not dependent on MHC-restricted processes, since the lack of MHC expression did not alter retinal pathology, neither did the lack of mature B and T cells. These findings leave macrophages activated by IFN-γ (CUTHBERTSON et al. 1990; GOUREAU et al. 1992) or toxic effects of the cytokine (BIRDSALL 1991; SCHNEIDER-SCHAULIES et al. 1991) as possible sources of neuronal damage (ANDERSON et al. 1988; LEM et al. 1991) in this model. The increased accumulation of neurofilaments in neuronal cells is clinically and experimentally connected with disturbances of the axonal transport system of affected neurons (AYEHUNIE et al. 1993; BRADY 1993; CÔTÉ et al. 1993); such disturbances could be a causative mechanism or a secondary participant in the observed pathological destruction.

4.6.2 Abrogation of the Intraocular Immune Privilege by Interferon-γ

The rhoγ model was used to study the influence of IFN-γ on the properties of the intraocular compartment, which characteristically lacks cellular inflammation or delayed type hypersensitivity (DTH) but manifests distinctive humoral responses. This system is termed anterior chamber associated immune deviation (ACAID). BALB/c-derived transgenic and nontransgenic mice were challenged intravitreally either with allogeneic splenocytes from C57B1/6 mice or with bovine serum albumin (BSA) in incomplete Freund's adjuvant (ICFA). The mice were then tested for the DTH reaction to BSA by intrapinnal injection of the same antigen in the ear 7 days later. Pathological changes of the eyes were evaluated by histology and immunhistochemistry. The rhoγ transgenic mice, which expressed IFN-γ, developed an increased amount of ocular inflammation in response to both antigens compared to the nontransgenic controls. Furthermore, these transgenic mice showed marked DTH reactions to BSA, whereas the nontransgenic mice did not

(GEIGER and SARVETNICK 1994a). After challenge with either antigen, the rhoγ mice had an increased number of cells expressing the adhesion molecule LFA-1 in both the retina and choroid. Furthermore, MHC class I and class II expression increased on retinal cells of the challenged transgenic mice, whereas their nontransgenic counterparts had virtually no such change. To determine whether IFN-γ would induce up-regulation of costimulatory molecules upon antigenic challenge, the expression of the costimulatory molecule B7 by immunostaining was tested with it's ligand CTLA-4 and with 16–10A1, an antibody against B7. Initially, few cells were positive for B7 within the ciliary bodies or choroids of either transgenic or nontransgenic mice. However, when challenged with BSA or splenocytes, transgenic mice responded with an expansion of the B7-positive cell population within the retina, but control BALB/c mice had little or no reaction.

IFN-γ apparently favors cellular infiltration by influencing the local immuno-suppressive properties of the eye. Disturbance of the intraocular microenvironment seems to be profound in transgenic mice that express IFN-γ, since they make immune responses to such mild stimuli as ICFA without additional protein. This up-regulation of B7 in transgenic mice after challenge is consistent with costimulatory events having a role in the development of uveitis under these conditions. Yet, IFN-γ expression alone is not sufficient to up-regulate expression of B7 in the retina, which may, instead, rely on stimuli linked to an apparent disturbance of the blood-retinal barrier induced by the cytokine, as demonstrated by staining for factor VIII-related antigen, and cellular infiltration of the retina. Consistently, the increased number of cells positive for the adhesion molecule LFA-1 in the retina and choroid of transgenic mice after challenge could indicate the presence of cytotoxic T cells within the eye. These findings indicate that local IFN-γ production disturbs the immunosuppressive properties of the eye and prevents the induction of ACAID in response to intraocularly presented antigen. Furthermore, these data confirm the extension of the intraocular immune privilege to the posterior chamber of the eye in normal mice.

4.6.3 Protective Effects of Intraocularly Produced Interferon-γ in Intraocular and Intracerebral Viral Infection

Transgenic rhoγ mice provided a tool for studying qualitatively and quantitatively the influence of IFN-γ produced in the eye on intraocular and cerebral viral infection. This model involved intravitreal injection of BALB/c and C57B1/6-based transgenic and nontransgenic mice with herpes simplex virus (HSV-1, strain F), which infected the recipients' brains and induced severe retinitis of the injected eyes, while the second uninoculated eyes of the transgenic mice were not affected, nontransgenic BALB/c mice developed bilateral retinitis. Nontransgenic mice died from HSV-1 infection at 3 weeks, whereas transgenic mice of the same age survived (GEIGER et al. 1994b). After infection, viral antigen and pathological changes were located mostly in the optic and trigeminal projection of the brain, including the geniculate and the colliculus superior. Subsequently, the viral antigen disappeared faster from the brains of transgenic mice than from those of nontransgenic controls, which retained vast quantities of antigen in their brains at day 10.

Although the transgenic mice had considerably more inflammation in infected eyes than nontransgenic animals, the brains of both groups had nearly the same amount of inflammation. This inflammatory response consisted of NK cells, macrophages and T cells in the eyes and brain. Quantitation of CD4 (L3T4) and CD8 (LY2) cells revealed that transgenic mice had three to five times higher total cell counts and relatively more CD8 cells from the beginning of infection than found in control animals. In both transgenic and nontransgenic animals, all affected tissues expressed an overall increase of MHC class I and class II antigens, presumably most abundant on glial cells and astrocytes.

Recovery of infectious virus from infected tissue homogenates served to determine whether the protection conferred on transgenic mice by HSV-1 infection resulted from a block of viral replication induced by IFN-γ. Virus was recoverable by day 1 from the injected (right) eyes of all animals studied and from the brains and the noninjected (left) eyes on day 2. No virus was recovered from noninfected control animals or from animals infected for more than 3 weeks. At days 1 and 2 after HSV-1 injection, tissues of transgenic and nontransgenic mice contained similar amounts of virus. Subsequently, the relative amount of viral yield decreased in the transgenic mice. By day 4, no virus could be recovered from the noninoculated left eyes of these mice.

The virus strain used for these experiments had characteristics different from strain KOS, used in well-known models of intraocular infection with HSV-1 (ANDERSON and FIELD 1984; ATHERTON and STREILEIN 1987; AZUMI and ATHERTON 1994; DIX et al. 1987; VANN and ATHERTON 1991; WHITTUM-HUDSON and PEPOSE 1988). Independent of the inoculation site (anterior chamber vs intravitreal infection route), all mice developed ipsilateral retinitis. The lethal susceptibility to HSV-1 of 3 week old BALB/c mice could be related to the selective defect of IFN-γ production in mice of this age (ADKINS et al. 1993). This defect probably correlates with the lack of IFN-γ production in response to viral infection in macrophages (LUCCHIARI and PEREIRA 1990) and the unresponsiveness of NK cells to stimulation with IFN-γ (PROVINCIALI et al. 1989). In contrast, 3 week old transgenic mice were protected from this lethal effect, presumably by the local production of IFN-γ. We do not know whether other factors, such as induction of resistance to virus-induced damage on a cellular level or an age-related increase in the susceptibility of neurons to HSV (McKENDALL and WOO 1987), might play a role.

IFN-γ could also act indirectly via increased MHC expression in adult transgenic mice (DAYTON et al. 1985; HAMEL et al. 1990; HUGHES et al. 1988), rendering the viruses more recognizable for the host. Properties of IFN-γ that favor a cellular immune response and lead to disturbance of the blood-tissue barrier might provide a mechanism responsible for the protection from HSV in transgenic mice. However, a direct viral replication block induced by IFN-γ is unlikely to be of major importance in HSV-1 infection, since the amounts of infections virus in transgenic and nontransgenic mice were nearly equal until day 2 after infection, a time when the virus was already present in the second, noninjected, eye. Any block of viral replication would occur too late to have an impact on its invasion of the brain. This result fits well earlier reports describing an 18 h delay between the presence of IFN-γ and the beginning of effects on viral replication in culture (KLOTZBÜCHER et al. 1990). Although these results do explain the failure of IFN-γ to prevent the neuroinvasion of HSV-1, they do not elucidate why the second eyes of transgenic mice did not develop pathology, despite the small amounts of virus recovered at day 2, or why these mice showed no clinical signs of encephalitis, unless the cytokine either directly influenced the propagation characteristics of the virus during its passage through the eye, or sent a systemic signal protecting infected neurons from death.

The lack of neurologic symptoms in young transgenic mice infected with HSV-1 could indicate the site of its protective effect. Since HSV-1, strain F, has been shown to develop latency (BATRA and BROWN 1990; GORDON 1990; MEIGNIER et al. 1983; NESBURN et al. 1976; TROUSDALE et al. 1991), it is not likely that the virus is ever cleared from surviving infected cells. Therefore, the more rapid loss of viral antigen from the brains of transgenic mice than from nontransgenic controls could mean that IFN-γ favors the pathway towards latency. Additional mechanisms of this scenario could include synergistic action with antibodies (HAMZAOUI et al. 1990; LOUSCH et al. 1991; MATTHIESES et al. 1988; SPIEZIA et al. 1990) and other cytokines, such as TNF-α (ROSSOL-VOTH et al. 1991), protection of infected neurons from death by cytotoxic T cells, (MARTZ and GAMBLE 1992) or IFN-γ-induced production of nitric oxide (CROEN 1993).

4.7 Expression of MHC Class I in the Brain

A less direct method for testing the influence of the host immune system on intracerebral viral infection was to use transgenic mice that express MHC class I on astrocytes in the brain under the control of the GFAP promoter (RALL et al. 1994) and infect them with lymphocytic choriomeningitis virus (LCMV). These transgenic mice did not develop spontaneous neuropathy, indicating that the expression of class I antigens in the brain does not, by itself, induce pathological changes. However, MHC antigens were expressed in a functional manner and induced antigen-specific lysis of virally infected astrocytes by cytotoxic T cells, rendering this model a useful tool for the study of viral infection in the brain. Experiments with mice expressing LCMV antigens on astrocytes under the GFAP promoter confirmed that antigen-specific activated cytotoxic T cells are capable of entering the brain tissue, despite an intact blood-brain barrier (OLDSTONE and SOUTHERN 1993).

5 Conclusion

From these studies, one must conclude that overexpression of inflammatory cytokines is usually harmful for the host animal. Many of the pathologic changes involved are reminiscent of those found in the inflammatory diseases of humans, confirming the involvement of host immune factors in the development of tissue damage. The expression of IL-6 revealed features present in diseases of different etiologies, such as Alzheimer's disease and MS, that may commonly underlie the development of damage in the CNS. The expression of IFN-γ yielded several unexpected results, demonstrating multiple facets of antiviral responses and implying that we know too little about the factors at play in viral infection of the CNS. The sites and extent of such damage are dependent on the regulatory mechanism used. For example, IFN-γ expressed under the α-crystallin promoter of the lens causes pathologic changes that do not occur when its expression is regulated by the rhodopsin promoter. Therefore, mice that acquire the expression of ectopic cytokines after birth may be better suited than transgenic animals, which express their transgenes during their fetal period, to the study of a cytokine's therapeutic potential. Nevertheless, transgenic animals represent a suitable means for studying the effects of cytokines in various diseases, providing a means to evaluate results previously obtained by in vitro studies for their relevancy in vivo. Similarly, perhaps the eye and the brain must be considered as quite different organs with disparate properties, despite their similar immunological functions. For example, expression of transgenes in the eyes may not automatically lead to expression in the brain. Therefore, data derived from these two sites may not be comparable. Still, the studies described here confirm the interaction and mutual regulation of cytokines and identify multiple factors shaped by specific properties of the CNS that contribute to disease.

References

Adams CWM (1977) Pathology of multiple sclerosis, Br. Med J 33: 15–20

Adkins B, Ghanei A, Hamilton K (1993) Developmental regulation of IL-4, IL-2 and IFN-gamma production by murine peripheral T lymphocytes. J Immunol 151(12): 6617–6626

Anderson DH, Williams DS, Neitz J, Fariss RN, Fliesler SJ (1988) Tunicamycin-induced degeneration in cone photoreceptors. Vis Neurosci 1: 153–158

Anderson JR, Field HJ (1984) An animal model of ocular herpes, keratitis, retinitis and cataract in the mouse. Br J Exp Pathol 65: 283–297

Arai K, Lee F, Miyajima A, Miyatake S, Arai N (1990) Cytokines: coordinators of immune and inflammatory responses. Annu Rev Biochem 59: 783–836

Arocker-Mettinger E, Steurer-Georgiew L, Steurer M, Huber-Spitzy V, Hoelzl E, Grabner G, Kuchar A (1992) Circulating ICAM-1 levels in serum of uveitis patients. Curr Eye Res 11 Suppl: 161–166

Atherton SS, Streilein JW (1987) Virus-specific DTH prevents contralateral retinitis following intra-cameral inoculation of HSV-2. Curr Eye Res 6: 133–139

Ayehunie S, Sonnerborg A, Yemane-Berhan T, Zewdie DW, Britton S, Strannegard O (1993) Raised levels of tumor necrosis factor-alpha and neopterin, but not interferon-alpha, in serum of HIV-1-infected from Ethiopia. Clin Exp Immunol 91: 37–42

Azumi A, Atherton SS (1994) Sparing of the ipsilateral retina after anterior chamber inoculation of HSV-1: Requirement for either CD4+ or CD8+ T cells. Invest Ophthalmol Vis Sci 35: 3251–3259

Batra SK, Brown SM (1990) Herpes simplex virus genes controlling reactivation from latency in rabbit eye model. Indian J Med Res 1991: 252–257

Benacerraf B (1981) Role of MHC gene products in immune regulation. Science 212: 1229–1238

Benveniste EN (1992) Inflammatory cytokines within the central nervous system: sources, function, and mechanism of action. Am J Physiol 263: C1-C16

Bernton EW, Bryant HU, Decoster MA, Orenstein JM, Ribas JL, Meltzer MS (1992) No direct neurotoxicity by HIV-1 virions or culture fluids from HIV-infected T cells or monocytes. AIDS Res Hum Retrovir 8: 495-503

Bignami A, Dahl D (1979) The radial glia of Muller in the rat retina and their response to injury. An immunofluorescence study with antibodies to the glial fibrillay acidic protein. Exp Eye Res 28: 63-69

Bignami A, Eng LF, Dahl D, Uyeda CT (1972) Localization of the glial fibrillary acidic protein in astrocytes by immunofluorescence. Brain Res 43: 429-435

Birdsall HH (1991) Induction of ICAM-1 on human neural cells and mechanisms of neutrophil-mediated injury. Am J Pathol 139: 1341-1350

Brady ST (1993) Motor neurons and neurofilaments in sickness and in health. Cell 73: 1-3

Campbell IL, Abraham CR, Masliah E, Kemper P, Inglis JD, Oldstone MBA, Mucke L (1993) Neurologic disease induced in transgenic mice by cerebral overexpression of interleukin-6. Proc Natl Acad Sci USA 90: 10061-10065

Campbell IL, Hobbs MV, Kemper P, Oldstone MBA (1994) Cerebral expression of multiple cytokine genes in mice with lymphocytic choriomeningitis. J Immunol 152: 716-723

Capecchi MR (1989) Altering the genome by homologous recombination. Science 244: 1288-1292

Chiang CS, Masliah E, Stalder A, Samimi A, Campbell IL (1994) Gliosis and neurodegeneration as a consequence of the cerebral expression of IL-3 in transgenic mice. Soc Neurosci Abstr 20: 176.14

Côté F, Collard JF, Julien JP (1993) Progressive neuronopathy in transgenic mice expressing the human neurofilament heavy gene: a mouse model of amyotrophic lateral sclerosis. Cell 73: 35-46

Croen KD (1993) Evidence for antiviral effect of nitric oxide. Inhibition of herpes simplex virus type 1 replication. J Clin Invest 91(6): 2446-2452

Cuthbertson RA, Lang RA, Coghlan JP (1990) Macrophage products IL-1 α, TNF-α and bFGF may mediate multiple cytopathic effects in the developing eyes of GM-CSF transgenic mice. Exp Eye Res 51: 335-344

Dalton DK, Pitts-Meek S, Keshav S, Figari IS, Bradley A, Steward TA (1993) Multiple defects of immune cell function in mice with disrupted interferon-γ genes. Science 259: 1739-1742

Dayton ET, Matsumoto-Kobayashi M, Perussia P, Trinchieri G (1985) Role of immune interferon in the monocytic differentiation of human promyelocytic cell lines by leukocyte conditioned medium. Blood 66(3): 583-594

Dix RW, Streilein JW, Cousins S, Atherton SS (1987) Histopathologic characteristics of two forms of experimental herpes simplex virus retinitis. Curr Eye Res 6: 47-52

Egwuagu CE, Sztein J, Chan CC, Reid W, Mahdi R, Nussenblatt RB, Chepelinsky AB (1994) Ectopic expression of gamma interferon in the eyes of transgenic mice induces ocular pathology and MHC class II gene expression. Invest Ophthalmol Vis Sci 35(2): 332-341

Elner SG, Elner VM, Pavilack MA, Todd RF, Mayo-Bond L, Franklin WA, Strieter RM, Kunkel SL, Huber AR (1992) Modulation and function of intercellular adhesion molecule-1 (CD54) on human retinal pigment epithelial cells. Lab Invest 66(2): 200-211

Geiger K, Sarvetnick N (1994a) Local production of IFN-γ abrogates the intraocular immune privilege in transgenic mice and prevents the induction of ACAID. J Immunol 153: 5239-5246

Geiger K, Sarvetnick N (1994b) MHC class I expression contributes to the development of cataract in transgenic mice with ectopic expression of IFN-γ. In: Nussenblatt RB, Whitcup SM, Caspi RR, Gery I (eds) Advances in ocular immunology. Elsevier, Amsterdam, pp 135-138

Geiger K, Howes E, Gallina M, Huang XJ, Travis GH, Sarvetnick N (1994a) Transgenic mice expressing IFN-γ in the retina develop inflammation and photoreceptor loss. Invest Ophthalmol Vis Sci 35(5): 2667-2681

Geiger K, Howes EI, Sarvetnick N (1994b) Ectopic expression of IFN-γ in the eye protects transgenic mice from intraocular HSV-1 infections. J Virol 69: 5556-5567

Giulian G, Wendt E, Vaca K, Noonan CA (1993) The envelope glycoprotein of human immunodeficiency virus type 1 stimulated release of neurotoxins from monocytes. Proc Natl Acad Sci USA 90: 2769-2773

Gordon JW (1989) Transgenic animals. Int Rev Cytol 115: 171-227

Gordon YJ (1990) Pathogenesis and latency of herpes simplex virus type 1 (HSV-1): an ophthalmologist's view of the eye as a model for the study of the virus-host relationship. Adv Exp Med Biol 278: 205-209

Goureau O, Lepoivre M, Courtois Y (1992) Lipopolysaccharide and cytokines induce a macrophage-type of nitric oxide synthase in bovine retinal pigmented epithelial cells. Biochem Biophys Res Commun 186(2): 854–859

Guiloff RJ, Tan SV (1992) Central nervous system opportunisitic infections in HIV disease: clinical aspects. Baillieres Clin Neurol 1: 103–154

Hamel CP, Detrick B, Hooks JJ (1990) Evaluation of Ia expression in rat ocular tissues following inoculation with interferon-gamma. Exp Eye Res 50(2): 173–182

Hamzaoui K, Slim AA, Hamza M, Touraine J (1990) Natural killer cell activity, interferon-gamma and antibodies to herpes viruses in patients with Behçet's disease. Clin Exp Immunol 79: 28–34

Hiscott PS, Grierson I, Trombetta CJ, Rahi AHS, Marshall J, McLeod D (1984) Retinal and epiretinal glia. An immunohistochemical study. Br J Ophthalmol 68: 698–707

Hogan RN, Baringer JR, Prusiner SB (1987) Scrapie infection diminishes spines and increases varicosities of dendrites in hamsters: a quantitative Golgi analysis. J Neuropathol Exp Neurol 46: 461–473

Huang S, Hendriks W, Althage A, Hemmi S, Bluethmann H, Kamijo R, Vilcek J, Zinkernagel RM, Aguet M (1993) Immune response in mice that lack the interferon-γ receptor. Science 259: 1742–1745

Hughes CC, Male DK, Lantos PL (1988) Adhesion of lymphocytes to cerebral microvascular cells: effects of interferon-gamma, tumor necrosis factor and interleukin-1. Immunology 64: 677–681

Ishida Y, Nishi M, Taguchi O, Inaba K, Hattori M, Minato N, Kawaichi M, Honjo T (1989) Expansion of natural killer cells but not T cells in human interleukin-2/interleukin-2 receptor (Tac) trangenic mice. J Exp Med 170: 1103–1115

Iwakura Y, Asano M, Nishimune Y, Kawade Y (1988) Male sterility of transgenic mice carrying exogenous mouse interferon-beta gene under the control of the metallothionein enhancer-promoter. EMBO J 7: 3757–3762

Janzer RC, Raff MC (1987) Astrocytes induce blood-brain barrier properties in endothelial cells. Nature 325: 253–257

John LD, Babcock G, Green D, Freedman M, Sriram S, Ransohoff RM (1992) Transforming growth factor-beta 1 differentially regulates proliferation and MHC class-II antigen expression in forebrain and brainstem astrocyte primary culture. Brain Res 585(1–2): 229–236

Johnston PV, Roots BI (1976) Neuron-glial relationships. In: Vinken PJ, Bruyn GW (eds) Handbook of clinical neurology. North-Holland, Amsterdam, pp 401–421

Klimpel GR, Asuncion M, Fons M, Norton JD, Albrecht T, Klimpel KD, Stein MD (1989) Interleukin 2 induced non MHC-restricted killing of herpes simplex type-1 (HSV-1) infected allogeneic and autologous lymphoblasts. J Clin Lab Immunol 29: 1–7

Klotzbücher A, Mittnacht S, Kirchner H, Jacobsen H (1990) Different effects of IFNγ and IFN α/β on "immediate early" gene expression of HSV-1. Virology 179: 487–491

Kroemer G, De Cid R, De Alboran IM, Gonzalo J-A, Iglesias A, Martinez AC, Gutiérrez-Ramos JC (1991) Immunological self-tolerance: an analysis employing cytokines or cytokine receptors encoded by transgenes or a recombinant vaccinia virus. Immunol Rev 122: 173–204

Lang RA, Metcalf D, Cuthbertson RA, Lyons I, Stanley E, Kelso A, Kannourakis G, Williamson DJ, Klintworth GK, Gonda TJ, Dunn AR (1987) Transgenic mice, expressing a hematopoetic growth factor gene (GM-CSF) develop accumulations of macrophages, blindness, and a fatal syndrome of tissue damage. Cell 51: 675–687

Lem J, Applebury ML, Falk JD, Flannery JG, Simon MI (1991) Tissue-specific and developmental regulation of rod-opsin chimeric genes in transgenic mice. Neuron 6(2): 201–210

Liversidge J, Forrester JV (1992) Antigen processing and presentation in the eye: a review. Curr Eye Res 11 Suppl: 49–58

Lousch RN, Staats H, Oakes JE, Cohen GH, Eisenberg RJ (1991) Prevention of herpes keratitis by monoclonal antibodies specific for discontinuous and continuous epitopes on glycoprotein D. Invest Ophthalmol Vis Sci 32: 2735–2740

Lucchiari MA, Pereira CA (1990) A major role of macrophage activation by interferon-gamma during mouse hepatitis virus type 3 infection. II. Age-dependent resistance. Immunobiology 181(1): 31–39

MacMillan V, Judge D, Wiseman A, Settles D, Swain J, Davis J (1993) Mice expressing a bovince basic fibroblast growth factor transgene in the brain show increased resistance to hypoxemic-ischemic cerebral damage. Stroke 24: 1735–1739

Marcinkiewicz J (1990) Cell-mediated immunity: roll of IL-3 and IL-6 in the regulation of contact sensitivity reaction. Folia Histochem Cytobiol 28: 107–119

Martz E, Gamble SR (1992) How do CTL control virus infections? Evidence for prelytic halt of herpes simplex Viral Immunol 5(1): 81–91

Massa PT, Hirschfeld S, Levi B-Z, Quigley LA, Ozato K, McFarlin DE (1992) Expression of major histocompatibility complex (MHC) class I genes in astrocytes correlates with the presence of nuclear factors that bind to constitutive and inducible enhancers. J Neuroimmunol 41: 35–42

Matthieses MA, Pesson A, Sundquist VA, Wahren B (1988) Neutralization capacity and antibody dependent cell mediated cytotoxicity of separated IgG subclasses 1, 3 and 4 against herpes simplex. J Immunol 72: 211–215

McKendall RR, Woo W (1987) Possible neural basis for age-dependent resistance to neurologic disease from herpes simplex virus. J Neurol Sci 81: 227–237

Meignier B, Norrild B, Roizman B (1983) Colonization of murine ganglia by a superinfecting strain of herpes simplex virus. Infect Immun 41(2): 702–708

Merril JE (1992) Tumor necrosis factor alpha, interleukin 1 and related cytokines in brain development: normal and pathological. Dev Neurosci 14: 1–10

Mitchell WJ, Gressens P, Martin JR, DeSanto R (1994) Herpes simplex virus type 1 DNA persistence, progressive disease and transgenic immediate early gene promoter activity in chronic corneal infections in mice. J Gen Virol 75: 1201–1210

Morell P, Norton WT (1980) Myelin. Sci Am 242: 88–118

Mosmann TR, Coffman RL (1989) TH1 and TH2 cells: different patterns of lymphokine secretion lead to different functional properties. Annu Rev Immunol 7: 145–173

Mountz JD, Zhou TJL (1990) Production of transgenic mice and application to immunology and autoimmunity. Am J Med Sci 300: 322–329

Neeley SP, Cross AJ, Crow TJ, Johnson JA, Taylor GR (1985) Herpes simplex virus encephalitis. Neuroanatomical and neurochemical selectivity. J Neurol Sci 71: 325–337

Nesburn AB, Dickinson R, Radnoti M, Green MJ (1976) Experimental reactivation of ocular herpes simplex in rabbits. Surv Ophthalmol 21(2): 185–190

Oldstone MBA, Southern PJ (1993) Trafficking of activated cytotoxic T lymphocytes into the central nervous system: use of a transgenic model. J Neuroimmunol 46: 25–32

Pepose JS, Holland GN, Nestor MS, Cochran AJ, Foos RY (1985) Acquired immune deficiency syndrome: Pathogenic mechanisms of ocular disease. Ophthalmology 92(1): 472–484

Perry VH, Gordon S (1988) Macrophages and microglia in the nervous system. Trends Neurosci 11: 273–277

Perry VH, Hume DA, Gordon S (1985) Immunohistochemical localization of macrophages and microglia in adult and developing mouse brain. Neuroscience 15: 313–326

Price RW, Brew B, Sidtis J, Rosenblum M, Scheck AC (1988) The brain in AIDS: central nervous system HIV-1 infection and AIDS dementia complex. Science 239 (4840): 586–592

Price RW, Brew BJ, Rosenblum M (1990) The AIDS dementia complex and HIV-1 brain infection: a pathogenetic model of virus-immune interaction. Res Publ Assoc Res Nerv Ment Dis 68: 269–290

Provinciali M, Muxxioli M, Fabris N (1989) Timing of appearance and disappearance of IFN and IL-2 induced natural immunity during ontogenetic development and aging. Exp Gerontol 24(3): 227–236

Raff MC, Mirsky R, Fields KL, Lisak RP, Dorfman SH, Silberberg DH, Gregson NA, Leibowitz S, Xennedy M (1978) Galactocerebroside is a specific cell surface antigenic marker for oligo-dendrocytes in culture. Nature 274: 813–815

Rall GF, Mucke L, Nerenberg M, Oldstone MBA (1994) A transgenic mouse model to assess the interaction of cytotoxic T lymphocytes with virally infected, class I MHC-expressing astrocytes. J Neuroimmunol 52: 61–68

Ramshaw IA, Ruby J, Ramsay A (1992) Cytokine expression by recombinant viruses- a new vaccine strategy. Tibtech 10: 424–426

Ranscht B, Clapshaw PA, Pride J, Noble M, Seifert W (1982) Development of oligodendrocytes and Schwann cells studied with a monoclonal antibody against galactocerebroside. Proc Natl Acad Sci USA 79: 2709–2713

Rock DL, Fraser NW (1983) Detection of the HSV-1 genome in central nervous system of latently infected mice. Nature 302: 523–525

Rossol-Voth R, Rossol S, Schutt KH, Corridori S, de Cian W, Falke D (1991) In vitro protective effect of tumor necrosis factor alpha against experimental infection with herpes simplex type 1. J Gen Virol 72: 143–147

Satoh T, Nakamura S, Taga T, Matsuda T, Hirano T, Kishimoto T, Kaziro Y (1988) Induction of neuronal differentiation in PC12 cells by B cell-stimulatory factor 2/interleukin-6. Mol Cell Biol 8: 3546–3549

Schlote W (1991) HIV-encephalopathy. Verh Dtsch Ges Pathol 17: 51–60

Schneider-Schaulies J, Kirchhoff F, Archelos J, Schachner M (1991) Down-regulation of myelin-asociated glycoprotein on Schwann cells by interferon-gamma and tumor necrosis factor-alpha affects neurite outgrowth. Neuron 7: 995–1005

Shankar V, Kao M, Hamir AN, Sheng H, Koprowski H, Dietzschold B (1992) Kinetics of virus level and changes in levels of several cytokine mRNAs in the brain after intranasal infection of rats with Borna disease virus. J Virol 66(2): 992–998

Sharer LR (1992) Pathology of HIV-1 infection of the central nervous system: a review. J Neuropathol Exp Neurol 51: 3–11

Sligh JE, Ballantyne CM, Rich SS, Hawkins HK, Smith CW, Bradley A, Beaudet AL (1993) Inflammatory and immune responses are impaired in mice deficient in intercellular adhesion molecule 1. Proc Natl Acad Sci USA 90: 8529–8533

Smith KA (1992) Interleukin-2. Curr Opin Immunol 4: 271–276

Spiezia KV, Dille BJ, Mushahwar IK, Kifle L, Okasinski GF (1990) Prevalence of specific antibodies to herpes simplex virus type 2 as revealed by an enzyme-linked immunoassay and Western blot analysis. Adv Exp Med Biol 278: 231–242

Streilein JW (1993) Immune privilege as the result of local tissue barriers immunosuppressive microenvironments. Curr Opin Immunol 5: 428–432

Suzumura A, Silberberg DH, Lisak RP (1986) The expression of MHC antigens on oligodendrocytes: induction of polymorphic H-2 expression by lymphokines. J Neuroimmunol 71: 179–190

Taverne J (1993) Transgenic mice in the study of cytokine function. Int J Exp Pathol 74: 525–546

Trousdale MD, Steiner I, Spivack JG, Deshmane SL, Brown SM, McLean A, Subak-Sharpe JH, Fraser NW (1991) In vivo and in vitro reactivation impairment of a herpes simplex virus type 1 latency-associated transcript variant in a rabbit eye model. J Virol 65(12): 6989–6993

Van Snick JV (1990) Interleukin-6: an overview. Annu Rev Immunol 8: 253–278

Vann VR, Atherton SS (1991) Neural spread of herpes simplex virus after anterior chamber inoculation. Invest Ophthalmol Vis Sci 32: 2462–2472

Vazeux R (1990) AIDS encephalopathy and tropism of HIV for brain monocytes/macrophages and microglial cells. Pathobiology 59: 214–218

Vitkovic L, Da Cunha A, Tyor WR (1993) Cytokine expression and pathogenesis in AIDS brain. In: Price PW, Perry SW (eds) HIV, AIDS, and the brain. Raven, New York

Wawrousek EF, Lai JC, Gery I, Chan CC (1994) Progressive inflammatory disease and neovascularization in the eyes of interleukin-1β transgenic mice. In: Nussenblatt RB, Whitcup SM, Caspi RR, Gery I (eds) Advances in ocular immunology. Elsevier, Amsterdam, pp 143–146

Weinstein DL, Walker DG, Akiyama H, McGeer PL (1990) Herpes simplex virus type I infection of the CNS, induces major histocompatibility complex antigen expression on rat microglia. J Neurosci Res 26: 55–65

Whittum-Hudson JA, Pepose JS (1988) Herpes simplex virus type 1 induces anterior chamber-associated immune deviation (ACAID) in mouse strains resistant to intraocular infection. Curr Eye Res 7(2): 125–130

Wong GHW, Bartlett PF, Clark-Lewis I, Battye F, Schrader JW (1984) Inducible expression of H-2 and Ia antigens on brain cells. Nature 310: 688–691

Zack DJ, Bennett J, Wang Y, Davenport C, Klaunberg B, Gaerhart J, Nathans J (1991) Unusual topography of bovine rhodopsin promoter-LacZ fusion gene expression in transgenic mouse retinas. Neuron 6(2): 187–199

Mx Transgenic Mice – Animal Models of Health*

H. Arnheiter[1], M. Frese[2], R. Kambadur[1,3], E. Meier[1], and O. Haller[2]

1 Introduction

If viruses, by their variety, distribution, potential virulence and sheer inexhaustibility, present a permanent threat to the well-being of organisms, one might wonder why good health still exists. Yet healthy organisms do exist, though not because they simply evade the tireless viral intruders, but because they constantly mobilize potent defense forces. In vertebrates, T and B cell-mediated immunity is usually credited with being of prime importance in antiviral defense, but often, the first hurdles viruses have to clear are not antibodies and T cells but the flow of mucus, the acidity of the stomach, the paucity of appropriate cellular

---

[1] Laboratory of Developmental Neurogenetics, National Institute of Neurological Disorders and Stroke, National Institutes of Health, Building 36, Room 5DO4, 36 Convent Drive, MSC 4160, Bethesda, MD 20892-4160, USA
[2] Department of Virology, Institute for Medical Microbiology and Hygiene, University of Freiburg, 79104 Freiburg, Germany
[3] Present Address: Department of Medicine, Johns Hopkins University School of Medicine, Baltimore, MD 21203, USA
*Dedicated to Prof. Jean Lindenmann on the occasion of his 70th birthday

virus receptors, or the inadequacy of a synthetic machinery that viruses need to usurp for their replication. These hurdles are host defense mechanisms no less important for the homeostasis of health than the specific immune system; they are placed at various levels and make the interplay between virus and host a notoriously complex affair. The method of choice to elucidate such defense mechanisms at the molecular level is the analysis of genetic variants in which these defense forces are altered. Although such variants abound from plants (FRASER 1990) to mammals (for mice, see GREEN 1989), only a few of the associated genes have been isolated or characterized (for example, DÉZÉLÉE et al. 1989; DVEKSLER et al. 1991; YOKOMORI and LAI 1992; NÉDELLEC et al. 1994; BUREAU et al. 1993), most likely because many of these variants are multigenic in origin and difficult to analyze.

One of the best-studied genetic variants of susceptibility to a viral infection in mammals is the resistance of A2G mice to influenza A and B viruses. This resistance was discovered by LINDENMANN (1962) and was attractive to study for many reasons. It was inherited as a single, dominant trait, was specific for orthomyxoviruses, effective against high virus doses, and independent of the route of infection and hence not a peculiarity of some particular organ (reviewed in HALLER 1981). During the past years we have learned that resistance is brought about by a single gene, *Mx1*, that is localized at 64 map units from the centromere on chromosome 16 and is structurally altered in influenza-susceptible mice. This gene is not normally expressed, but is induced upon viral infection through the action of interferon α and β, not γ, and it encodes a single nuclear GTPase, the Mx1 protein (reviewed in ARNHEITER and MEIER 1990; STAEHELI et al. 1993). This protein interferes with the accumulation of primary influenza viral transcripts (KRUG et al. 1985; PAVLOVIC et al. 1992) probably by inhibiting the transcriptional elongation of viral mRNAs by the viral RNA-dependent RNA polymerase (HUANG et al. 1992; STRANDEN et al. 1993; LANDIS et al. 1994). *Mx1*-dependent resistance is found in at least one other inbred mouse strain, SL/NiA (HALLER et al. 1986), and also in wild mice, although only in approximately three-quarters of them (HALLER et al. 1987). This finding is exciting because it shows that resistance is not a laboratory artifact, and at the same time it is puzzling because wild mice are not natural hosts for orthomyxo viruses. Still more puzzling is the observation that virtually all vertebrates, including many for which no pathogenic influenza viruses have been found, possess interferon-inducible *Mx* genes. Furthermore, *Mx* genes show sequence relationships with genes encoding dynamins, a family of constitutive GTPases involved in endocytosis (reviewed in ROBINSON et al. 1994), and with yeast genes encoding GTPases that are also involved in basic cellular processes (reviewed in ARNHEITER and MEIER 1990; STAEHELI et al. 1993). Thus, it is possible that the primary role of *Mx* in interferon-exposed cells is not in inhibiting influenza virus but in some other, yet uncharacterized cellular process. Alternatively, the full antiviral profile of Mx proteins may not have been discovered so far, and *Mx* may in fact inhibit in each species a set of species-specific pathogens. The recent discoveries that mouse Mx1 and human MxA protein inhibit tick-born influenza-like viruses (HALLER et al. 1993, 1995; PAVLOVIC et al. 1995; FRESE et al. 1995), which produce sporadic infections in rodents and humans, and that human MxA protein affects the synthesis of measles virus glycoproteins in cultured cells (SCHNORR et al. 1993) would seem to add weight to the argument that Mx is primarily an antiviral system; it might be a long way, however, until each Mx protein will have been assigned its specific viral target.

The dominant and efficient anti-influenza virus activity of some Mx proteins and the existence of mice unable to synthesize antivirally active Mx proteins have prompted studies in which *Mx* genes were used as transgenes for germline transformation and hence conversion of a susceptible host into a resistant one. The goals of these studies were: (a) to close the chain of observations that have linked *Mx* with resistance by rescuing an *Mx*1-deficient (*Mx*1⁻) host with a cDNA encoding a single protein, (b) to test whether constitutive expression of an antiviral protein which is normally expressed only during infection is possible and would give an organism a head start in the fight against infections, (c) to establish a dose/response curve in vivo by manipulating the amount of Mx protein expressed, (d) to test in vivo the efficacy of Mx proteins that are derived from organisms for which no genetic variants in *Mx* are available, and (e) to establish the ground for future "germline vaccination" of farm animals.

The successful pilot study with mouse Mx1 in transgenic mice (ARNHEITER et al. 1990) was followed by three types of subsequent studies. First, there was the introduction of human MxA protein into mice (HALLER et al. 1993; Pavlovic et al. submitted). Human MxA is a protein whose antiviral activity has been studied both in mouse and human cells (AEBI et al. 1989; PAVLOVIC et al. 1990, 1992; STAEHELI and PAVLOVIC 1991; PITOSSI et al. 1993; SCHNORR et al.; reviewed in ARNHEITER and MEIER 1990; STAEHELI et al. 1993), but whose in vivo relevancy could not be assessed, since, until now, no humans with mutations in this gene have been found. [In fact, the only genetic abnormalities associated with human *Mx* may be gene dosage alterations in patients with aneuploidy of chromosome 21 on which human *Mx* resides.] Second, when it was recently discovered that Thogoto virus, a tick-born influenza-like virus, was inhibited by mouse Mx1 and human MxA protein, transgenic mice expressing either of these proteins were used to prove that it is *Mx* and not some other gene that mediates inhibition of this virus in vivo (HALLER et al. 1993, 1995; PAVLOVIC et al. 1995; FRESE et al. 1995). Third, the first introduction of mouse Mx1 transgenes into pigs has been reported (MÜLLER et al. 1992b). Pigs have *Mx* genes of their own, but they are susceptible to influenza and provide a substantial reservoir for swine influenza viruses. By introducing the potent mouse Mx1 transgene, it was thought that pigs with increased influenza resistance could be obtained. None of these pigs expressed detectable amounts of transgenic Mx1 protein, however, but they represent, nevertheless, a first step in the direction of generating virus-resistant livestock. This may be of particular relevance since several other domestic animals are also susceptible to influenza, and since new human influenza pandemics may have their origin in domestic animals (for review, see WEBSTER et al. 1993). This review is intended to bring these more recent developments into a wider perspective.

2 The Mx Proteins

Sequence data, now available for at least eight different vertebrate species, reveal a considerable conservation among Mx proteins (Fig. 1) which allows us to depict their common structural features in a schematic way (Fig. 2). All Mx proteins share three sequence elements characteristic of conventional GTP-binding proteins. These proteins comprise a family of an estimated 50–100 members per genome and include proteins involved in signal transduction, intracellular protein trafficking, and polypeptide chain elongation (reviewed in BOURNE et al. 1990, 1991). Consistent with the presence of these elements, several Mx proteins have been found to display GTPase activity (Table 1). Surrounding these three elements, there are additional blocks of near perfect sequence conservation. One of these blocks overlaps the "dynamin family signature" (Figs 1–3) and has been implicated in the polymerization of mouse Mx1 protein in vitro (NAKAYAMA et al. 1993). At the NH$_2$-terminals, however, Mx sequences diverge considerably.

Compared with the sequence conservation in the regions of the GTP-binding elements, there is a greater sequence variability in the COOH-terminal portions of the molecules. Near their carboxyl ends, one finds two blocks, each containing a set of leucines (or rarely another hydrophobic amino acid) that are spaced seven amino acids apart and may assume the conformation of leucine zippers; these blocks have been shown to mediate oligomerization of a heterologous monomeric protein and may be involved in the oligomerization of Mx proteins as well (MELEN et al. 1992). Even though some alternatively spliced Mx mRNAs have been found, alternative splicing has not so far been shown to affect coding exons.

In their NH$_2$-terminal thirds, Mx proteins share sequence similarities with dynamins, proteins that in vertebrates and Drosophila are involved in endocytosis and come in various alternatively spliced forms. Mx proteins also share sequence similarities with two yeast proteins, VPS1 and MGM1. VPS1 is a cytoplasmic protein involved in exocytotic protein transport (ROTHMAN et al. 1990) and MGM1 is a protein necessary for the maintenance of the mitochondrial genome (JONES and FRANGMAN 1992). As shown in Fig. 3, the sequence similarities are again striking in the blocks surrounding the conserved GTP-binding elements; they exceed by far the extent of similarities with other GTPases such as the prototype GTPase p21H^{-ras} and justify grouping the Mx proteins, the dynamins, and the two yeast proteins together in an Mx or dynamin family of large GTPases. The

Fig. 1. Alignments of Mx protein sequences. The sequences were aligned using the Clustal V program (HIGGINS and SHARP 1988) through SeqApp (Don Gilbert, University of Washington). *Shaded areas,* mark identities in at least nine of the 12 aligned sequences. *Boxes* identify the three consensus elements of GTP-binding proteins, and the *horizontal bar* the conserved "dynamin family signature" from the Prosite Database (PS00410). The database accession numbers are from Swissprot: huMxA (P20591); huMxB (P20592); oaMx (P33237); ssMx1 (P27594); muMx1 (P09922); rnMx1 (P18588); rnMx2 (P18589); rnMx3 (P18590); apMx (P33238); pfMx (301 amino acid fragment, P20593), and from GenBank: muMx2 (J03368) (in this sequence the supernumerary C1366 was removed to make a 655 amino acid open reading frame: the corresponding protein has been tested for antiviral activity in tissue culture, see ZÜRCHER et al. 1992b); gdMx (Z23168)

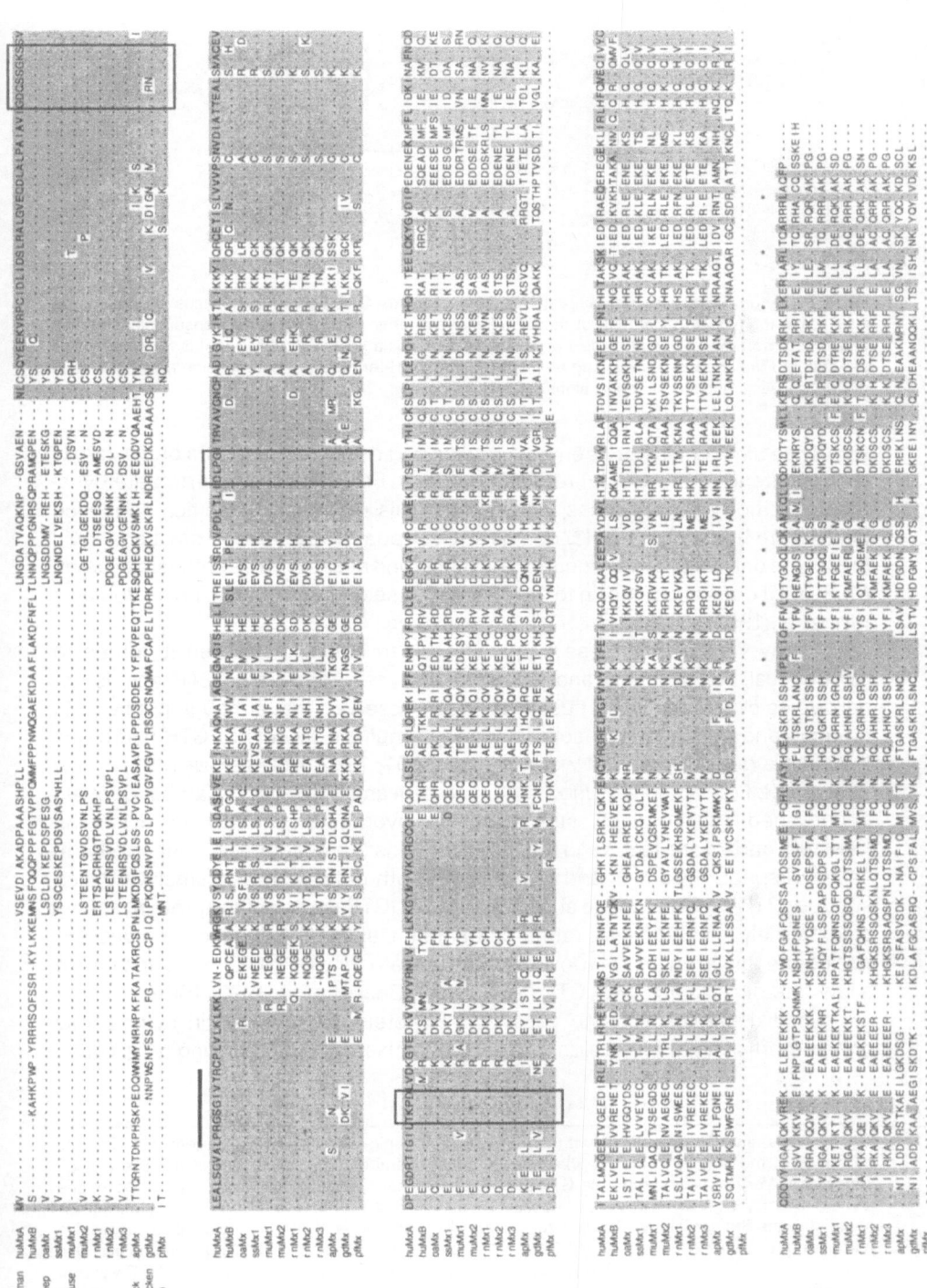

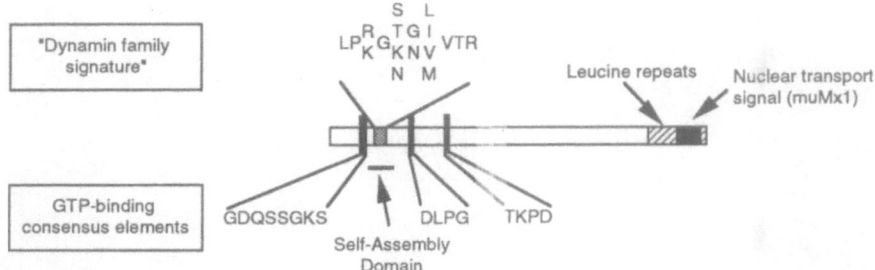

Fig. 2. Structural features of Mx proteins. The tripartite GTP-binding consensus elements are localized in the NH₂-terminal third of the proteins. A "dynamin family signature" consensus sequence that is common to all Mx and Mx-like proteins and overlaps a "self-assembly" motif is found between the first and the second GTP-binding elements. The COOH-terminal contains leucine repeats, and, in muMx1, a basic nuclear transport signal

sequence similarities between Mx proteins and the other members of this family do not extend to the COOH-terminal portions. For instance, in vertebrate dynamins, but not in Mx proteins, one finds in this region a proline-rich sequence binding Src homology 3 (SH3) domains (Gout et al. 1993; Herskovits et al. 1993). In Mx, the only proline-rich sequence is that found near the NH₂-terminal of human MxB, but it does not conform to the consensus sequence of SH3-binding domains (Ren et al. 1993).

It may well be for these sequence dissimilarities that when it comes to biochemical, cell biological and functional analyses, these large GTPases do not behave as close relatives. Thus, it has been observed that although purified Mx proteins and dynamins are able to undergo multiple rounds of GTP hydrolysis (hydrolysis rates >3 min⁻¹) (Nakayama et al. 1991, 1992; Horisberger 1992; Pitossi et al. 1993; Melen and Julkunen 1994; Shpetner and Valle 1992; Kambadur et al., unpublished), microtubles stimulate this hydrolysis potently in dynamins (Shpetner and Vallee 1992) and not in Mx proteins (Kambadur et al., unpublished). In this respect, dynamins and Mx proteins both differ from the small GTPases such as p21H⁻ʳᵃˢ, which in the absence of a GAP (GTPase activating protein) shows an extremely low hydrolysis rate (~0.008 min⁻¹), and in the absence of a guanine nucleotide releasing factor, a low dissociation rate for GDP (0.02 min⁻¹) and hence a low rate of reloading with GTP (reviewed in Bourne et al. 1990, 1991).

Since dynamins and the yeast VPS1 protein are involved in cytoplasmic protein trafficking, their localization in the cytoplasm is not surprising (Vater et al.

Fig. 3. Alignment of mouse Mx1 with dynamin and dynamin-like proteins. Alignment was done as described in the legend to Fig. 1. *Shaded areas* mark identities in at least four of the five aligned sequences. *Vertical boxes* mark the GTP-binding consensus elements; the *horizontal bar* marks the "dynamin family signature" (Prosite Database PS00410), and the *horizontal box* the region corresponding to the Src homology 3-binding consensus element (Ren et al. 1993). The database accession numbers are: muMx1 (Swissprot P09922), mouse Dyn1 (GenBank L31397); Drosophila Dyn (Swissprot p27619); C. Elegans Dyn (GenBank L29031); VPS1 (Swissprot P21576); MGM1 (Swissprot P32266).

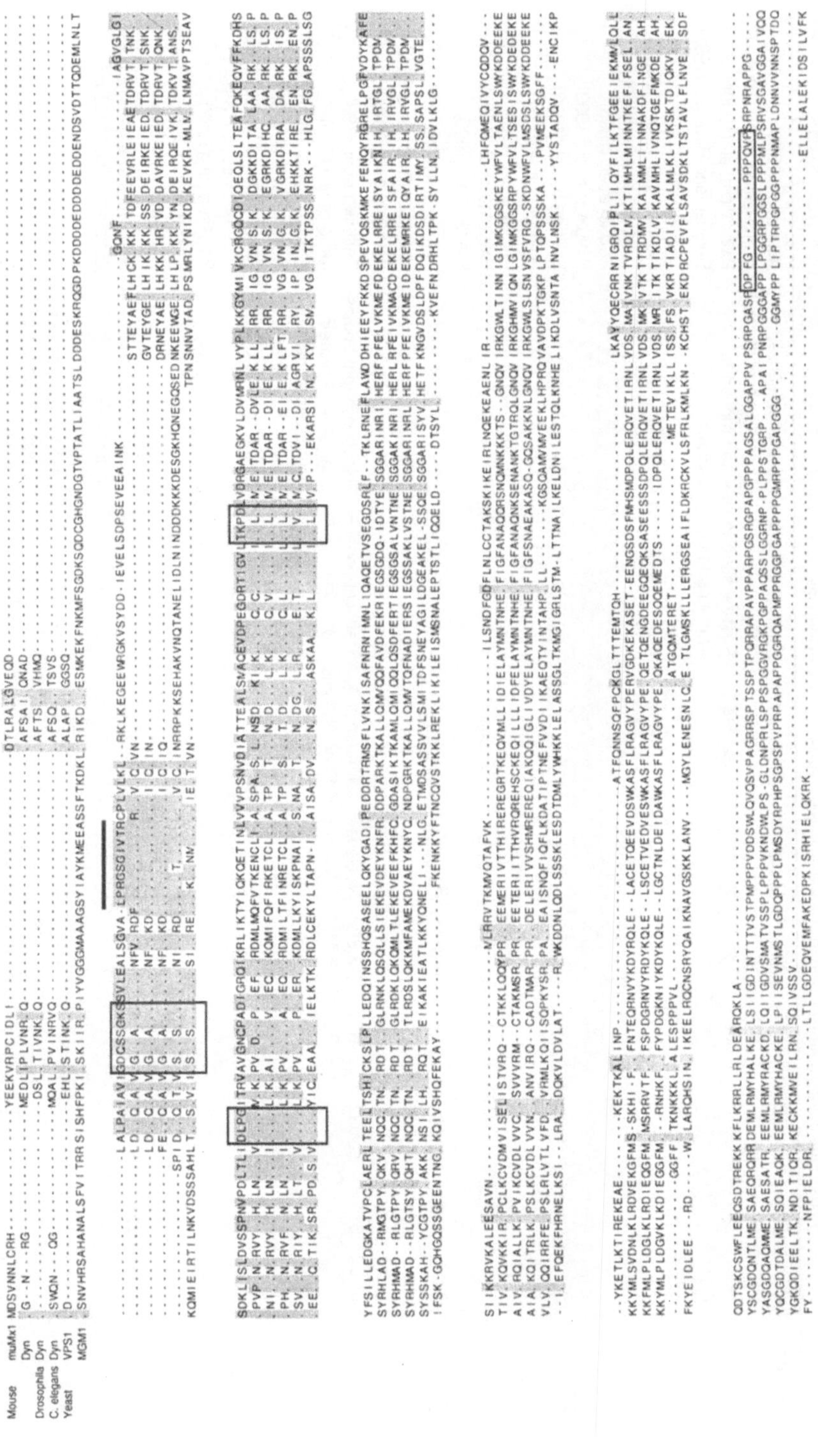
Mouse muMx1
Dyn
Drosophila Dyn
C. elegans Dyn
Yeast VPS1
MGM1

Table 1. Characteristics of natural Mx proteins

Species	Name[a]	Amino acids	M_r (calculated)	GTPase activity		Subcellular localization	Antiviral activity in cell culture		Remarks	References
				k_m (μM)	V_{max} (min⁻¹)		Virus	Mechanism		
Human	huMxA	661	75 403	100[b]; 62[c]; 768[d]	70[b]; 3.1[c]; 35[d]	Cytoplasm	Influenza A; Thogoto; measles; VSV	Posttranscriptional; n.d; Posttranscriptional; Transcriptional	Inhibits influenza A virus transcriptionally when moved to nucleus	AEBI et al. (1989); HORISBERGER et al. (1990); PAVLOVIC et al. (1990); STAEHELI and PAVLOVIC (1991); PAVLOVIC et al. (1992); HORISBERGER et al. (1992); ZÜRCHER et al. (1992a); PITOSSI et al. (1993); PAVLOVIC et al. (1993); SCHNORR et al. (1993); HALLER et al. (1993); MELEN et al. (1994); PAVLOVIC et al. (1995); FRESE et al. (1995)
	huMxB	715	82 088	n.d	n.d	Cytoplasm	(inactive)	–		AEBI et al. (1989); HORISBERGER et al. (1990); PAVLOVIC et al. (1990)
Mouse	muMx1	631	72 037	667[e]; 65[c]; 986[d]	13.8[e]; 7.1[c]; 47[d]	Nucleus	Influenza A; Thogoto virus	Transcriptional; n.d	Injection of antibodies against Mx1 protein neutralizes antiviral activity in interferon-treated Mx1+ cells	DREIDING et al. (1985); KRUG et al. (1985); STAEHELI et al. (1986); NOTEBORN et al. (1987); ARNHEITER and HALLER (1988); NAKAYAMA et al. (1991); PITOSSI et al. (1993); MELEN et al. (1994); HALLER et al. (1995)
	muMx2	423 (ORF1)	46 742	n.d	n.d	See remarks	(Inactive)	–	Truncated protein; inhibits VSV when supernumerary basepair is removed to allow synthesis of a	STAEHELI and SUTCLIFFE (1988); ZÜRCHER et al. (1992b)

Species	Protein	No. of amino acids[a]	Mr			Nucleus	Influenza A[g]	Thogoto	Comments	References
Rat	r nMx1	652	74 469	n.d	n.d	Nucleus	n.d	n.d	cytoplasmic protein of 655 amino acids	MEIER et al. (1990); HALLER et al. (1995)
	r nMx2	659	75 073	130[f]	5[f]	Cytoplasm	VSV		When moved to nucleus by NH_2-terminal nuclear transport signal, inhibits influenza A virus and loses anti-VSV activity	MEIER et al. (1990); JOHANNES et al. (1993); KAMBADUR et al. (unpublished)
	r nMx3	659	74 951	150[f]	5[f]	Cytoplasm	inactive	–	Inactive, even when moved to nucleus	MEIER et al. (1990); JOHANNES et al. (1993); KAMBADUR et al. (unpublished)
Sheep	oaMx	654	75 573	n.d	n.d	n.d	n.d			CHARLESTON and STEWART (1993)
Pig	ssMx	663	75 587	n.d	n.d	n.d	n.d			MÜLLER et al. (1992a)
Chicken	gdMx	705	79 493	n.d	n.d	Cytoplasm	Inactive	–		BERNASCONI et al. (1995)
Duck	apMx	721	81 953	n.d	n.d	Nucleus and cytoplasm	Inactive	–	Two allelic forms, both inactive in avian cells	BAZZIGHER et al. (1993)
Fish	pfMx	–	–	–	–	–	–	–	Partial DNA sequence	STAEHELI et al. (1989)

n.d., not determined; VSV, vesicular stomatitis virus; ORF, open reading frame.

[a] Sequences from the databases as indicated in Fig. 1.

[b] Natural protein (HORISBERGER 1992).

[c] Protein from recombinant baculovirus (MELEN et al. 1994).

[d] E. coli-derived (PAVLOVIC et al. 1993).

[e] E. coli-derived (NAKAYAMA et al. 1992).

[f] E. coli-derived (KAMBADUR et al., unpublished)

[g] A weak activity against VSV has been observed in cells expressing rat Mx1 transiently. This activity may be due to residual cytoplasmic rat Mx1 protein (MEIER et al. 1990).

1992; for dynamins, see Robinson et al. 1994). Many Mx proteins are also cytoplasmic, but some, including mouse Mx1, accumulate in the nuclei of cells in a characteristic punctate staining pattern (Dreiding et al. 1985; Meier et al. 1988). The hallmark of Mx proteins is antiviral activity, and it is implied (though it has never been formally tested) that dynamins, which are constitutive, are inactive against viruses, and that Mx proteins, which are interferon-induced, are inactive in vesicular protein trafficking. At first sight, the antiviral activities of Mx proteins would seem to follow a clear pattern: nuclear forms such as mouse or rat Mx1 inhabit viruses whose replication depends on a nuclear phase, namely influenza viruses (Staeheli et al. 1986; Noteborn et al. 1987; Meier et al. 1990) and their distant relatives, tick born influenza-like viruses (Haller et al. 1993, 1995; Haller and Thimme, unpublished). In contrast, cytoplasmic Mx proteins such as human MxA or rat Mx2 inhibit viruses whose replication takes place entirely in the cytoplasm, namely, vesicular stomatitis virus (VSV) (Meier et al. 1990; Staeheli and Pavlovic 1991) and, in the case of human MxA, measles virus (Schnorr et al. 1993). But that would be making things too simple: for instance, like nuclear Mx1, the cytoplasmic MxA protein also inhibits influenza virus, albeit at a posttranscriptional, presumably cytoplasmic step, in contrast to nuclear mouse Mx1 which inhibits a nuclear, transcriptional step (Pavlovic et al. 1992). When a mutant human MxA protein is moved into the nucleus by attachment of a nuclear transport signal, it still inhibits influenza virus, but now at a transcriptional step (Zürcher et al. 1992a), and when the same is done with rat Mx2, it now inhibits influenza virus and no longer VSV (Johannes et al. 1993). And then, some Mx proteins seem to be devoid of any antiviral activity (while still being active as GTPases) (Pavlovic et al. 1990; Meier et al. 1990; Bazzigher et al. 1993; Bernasconi et al. 1995; Kambadur et al. unpublished). For instance, the Mx proteins of both ducks and chickens have apparently no activity against influenza virus (Bazzigher et al. 1993; Bernasconi et al. submitted), even though ducks, in contrast to chickens, usually resist the lethal effects of influenza (for review, see Murphy and Webster 1990). In sum, there is no strict correlation between the intracellular localization of an Mx protein and the type of virus that is, or is not, inhibited.

Mx proteins have been subjected to extensive mutational analyses, summarized in Table 2. These studies have revealed that, for antiviral activity, an intact GTP-binding domain is required, though not sufficient. A proper subcellular localization is also critical, lest the protein may change its specificity, mechanism of action, or lose activity altogether. A particularly interesting mutation in human MxA, $E_{645} \rightarrow R$, has lost activity against VSV while retaining activity against influenza and Thogoto virus (Zürcher et al. 1992a; Haller et al. 1993; Frese et al. 1995). This observation may suggest a direct interaction between MxA and the different viruses (Staeheli et al. 1993), even though it does not exclude the possibility that different domains of MxA interact with distinct cellular auxiliary factors, one of them involved in inhibition of VSV, another in inhibition of influenza and Thogoto virus, similar to what has been discussed for rat Mx proteins (Johannes et al. 1993). Alterations in additional parts of the molecules (that is, various deletions and linker insertions) all have resulted in a loss of activity (Garber et al. 1993; Melen and

Table 2. Examples of mutational analyses of Mx proteins

Protein mutated	Domain affected	Specific mutation	Nature of mutation	GTPase activity[a]	intracellular localization	Antiviral activity	Remarks	References
muMx1	—	—	Wild type	Yes	Nucleus, punctate	Yes[b]		DREIDING et al. (1985) STAEHELI et al. (1986) NAKAYAMA et al. (1991)
muMx1	"P-loop"	Δ 23–95	Deletion, including "dynamin family signature" and self-assembly domain	n.d	Nucleus, diffuse	No[d]	Many different deletions also result in loss of antiviral activity	GARBER et al. (1993)
muMx1	"P-loop"	$K_{49} \rightarrow M$	Conserved lysine interacting with β-and γ-P of GTP	No	n.d	No[b]	Similar result with equivalent mutation in huMxA	PITOSSI et al. (1993)
muMx1	"P-loop"	$S_{47} \rightarrow C$	Nonconserved serine in "P-loop"	Yes	n.d	Yes[b]	Equivalent result with mutation in huMxA	PITOSSI et al. (1993)
muMx1	GTP-binding element 2	$D_{144} \rightarrow L$ $L_{145} \rightarrow E$	Consensus aspartate	No	Nucleus	Reduced[c]		MELEN and JULKUNEN (1994)
muMx1	GTP-binding element 3	$T_{213} \rightarrow L$ $K_{214} \rightarrow E$	Consensus threonine and lysine	No	Nucleus	Reduced[c]		MELEN and JULKUNEN (1994)
muMx1	Leucine repeats	$L_{612} \rightarrow K$	Introduction of charged amino acid	n.d	Nucleus, punctate	No[d]		GARBER et al. (1993)
muMx1	Leucine repeats	$L_{619} \rightarrow P$	Introduction of α helix-breaking proline	n.d	Nucleus, diffuse	No[d]		GARBER et al. (1993)

Table 2. (Contd.)

Protein mutated	Domain affected	Specific mutation	Nature of mutation	GTPase activity[a]	intracell localization	Antiviral activity	Remarks	Reference
muMx1	Leucine repeats	Δ531–602	Deletion of part of the leucine repeats	n.d	Nucleus	No[c]	Domain may serve as dimerization domain on heterologous proteins	MELEN and JULKUNEN (1994)
muMx1	Nuclear transport signal	Δ621–631	Partial deletion	n.d	Nucleus and cytoplasm	reduced[c]		NOTEBORN et al. (1987)
muMx1	Nuclear transport signal	Δ617–631	Deletion	n.d	Cytoplasm	No[d]		GARBER et al. (1993)
muMx1	Nuclear transport signal	$R_{614} \rightarrow E$	R in nuclear, E in cytoplasmic mammalian Mx	n.d	Cytoplasm	No[b]	Active against influenza virus when moved back to nucleus	ZÜRCHER et al. (1992c)
huMxA		$E_{645} \rightarrow R$	Analogous to R_{614} in muMx1	n.d	Cytoplasm	See remarks	No activity against VSV, but retains activity against influenza and Thogoto virus	ZÜRCHER et al.(1992a) HALLER et al. (1993) FRESE et al. (1995)

[a]"no", less than 10% of wild type.
[b]Tested in stably transfected cell lines.
[c]Tested in transiently transfected cells.
[d]Tested in recombinant retrovirus-infected cells.

JULKUNEN 1994). Thus, it seems unlikely that Mx proteins are simply organized in two functional domains, for instance an NH_2-terminal regulatory domain and a COOH-terminal effector domain, as suggested for VPS1 (VATER et al. 1992).

Despite the many biochemical and functional differences between Mx proteins and dynamins, recent evidence suggests a potential functional link between these molecules. In vitro, dynamin has been found to self-assemble into long helical structures with an outer diameter of ~50 nm and an inner diameter of ~30 nm (HINSHAW and SCHMID 1995). These structures seem to correspond, both in shape and dimension, to helical, dynamin-containing rings which were seen to form around the necks of tubular membrane invaginations induced by incubating nerve terminals with GTPγS (TAKEI et al. 1995). Interestingly, Mx1 can also self-assemble into C-shaped and ring-shaped polymers with similar dimensions (NAKAYAMA et al. 1993). Thus, it is tempting to speculate that at least some Mx proteins might inhibit viruses by spiraling around structures with a diameter of ~25 nm, such as nucleocapsids.

3 The Viruses Inhibited in Mx Transgenic Mice

Many viruses have been tested for susceptibility to Mx, both in vivo and in vitro. Those that were found to be inhibited by Mx do not have many properties in common, except that they are all enveloped negative-standed RNA viruses and that for none of them viral variants exist that overcome the Mx block. But in many respects, these viruses are strikingly dissimilar: they replicate in the cytoplasm or the nucleus, have a segmented or a non-segmented RNA genome, show RNA splicing or not, have overlapping open reading frames or not, have bicistronic mRNAs or not, employ RNA "editing" or not, and show little or no sequence similarities.

3.1 Influenza Virus

Influenza viruses exist in three varieties, A, B, and C, of which one, influenza A, is a major human pathogen. Natural isolates of these viruses are not readily pathogenic for mice; only upon serial passage do they become virulent for this species. Thus, for most work on Mx resistance, influenza viruses have been used whose history includes many mouse passages. Of particular interest were pneumotropic viruses that were derived from the human influenza A/PR/8/34 isolate, neurotropic viruses such as influenza A/NWS derived from the original human A/WS isolate, and a hepatotropic variant (TURH) (HALLER 1981) derived from the avian influenza A/TUR/England/63 virus that causes fowl plague. All passages were prepared in susceptible mice. Although some exceptional A2G-virulent influenza A variants exist (see below), they have not been arrived at by serial passage in

Mx1⁺ mice; in fact, serial passage of an A2G-avirulent influenza virus in A2G mice has never yielded an A2G-virulent one.

All influenza viruses have a single stranded RNA genome that is separated into eight (A and B) or seven (C) segments. It is this genome segmentation that allows for the generation of reassortants, that is, viruses that are composed of segments from different parental strains and that may be at the source of new influenza pandemics (for review, see WEBSTER et al. 1993). With influenza A virus, the segments encode three proteins that are part of the transcriptase complex, a nucleocapsid protein, two envelope glycoproteins, two membrane proteins, and two infected cell proteins. Functionally equivalent proteins are found with the other influenza viruses, although their number may vary. The genomic segments have conserved 5' and 3' terminals that are self-complementary and serve as promoter elements for mRNA and viron RNA synthesis. Viral mRNA synthesis is an unusual process that involves the cleavage of a subset of host nuclear RNAs by a virus-encoded cap-dependent endonuclease which generates 10–13 nucleotide-long capped primers that are needed initiation of transcription. Some viral mRNAs are spliced (the only known examples of splicing of RNAs not transcribed by RNA polymerase II), and some are bicistronic, yielding more than one protein product. However, additional coding strategies that viruses may use to increase their coding capacity such as non-AUG initiators, ribosomal frame-shifting, suppression of termination codons or RNA editing have not been found with influenza viruses. Mouse Mx1 inhibits influenza A and B viruses equally well in vivo, suggesting that the target of Mx is not one of the features that differ between these two viruses. Influenza C, however, has not so far been tested for susceptibility to Mx.

3.2 Tick-Born Influenza-Like Viruses

Among the many tick-born viruses, two have been recognized to be distantly related to orthomyxoviruses. One of them, Dhori virus (reviewed in NUTTALL et al. 1994), was found to be inhibited in its replication in A2G mice (Thimme and Haller, unpublished), but is not further considered here as it has not yet been tested for inhibition by Mx in transgenic mice. The other, Thogoto virus, was likewise found to be inhibited in A2G mice and has now been tested in transgenics (HALLER et al. 1993, 1995; PAVLOVIC et al. 1995). It was first isolated in Thogoto forest near Nairobi, Kenya (HAIG et al. 1965) and has since been found in a variety of ticks and their vertebrate hosts in central Africa, Egypt, Iran, Sicily and Portugal (reviewed in DAVIES et al. 1986). The virus replicates both in the tick and in the vertebrate and antibodies have been found in rats, buffaloes, camels, donkeys, cattle, sheep and humans. In sheep, infections may be associated with febrile illness and possibly abortions (DAVIES et al. 1984), an interesting observation in light of the fact that normal pregnant sheep express Mx mRNA in the endometrium (CHARLESTON and STEWART 1993). Thogoto virus infections in humans have also been reported but may be rare (MOORE et al. 1975).

The similarities of these tick-born viruses with influenza virus are based on several molecular characteristics. For instance, Thogoto virus comprises six segments of single-stranded, negative sense RNAs and, like influenza virus, is capable of producing reassortants in vivo (DAVIES et al. 1987). Two segments have been cloned: one encodes a protein with some low sequence similarity to the influenza PA protein, the other a glycoprotein with similarity to the baculovirus glycoprotein gp64 (reviewed in NUTTALL et al. 1990). The ends of the segments have partial sequence complementarities and resemble those of influenza viruses. Thogoto virus has a nuclear phase in its replication, but viral mRNA synthesis may not involve the stealing of host mRNA caps as is observed with influenza viruses.

3.3 Vesicular Stomatitis Virus

Vesicular Stomatitis Virus is a member of the rhabdoviridae and thus a relative of the rabies virus, an important human pathogen. VSV is of agricultural importance, primarily infecting horses. Its genome is a single strand of a negative sense RNA, 11 162 nucleotides in length (Indiana serotype), which is transcribed, through a stop/start mechanism, into five individual mRNAs. They encode the nucleoprotein N, the phosphoprotein NS, the matrix protein M, the glycoprotein G, and the RNA-dependent RNA polymerase L. The virus replicates entirely in the cytoplasm, and no RNA splicing occurs. Also, there is no RNA editing, no utilization of overlapping open reading frames or other mechanisms that could increase the coding capacity of the virus. While certain Mx proteins inhibit VSV, experiments with the related virus, rabies, have not so far been published.

Besides these viruses, measles virus glycoprotein synthesis has been found to be inhibited in a human monocytic cell line constitutively expressing human MxA protein (SCHNORR et al. 1993). Measles virus is a human pathogen, causing acute infections and rarely subacute sclerosing panencephalitis and inclusion body encephalitis. It employs editing of its P mRNA as an additional coding strategy, yielding a protein that shares its NH_2-terminal but not its COOH-terminal with the phosphoprotein encoded by the unedited mRNA. However, measles virus inhibition by human MxA protein has not so far been tested in transgenic mice.

4 Mx Transgenic Mice

Most inbred laboratory mice are $Mx1^-$, that is, they are homozygous for either one of two $Mx1^-$ alleles with the coding capacities for only the NH_2-terminal half of the full size protein. They are susceptible to influenza, consistent with the observation that deletions in the COOH-terminal half produce Mx1 proteins unable to protect fibroblasts against influenza virus in culture (GARBER et al. 1993; MELEN and

JULKUNEN 1994). Thus, any one of the standard mouse strains may be used as recipients for Mx cDNA constructs.

4.1 Inducible Mouse Mx1 Promoter

The initial goal was to test whether virus-resistant transgenic mice could be generated in which the need for interferon induction by the infecting virus or the experimenter would be circumvented and in which resistance would thus be reduced to its single most important element, the Mx1 protein. However, using the constitutive SV40 early enhancer/promoter or a metallothionein promoter in front of a mouse Mx1 cDNA, transgenic mice were obtained which expressed Mx1 protein at insufficient levels to render them resistant, or which did not express Mx1 protein at all (Arnheiter et al., unpublished; Meier et al., unpublished). Thus, it was possible that constitutive expression of this normally interferon-induced protein was deleterious, eliminating high-level expressors before birth. Therefore, other transgenics were generated (ARNHEITER et al. 1990) in which Mx1 expression was subjected to induction by the infecting virus. This was achieved by using as promoter the interferon-inducible mouse Mx1 promoter itself (HUG et al. 1988). Not surprisingly, this approach was successful. At least one line, line 979, expressed Mx1 protein at a high level (63% as compared to A2G mice when injected with interferon (ARNHEITER et al. 1990)). This line was bred to homozygosity at the transgene locus and individual transgenics and non-transgenic control mice were infected with graded doses of a pneumotropic, mouse-adapted human influenza virus (influenza A/PR/8/34). As shown in Table 3 and summarized in Table 4, all transgenics survived and almost all sero-converted while

Table 3. Resistance of transgenic mice of line 979 to pneumotropic influenza virus

Line[a]	Virus dose[b]	Number dead/ number infected[c]	Day of death[d]	HAI titer of survivors[e]
979	10–1	0/5	–	320, 80, 80, 80, 40
	10–2	0/5	–	320, 160, 160, 160, 160, 80
	10–3	0/5	–	640, 320, 320, 320, 40
	10–4	0/5	–	40, 40, 20, <20, <20
	10–5	0/4	–	40, 40, 20, <20
Control	10–1	5/5	4, 5, 5, 6, 6	–
	10–2	5/5	5, 5, 5, 6, 6	–
	10–3	5/5	6, 6, 6, 7, 7	–
	10–4	5/5	7, 7, 7, 8, 8	–
	10–5	4/4	8, 8, 10, 12	
	10–6	0/5	–	640, 160, 80, 40, 40

HAI, Hemagglutinin inhibition assay.
[a]Adult mouse Mx1 transgenics of line 979 (ARNHEITER et al. 1990), homozygous at the transgene locus, and control nontransgenic mice.
[b]Influenza A/PR/8, dilutions of stock virus (allantoic fluid), 50 μl intranasally.
[c]Indicates dead mice 21 days after infection.
[d]Deaths of individual infected mice.
[e]Sera of survivors collected 21 days after infection and tested for antibodies to influenza A/PR/8 by a hemagglutinin inhibition assay.

Table 4. Transgenic animals expressing Mx proteins

Species	Transgene Promoter	cDNA	Trans-genic founders obtained	Express-ing trans-gene[b]	Analyzed for resis-tance	Level of expres-sion[b]	Antiviral Resistance[a] Influenza A Pneumo-tropic[c]	Neuro-tropic[d]	Hepato-tropic[e]	Thogoto[f]	VSV[g]	Remarks	References
Mouse	muMx1[h]	muMx1	9	6	Line 979	63%	High	High	Low	High	n.d		ARNHEITER et al. (1990) and this paper; see Fig. 4 and 5 and Table 3)
					Line 964	11%	n.d	Medium	n.d	n.d	n.d	Paradox titration curve with neurotropic Influenza A	HALLER et al. (1995)
	muHMG[i]	muMx1	13	3	Line 1009 K(hemi-zygous)	3%	n.d	None	n.d	n.d	n.d		KOLB et al. (1992)
						n.d	n.d	Medium	n.d	n.d	n.d	Line retains influenza A resistance when treat-ed with antibodies to interferon αβ	
					KK(homo-zygous)	n.d	n.d	Medium	n.d	n.d	n.d		
	muAlb[k]	muMx1	9	>2	None	–	–	–	–	–	–	In some transgenic lines, expression liver-specific. However, postnatal expression is reduced over the course of 6 months	KOLB et al. (1992)
	muMx1[h]	huMxA	10	2	Line 58	Low	n.d	None	n.d	Medium	n.d.		PAVLOVIC et al. (1995 and unpublished results)
					Line 32	Low	n.d.	None	n.d	Low	n.d.		HALLER et al. (unpublished)
	muHMG[i]	huMxA	15	3	G(hemi-zygous)	Low	Low	Low	n.d	High	Low		HALLER et al. (1993) PAVLOVIC et al. (1995 and unpublished results) FRESE et al. (1995)
					L(homo-zygous)	Low	Low	Low	Low	High	Low		

Table 4. (Contd.)

Species	Transgene						Antiviral Resistance[a]					Remarks	References
							Influenza A						
	Promoter	cDNA	Trans- genic founders obtained	Express- ing trans gene[b]	Analyzed for resis- tance	Level of expres- sion[b]	Pneumo- tropic[c]	Neuro- tropic[d]	Hepato- tropic[e]	Thogoto[f]	VSV[g]		
Pig	huMTII$_A$[l]	muMx1	4	None	None	–	–	–	–	–	–		Muller et al. (1992b)
	SV40[m]	muMx1	1	None	None	–	–	–	–	–	–		Muller et al. (1992b)
	muMx1[n]	muMx1	8	None	None	–	–	–	–	–	–		Muller et al. (1992b)

[a] High: 50%–100% of mice hemizygous at the transgene locus survive, e.g., up to 10 000 LD_{50} for neurotropic influenza A; medium: 30%–70% of the mice survive, depending on virus dose; low: 0%–30% survive, resistance often only manifested as a delay in the time of death.

[b] Determined at protein level and compared to A2G mice after injection of interferon or interferon inducer.

[c] Influenza A/PR/8, at least 100 LD_{50} intranasally.

[d] Influenza A/NWS, at least 100 LD_{50} intracerebrally.

[e] Influenza A/TURH, at 500 LD_{50} intraperitoneally.

[f] At least 100 pfu intracerebrally or 10 pfu intraperitoneally.

[g] VSV, serotype Indiana, 300 pfu, intracerebrally.

[h] Mouse Mx1 promoter, 1.8 kb.

[i] Mouse 3-hydroxy-3-methylglutaryl coenzyme A reductase promoter, noncoding exon 1 and complete intron 1 (5.5 kb).

[k] Mouse albumin enhancer/promoter, 2.3 kb.

[l] Human metallothionein IIA promoter/enhancer, 836 bp.

[m] SV40 early enhancer/promoter, 400 bp.

[n] Mouse Mx1 promoter, 2.3 kb.

among the controls, only the ones infected with the highest virus dilution survived. Immunohistochemical analysis of the lungs of another set of paired transgenic and nontransgenic mice that were infected with 1000 LD$_{50}$ of this virus showed the following: Controls had wide-spread influenza virus-specific staining in their lungs (Fig. 4a), and no Mx1-specific staining (not shown). Transgenics had very restricted influenza staining in a few bronchioles (Fig. 4b,c), and showed expression of Mx1 protein in a few cells in the immediate vicinity of the infected cells (Fig. 4d,e). Similar observations were made earlier with a neurotropic influenza virus injected intracerebrally (ARNHEITER et al. 1990). These analyses explain by what principle the mice become resistant. There is evidently no need for a

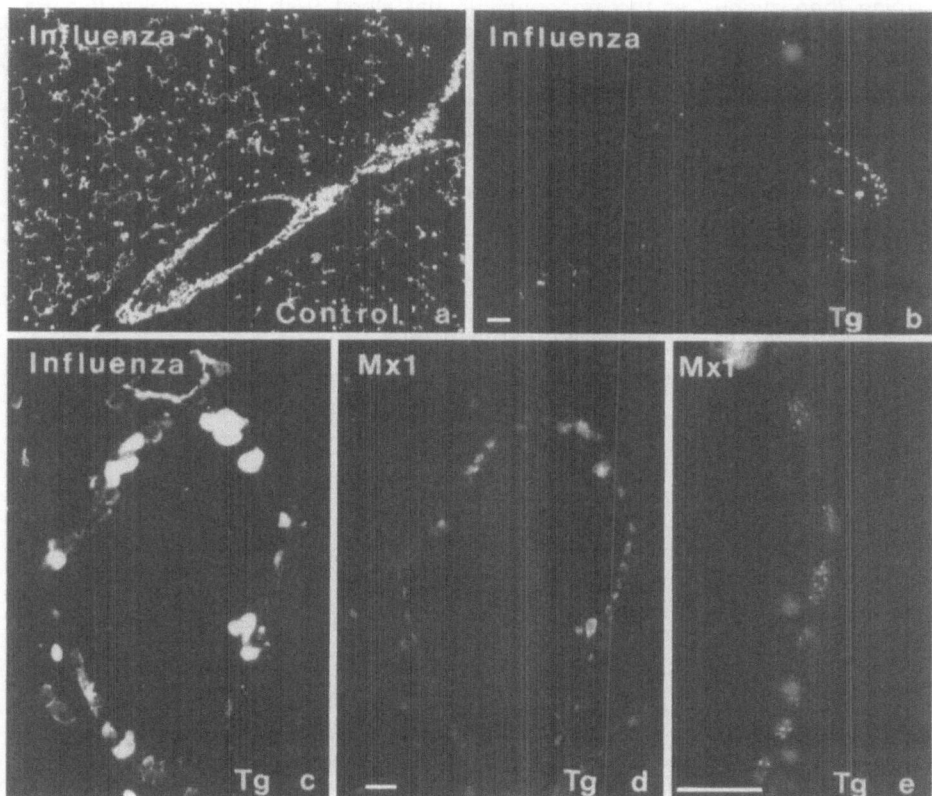

Fig. 4a–e. Induction of Mx1 protein in lungs of Mx1 transgenic mice by pneumotropic influenza virus. Adult line 979 mice, homozygous at the transgenic locus, and control mice were infected intranasally with 1000 LD$_{50}$ of a mouse-adapted influenza A/PR8 virus. Four days later, lungs were harvested and cryostat sections prepared and double labelled by indirect immunofluorescence with antibodies to influenza A virus and antibodies to Mx1 protein. **a** Widespread influenza virus immunofluorescence in susceptible control mice. **b** Restricted immunofluorescence in transgenic mice. **c** Higher magnification of influenza antigens present in transgenic mice. **d** Coresponding field showing Mx1 immunofluorescence. **e** Higher magnification of cells shown in **d** expressing the characteristically punctate, nuclear Mx1 protein. *Bars*: for **a** and **b**: 100 µm; for **c–e** 20 µm

complete block of virus replication, nor is there a need for massive Mx1 induction. What is important is a rapid expression of the antiviral protein Mx1 at precisely the anatomical sites where the virus initially replicates. The infected cells thus become demarcated by a barrier of protected cells, virus spread is blocked, and valuable time is gained for other defense mechanisms to complete the job of virus elimination and wound repair. One might imagine, then, that a strain of virus capable of growing exceptionally fast might outpace the induction of Mx1 and kill the host before there is time to build up protective levels of Mx1 protein. Such a fast growing virus is the hepatotropic, mouse-adapted influenza A virus TURH, which usually kills $Mx1^-$ mice in less than 72 h (HALLER 1975). Although avirulent for A2G mice, infection with this virus killed three of four infected line 979 transgenic mice (hemizygous at the transgene locus) albeit with some delay (Fig. 5a, Table 4). Thus, compared with A2G mice, the transgenics with their relatively fewer Mx1-positive cells did not cope well with this rapidly growing virus. This observation reminds us of earlier experiments with an even more virulent influenza A virus variant that was in fact as virulent for $Mx1^+$ mice as it was for $Mx1^-$ mice. The reason for this virus' virulence was not inability to induce interferon or insensitivity to the action of Mx1 protein; it was the fact that it grew more rapidly than other variants (HALLER 1981).

With the above observations in mind, it was interesting to determine what would be the minimal amount of Mx1 protein that might still lead to protection against less rapidly growing influenza viruses. As published earlier and summarized in Table 4, a line expressing at only 3% of the level of A2G mice was as susceptible to the neurotropic influenza A virus NWS as nontransgenic control mice (ARNHEITER et al. 1990). However, a line expressing at an intermediate level, 11%, showed a paradoxical titration curve: transgenic mice were protected after *high* dose

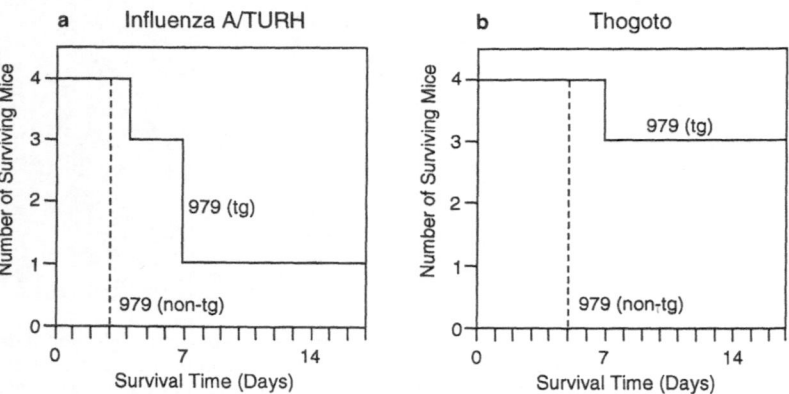

Fig. 5a,b. Resistance of transgenic mice expressing mouse Mx1 protein under the control of the mouse Mx1 promoter to hepatotropic influenza and Thogoto virus. **a** Four transgenic mice (hemizygous at the transgene locus) and four nontransgenic litter mates were infected intraperitoneally with 500 LD_{50} of influenza A/TURH (HALLER 1975), and deaths were checked on each day after infection. **b** Four transgenic mice and control litter mates were injected intraperitoneally with 1000 pfu of Thogoto virus, and deaths were checked on each day after infection

infection, but not after *low* dose infection. It was possible that this susceptibility to low dose infection was in fact due to the inability of the virus to induce Mx1 rapidly enough and in amounts high enough to protect the mice against the low virus dose. That this interpretation was correct was indirectly suggested by the following experiment. When low-dose infected animals, normally destined to die of the infection, were given a single shot of interferon along with the virus challenge, they survived, whereas low-dose infected non-transgenic litter mates succumbed despite such interferon treatment. This suggests that high dose infection (as is usually used for experimental work) was able to induce enough Mx1 on its own, enough even to protect against the particular virus dose. Similar observations have been made in other systems (cited in ARNHEITER et al. 1990; SMITH and DELUCA 1992). These observations need careful consideration, since natural infections usually occur with low amounts of virus that may not induce much interferon and hence may not immediately trigger early defense mechanisms such as Mx.

Recent experiments showed a high degree of resistance of line 979 transgenics to Thogoto virus (Fig. 5b, Table 4). Independent of pretreatment with an interferon inducer, poly(I)-poly(C), of ten mice infected intraperitoneally with 1000 plaque forming units of Thogoto virus, all survived, whereas of 15 control mice, all but one succumbed (HALLER et al. 1995).

Transgenic mice expressing human MxA under the same mouse Mx1 promoter (HUG et al. 1988) have also been generated. Since human MxA protein was shown to inhibit influenza virus, influenza-like tick born viruses, and VSV in cultured cells (see Table 1), such mice would be expected to display resistance to infection with either one of these viruses. Upon infection with a neurotropic influenza virus, however, the mice showed little more than marginal protection, manifested by a slight delay in the time point of death, and following Thogoto virus infection, only one of two lines was moderately protected (Table 4). This low level protection was likely due to a low level of MxA expression following infection or interferon treatment, and not to an intrinsic inability of MxA to inhibit these viruses in vivo, since constitutive expression of MxA rendered mice more readily resistant to challenge with these viruses (see below).

A similar construct as present in line 979 transgenic mice was used to generate transgenic pigs, but Mx1 protein remained undetectable as far as tested in the founder pigs and their transgenic offspring (MÜLLER et al. 1992a). The underlying reasons may be transgene rearrangement, as found in some of the pigs, and the possibility that the mouse Mx1 promoter is not optimally operating in pig cells.

4.2 Constitutive Promoters

Even though initial experiments to obtain transgenic mice constitutively expressing Mx1 failed, subsequent experiments using the murine 3-hydroxy-3-methylglutaryl coenzyme A reductase promoter (muHMG), the noncoding exon 1 and adjacent intron, were more successful (KOLB et al. 1992). Thus, transgenic mice were obtained that expressed mouse Mx1 constitutively in a few organs, notably the brain, and that were partially resistant to a neurotropic influenza virus. Most importantly, these mice retained their resistance when coinjected with antibodies to interferon at levels sufficient to render A2G mice more susceptible, thus proving that constitutive Mx expression can circumvent the need for rapid interferon induction. This experiment appears to reflect an earlier observation made with newborn *Mx1+* and *Mx1-* mice. Both newborn *Mx1+* and *Mx1-* mice are susceptible to influenza, but when pretreated with interferon, only *Mx1+* mice become resistant. This suggests that it is the Mx1 protein, and not an interferon-induced "general" antiviral state, that protects the interferon-treated newborn *Mx1+* mice (HALLER et al. 1981).

Using the above mentioned muHMG promoter/enchancer, constitutive expression of human MxA has also recently been achieved (HALLER et al. 1993; PAVLOVIC et al. 1995). Such transgenic mice are protected, albeit at a low level, against pneumotropic and neurotropic influenza virus and VSV, and at a high level against Thogoto virus (Table 4). These results are particularly interesting, as they demonstrate that the human MxA protein, well characterized in vitro, in fact also operates in vivo and precisely against those viruses that proved susceptible to MxA in cultured cells.

However, with respect to transgenic pigs, constitute expression of Mx1 protein has not yet been observed (MÜLLER et al. 1992b).

5 Problems with Mx Transgenic Mice

5.1 Constitutive Expression of Mx

As mentioned above, Mx proteins are not usually expressed in normal mice, except in a few cell types such as a fraction of peritoneal macrophages. Thus, it was not a priori evident that constitutive Mx1 expression would be tolerated in cells of many organs and beginning early during development. A first indication that such constitutive Mx1 expression might be tolerated in cells other than macrophages came from observations made in several lines of mice harboring the Mx1 cDNA under the control of the Mx1 promoter (ARNHEITER et al. 1990). These mice showed, for reasons not investigated further, constitutive expression in their skeletal muscles, not only of transgenic RNA but also of transgenic Mx1 protein. Interferon treatment of these animals raised the levels of expression slightly, whereas in A2G mice, it raised it from undetectable levels to moderate levels (Fig. 6). The muscles of the transgenic animals did not show alterations, neither anatomical nor functional, so that it was concluded that constitutive transgene expression in this cell type was tolerated. This observation contrasts in interesting ways with those made in different transgenics in which Mx1 was expressed under the control of the albumin promoter (KOLB et al. 1992). In these animals,

A) mRNA   B) Protein
                             - Interferon   + Interferon

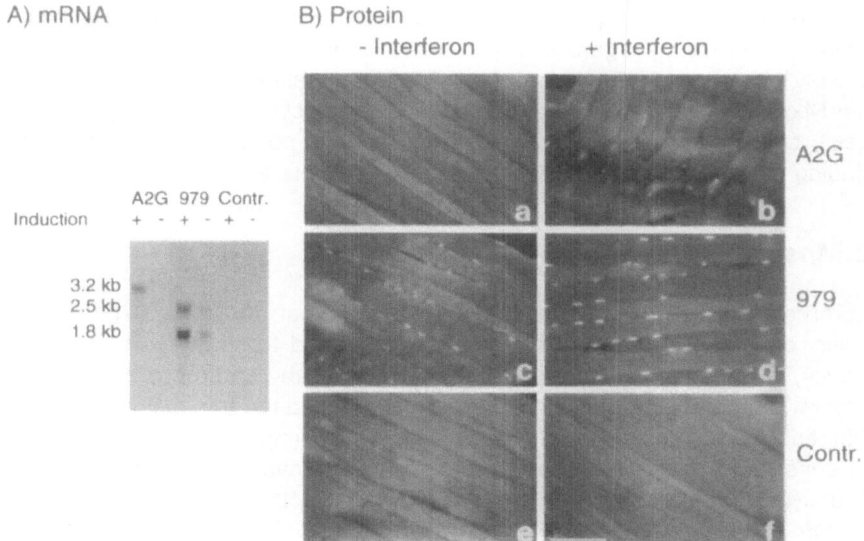

Fig. 6A,B. Constitutive expression of Mx1 in skeletal muscle of line 979 transgenics. **A** Northern blot analysis. Adult A2G mice, line 979 transgenic mice, or control litter mates were injected intraperitoneally with poly(I)-poly(C) (+) or were left untreated (–). After 6 hours, quadriceps mRNA was prepared and subjected to Northern analysis using a probe that detects sequences corresponding to exons 9, 10, and 11 of Mx1 which are present in A2G and Mx1 transgenic mRNA but absent in Mx1⁻ mRNA from control mice. In the transgenic mice, one mRNA of expected size (2.5 kb) and one of aberrant size (1.8 kb) are found. These mRNAs comigrate with those found in other organs of the transgenic mice (ARNHEITER et al. 1990). **B** Immunocytochemistry. Adult mice as indicated were either left untreated (–interferon, *a, c, e*) or injected intravenously with 10⁶ units of recombinant interferon (+interferon *b, d, f*). After 18 hours, quardriceps muscles were excised and cryostat sections prepared. The sections were stained by indirect immunofluorescence with antibodies to Mx1. Muscle cell nuclei are brightly fluorescent in line 979 muscle, independent of interferon pretreatment. *Bar:* 10 µm

Mx1 was expressed in hepatocytes early after birth, but this expression decreased as the mice became older. At face value, this latter phenomenon resembles an earlier observation made with transgenic mice expressing the urokinase-type plasminogen activator (uPA) under the control of the albumin promoter. Although most of the neonates harboring this construct died because of toxicity to the liver and a disruption of normal hemostasis, there were in two lines some animals which survived and in which plasma uPA levels gradually returned to normal. The reason for this phenomenon was somatic transgene rearrangement, accompanied by the abolishment of transgene expression, in a few hepatocytes which were able to proliferate and ultimately replace the entire liver. Consequently, the restored liver tissue showed all anatomical signs of regeneration (SANDGREN et al. 1991). Nothing similar was found in the albumin promoter-Mx1 transgenic mice, however, and expression of the transgene in these mice must have been reduced by a different mechanism that has not so far been elucidated. It is possible, of course, that reduction with age in liver expression of Mx1 is independent of the transgenic Mx1 protein but dependent on other properties intrinsic to the particular Mx1 transgene. However, it is interesting to note that other Mx transgenics prepared with constitutive promoters also show relatively low expression in hepatocytes. Thus, it is possible that Mx1 and perhaps other Mx proteins are not as well tolerated in hepatocytes when compared with muscle cells, macrophages, lung cells and several populations of brain cells, including neurons and migratory glial cells (GOUT and DUBOIS-DALCQ 1993).

5.2 Mosaic Expression

A common feature of many transgenic models is that transgene expression is mosaic, only seen in a subset of the cells expected to show expression. For instance, in transgenic mice expressing Mx1 under the control of the muHMG promoter, only a subset of cells in the brain expressed the transgene (KOLB et al. 1992), and the same was true for several lines harboring the interferon-inducible Mx1 constructs, even when they were treated with massive doses of interferon (10^6 units per 20 g mouse) (ARNHEITER et al. 1990). Despite this mosaicism of expression, transgenic mice were obtained that displayed resistance to viral infections, albeit never quite at the level of the naturally resistant A2G mice. Thus, if transgenic animals could be generated whose transgene expression was more uniform and less mosaic, the degree of antiviral resistance might be increased.

The mechanisms behind mosaic expression are not clear. One interpretation is that of position effect. The transgenes, usually integrated as tandem copies in a random localization in a chromosome, could become subject to influence by neighboring sequences, generally lowering expression levels, so that on average less cells show a detectable expression when compared to expression from the corresponding endogenous promoters. The possibility of transgene methylation as a reason for reduced transgene expression has been invoked as well, as has the local formation of heterochromatin due to the presence of repeat units (DORER and HENIKOFF 1994). In any event, it would seem preferable in future experiments to place transgenes as single copies under the transcriptional control of endogenous promoters, as is now possible by homologous recombination in embryonic stem cells. For interferon-inducible expression, for instance, the endogenous Mx1 promoter, either of *Mx1⁻* or *Mx1⁺* mice, could be exploited, since both the Mx1⁻ or *Mx1⁺* promoters are active in vivo (ARNHEITER et al. 1990) and since the Mx1 gene itself seems to be dispensable for normal life.

6 Concluding Remarks

The discovery of the Mx system, its genes and antiviral proteins, started with the chance observation of a strain of influenza-resistant mice by Lindenmann, an observation that gave Mx its firm biological foundation. Without this observation, Mx may have been discovered nonetheless, be it the result of screening cDNA libraries for interferon-induced sequences or of genomic libraries for sequences cross-hybridizing with dynamin probes, but the subsequent course of events would have been different. One might have found rapidly that in normal laboratory mice, *Mx1* alleles take the shape of "irrelevant" pseudogenes incapable of encoding full size *Mx1* proteins and that despite the absence of such full size proteins, mice are evidently healthy. Because of potential physiological roles of the still intact GTPase domain of the otherwise crippled Mx1⁻ protein, one might have gone on to generate by gene targeting true null mutants in *Mx1* (and for completion, also in *Mx2*), only to find, most likely, that not even now there was a phenotype. That, then, might have been the end of the story, unless an observation like that by Lindenmann would have helped us on the right path. One wonders, thus, how many genetic models, including gene-targeted mice, lie dormant because the right challenge has not been found to uncover their deficiencies. Yet, as the Mx model shows, even if the "right" challenge is discovered, it may still take its time until we can say with confidence that the newly found deficiency is biologically relevant, not only under the experimental conditions tested but for life in nature.

Mx transgenic mice have in many ways furthered our understanding and appreciation of the role of early host defense mechanisms in experimental viral pathogenesis. When influenza virus-susceptible mice were rescued with an Mx1 cDNA, all previous evidence that resistance associated with the *Mx1* locus was due to the single Mx1 protein became indisputable. Until then, the possibility, however remote, existed that protection of the living mouse could have been due to the action of proteins encoded by other genes situated in the vicinity of the *Mx1* gene or even within the gene itself. Likewise, when Thogoto virus was found to be inhibited in *Mx1⁺* congenic mice, it was theoretically possible that this novel influenza virus might have been inhibited by a different, novel protein. Such thoughts could be put aside, however, when the first Mx1 transgenic mice were shown to survive lethal doses of influenza viruses and more recently, Thogoto virus. The constitutive, interferon-independent expression of Mx1 in transgenic mice, in conjunction with the administration of anti-interferon antibodies, has also shown that the interferon-induced antiviral state can be dissected into its components in the intact, living organism, and not just in cultured cells. By crossing Mx transgenic mice expressing Mx proteins constitutively with mice whose type I interferon receptor gene has been rendered nonfunctional by gene targeting (MÜLLER et al. 1994), one can now proceed to test the efficacy of Mx proteins in vivo in the complete absence of a functional type I interferon system.

Transgenic mice have also been instrumental in the demonstration that the human MxA protein shows in vivo antiviral activities. This demonstration provides compelling evidence that MxA protein is potentially as powerful in humans as is Mx1 protein in mice. In fact, MxA is induced efficiently in humans after vaccination with live virus (ROERS et al. 1994). It remains to be demonstrated, though, that MxA in humans would be induced with similar efficiency during natural infections.

Deliberate expressions of antiviral molecules within cells has become known as "intracellular immunization" (BALTIMORE 1988). The approach has been particularly successful in higher plants in which virus-derived genes are used to express proteins capable of interfering with viral replication. Mx proteins, in contrast, are host-derived antiviral proteins whose corresponding genes are present in most vertebrates, including domestic animals. Thus, deliberate expression of Mx proteins might only lead to increased resistance if susceptibility of an animal to infection is in fact due to a comparably weak Mx system. This seems to be the case in chickens which have an Mx protein that is devoid of antiviral activity and in which avian influenza viruses cause devasting epidemics. Here, expression of an active Mx protein (for instance, an Mx protein from another species or a chicken Mx protein engineered to display higher antiviral activity) may yield chickens with increased resistance. If used widely, such transgenic chickens and other domestic animals with increased resistance might in turn help to control the emergence of new human pandemic strains of influenza virus that may originate in these animal reservoirs.

Mx transgenic mice have taught us in a clear way that a subtle balance exists between the infecting virus and the mobilization of early defense forces, and that the localized and timely mobilization of these defense forces is by far more important than the number of cells in which they are induced: a few Mx positive cells at the right place help to defy death or disease and are able to tip the balance in favor of health. Thus, a northern or western analysis performed with tissue extracts from an infected transgenic animal would show little if any signs of the very factor that is the cornerstone of resistance. Such is at times the economy of nature that what is produced in just a few cells may be of utmost importance.

Acknowledgments. The authors thank Drs. M. Dubois-Dalcq, L. Hudson and P. Staeheli for critical comments, Drs. K. Melen, I. Julkunen, D. Bernasconi and J. Pavlovic for sharing results prior to publication, and S. Skuntz for excellent technical assitance. Part of these studies were supported by grants to O.H. from the Deutsche Forschungs-Gemeinschaft and Land Baden-Württemberg.

References

Aebi M, Fäh J, Hurt N, Samuel CE, Thomis D, Bazzigher L, Pavlovic J, Haller O, Staeheli P (1989) cDNA structures and regulation of two interferon-induced human Mx proteins. Mol Cell Biol 9: 5062–5072

Arnheiter H, Haller O (1988) Antiviral state against influenza virus neutralized by microinjection of antibodies to interferon-induced Mx proteins. EMBO J 7: 1315–1320

Arnheiter H, Meier E (1990) Mx proteins: antiviral proteins by chance or by necessity? New Biol 2: 851–857

Arnheiter H, Skuntz S, Noteborn M, Chang S, Meier E (1990) Transgenic mice with intracellular immunity to influenza virus. Cell 62: 51–61

Baltimore D (1988) Intracellular immunization. Nature 335: 395–396

Bazzigher L, Schwartz A, Staeheli P (1993) No enhanced influenza resistance of murine and avian cells expressing cloned duck Mx protein. Virology 195: 100–112

Bernasconi D, Schultz U, Staeheli P (1995) The interferon-induced Mx protein of chickens lacks antiviral activity. J Interferon Cytokine Res 15: 47–53

Bourne HR, Sanders DA, McCormick F (1990) The GTPase superfamily: a conserved switch for diverse cell functions. Nature 348: 125–132

Bourne HR, Sanders DA, McCormick F (1991) The GTPase superfamily: conserved structure and molecular mechanism. Nature 349: 117–127

Bureau J-F, Montagutelli X, Bihl F, Lefebvre S, Guenet J-L, Brahic M (1993) Mapping loci influencing the persistence of Theiler's virus in the murin central nervous system. Nature Genet 5: 87–91

Charleston B, Stewart HJ (1993) An interferon-induced Mx protein: cDNA sequence and high-level expression in the endometrium of pregnant sheep. Gene 137: 327–331

Davies CR, Jones LD, Nuttall PA (1986) Experimental studies on the transmission cycle of Thogoto virus, a candidate orthomyxovirus, in *Rhipicephalus appendiculatus*. Am J Trop Hyg 35: 1256–1262

Davies CR, Jones LD, Green BM, Nuttall PA (1987) In vivo reassortment of Thogoto virus (a tick-born influenza-like virus) following oral infection of *Rhipicephalus appendiculatus*. J Gen Virol 68: 2331–2338

Daviess FG, Soi RK, Wairu BN (1984) Abortion in sheep caused by Thogoto virus. Vet Rec 115: 654

Dézélée S, Bras F, Contamine D, Lopez-Ferber M, Segretin D, Tenninges D (1989) Molecular analysis of *ref(2)P*, a *Drosophila* gene implicated in sigma rhabdovirus multiplication and necessary for male fertility. EMBO J 8: 3437–3446

Dorer D, Henikoff S (1994) Expansion of transgene repeats cause heterochromatin formation and gene silencing in drosophila. Cell 77: 993–1002

Dreiding P, Staeheli P, Haller O (1985) Interferon-induced protein Mx accumulates in nuclei of mouse cells expressing resistance to influenza viruses. Virology 140: 192–196

Dveksler GS, Pensiero MN, Cardellichio CB, Williams RK, Jiang G-S, Holmes KV, Dieffenbach CV (1991) Cloning of the mouse hepatitis virus receptor: expression in human and hamster cell lines confers susceptibility to MHV. J Virol 65: 6881–6891

Fraser RSS (1990) The genetics of resistance to plant viruses. Annu Rev Phytopathol 28: 179–200

Frese M, Kochs G, Meier-Dieter U, Siebler J, Staeheli P, Haller O (1995) Human MxA protein inhibits tick-born Thogoto but not Dhori virus. J Virol 69: 3904–3909

Garber EA, Hreniuk DL, Schiedel LM, van der Plog LHT (1993) Mutations in murine Mx1: effects on localization and antiviral activity. Vriology 194: 715–723

Gout O, Dubois-Dalcq M (1993) Directed migration of transplanted glial cells toward a spinal cord demyelinating lesion. Int J Dev Neurosci 11: 613–623

Gout I, Dhand R, Hiles ID, Fry MJ, Panayotou G, Das P, Truong O, Totty NF, Hsuan J, Booker GW, Campbell ID, Waterfield MD (1993) The GTPase dynamin binds to and is activated by a subset of SH3 domains. Cell 75: 25–36

Green MC (1989) Catalog of mutant genes and polymorphic loci. In: Lyon MF, Searle AG (eds) Genetic variants and strains of the laboratory mouse. Oxford University Press, New York

Haig DA, Woodall JP, Danskin D (1965) Thogoto virus: a hitherto undescribed agent isolated from ticks in Kenya. J Gen Microbiol 38: 389–394

Haller O (1975) A mouse hepatotropic variant of influenza virus. Arch Virol 49: 99–116

Haller O (1981) Inborn resistance of mice to orthomyxoviruses. Curr Top Microbiol Immunol 92: 25–52

Haller O, Arnheiter H, Gresser I, Lindenmann J (1981) Virus-specific interferon action. Protection of newborn Mx carriers against lethal infection with influenza virus. J Exp Med 154: 199–203

Haller O, Acklin M, Staeheli P (1986) Genetic resistance to influenza virus in wild mice. Curr Top Microbiol Immunol 127: 331–337

Haller O, Acklin M, Staheli P (1987) Influenza virus resistance of wild mice : wild type and mutant *Mx* alleles occur at comparable frequencies. J Interferon Res 7: 647–656

Haller O, Frese M, Kochs G, Arzet H, Hefti H, Pavlovic J (1993) Human MxA Protein protects transgenic mice from infection with Thogoto virus. J Interferon Res 13: S121

Haller O, Frese M, Rost D, Nuttall PA, Kochs G (1995) Tick-born Thogoto virus infection in mice is inhibited by the orthomyxovirus resistance gene product Mx1. J Virol 69: 2596–2607

Herskovits JS,Shpetner HS, Burges CC, Vallee RB (1993) Microtubules and src homology 3 domains stimulate the dynamin GTPase via its C-terminal domain. Proc Natl Acad Sci USA 90: 11468–11472

Higgins DG, Sharp PM (1988) CLUSTAL: a package for performing multiple sequence alignments on a microcomputer. Gene 73: 237–244

Hinshaw JE, Schmid SL (1995) Dynamin self-assembles into rings suggesting a mechanism for coated vesicle budding. Nature 374: 190–192

Horisberger MA (1992) Interfron-induced human protein MxA is a GTPase which binds transiently to cellular proteins. J Virol 66: 4705–4709

Horisberger MA, McMaster GK, Zeller H, Wathelet MG, Dellis J, Content J (1990) Cloning and sequences analysis of cDNAs for interferon- and virus-induced human Mx proteins reveal that they contain putative guanine nucleotide-binding sites: functional study of the corresponding gene promoter. J Virol 64: 1171–1181

Huang T, Pavlovic J, Staeheli P, Krystal M (1992) Overexpression of the influenza virus polymerase can titrate out inhibition by the murine Mx1 protein. J Virol 66: 4705–4709

Hug H, Costas M, Staeheli P, Aebi M, Weissmann C (1988) Organization of the murine Mx gene and characterization of its interferon- and virus-inducible promoter. Mol Cell Biol 8: 3065–3079

Johannes L, Arnheiter H, Meier E (1993) Switch in antiviral specificity of a GTPase upon translocation from the cytoplasm to the nucleus. J Virol 67: 1653–1657

Jones B, Fangman W (1992) Mitochondrial DNA maintenance in yeast requires a protein containing a region related to the GTP binding domain of dynamin. Genes Dev 6: 380–389

Kolb E, Laine E, Strehler D, Staeheli P (1992) Resistance to influenza virus infection of Mx transgenic mice expressing Mx protein under the control of two constutive promoters. J Virol: 1709–1716

Krug RM, Shaw M, Broni B, Shapiro G, Haller O (1985) Inhibition of influenza viral mRNA synthesis in cells expressing the interferon-induced Mx gene product. J Virol 56: 201–206

Landis H, Hefti HP, di Paolo C, Pavlovic J (1994) Mx1 inhibits influenza A virus mRNA synthesis at the level of elongation. Experientia 50: A50

Lindenmann J (1962) Resistance of mice to mouse adapted influenza A virus. Virology 16: 203–204

Meier E, Fäh J, Grob MS, End R, Staeheli P, Haller O (1988) A family of interferon-induced Mx-related mRNAs encodes cytoplasmic and nuclear proteins in rat cells. J Virol 62: 2386–2393

Meier E, Kunz G, Haller O, Arnheiter H (1990) Activity of rat Mx proteins against a rhabdovirus. J Virol 64: 6263–6269

Melen K, Julkunen I (1994) Mutational analysis of murine Mx1 protein. GTP binding core domain is essential for anti-influenza A activity. Virology 205: 269–279

Melen K, Ronni T, Lotta T, Julkunen I (1994) Enzymatic characterization of interferon-induced antiviral GTPases murine Mx1 and human MxA protein. J Biol Chem 269: 2009–2015

Melen K, Ronni T, Broni B, Krug RM, von Bonsdorff C-H, Julkunen I (1992) Interferon-induced Mx proteins form oligomers and contain a putative leucine zipper. J Biol Chem 267: 25898–25907

Moore DL, Causey OR, Carey DE, Reddy S, Cooke AR, Akinkugbe FM, David-West TS, Kemp GE (1975) Arthropod-born viral infections of man in Nigeria, 1964–1970. Ann Trop Med Parasitol 69: 49–64

Müller M, Winnaker E-L, Brem G (1992a) Molecular cloning of porcine Mx cDNAs: new members of a family of interferon-inducible proteins with homology to GTP-binding proteins. J Interferon Res 12: 119–129

Müller M, Brenig B, Winnacker E-L, Brem G (1992b) Transgenic pigs carrying cDNA copies encoding the murin Mx1 protein which confers resistance to influenza virus infection. Gene 121: 263–270

Müller U, Steinhoff U, Reis LFL, Hemmi S, Pavlovic J, Zinkernagel RM, Aguet M (1994) Functional role or type I and type II interferons in antiviral defense. Science 264: 1918–1921

Murphy BR, Webster RG (1990) Orthomyxoviruses. In: Fields BN, Knipe DM (eds) Fields Virology. Raven, New York, pp 1091–1152

Nakayama M, Nagata K, Kato A, Ishihama A (1991) Interferon-inducible mouse Mx1 protein that confers resistance to influenza virus is GTPase. J Biol Chem 266: 21404–21408

Nakayama M, Nagata K, Ishihama A (1992) Enzymatic properties of the mouse Mx1 protein-associated GTPase. Virus Res 22: 227–234

Nakayama M, Yazaki K, Kusano A, Nagata K, Hanai N, Ishihama A (1993) Structure of mouse Mx1 protein: molecular assembly and GTP-dependent conformational change. J Biol Chem 268: 15033–15038

Nédellec P, Dveksler GS, Daniels E, Turbide C, Chow B, Basil AA, Holmes KV, Beauchemin N (1994) Bgp2, a new member of the carcinoembryonic antigen-related gene family, encodes an alternative receptor for mouse hepatitis virus. J Virol 68: 4525–4537

Noteborn M, Arnheiter H, Richter-Mann L, Browning H, Weissmann C (1987) Transport of the murine Mx protein into the nucleus is dependent on a basic carboxy-terminal sequence. J Interferon Res 7: 657– 669

Nuttall PA, Booth TF, Carey D, Davies CR, Jones LD, Morse MA, Moss SR (1990) Biological and molecular characteristics of orbiviruses and orthomyxoviruses isolated from ticks. Arch Virol 1: 219–225 Suppl

Nuttall PA, Morse MA, Jones LD, Portela A (1995) Orthoacariviruses. In: Gibbs AJ, Calisher CM (eds) Molecular basis of viral evolution. Cambridge University Press, Cambridge (in press)

Pavlovic J, Zürcher T, Haller O, Staeheli P (1990) Resistance to influenza virus and vesicular stomatitis virus conferred by expression of human MxA protein. J Virol 64: 3370–3375

Pavlovic J, Haller O, Staeheli P (1992) Human and mouse Mx-proteins inhibit different steps of the influenza virus multiplication cycle. J Virol 66: 2564–2569

Pavlovic J, Schröder A, Blank A, Pitossi F, Staeheli P (1993) Mx proteins: GTPases involved in the interferon-induced antiviral state. Ciba Found Symp 176: 233–247

Pavlovic J, Arzet HA, Hefti H, Frese M, Rost D, Ernst B, Kolb E, Staeheli P, Haller O (1995) Enhanced virus resistance of transgenic mice expressing the human MxA protein. J Virol: 4506–4510

Pitossi F, Blank A, Schröder A, Schwarz A, Hussi P, Schwemmle M, Pavlovic J, Staeheli P (1993) A functional GTP-binding motif is necessary for antiviral activity of Mx proteins. J Virol 67: 6726–6732

Ren R, Mayer BJ, Cicchetti P, Baltimore D (1993) Identification of a ten-amino acid proline-rich SH3 binding site. Science 259: 1157–1161

Robinson PJ, Liu J-P, Powell KA, Fykse EM, Südhof TC (1994) Phosphorylation of dynamin I and synaptic-vesicle recycling. Trends Neurosci 17: 348–353

Roers A, Hochkeppel HK, Horisberger MA, Hovanessian A, Haller O (1994) MxA gene expression after live virus vaccination: a sensitive marker for endogenous type I interferon. J Infec Dis 169: 807–813

Rothman JH, Raymond CK, Gillbert T, O'Hara PJ, Stevens TH (1990) A putative GTP binding protein homologous to interferon-inducible Mx proteins performs an essential function in yeast protein sorting. Cell 61: 1063–1074

Sandgren EP, Palmiter RD, Heckel JL, Daugherty CC, Brinster RL, Degen JL (1991) Complete hepatic regeneration after somatic deletion of an albumin-plasminogen activator transgene. Cell 66: 245–256

Schnorr J-J, Schneider-Schaulies S, Simon-Jödicke A, Pavlovic J, Horisberger MA, ter Meulen V (1993) MxA-dependent inhibition of measles virus glycoprotein synthesis in a stably transfected human monocytic cell line. J Virol 67: 4760–4768

Shpetner HS, Vallee RB (1992) Dynamin is a GTPase stimulated to high levels of activity by microtubules. Nature 355: 733–735

Smith CA, DeLuca NA (1992) Transdominant inhibition of herpes simplex virus growth in transgenic mice. Virology 191: 581–588

Staeheli P, Pavlovic J (1991) Inhibition of vesicular stomatitis virus mRNA synthesis by human MxA protein. J Virol 65: 4498–4501

Staeheli P, Sutcliffe JG (1988) Identification of a second interferon-regulated murine Mx gene. Mol Cell Biol 8: 4524–4528

Staeheli P, Haller O, Boll W, Lindenmann J, Weissmann C (1986) Mx protein: constitutive expression in 3T3 cells transformed with cloned Mx cDNA confers selective resistance to influenza virus. Cell 44: 147–158

Staeheli P, Yu Y-X, Grob R, Haller O (1989) A double-stranded RNA inducible fish gene homologous to the murine influenza virus resistance gene Mx. Mol Cell Biol 9: 3117–3121

Staeheli P, Pitossi F, Pavlovic J (1993) Mx proteins: GTPases with antiviral activity. Trends Cell Biol 3: 268–272

Stranden AM, Staeheli P, Pavlovic J (1993) Function of the mouse Mx1 protein is inhibited by overexpression of the PB2 protein of influenza virus. Virology 197: 642–651

Takei K, McPherson PS, De Camilli P (1995) Tubular membrane invaginations coated by dynamin rings are induced by GTP-γS in nerve terminals. Nature 374: 186–189

Vater CA, Raymond CK, Ekena K, Howald-Stevenson I, Stevens TH (1992) The VPS1 protein, a homolog of dynamin required for vaculor protein sorting in Saccharomyces cerevisiae, is a GTPase with two functionally separable domains. J Cell Biol 119: 773–86

Webster RG, Wright SM, Castrucci MR, Bean WJ, Kawaoka Y (1993) Influenza – a model of an emerging virus disease. Intervirology 35: 16–25

Yokomori K, Lai MMC (1992) The receptor for mouse hepatitis virus in the resistant mouse strain SJL is functional: implications for the requirement for a second factor for viral infection. J Virol 66: 6931–6938

Zürcher T, Pavlovic J, Staeheli P (1992a) Mechanism of human MxA protein action – variants with changed antiviral properties. EMBO J 11: 1657–1661

Zürcher T, Pavlovic J, Staeheli P (1992b) Mouse Mx2 protein inhibits vesicular stomatitis virus but not influenza virus. Virology 187: 796–800

Zürcher T, Pavlovic J, Staeheli P (1992c) Nuclear localization of Mx protein is necessary for inhibition of influenza virus. J Virol 66: 5059–5066

Hepatitis B Virus Transgenic Mice: Models of Viral Immunobiology and Pathogenesis

F.V. Chisari

Department of Molecular and Experimental Medicine, The Scripps Research Institute, 10666 North Torrey Pines Road, La Jolla, CA 92037, USA

1 Introduction

The hepatitis B virus (HBV) is a noncytopathic, enveloped virus with a circular, double-stranded DNA genome that causes acute and chronic necroinflammatory liver disease and hepatocellular carcinoma (reviewed in CHISARI and FERRARI 1994). HBV infection acquired in adult life is often clinically inapparent and the vast majority of acutely infected adults recover completely from the disease and clear the virus. Rarely, however, the acute liver disease may be so severe that the patient can die from fulminant hepatitis. Approximately 5%–10% of acutely infected adults become persistently infected by the virus and develop chronic liver disease of varying severity. Neonatally transmitted HBV infection, however, is rarely cleared, and over 90% of such children become chronically infected. Because HBV is commonly spread from infected mother to newborn infant in highly populated areas of Africa and Asia, several hundred million people throughout the world are persistently infected by HBV for most of their lives and suffer varying degrees of chronic liver disease which greatly increases their risk of developing cirrhosis and hepatocellular carcinoma (HCC). Indeed, the risk of HCC is increased 100-fold in patients with chronic hepatitis, and the lifetime risk of HCC in males infected at birth approaches 40% (BEASLEY et al. 1981). Accordingly, a large fraction of the world's population suffers and dies from these late complications of HBV infection.

It is generally thought that HBV is not directly cytopathic for the hepatocyte but that the associated liver diseases are caused by the immune response to viral antigens expressed by infected hepatocytes. HBV induced hepatocarcinogenesis has been ascribed to a variety of causes including: (a) the insertional deregulation of cellular growth control genes by the integrated viral DNA sequences that are commonly found in HCC; (b) the random deregulation of cellular growth control genes by chromosomal and genetic lesions caused by increased hepatocellular turnover and the mutagenic environment of the inflamed liver; (c) the transcriptional deregulation of cellular growth control genes by the viral transactivating protein (the X protein) that may itself be deregulated secondary to structural changes in the viral genome that occur during integration.

It has been difficult to examine these pathogenetic mechanisms in great detail however, because of the limited host range of this and by the lack of in vitro culture systems to propagate it. For this reason, until recently most studies of HBV pathogenesis were limited to the analysis of HBV infected patients and chimpanzees or of HBV-related hepadnavirus infections in the woodchuck, ground squirrel and Pekin duck, outbred species whose immune systems have not been characterized and have limited value for analysis of the antiviral immune response.

While a great deal has been learned from the those studies, definitive analysis of the immunological mechanisms involved in HBV pathogenesis required the development of an inbred animal model with a well defined immune system, i.e., the HBV transgenic mouse. In the course of those studies many other previously unknown aspects of HBV biology and pathogenesis have been elucidated because of the unique power of the transgenic mouse system to express each of the viral genes and the entire viral genome in its most physiological environment, the primary hepatocyte in vivo.

In the remainder of this chapter we will review the salient characteristics of the various HBV transgenic mouse models produced thus far, with an emphasis on those aspects of HBV biology and pathogenesis that they have played an especially key role in elucidating. Before describing these models, however, it is necessary to briefly review some of the features of the HBV genome and life cycle, in order to place the transgenic mouse models in perspective.

2 The Hepatitis B Virus Genome and Life Cycle

As previously reviewed (CHISARI 1991), the HBV virion contains a circular, partially double-stranded, DNA genome approximately 3200 base pairs in length. The long (minus) strand of the viral DNA encodes a greater than genome length 3.5 kb pregenomic RNA which is reverse transcribed as an early step in viral replication. The minus strand also encodes 3.5, 2.4, 2.1 and 0.7 kb mRNA species that are translated into the structural (envelope, nucleocapsid) and nonstructural (polymerase, transactivating protein) proteins of the virus. Because the experimental systems and concepts that have led to our current understanding of HBV pathogenesis touch on many aspects of the structure and function of each of these gene products, we will provide a brief review of their characteristics and the role they play in the life cycle of this virus.

2.1 The Hepatitis B Virus Proteins

The nucleocapsid open reading frame contains two in-phase start codons that define two overlapping polypeptides the shorter of which (core, HBcAg) is a cytoplasmic and nuclear protein which self-assembles to form the viral nucleocapsid. The COOH-terminal region of the core protein includes an arginine-rich domain that contains nuclear localization sequences (YEH et al. 1990; ECKHARDT et al. 1991) and an overlapping sequence required for encapsidation of the 3.5 kb viral RNA pregenome (NASSAL 1992). Core particles are assembled from core protein dimer precursors in a spontaneous, concentration-dependent process that can occur in the absence of other viral components (ZHOU and STANDRING 1991, 1992; SEIFER et al. 1993; ZHOU et al. 1992). It is thought that core particles transport the viral replication complex to the nucleus for amplification and that they also associate with viral envelope at the endoplasmic reticulum membrane for virion assembly and export (see below).

A longer counterpart (precore) of the core protein is translocated, via a signal peptide at its extreme NH_2-terminal (STANDRING et al. 1988), into the endoplasmic reticulum from which it is secreted as HBeAg following truncation of NH$_2$- and COOH-terminal residues, although traces of this protein have been also identified in cytoplasmic, nuclear and membrane compartments (Ou et al. 1989; SCHLICHT and SCHALLER 1989). While the presence of HBeAg in the serum of infected patients is a good serological marker of viral replication, the function of the precore protein in the viral life cycle is currently undefined and it is clearly not required for viral replication (CHEN et al. 1992a).

The envelope open reading frame contains three in-phase translation start codons which define the NH$_2$-terminals of three overlapping polypeptides, the expression of which is transcriptionally regulated. The relative abundance of these three proteins plays an important role in viral particle morphogenesis. The major and middle envelope polypeptides assemble into small, 22 nm spherical particles by budding into the lumen of the endoplasmic reticulum, and they are rapidly secreted after passage and glycosylation in the Golgi apparatus (PATZER et al. 1984, 1986; EBLE et al. 1986, 1987). The large envelope polypeptide appears to be an essential component of the infectious virion (Dane particle) and may play an important structural role in complete virus particle assembly. The large envelope polypeptide also exerts an important structural influence on the formation of the abundant, noninfectious, subviral particles that are characteristic of HBV infection (CHISARI et al. 1986; STANDRING et al. 1986; McLACHLAN et al. 1987; OU and RUTTER 1987; MOLNAR-KIMBER et al. 1988). Depending on the relative molar ratio of the large envelope polypeptide to the major and middle envelope polypeptides within a given cell, either short secretable filaments are formed or long, nonsecretable branching filaments are produced that accumulate within the endoplasmic reticulum and lead to the development of "ground glass" hepatocytes and may ultimately contribute to the death of the cell (CHISARI et al. 1987; GERBER et al. 1974a,b).

The X open reading frame encodes a transcriptional transactivating protein that positively regulates transcription from HBV and other viral, and cellular, promoters (TWU and SCHLOEMER 1987; SPANDAU and LEE 1988; SETO et al. 1988; TWU et al. 1989; COLGROVE et al. 1989) in vitro. While X does not appear to be required for virus replication in transformed cell lines (BLUM et al. 1992), it has recently been shown that X is important for the establishment woodchuck hepatitis virus of (WHV) infection in woodchucks (CHEN et al. 1994; ZOULIM et al. 1994). Current data suggest that HBx is not a DNA binding protein; instead, it complexes with cellular transcription factors and modifies their ability to bind to regulatory elements, thereby influencing the transcription process in a general fashion rather than by directly binding DNA target sequences itself (COLGROVE et al. 1989; MAGUIRE et al. 1991; SIDDIQUI et al. 1989; UNGER and SHAUL 1990). Furthermore, it has been recently shown that the X protein complexes with the p53 tumor suppressor protein and inhibits its sequence-specific DNA binding capacity, transcriptional activation function and, perhaps, its indirect DNA repair function in vitro (WANG et al. 1994). Because of these properties, it has been proposed that dysregulated expression of the X gene product may be involved in hepatocarcinogenesis in chronic HBV infection (WOLLERSHEIM et al. 1988). Importantly, it has recently been shown that high level expression of the X gene product is associated with the development of HCC in the absence of any associated liver disease in some lineages of transgenic mice (KOIKE et al. 1994).

The polymerase open reading frame encodes the viral polymerase protein which is translated by internal ribosomal initiation of the 3.5 kb core mRNA (KAWAMOTO et al. 1990; JEAN-JEAN et al. 1989). The polymerase protein contains DNA-dependent DNA polymerase, reverse transcriptase and RNAse H domains as well as a 5' DNA binding protein which serves as a primer for reverse transcription of the viral pregenome (MACK et al. 1988; BOSCH et al. 1988; BAVAND and LAUB 1988; BAVAND et al. 1989; SCHLICHT et al. 1989; WANG and SEEGER 1992). The polymerase gene products also play essential roles in encapsidation and replication of the viral genome (HIRSCH et al. 1990; BARTENSCHLAGER et al. 1990; LAVINE and HIRSCH 1989) by binding to a hairpin structure at the 5' end of the 3.5 kb pregenomic RNA which serves both as a packaging signal (HIRSCH et al. 1990; JUNKER-NIEPMANN et al. 1990) and as the template for initiation of reverse transcription of the viral pregenome (WANG and SEEGER 1993).

2.2 The Viral Life Cycle

As previously reviewed (CHISARI et al. 1989), the mechanism of viral entry into the hepatocyte is not known although there is evidence that attachment may be mediated by the interaction of a domain in the preS (1) region of the large envelope polypeptide (NEURATH et al. 1986) with one or more structures on the hepatocyte membrane. Several candidate receptors for HBV have been suggested, including the polyalbumin receptor (MACHIDA et al. 1983), the transferrin receptor (FRANCO et al. 1992), the interleukin-6 receptor (NEURATH et al. 1992), endonexin II (HERTOGS et al. 1993) and apolipoprotein H (MEHDI et al. 1994), among others. At present, however, the molecular basis for viral attachment and entry is not known for HBV or for any of the other hepatitis viruses.

The replication of HBV is unique among DNA viruses in that it involves reverse transcription of an RNA pregenome (SUMMERS and MASON 1982). Based on this seminal observation and subsequent studies from several laboratories (SUMMERS and MASON 1982; MASON et al. 1982; ROSENTHAL et al. 1983; MILLER et al. 1984a,b; MOLNAR-KIMBER et al. 1984; SEEGER et al. 1986; LIEN et al. 1986; TUTTLEMAN et al. 1986; WILL et al. 1987) the following model of the HBV life cycle has emerged (ROBINSON et al. 1987).

Following entry and presumptive uncoating, viral plus strand DNA synthesis is completed within the nucleocapsid particle which delivers the open circular viral genome to the nucleus via nuclear localization signals located at the COOH-terminal of HBcAg (see above). Recent data from a transgenic mouse model suggest that nucleocapsid particles do not enter the nucleus (see below). This suggests that capsid disassembly probably occurs on the cytoplasmic face of the nuclear membrane and that the viral genome is released into the nucleus where DNA repair enzymes process the viral minus and plus strands to produce the covalently closed circular DNA molecule that serves as the template for transcription of the viral pregenomic and messenger RNAs (SUMMERS and MASON 1982).

The RNA pregenome is transported into the cytoplasm where it becomes incorporated into a nucleocapsid particle by interacting with core and polymerase proteins that have been translated from their respective mRNAs. Within these nucleocapsid particles new DNA minus strands are synthesized by reverse transcription of the pregenomic RNA. Newly formed DNA minus strands serve as the template for DNA plus strand synthesis, and plus strand elongation converts the linear DNA intermediates into a relaxed, circular, double-stranded molecule.

Some of these core particles transport the developing viral genome back to the nucleus in a process that effectively amplifies the HBV copy number in the cell (TUTTLEMAN et al. 1986). Other core particles associate with viral envelope proteins that exist as integral membrane proteins in the endoplasmic reticulum into which they bud and are ultimately secreted as infectious virions to initiate new rounds of infection in susceptible cells. After many such cycles (chronic infection) the viral DNA may integrate into the host genome. Integration may be associated with extensive rearrangement of viral and host flanking sequences with attendant modulation of expression of viral and host genes.

3 Hepadna Virus Models

The experimental approaches to HBV pathogenesis have been severely hampered because the host range of HBV is limited to humans and chimpanzees, and since in vitro culture systems for the propagation of HBV do not exist. Many questions have been answered, however, thanks to the existence of several related viruses (hepadnaviruses) in woodchucks, ground squirrels and Pekin ducks.

3.1 Woodchuck Hepatitis Virus

The discovery of a naturally occurring hepadnavirus in woodchucks (SNYDER and SUMMERS 1980), and its association with acute and chronic liver disease and HCC (SNYDER and SUMMERS 1980; POPPER et al. 1987) laid the groundwork for much of our current understanding of hepadnavirus biology and pathogenesis. For example, similar to HBV, neonatal infection by the WHV invariably leads to persistent infection and HCC, while adult onset infection leads to acute self-limited hepatitis and viral clearance (KORBA et al. 1989). Discovery of the extrahepatic replication of WHV (KORBA et al. 1990), especially its ability to replicate efficiently in lympho-mononuclear cells (KORBA et al. 1986, 1987, 1989; POTTS et al. 1981; CHEMIN et al. 1992), reinforced the concept that HBV is not strictly hepatotropic and that extrahepatic reservoirs of virus may exist that can contribute to viral persistence and serve as a continuing source of virus and viral antigens to maintain the immune response long after seroconversion and recovery from acute viral hepatitis (see below).

The WHV model has also greatly strengthened the concept that the antiviral T cell response plays a critical role in viral clearance and disease pathogenesis, since cyclosporin A treated woodchucks with suppressed T cell function fail to terminate WHV infection when infected as adults (COTE et al. 1991). This model also documented the dependence of the hepatitis delta virus (HDV) on coincident or preceding HBV infection (NEGRO et al. 1989). Furthermore, due to the ability to infect the woodchuck liver by direct intrahepatic injection of cloned WHV genomes, it has been shown that the precore protein is dispensable for viral replication (CHEN et al. 1992b) but that the X protein is not (CHEN et al. 1993; ZOULIM et al. 1994).

Recently, the woodchuck model has been used to examine the physiological basis for viral clearance during acute WHV infection (KAJINO et al., submitted). The results of these studies are compatible with a hypothesis that has recently been forthcoming from a transgenic mouse model of viral hepatitis (GUIDOTTI et al. 1994a), that, in addition to destroying infected hepatocytes, the immune response can also deliver a noncytolytic signal that eliminates the virus from the hepatocyte without killing it (see below).

Perhaps the most important contribution of the woodchuck model was in the area of hepatocarcinogenesis. Not only was it shown that virtually 100% of neonatally woodchucks develop persistent WHV infection and chronic hepatitis that progresses to HCC, but the insertional or transcriptional activation of the *myc* family of oncogenes was established as a critical early element in hepatocarcinogenesis in these animals (ETIEMBLE et al. 1994).

3.2 Other Hepadnaviruses

The discovery of a related hepadnavirus in Beechey ground squirrels (GSHV) (MARION et al. 1980) and the demonstration that chronic infection by this virus also leads to HCC (MARION et al. 1986) confirmed the previous observations in chronically WHV infected woodchucks, thereby establishing the oncogenic potential of HBV in humans. These studies further corroborated that hepadnaviruses do not carry an acutely transforming oncogene but, rather, that HCC develops after many years of antecedent infection and liver disease.

Extension of the hepadnavirus family to Pekin ducks (DHBV) (MASON et al. 1980) created the opportunity to define the molecular aspects of the viral life cycle (see above, reviewed in SUMMERS 1988) since it is possible to transmit this virus readily to adult and unborn animals (MASON et al. 1983) and to infect cultured primary hepatocytes (TUTTLEMEN et al. 1986) and a transformed avian hepatoma cell line (CONDREAY et al. 1990). The duck model has been particularly useful to study antiviral drug activity in vivo and in vitro (PETCU et al. 1988; FUKUDA et al. 1988; CONDREAY et al. 1990; HARITANI et al. 1989; YOKOTA et al. 1990; NIU et al. 1990) and to examine the role of certain physiological events, especially hepatocellular turnover, in viral clearance (JILBERT et al. 1992).

Despite the enormous strides that were made possible by the availability of the foregoing animal models, the outbred nature of these species and the lack of reagents to define their immune responses to these viral antigens have severely hampered the analysis of hepadnaviral immunobiology and immuno-pathogenesis.

4 Hepatitis B Virus Transgenic Mice

With the advent of embryo microinjection technology it became evident that many questions relating to HBV immunobiology and pathogenesis might be directly examined by introduction of partial or complete copies of the HBV genome into transgenic mice (summarized in Table 1). The major issues that have been approachable with these models will be summarized in the remainder of this review.

4.1 Determinants of Hepatitis B Virus Host Range and Tissue Specificity

Using constructs containing only HBV derived regulatory sequences, several laboratories (BABINET et al. 1985; CHISARI et al. 1985; BURK et al. 1988; FARZA et al. 1988; ARAKI et al. 1989; KIM et al. 1991; GILLES et al. 1992a; CHISARI et al., unpublished) have produced transgenic mice that preferentially express all of the viral gene products, and even replicate the virus, in the hepatocyte (reviewed in CHISARI and FERRARI 1994). Interestingly, these mice also express the viral gene products in kidney tubular epithelial cells, sometimes preferentially, and they also display sporadic and unpredictable expression in miscellaneous other tissues that are unique to each transgenic lineage, presumably reflecting integration site influences. Importantly, the supercoiled form of HBV DNA (cccDNA) has not been detected in any of these lineages, suggesting that species-specific differences at this level may play a role in determining the host range of this virus.

These studies demonstrated that HBV has the potential to be expressed and to replicate in many cells besides the hepatocyte, that viral gene expression is developmentally regulated (DeLoIA et al. 1989) and that it is positively regulated by androgens (FARZA et al. 1987) and glucocorticoids (FARZA et al. 1987), providing insight into the male predominance of HBV infection and the increased viral burden associated with steroid therapy of chronic hepatitis B in humans. Together with evidence of extrahepatic viral DNA and virus expression in infected patients and the various hepadnavirus models (reviewed in KORBA et al. 1986), these data strongly suggest that the relative liver specificity of HBV must reflect multiple constraints at the levels of viral entry, replication and gene expression, and that none of these constraints is absolutely specific for the human hepatocyte.

Table 1. Hepatitis Virus Transgenic Mice

Promoter	Genes expressed	Characteristics	Reference
HBV	All HBV ORFs	Viral replication	ARAKI et al. (1989); FARZA et al. (1987)
		CTL inhibit viral replication	GUIDOTTI et al. (unpublished)
		CTL induced hepatitis	GUIDOTTI et al. (1994a)
HBV	Small S	Developmental and hormonal regulation	FARZA et al. (1987); DELOIA et al. (1989); BURK et al. (1988)
		CTL and cytokine regulation of HBV gene expression	GUIDOTTI et al. (1994a,b); GILLES et al. (1992a); GUILHOT et al. (1993)
		Small S is not cytopathic	CHISARI et al. (1985)
		CTL induced hepatitis	MORIYAMA et al. (1990); ANDO et al. (1993)
MT + HBV	Large and small S	Small S forms rapidly secretable spherical particles	CHISARI et al. (1986)
		Large S forms HBsAg filaments that are retained in the ER and cause ground glass cell formation	CHISARI et al. (1986)
ALB + HBV	Large and small S	HBsAg storage disease	CHISARI et al. (1989)
		Cytokine induced hepatitis	GILLES et al. (1992b)
		CTL induced hepatitis	MORIYAMA et al. (1990); ANDO et al. (1993)
		Immunological tolerance	WIRTH et al. (submitted)
		HCC	CHISARI et al. (1989)
HBV	X	HCC without prior liver disease	KOIKE et al. (1994)
		No HCC	
α1AT	X	No HCC	LEE et al. (1990)
MUP	X	No HCC	CHISARI et al. (unpublished)
MUP	Core	Core particles do not cross nuclear membrane	GUIDOTTI et al. (1994c)
		HBcAg is not cytopathic	GUIDOTTI et al. (1994c)
MUP	Precore	Precore protein is secreted, not cytoplasmic or nuclear	CHISARI et al. (unpublished)
		Precore is not cytopathic	CHISARI et al. (unpublished)
MT	Core	Immunological tolerance	MILICH et al. (1994)
MT	Precore	Immunological tolerance	MILICH et al. (1990)
ALB	Delta	Delta antigen is not cytopathic	GUILHOT et al. (1994)

HBV, hepatitis B virus; ORF, open reading frame; CTL, cytotoxic T lymphocytes; HBsAg, subviral filamentous particles; MT, metallothionein; ALB, albumin; α_1 AT, α_1-anti-trypsin; MUP, mouse major urinary protein; HCC, hepatocellular carcinoma

4.2 Assembly, Transport and Secretion of the Hepatitis B Virus Structural Proteins

An important by-product of these studies was the demonstration that most of the HBV gene products, and the process of viral replication itself, is not directly cytopathic for the hepatocyte, at least at the levels attained in animals containing the complete viral genome (ARAKI et al. 1989; FARZA et al. 1988). To examine this issue further, we and others produced an assortment of transgenic lineages that express each of the HBV gene products under the control of the native viral regulatory elements or liver specific cellular promoters.

4.3 The Envelope Proteins

To examine the behavior of the three HBV envelope proteins, transgenic mice were produced in which the envelope coding region was controlled either by the native HBV regulatory elements (Table 1) or in which the preS promoter was replaced by the inducible, liver-specific mouse metallothionein promoter, or the constitutively active mouse albumin promoter. In these constructs, expression of the large envelope polypeptide is controlled either by the HBV preS promoter or by the exogenous promoters, while expression of the middle and small envelope polypeptides is controlled by the native HBs promoter present within the fragment.

In these studies, it was shown that the middle and major envelope proteins assemble into small 22 nm spherical particles that bud into the endoplasmic reticulum (ER) and are rapidly secreted by the cell (CHISARI et al. 1986, 1987). In contrast, the HBV large envelope protein assembles into long, branching, filamentous HBsAg particles that become trapped in the ER and are not secreted (CHISARI et al. 1986, 1987). It was subsequently shown that the progressive accumulation of these subviral filamentous particles leads to a dramatic expansion of the ER in the hepatocyte (CHISARI et al. 1987), eventually causing ultrastructural and histologic changes that are characteristic of the ground glass hepatocytes found in the liver of chronically infected patients with integrated HBV DNA (GERBER et al. 1974a,b).

4.4 The Nucleocapsid Proteins

To examine factors that influence the intracellular localization of nucleocapsid proteins and particles in the primary hepatocyte in vivo, transgenic mice that express the HBV core and precore proteins under the transcriptional control of the liver specific mouse major urinary protein (MUP) promoter were produced. In these studies it was learned that the precore protein is strictly secreted into the blood as HBeAg and that it is not detectable within any compartment in the hepatocyte by immunohistochemical techniques.

Importantly, the viral core protein, synthesized at a low rate by the hepatocytes in these animals, was shown to be rapidly transported into the nucleus where it accumulated until it reached a critical concentration threshold for nucleocapsid particle assembly to occur. Subsequently, the nucleocapsid particles remained in the nucleus until the nuclear membrane dissolved during M-phase of cell division, whereupon they entered the cytoplasm (GUIDOTTI et al. 1994c). The cytoplasmic nucleocapsid particles could not reenter the nucleus, however, when the nuclear membrane reformed. Furthermore, the nucleocapsid particles appeared to be unstable in the cytoplasm after cell division, since HBcAg completely disappeared from the cell shortly thereafter, only to gradually reappear (once again in the nucleus) several days later.

The data suggest that the rate of nuclear transport of nascent core monomers and dimers is quite brisk in the hepatocyte; in fact it is faster than the rate of core protein synthesis in these animals, precluding the assembly of nucleocapsid particles in the cytoplasm. This feature made it possible to demonstrate that nucleocapsid assembly can occur de novo in the nucleus and that preformed nucleocapsid particles are not transported across the intact nuclear membrane in either direction. Since core particles are not transported into the nucleus, the data suggest that nucleocapsid disassembly must occur at the cytoplasmic face of the nuclear membrane to permit entry of the hepadnaviral genome into the nucleus during the initial infection and for subsequent genome amplification within the infected cell. If this is confirmed, it raises questions about the relevance of the intranuclear core particles during the viral life cycle.

The data also demonstrate that nuclear HBcAg turnover is very slow relative to the rate of HBcAg turnover in the cytoplasm, implying the existence of an active cytoplasmic HBcAg degradative pathway, at least during the early stages of G_0 soon after mitosis. This is an important new concept because if HBcAg degradation is accelerated during the hepatocellular regeneration that occurs secondary to liver cell injury in patients with viral hepatitis, it is theoretically possible that this process could interrupt viral replication and contribute to viral clearance from the infected cell without killing it.

Finally, the data suggest that when HBcAg is detectable immunohistochemically in the cytoplasm during HBV infection, it probably reflects either recent cell division with the release of nuclear HBcAg into the cytoplasm (as shown in the core transgenic mice), or very high rates of core protein synthesis with de novo assembly of core particles in that compartment (as would occur during high level viral replication). This is reinforced by the observation that the antigenically cross-reactive precore protein is so rapidly secreted that it is never detected in the cytoplasm by this technique in the corresponding transgenic mice, despite the fact that steady state levels of HBeAg reach several micrograms per ml of serum in those animals (see above).

4.5 Pathogenetic Potential of the Viral Proteins

While overexpression of the HBV core, precore, X, small and middle envelope proteins, and viral replication are not associated with any evidence of liver disease in the transgenic mouse model, nor for that matter is the expression of the small or large delta antigens (GUILHOT et al. 1994), the HBV large envelope protein has the potential to compromise hepatocellular function and kill the hepatocyte when it is expressed at high levels, accumulates in the ER as filamentous HBsAg particles, and causes the formation of ground glass hepatocytes. Unexpectedly, the HBsAg laden ground glass hepatocyte was shown to be hypersensitive to certain endogenous stimuli that are normally not toxic to the HBsAg negative hepatocyte. Specifically, it has been shown that bacterial lipopolysaccharide (LPS) and interferon gamma (IFNγ) cause severe hepatitis in animals that retain HBsAg and that the degree of hepatocellular injury is a direct function of the amount of HBsAg retained in the ER (GILLES et al. 1992b). The pathophysiological basis for this effect is not known at this time.

Also unexpectedly, it was shown that prolonged storage of high concentrations of these long subviral filaments in the ER of ground glass hepatocytes is directly cytotoxic to the hepatocyte, initiating a storage disease characterized by chronic hepatocellular necrosis and a secondary inflammatory and regenerative response (CHISARI et al. 1987; FILIPPI et al. 1988) that inexorably leads to HCC (CHISARI et al. 1989; DUNSFORD et al. 1990; SELL et al. 1991). Thus, the inappropriate expression of a single structural viral gene is sufficient to set in motion a complex series of events that ultimately leads to malignant transformation. The incidence of hepatocellular carcinoma in this model corresponds to the frequency, severity and age of onset of liver cell injury which itself corresponds to the intrahepatic concentration of HBsAg and is influenced by genetic background and sex.

A more detailed description of this model will be presented later in this chapter. In the next several sections, we will review the immunological studies performed in transgenic mice that have led to new insights into HBV immunobiology and pathogenesis.

4.6 Pathogenetic Potential of the Cellular Immune Response to Hepatitis B Virus

4.6.1 Class I-Restricted, HBsAg-Specific CTLs Cause Acute Hepatitis When Injected into HBV Transgenic Mice

In recent studies it was shown that transgenic mice that express HBV envelope antigens in their hepatocytes (WIRTH et al., submitted) develop a liver disease resembling acute viral hepatitis following adoptive transfer of CD8-positive, MHC class I (Ld)-restricted, HBsAg-specific CTLs (MORIYAMA et al. 1990). In these studies, the severity of the CTL induced necroinflammatory liver disease depended on the route of CTL administration, the production of interferon-γ (IFN-γ) by the CTL when they recognize antigen, the number of HBsAg-positive hepatocytes present within the liver, the amount of HBsAg within each hepatocyte, the sex of the recipient and the route of CTL administration. Furthermore, by quantitative, morphometric, histopathological analysis of the liver at different time points after the injection of CTL, it was shown that the disease progresses through a series of clearly definable steps in an orderly fashion.

4.6.2 To Initiate Disease, CTLs Bind to the HBsAg-Positive Hepatocytes and Trigger Apoptosis

The earliest detectable pathological event that occurs following the entry of these CTL into the liver is their attachment to HBsAg-positive hepatocytes, which they trigger to undergo apoptosis (step 1) (ANDO et al. 1994b). The direct CTL-target cell interaction results in the appearance of widely scattered, acidophilic, Councilman bodies (apoptotic hepatocytes that are characteristic of acute viral hepatitis in humans. Importantly, however, the direct cytopathic effect of the CTL is limited to very few hepatocytes, possibly because the effector: target (E: T) cell ratio in the liver is very low (it never exceeds 0.03) in these experiments, and because free-rangin gCTL movement is severely limited by the architectural constraints of solid tissue. These factors are probably operative in other examples of T cell-mediated immunopathology, although they are not widely appreciated because most of our concepts of CTL-induced disease are based on in vitro studies using much higher E:T ratios and cell mixtuers in which free ranging CTL movement is possible.

4.6.3 Antigen-Nonspecific Amplification Mechanisms Cause Most of the Damage Initiated by the CTLs

Between 4 and 12 h after injection, the CTL recruit many host-derived, antigen-nonspecific inflammatory cells into their immediate vicinity (step 2), resulting in the formation of necroinflammatory foci, in which hepatocellular necrosis extends to the periphery of necroinflammatory foci, well beyond the CTL, indicating that most of the hepatocytes were killed by cells other than the CTL themselves (ANDO et al. 1993). In most HBsAg-positive transgenic mouse lineages, the liver disease does not progress beyond this point.

Like most cases of acute viral hepatitis in humans, the disease is transient in these mice, relatively mild (destroying no more than 5% of the hepatocytes) and non-fatal, unless the hepatocytes retain HBsAg, as discussed above. In that case the disease process extends to step 3, which kills nearly half of the mice due to liver failure within 48–72 h of CTL injection. This process is characterized by the widespread necrosis of HBsAg-laden hepatocytes with a marked diffuse lympho-mononuclear inflammatory cell infiltrate and Kupffer cell hyperplasia, thereby resembling the histopathological features HBV-induced fulminant hepatitis in humans (ANDO et al. 1993). The inflammatory cells, especially the macrophages, outnumber the injected CTL by at least 100-fold at this point in the disease process, suggesting that most of the destruction is caused by the inflammatory cells that the CTL recruit and/or activate and not by the CTL themselves. This hypothesis is strengthened by the fact step 3 can be completely prevented by the prior administration of neutralizing antibodies to IFN-γ or by the inactivation of macrophages by multiple injections of carrageenan (ANDO et al. 1993).

Collectively, these observations suggest that most of the histopathological manifestations of the liver disease in this model, except hepatocellular apoptosis, are mediated by antigen-nonspecific cytokines and effector cells that have been activated by the virus-specific T cells, and not by the CTL themselves. The striking similarities between the immunopathological and histopathological features of the current model and acute viral hepatitis in humans suggest that similar events probably contribute importantly to the pathogenesis of the human diseases well.

4.6.4 CTLs Do Not Have Access to HBsAg-Positive Parenchymal Cells in Most Extrahepatic Tissues

While intravenously administered HBsAg-specific CTL can recognize and destroy HBsAg-positive hepatocytes in these transgenic mice, the same CTL are unable to recognize antigen expressed in other tissues (e.g., brain, kidney, testis, pancreas, gastrointestinal tract) in these animals (ANDO et al. 1994a). However, when they are injected beneath the kidney capsule or intracerebrally, i.e., extravascularly, the CTL are highly cytopatic for HBsAg-positive renal tubules and choroid plexus epithelial cells. These data suggest that the CTL ignore the extrahepatic viral antigen because their access to antigen in these sites is precluded by microvascular anatomical barriers that are present in these tissues and do not exist in the sinusoidal structure of the liver, not because their capacity to process and present the antigen to the CTL is defective. This raises the possibility that extrahepatic reservoirs of HBV may be more difficult to eradicate than infected hepatocytes during natural HBV infection, and thereby contribute to viral persistence.

4.7 Noncytopathic Antiviral Potential of the Immune Response to Hepatitis B Virus

In a recent series of experiments, certain soluble products of the immune response, especially IFN-γ (GUIDOTTI et al. 1994a), tumor necrosis factor (TNFα) (GILLES et al. 1992a; GUIDOTTI et al. 1994a) and interleukin (IL-2) (GUIDOTTI et al. 1994b), have been shown to suppress the steady state content of HBV mRNA in the hepatocyte of transgenic mice. Furthermore, it has been shown that these effects are mediated by a post-transcriptional mechanism that selectively accelerates the degradation of cytoplasmic HBV mRNA (GUILHOT et al. 1993). The same events are set in motion when HBsAg-specific CTL secrete IFNγ and induce TNFα following antigen recognition (GUIDOTTI et al. 1994a). In recent studies of newly derived transgenic mouse lineages that replicate the HBV genome (GUIDOTTI and CHISARI, unpublished observations), it has been shown that all of the viral gene products, including the intrahepatic nucleocapsids and the viral replicative intermediates, are eliminated, noncytolytically, by the same CTL. In the same studies it was shown that the intrahepatic nucleocapsid particles and replicative intermediates are also eliminated during hepatocellular regeneration following partial hepatectomy, which, in contrast to the response to CTL administration, had no effect on the hepatic HBV mRNA content.

These observations suggest that, in addition to destroying HBV infected hepatocytes, a strong intrahepatic immune response to HBV might be able to suppress viral replication and perhaps even "cure" infected hepatocytes of the virus without killing them. The data suggest that at least two intracellular inactivation pathways might be activated by the CTL, one that degrades the viral RNA in response to the release of IFNγ and TNFα from antigen-activated CTL, and another that degrades the nucleocapsid particles and their contents in regenerating hepatocytes. In support of this hypothesis, it has been shown that the WHV is cleared from all of the hepatocytes in the acutely infected woodchuck liver without massive hepatic necrosis (JILBERT et al. 1992), suggesting that noncytolytic clearance mechanisms may be operative in that model also.

4.8 Tolerogenic Potential of Hepatitis B Virus Nucleocapsid and Envelope Proteins

Clonal deletion of HBV-specific T cells as a consequence of transplacental infection of the developing fetus, or transplacental passage of subviral antigens, could play an important role in the chronic infection that develops in neonates born to infected mothers. In line with this possibility, it has been shown that nontransgenic progeny of HBeAg positive transgenic mothers are tolerant to both HBeAg and HBcAg at the T cell level, presumably due to the thymic deletion of MHC class II-restricted HBV nucleocapsid-specific helper T cells as a result of transplacental exposure to HBeAg (MILICH et al. 1990). Not surprisingly, transgenic mice that express the viral nucleocapsid and envelope antigens have been shown by many groups to be tolerant to these antigens at the T cell level (MILICH et al. 1990, 1994; MANCINI et al. 1993; WILL 1991). Since intrauterine infection of the fetal liver by HBV has been described by many investigators (ALEXANDER and EDDLESTON 1986), it is likely that tolerance to many of the viral proteins can contribute to viral persistence by negative selection of the responding cells in the thymus during fetal development.

4.9 Hepatocellular Carcinoma

4.9.1 Overexpression of the HBV X Gene Can Cause Hepatocellular Carcinoma in Transgenic Mice

Since the HBV X gene product displays transcriptional transactivating properties and can transactivate cellular genes (e.g., c-myc and c-jun) associated with cellular growth control, it has been suggested that unregulated expression of this gene may contribute to the malignant transformation of the infected hepatocyte. Interestingly, it has recently been reported that the HBV X protein inhibits p53 gene function in vitro (WANG et al. 1994), strengthening the evidence for a role of this viral transactivator in hepatocarcinogenesis.

In support of this hypothesis it has been shown that high level expression of the HBV X gene can lead to HCC in transgenic mice (KOIKE et al. 1994), although others have not observed the induction of HCC in independently derived X gene transgenic mouse strains (LEE et al. 1990). This discrepancy might be related to differences in the strength and duration of HBx protein expression. Importantly, it might also be related to the different genetic backgrounds on which the various transgenic models were produced. In this respect, the mice that develop HCC were produced and maintained on a CD-1 background which displays a high spontaneous rate of HCC (HOMBURGER et al. 1975). This might suggest that the X protein functions as a cofactor in the process of hepatocarcinogenesis and that it may not be sufficient to induce HCC by itself. This is supported by the ability of the transfected X gene product to promote the transformation of previously immortalized but not primary cell lines in vitro (HÖHNE et al. 1990). It is further strengthened by the long interval before tumor development in the HCC susceptible transgenic mouse lineages, indicating that X gene expression alone is not sufficient for carcinogenesis and that other genetic and/or epigenetic events are necessary for HCC to develop.

4.9.2 Chronic Liver Cell Injury Can Cause Hepatocellular Carcinoma in HBsAg Transgenic Mice

Although it has not yet been possible to develop a model of immunologically induced chronic hepatitis in any of the HBV transgenic mice available at this point in time, chronic hepatitis does develop in certain lineages that greatly overexpress the HBV large envelope polypeptide and accumulate nonsecretable HBsAg filamentous particles in the ER (CHISARI et al. 1986). As mentioned earlier, lineages that accumulate moderate amounts of these particles display normal baseline liver function but their hepatocytes are hypersensitive to the cytotoxic effects of IFNγ (GILLES et al. 1992b; ANDO et al. 1993). However, excessive storage of HBsAg apparently compromises ER function sufficiently that the hepatocytes die spontaneously, causing a secondary inflammatory (mutagenic) and regenerative (mitogenic) response that leads to HCC in mice from strains that have a very low incidence of spontaneous HCC (CHISARI et al. 1989).

Importantly, hepatocellular turnover in these mice, relative to nontransgenic controls, is increased nearly 100-fold for at least a year before the onset of HCC (HUANG and CHISARI, unpublished observations). Additionally, oxygen radical production is greatly increased and the antioxidant (glutathione and catalase) content of the liver is greatly decreased in the liver of these mice. Importantly, these changes are associated with a dramatic increase in oxidative DNA damage to hepatocellular genomic DNA (HAGEN et al. 1994). It is reasonable to assume that these events could lead to the development of random mutations throughtout the liver cell genome that eventually contribute to the development of HCC.

In view of the prolonged antecedent hepatocellular injury, regeneration and oxidative DNA damage, it is likely that transformation occurs via a multistep process in this transgenic model. Since transformation was observed in two independent lineages, without evidence of transgene rearrangement or instability, direct insertional activation of a cellular oncogene is not likely. Rather, it likely involves activating or inactivating mutations in multiple cellular genes that are spatially and functionally independent of the integrated HBV sequences.

Interestingly, insulin-like growth factor-II (IGF-II) and mdr III gene expression are transcriptionally activated in the majority of the transgenic mouse hepatomas

(SCHIRMACHER et al. 1992; KUO et al. 1992). These genes are not overexpressed during the preneoplastic phase of the disease, however, suggesting that they represent late changes associated with tumor progression but not with tumor initiation in this model. The importance of these findings is underscored by the contrasting fact that no changes in p53, RB-1, Ha-*ras*, Ki-*ras*, N-*ras*, c-*myc*, N-*myc*, erb-A, erb-B, *src, mos, abl, sis, fms, fes, fos, jun*, TGF-α, TGF-β, PDGF-α, PDGF-β, EGF receptor, retinoic acid receptor-β, HNF-1, c/EBP, or CREB DNA copy number, gene structure, steady state RNA levels or protein content were detected in any of the tumors (PASQUINELLI et al. 1992). Obviously, the cellular genome is vast, as are the opportunities for growth promoting mutations and chromosomal abnormalities outside of these loci.

The link between injury and transformation is strengthened by the fact that HCC has not been observed in any HBV transgenic lineage that doesn't diaplay liver cell injury (CHISARI et al. 1985; BABINET et al. 1985; BURK et al. 1988; ARAKI et al. 1989; FARZA et al. 1987, 1988; HINO et al. 1989; YAMAMURA et al. 1987; DELOIA et al. 1989) except for the mice that express the transactivating X gene described above. The pathogenetic importance of injury in this regard is further strengthened by the development of HCC in livers of transgenic mice with neonatal hepatitis as a consequence of the hepatocellular retention of α 1-antitrypsin (DYCAICO et al. 1988). Additional support for this hypothesis is the fact that human HCC occurs in the context of necrosis, inflammation and regeneration (cirrhosis) in several diseases other than hepatitis B, such as chronic hepatitis C (reviewed in ALTER 1988), alcoholism (LIEBER et al. 1986), hemochromatosis (NIEDERAU et al. 1985), glycogen storage disease (LIMMER et al. 1988), α-1-antitrypsin deficiency (CARLSON and ERIKSSON 1985; ERIKSSON et al. 1986), and primary biliary cirrrhosis (MELIA et al. 1984).

These results suggest that severe, prolonged hepatocellular injury induces a preneoplastic proliferative and inflammatory response that places the dividing hepatocyte at risk of developing multiple random mutations or other chromosomal changes, including viral integration, some of which program the cell for unrestrained growth.

In the transgenic mouse model described above, injury is secondary to the overproduction of the HBV large envelope protein. Although this process may occur during viral infection, it is much more likely that immunological mechanisms are principally responsible for liver disease in HBV infected humans. While the initiating events are different in the human and mouse systems, the downstream processes that place the dividing hepatocyte at risk of developing the mutations and chromosomal damage that ultimately transform the hepatocyte are likely to be quite similar. It is very possible therefore, that the most devastating complication of HBV infection may be caused by a chronic antiviral immune response that destroys some, but not all, of the infected hepatocytes, one of which eventually develops into a malignant neoplasm that kills the host.

5 Summary

It should be apparent from the foregoing that the transgenic mouse model system has contributed substantially to our understanding of many aspects of HBV biology, immunobiology and pathogenesis in the past several years.

We have learned that HBV can replicate within the mouse hepatocyte, as well as other mouse cell types, suggesting that there are probably no strong tissue or species specific constraints to viral replication once the viral genome enters the cell. However, the failure thus far to detect viral cccDNA in the hepatocyte nucleus in several independently derived transgenic lineages suggests that other, currently undefined, constraints on host range and tissue specificity may also be operative.

Thanks to the transgenic mouse model we now understand the pathophysiological basis for HBsAg filament formation and ground glass cell production, and we have learned that at least this viral gene product can be toxic for the hepatocyte, first by compromising its ability to survive the hepatocytopathic effects of LPS and IFNα and eventually by causing it to die in the absence of any obvious exogenous stimulus.

In recent studies, it has been shown that preformed nucleocapsid particles do not cross the nuclear membrane in either direction at least in the mouse hepatocyte. If this is confirmed, it will have two important implications: first, that nucleocapsid disassembly must occur in the cytoplasm before the nascent viral genome can enter the nucleus; second, that the intranuclear nucleocapsid particles are empty, and therefore serve no currently defined purpose in the viral life cycle. This should stimulate new interest in the analysis of the function of these particles that are a prominent feature of mammalian hepadnavirus infection.

The transgenic mouse model has also established definitively that HBV-induced liver disease has an immunological basis, and that the class I-restricted CTL response plays a central role in this process. Additionally, the mouse studies have taught us that when the CTL recognize their target antigen on the hepatocytes they cause them to undergo apoptosis, forming the acidophilic, Councilman bodies that are characteristic of viral hepatitis. Further, we have learned that although the CTL initiate the liver disease, they actually contribute more to disease severity indirectly by recruiting antigen nonspecific effector cells into the liver than by directly killing the hepatocytes themselves. In addition, by releasing IFNγ when they recognize antigen, the CTL can destroy enough of the liver to cause fulminant hepatitis in mice whose hepatocytes overproduce the large envelope protein and are hypersensitive to the cytopathic effects of this cytokine.

We have also learned that the CTL are unable to recognize HBV-positive parenchymal cells outside of the liver, apparently because they cannot traverse the microvascular barriers that exist at most extrahepatic tissue sites. This important new discovery may permit the virus to survive a vigorous CTL response and contribute not only to the maintenance of memory T cells following acute hepatitis but also to serve as a reservoir to reseed the liver in patients with chronic hepatitis.

The transgenic mouse model has also revealed that activated CTL and the cytokines they secrete can down-regulate HBV gene expression, and possibly even control viral replication, by noncytotoxic intracellular inactivation mechanisms involving the degradation of viral RNA and, perhaps, the degradation of viral nucleocapsids and replicative DNA intermediates without killing the cell. If HBV replication is indeed interrupted by this previously unsuspected activity, it could contribute substantially to viral clearance during acute infection when the immune response to HBV is vigorous. Alternatively, it could also contribute to viral persistence, by only partially down-regulating the virus during chronic infection when the immune response is weak.

Our understanding of the events that contribute to neonatal tolerance in HBV infection has been greatly expanded by the demonstration that transgenic maternal HBeAg can cross the placenta and tolerize the nontransgenic fetus to both HBeAg and HBcAg at the T cell level. Additionally, the demonstration that HBV transgenic mice are tolerant to the viral nucleocapsid and envelope antigens at the T cell level, even if they are not secreted, confirms that tolerance due to intrauterine infection by HBV can be profound and extend to all of the structural viral proteins.

Finally, the transgenic mouse model has taught us that insertional activation or inactivation of hepatocellular growth control genes is not required for hepatocarcinogenesis, even though it plays an important role in WHV-induced HCC. We have also learned from transgenic mice that the HBV X protein can play a role in hepatocarcinogenesis, possibly by cooperating with other currently undefined viral, environmental or host-derived cofactors, when it is expressed at high levels for a sustained period of time in the liver. Additionally, it is clear that chronic liver cell injury sets in motion a cascade of pathophysiological events including inflammation, oxygen radical production, hepatocellular regeneration, and oxidative DNA damage that produce random genetic and chromosomal changes that ultimately cause HCC. All of these events, and perhaps others, are presumably operative during chronic HBV infection to explain the extraordinarily high incidence of HCC in patients with chronic hepatitis.

Collectively, the foregoing observations illustrate the power of transgenic mice to elucidate aspects of viral biology and pathogenesis, especially when they are used to address questions that cannot be otherwise approached in simpler systems. Based on the breadth of these contributions, we are optimistic that transgenic mouse technology will clarify many additional aspects of hepadnaviral biology and pathogenesis in the future.

Acknowledgments. I am indebted to Drs. Luca Guidotti, Claudio Pasquinelli, Stephane Guilhot, Pierre Filippi, Patrick Gilles, Takashi Moriyama, Kazuki Ando and Tetsuya Ishikawa who have contributed to the analysis of the transgenic models described in this paper. I am also grateful to Drs. Ralph Brinster, Richard Palmiter, Alan McLachlan and David Milich who helped me to initiate our HBV transgenic model program several years ago, to Drs. Christine Pourcel, Kenichi Yamamura and Gilbert Jay for sharing tissues and animals from their transgenic lineages, and to Drs. Shaonan Huang, Heinz Schaller,

Charles Rogler, Stewart Sell, M. Tien Kuo, John Curnutte, Tory Hagen and Bruce Ames for their active collaboration in recent years. Finally, I want to thank Jenny Price (Scripps) and Rick Huntress (DNX Corp, Princeton, NJ) for embryo microinjection, John Shutter, Kathy Klopchin, Patricia Fowler, Margie Pagels, Ying Tsu Loh, Jan Shoenberger, and Violet Martinez for outstanding technical assistance, and Bonnie Weier for preparation of this manuscript. These studies were supported by grants CA40489 and CA54560 from the National Institutes of Health. This is manuscript number 8782-MEM from The Scripps Research Institute.

References

Alexander GJ, Eddleston AL (1986) Does maternal antibody to core antigen prevent recognition of transplacental transmission of hepatitis-B-virus infection[2]. Lancet 1: 296–297

Alter HJ (1988) Transfusion-associated non-A, non-B hepatitis: the first decade. In: Zuckerman AJ (ed) Viral hepatitis and liver disease. Liss, New York, pp 534–542

Ando K, Moriyama T, Guidotti LG, Wirth S, Schreiber RD, Schlicht HJ, Huang S, Chisari FV (1993) Mechanisms of class I restricted immunopathology. A transgenic mouse model of fulminant hepatitis. J Exp Med 178: 1541–1554

Ando K, Guidotti LG, Cerny A, Ishikawa T, Chisari FV (1994a) CTL access to tissue antigen is restricted in vivo. J Immunol 153: 482–488

Ando K, Guidotti LG, Wirth S, Ishikawa T, Missale G, Moriyama T, Schreiber RD, Schlicht HJ, Huang S, Chisari FV (1994b) Class I restricted cytotoxic T lymphocytes are directly cytopathic for their target cells in vivo. J Immunol 152: 3245–3253

Araki K, Miyazaki J-I, Hino O, Tomita N, Chisaka O, Matsubara K, Yamamura K-I (1989) Expression and replication of hepatitis B virus genome in transgenic mice. Proc Natl Acad Sci USA 86: 207–211

Babinet C, Farza H, Morello D, Hadchouel M, Pourcel C (1985) Specific expression of hepatitis B surface antigen (HBsAg) in transgenic mice. Science 230: 1160–1163

Bartenschlager R, Junker-Niepmann M, Schaller H (1990) The P gene product of hepatitis B virus is required as a structural component for genomic RNA encapsidation. J Virol 64: 5324–5332

Bavand MR, Laub O (1988) Two proteins with reverse transcriptase activities associated with hepatitis B virus-like particles. J Virol 62: 626–628

Bavand Mr, Feitelson M, Laub O (1989) The hepatitis B virus-associated reverse transcriptase is encoded by the viral pol gene. J Virol 63: 1019–1021

Beasley RP, Lin C-C, Hwang LY, Chen C-S (1981) Hepatocellular carcinoma and hepatitis B virus: a prospective study of 22, 707 men in Taiwan. Lancet 2: 1129–1133

Blum HE, Zhang Z-S, Galun E, von Weizsäcker F, Garner B, Liang TJ, Wands JR (1992) Hepatitis B virus X protein is not central to the viral life cycle in vitro. J Virol 66: 1223–1227

Bosch V, Bartenschlager R, Radziwill G, Schaller H (1988) The duck hepatitis B virus P-gene codes for protein strongly associated with the 5'-end of the viral DNA minus strand. Virology 166

Burk RD, DeLoia JA, ElAwady MK, Gearhart JD (1988) Tissue preferential expression of the hepatitis B virus (HBV) surface antigen gene in two lines of HBV transgenic mice. J Virol 62: 649–654

Carlson J, Eriksson S (1985) Chronic "cryptogenic" liver disease and malignant hepatoma in intermediate alpha-1-antitrypsin deficiency identified by a pi 2-specific monoclonal antibody. Scand J Gastroenterol 20: 835–841

Chemin I, Baginski I, Vermot-Desroches C, Hantz O, Jacquet C, Rigal D, Trepo C (1992) Demonstration of woodchuck hepatitis virus infection of peripheral blood mononuclear cells by flow cytometry and polymerase chain reaction. J Gen Virol 73: 123–129

Chen HS, Knew MC, Hornbuckle WE, Tenant BC, Cote PJ, Gerin JL, Purcell RH, Miller RH (1992) The precore gene of the woodchuck hepatitis virus genome is not essential for viral replication in the natural host. J Virol 66: 5682–5684

Chen HS, Kaneko S, Girones R, Anderson RW, Hornbuckle WE, Tenant BC, Cote PJ, Gerin JL, Purcell RH, Miller RH (1993) The woodchuck hepatitis virus X gene is important for establishment of virus infection in woodchucks. J Virol 67: 1218–1226

Chen HS, Kaneko S, Girones R, Anderson RW, Hornbuckle WE, Tennant BC, Cote PJ, Gerin JL, Purcell RH, Miller RH (1994) The woodchuck hepatitis virus X gene is important for establishment of virus infection in woodchucks. J Virol 67: 1218–1226

Chisari FV (1991) Analysis of hepadnavirus gene expression, biology and pathogenesis in the transgenic mouse. Curr Top Microbiol Immunol 168: 85–101

Chisari FV, Ferrari C (1994) Immunobiology and pathogenesis of viral hepatitis. In: Nathenson N Ahmed R, Gonzatez-Scarano F, Griffin D, Holmes K, Murphy FA, Robinson H (eds) Viral pathogenesis. Raven New York, (in press)

Chisari FV, Pinkert CA, Milich DR, Filippi P, McLachlan A, Palmiter RD, Brinster RL (1985) A transgenic mouse model of the chronic hepatitis B surface antigen carrier state. Science 230: 1157–1160

Chisari FV, Filippi P, McLachlan A, Milich DR, Riggs M, Lee S, Palimiter RD, Pinker CA, Brinster RL (1986) Expression of hepatitis B virus large envelope polypeptide inhibits hepatitis B surface antigen secretion in transgenic mice. J Virol 60: 880–887

Chisari FV, Filippi P, Buras J, McILlachlan A, Popper H, Pinkert CA, Palmiter RD, Brinstaer RL (1987) Structural and pathological effects of synthesis of hepatitis B virus large envelope polypeptide in transgenic mice. Proc Natl Acad Sci USA 84: 6909–6913

Chisari FV, Klopchin K, Moriyama T, Pasquinelli C, Dunsford HA, Sell S, Pinkert CA, Brinster RL, Palmiter RD (1989) Molecular pathogenesis of hepatocellular carcinoma in hepatitis B virus transgenic mice. Cell 59: 1145–1156

Colgrove R, Simon G, Ganem D (1989) Transcriptional activation of homologous and heterologous genes by the hepatitis B virus X gene product in cells permissive for viral replication. J Virol 63: 4019–4026

Condreay LD, Aldrich CE, Coates L, Mason WS, Wu T-T (1990) Efficient duck hepatitis B virus production by an avian liver tumor cell line. J Virol 64: 3249–3258

Cote PJ, Korba BE, Steinberg H, Ramirez-Mejia C, Baldwin B, Hornbuckle WE, Tennant BC, Gerin JL (1991) Cyclosporin A modulates the course of woodchuck hepatitis virus infection and induces chronicity. J Immunol 146: 3138–3144

DeLoia JA, Burk RD, Gearhart JD (1989) Developmental regulation of hepatitis B surface antigen expression in two lines of hepatitis B virus transgenic mice. J Virol 63: 4069–4073

Dunsford HA, Sell S, Chisari FV (1990) Hepatocarcinogenesis due to chronic liver cell injury in hepatitis B virus transgenic mice. Cancer Res 50: 3400–3407

Dycaico MJ, Grant SG, Felts K, Nichols WS, Geller SA, Hager JH, Pollard AJ, Kohler SW, Short HP, Jirik FR et al. (1988) Neonatal hepatitis induced by alpha 1-antitrypsin: a transgenic mouse model. Science 242: 1409–1412

Eble BE, Lingappa VR, Ganem D (1986) Hepatitis B surface antigen: an unusual secreted protein initially synthesized as a transmembrane polypeptide. Model Cell Biol 6: 1454–1463

Eble BE, MacRae DR, Lingappa VR, Ganem D (1987) Multiple topogenic sequences determine the transmembrane orientation of hepatitis B surface antigen. Mol Cell Biol 7: 3591–3601

Eckhardt SG, Milich DR, McLachlan A (1991) Hepatitis B virus core antigen has two nuclear localization sequences in the arginine -rich carboxyl terminus. J Virol 65: 575–582

Eriksson S, Carlson J, Velez RN (1986) Risk of cirrhosis and primary liver cancer in alpha-1-antrypsin deficiency. N Engl J Med 314: 736–740

Etiemble J, Degott C, Renard CA, Fourel G, Shamoon B, Vitvitski-Trepo L, Hsu TY, Tiollais P, Babinet C, Buendia MA (1994) Liver-specific expression and high oncogenic efficiency of a c-myc transgene activated by woodchuck hepatitis virus insertion. Oncogene 9: 727–737

Farza H, Salmon AM, Hadchouel M, Moreau, Babinet C, Tiollais P, Pourcel C (1987) Hepatitis B surface antigen gene expression is regulated by sex steroids and glucocorticoids in transgenic mice. Proc Natl Acad Sci USA 84: 1187–1191

Farza H, Hadchouel M, Scotto J, Tiollais P, Babinet C, Pourcel C (1988) Replication and gene expression of hepatitis B virus in a transgenic mouse that contains the complete viral genome. J Virol 62: 4144–4152

Filippi P, Buras J, McLachlan A, Popper H, Pinkert CA, Palmiter RD, Brinster RL, Chisari FV (1988) Overproduction of hepatitis B virus large envelope polypeptide causes filament storage, ground glass cell formation, hepatocellular injury and nodular hyperplasia in transgenic mice, In: Zuckerman AJ (ed) Viral hepatitis and liver disease. Liss, New York, PP 632–640

Franco A, Paroli M, Testa U, Benvenuto R, Peschle CM, Balsano F, Barnaba V (1992) Transferrin receptor mediates uptake and presentation of hepatitis B envelope antigen by T lymphocytes. J Exp Med 175: 1195–1205

Fukuda R, Okinaga S, Akagi S, Hidaka M, Ono N, Fukumoto S, Shimada Y (1988) Alteration of infection pattern of duck hepatitis B virus by immunomodulatory drugs. J Med Virol 26: 387–396

Gerber MA, Hadziyannis S, Vissoulis C, Schaffner F, Paronetto F, Popper H (1974a) Hepatitis B antigen: nature and distribution of cytoplasmic antigen in hepatocytes of carriers (37912). Proc Soc Exp Biol Med 145: 863–867

Gerber MA, Hadziyannis S, Vissoulis C, Schaffner F, Paronetto F, Popper H (1974b) Electron microscopy and immunoelectron microscopy of cytoplasmic hepatitis B antigen in hepatocytes. Am J pathol 175: 489–502

Gilles PN, Fey G, Chisari FV (1992a) Tumor necrosis factor-alpha negatively regulates hepatitis B virus gene expression in transgenic mice. J Virol 66: 3955–3960

Gilles PN, Guerrette DL, Ulevitch RJ, Schreiber RD, Chisari FV (1992b) Hepatitis B surface antigen retention sensitizes the hepatocyte to injury by physiologic concentrations of gamma interferon. Hepatology 16: 655–663

Guidotti LG, Ando K, Hobbs MV, Ishikawa T, Runkel RD, Schreiber RD, Chisari FV (1994a) Cytotoxic T lymphocytes inhibit hepatitis B virus gene expression by a noncytolytic mechanism in transgeneic mice. Proc Natl Acad Sci USA 91: 3764–3768

Guidotti LG, Guilhot S, Chisari FV (1994b) Interleukin 2 and interferon alpha/beta downregulate hepatitis B virus gene expression in vivo by tumor necrosis factor dependent and independent pathways. J Virol 68: 1265–1270

Guidotti LG, Martinez V, Loh YT, Rogler CE, Chisari FV (1994c) Hepatitis B virus nucleocapsid particles do not cross the hepatocyte nuclear membrane in transgenic mice. J Virol 68: 5469–5475

Guilhot S, Guidotti LG, Chisari FV (1993) Interleukin-2 downregulates hepatitis B virus gene expression in transgenic mice by a post-transcriptional mechanism. J Virol 67: 7444–7449

Guilhot S, Huang S, Xia YP, LaMonica N, Lai MMC, Chisari FV (1994) Expression of the hepatitis delta virus and small antigens in transgenic mice. J Virol 68: 1052–1058

Hagen TM, Wehr C, Huang SN, Fowler P, Martine ZV, Curnutte J, Anes BN, Chisari FV (1994) Extensive oxidative DNA damage in hepatocytes of transgenic mice with chronic active hepatitis destined to develop hepatocellular carcinoma. Proc Natl Acad Sci USA 97: 12808–12812

Haritanin H, Uchida T, Okuda Y, Shikata T (1989) Effect of 3'-azido-3-deoxythmidine on replication of duck B virus in vivo and in vitro. J Med Virol 29: 244–248

Hertogs K, Lenders WPJ, Depla E, DeBruin WCC, Meheus L, Raymackers J, Moshage H, Yap SH (1993) Endonexin II, present on human liver plasma membranes, is a specific binding protein of small hepatitis B virus envelope protein. Virology 197: 549–557

Hino O, Nomura K, Ohtake K, Kawaguchi T, Sugano H, Kitagawa T (1989) Instability of integrated hepatitis B virus DNA with inverted repeat structure in a transgenic mouse. Cancer Genet Cytogenet 37: 273–278

Hirsch RC, Lavine JE, Chang L, Varmus HE, Ganem D (1990) Polymerase gene products of hepatitis B viruses are required for genomic RNA packaging as well as for reverse transcription. Nature 344: 552–555

Höhne M, Schaefer S, Seifer M, Feitelson MA, Paul D, Gerlich WH (1990) Malignant transformation of immortalized transgenic hepatocytes after transfection with hepatitis B virus DNA. EMBO J 9: 1137–1145

Homburger F, Russfield AB, Weisburger JH, Lim S, Chak SP, Weisburger EK (1975) Aging changes in CD1 HaM/ICR mice reared under standard laboratory conditions J Natl Cancer Inst 55: 37–45

Jean-Jean O, Weimer T, De Recondo AM, Will H, Rossignol JM (1989) Internal entry of ribosomes and ribosomal scanning involved in hepatitis B virus P gene expression. J Virol 63: 5451–5454

Jilbert AR, Wu T-T, England JM, De La M, Hall P, Carp NZ, O'Connell AP, Mason WS (1992) Rapid resolution of duck hepatitis B virus infections occurs after massive hepatocellular involvement. J Virol 66: 1377–1388

Junker-Niepmann M, Bartechlager R, Schaller H (1990) A short cis-acting sequence is required for hepatitis B virus progenome encapsidation and sufficient for packaging of foreign RNA. EMBO J 9: 3389–3396

Kawamoto S, Yamamoto S, Ueda K, Nagahata T, Chisaka O, Matsubara K, (1990) Translation of hepatitis B virus DNA polymerase from the internal AUG codon, not from the upstream AUG codon for the core protein. Biochem Biophys Res Commun 171: 1130–1136

Kim C-M, Koike K, Saito I, Miyamura T, Jay G (1991) HBx gene of hepatitis B virus induces liver cancer in transgenic mice. Nature 351: 317–320

Koike K, Moriya K, Iino S, Yotsuyanagi H, Endo Y, Miyammura T, Hurokawa K (1994) High level expression of hepatitis B virus HBx gene and hepatocarcinogenesis in transgenic mice. Hepatology. 19: 810–819

Korba BE, Wells F, Tennant BC, Yoakum GH, Purcell RH, Gerin JL (1986) Hepadnavirus infection of peripheral blood lymphocytes in vivo: woodchuck and chimpanzee models of viral hepatitis. J Virol 58: 1–8

Korba BE, Wells F, Tennant BC, Cote PJ, Gerin JL (1987) Lymphoid cells in the spleens of woodchuck hepatitis virus-infected woodchucks are a site of active viral replication. J Virol 61: 1318–1324

Korba BE, Cote PJ, Wells FV, Baldwin B, Popper H, Purcell RH, Tennant BC, Gerin JI (1989) Natural history of woodchuk hepatitis virus infections during the course of experimental viral infection: molecular virologic features of the liver and lymphoid tissues. J Virol 63: 1360–1370

Korba BE, Brown TL, Wells FV, Baldwin B, Cote PJ, Steinberg H, Tennant BC, Gerin JL (1990) Natural history of experimental woodchuck hepatitis virus infection: molecular virologic features of the pancreas, kidney, ovary, and testis. J Virol 64: 4499–4506

Kuo MT, Zou J-Y, Teeter LD, Ikeguchi M, Chisari FV (1992) Activation of multidrug resistance (P-glycoptotein) mdr3/mdrla gene during the development of hepatocellular carcinoma in hepatitis B virus transgenic mice. Cell Growth Differ 3: 531–540

Lavine J, Hirsch R (1989) A system for studying the selective encapsidation of hepadnavirus FNA. J Virol 63: 4257–4263

Lee T-H, Finegold MJ, Shen R-F, DeMayo JL, Woo SL, Butel JS (1990) Hepatitis B virus transactivator X protein is not tumorigenic in transgenic mice. J Virol 64: 5939–5947

Lieber CS, Garro A, Leo MA, Mak KM, Worner T (1986) Alcohol and cancer. Hepatology 6: 1005–1019

Lien J-M, Aldrich CE, Mason WS (1986) Evidence that a capped oligoribonucleotide is the primer for duck hepatitis B virus plus-strand DNA synthesis. J Virol 57: 229–236

Limmer J, Fleig WE, Leupold D, Bittner R, Ditscheunest H, Berger H-G (1988) Hepatocellular carcinoma in type 1 glycogen storage disease. Hepatology 8: 531–537

Machida A, Kishimoto S, Ohnuma H, Miyamoto H, Baba K, Oda K, Nakamura T, Miyakawa Y, Mayumi M (1983) A hepatitis B surface antigen polypeptide (P31) with the receptor for polymerized human as well as chimpanzee albumins. Gastroenterology 85: 268–274

Mack DH, Bloch W, Nath N, Sninsky JJ (1988) Hepatitis B virus particles contain a polypeptide encoded by the largest open reading frame: a putative reverse transcriptase. J Virol 62: 4768–4790

Maguire HF, Hoeffler JP, Siddiqui A (1991) HBV X protein alters the DNA binding specificity of CREB and ATF-2 by protein-protein interactions. Science 252: 842–844

Mancini M, Hadchouel M, Tiollais P, Pourcel C, Michel ML (1993) Induction of anti-hepatitis B surface antigen (HBsAg) antibodies in HBsAg producing transgenic mice: a possible way of circumventing "nonresponse" to HBsAg. J Med Virol 39: 67–74

Marion PL, Oshiro LS, Regnery DC, Scullard GH, Robinson WS (1980) A virus in Beechey ground squirrels which is related to hepatitis B virus of man. Proc Natl Acad Sci USA 77: 2941–2945

Marion PL, Van Davelaar MJ, Knight SS, Salazar FH, Garcia G, Popper H, Robinson WS (1986) Hepatocellular carcinoma in ground squirrels persistently infected with ground squirrel hepatitis virus. Proc Natl Acad Sci USA 83: 4543–4546

Mason WS, Seal S, Summers J (1980) Virus of Pekin ducks with structural and biological relatedness to human hepatitis B virus. J Virol 36: 829–836

Mason WS, Aldrich C, Summers J et al. (1982) Asymmetric replication of duck hepatitis B virus DNA in liver cells (free minus-strand DNA). Proc Natl Acad Sci USA 79: 3997–4001

Mason WS, Halpern MS, England JM, Seal G, Egan J, Coates L, Aldrich C, Summers J (1983) Experimental transmission of duck hepatitis B virus. Virology 131: 375–384

McLachlan A, Milich DR, Raney AK, Riggs MG, Hughes JL, Sorage J, Chisari FV (1987) Expression of hepatitis B virus surface and core antigens: influences of pre-S and precore sequences. J Virol 61: 683–780

Mehdi H, Kaplan MJ, Anlar FY, Yang X, Bayer R, Sutherland K, Peeples ME (1994) Hepatitis B virus surface antigen binds to apolipoprotein H. J Virol 68: 2415-2424

Melia WM, Wilkinson ML, Portmann BC, Johnson PJ, Williams R (1984) Hepatocellular carcinoma in the non-cirrhotic liver: a comparison with that complicating cirrhosis. Q J Med 53: 391–400

Milich DR, Jones JE, Hughes JL, Price J, Raney AK, McLachlan A (1990) Is a funtion of the secreted hepatitis B antigen to induce immunologic tolerance in utero? Proc Natl Acad Sci USA 87: 6599–6603

Milich DR, Jones JE, Hughes JL, Maruyama T, Price J, Melhado I, Jirik F (1994) Extrathymic expression of the intracellular hepatitis B core antigen results in T cell tolerance in transgenic mice. J Immunol 152: 455–466

Miller RH, Marion PL, Robinson WS (1984a) Hepatitis B viral DNA-RNA hybrid molecules in particles from infected liver are converted to viral DNA molecules during an endogenous DNA polymerase reaction. Virology 139: 64–72

Miller RH, Tran C-T, Robinson WS (1984b) Hepatitis B virus particles of plasma and liver contain viral DNA-RNA hybrid molecules. Virology 139: 53–63

Molnar-Kimber KL, Summers JW, Mason WS (1984) Mapping of the cohesive overlap of duck hepatitis B virus DNA and of the site of initiation of reverse transcription. J Virol 51: 181–191

Molnar-Kimber KL, Jarocki-Witek V, Dheer SK, Vernon SK, Conley AJ, Davis AR, Hung PP (1988) Distinctive properties of the hepatitis B virus envelope proteins. J Virol 62: 407–416

Moriyama T, Guilhot S, Klopchin K, Moss B, Pinkert CA, Palmiter RD, Brinster RL, Kanagawa O, Chisari FV (1990) Immunobiology and pathogenesis of hepatocellular injury in hepatitis B virus transgenic mice. Science 248: 361–364

Nassal M (1992) The arginine-rich domain of the hepatitis B virus core protein is required for pregenome encapsidation and productive viral positive-strand DNA synthesis but not for virus assembly. J Virol 66: 4107–4116

Negro F, Korba BE, Forzani B, Baroudy BM, Brown TL, Gerin JL, Ponzetto A (1989) Hepatitis delta virus (HDV) and woodchuck hepatitis virus (WHV) nucleic acids in tissues of HDV-infected chronic WHV carrier woodchucks. J Virol 63: 1612–1618

Neurath AR, Kent SB, Strick N, Parker K (1986) Identification and chemical synthesis of a host cell receptor binding site on hepatitis B virus. Cell 46: 429–436

Neurath AR, Strick N, Sproul P (1992) Search for hepatitis B virus cell receptors reveals binding sites for interleukin 6 on the virus envelope protein. J Exp Med 175: 461–469

Niederau C, Fischer R, Sonnenberg A, Stremmel W, Trampisch HJ, Strohmeyer G (1985) Survival and causes of death in cirrhotic and in noncirrhotic patients with primary hemochromatosis. N Engl J Med 313: 1256–1262

Niu J, Wang Y, Qiao M, Gowans E, Edwards P, Thyagarajan SP, Gust I, Locarnini S (1990) Effect of Phyllanthus amarus on duck hepatitis B virus replication in vivo. J med Virol 32: 212–218

Ou J-H, Rutter WJ (1987) Regulation of secretion of the hepatitis B virus major surface antigen by the preS-1 protein. J Vitrol 61: 782–786

Ou J-H, Yeh CT, Yen TSB (1989) Transport of hepatitis B virus precore protein in to the nucleus after cleavage of its signal peptide. J Virol 63: 5238–5243

Pasquinelli C, Bhavani K, Chisari FV (1992) Multiple oncogenes and tumor suppressor genes are structurally and functionally intact during hepatocarcinogenesis in hepatitis B virus transgenic mice. Cancer Res 52: 2823–2829

Patzer EJ, Nakamura GR, Yaffe A (1984) Intracellular transport and secretion of hepatitis B surface antigen in mammalian cells. J Virol 51: 346–353

Patzer EJ, Nakamura GR, Simonsen CC, Levinson AD, Brands R (1986) Intracellular assembly and packaging of hepatitis B surface antigen particles occur in the endoplasmic reticulum. J Virol 58: 884–892

Petcu DJ, Aldrich CE, Coates L, Taylor JM, Mason WS (1988) Suramin inhibits in vitro infection by duck hepatitis B virus, Rous sarcoma virus, and hepatitis delta virus. Virology 167: 385–392

Popper H, Roth L, Purcell RH, Tennant BC, Gerin JL (1987) Hepatocarcinogenicity of the woodchuck hepatitis virus. Proc Natl Acad Sci USA 84: 866–870

Potts RC, Sherif MM, Robesrton AJ, Gibbs JH, Brown RA, Beck JS (1981) Serum inhibitory factor in lepromatous leprosy: its effect on the pre-S-phase cell-cycle kinetics of mitogen-stimulated normal human lymphocytes. Scand J Immunol 14: 269–280

Robinson WS, Miller RH, Marion PL (1987) Hepadnaviruses and retroviruses share genome homology and features of replication. Hepatology 7: 64S–73S

Rosenthal N, Kress M, Gruss P et al. (1983) BK, viral enhancer element and a human cellular homology. Science 222: 749–755

Schirmacher P, Held WA, Chisari FV, Yang D, Rogler CE (1992) Reactivation of insulin-like growth factor II during hepatocarcinogenesis in transgenic mice suggests a role in malignant growth. Cancer Res 52: 2549–2556

Schlicht HJ, Schaller H (1989) The secretory core protein of human hepatitis B virus is expressed on the cell surface. J Virol 63: 5399–5404

Schlicht HJ, Radziwill G, Schaller H (1989) Synthesis and encapsidation of duck hepatitis B virus reverse transcriptase do not require formation of core-polymerase fusion proteins. Cell 56: 85–92

Seeger C, Ganem D, Varmus HE (1986) Biochemical and genetic evidence for the hepatitis B virus replication strategy. Science 232: 477–484

Seifer M, Zhou S, Standring DN (1993) A micromolar pool of antigenically distinct precursors is required to initiate cooperative assembly of hepatitis B virus capsids in Xenopus oocytes. J Virol 67: 249–257

Sell S, Hunt JM, Dunsford HA, Chisari FV (1991) Synergy between hepatitis B virus expression and chemical hepatocarcinogens in transgenic mice. Cancer Res 51: 1278–1285

Seto E, Yen TSB, Peterlin BM, Ou J-H (1988) Trans-activation of the human immunodeficiency virus long terminal repeat by the hepatitis B virus X protein. Proc Natl Acad Sci USA 85: 8286–8290

Siddiqui A, Gaynor R, Srinivasan A, Mapoles J, Farr RW (1989) Transactivation of viral enhancers including long terminal repeat of the human immunodeficiency virus by the hepatitis B virus X protein. Virology 173: 764–766

Snyder RL, Summers J (1980) Woodchuck hepatitis virus and hepatocellular carcinoma. Cold Spring Harbor Conf Cell Proliferation 7: 447–458

Spandau DF, Lee CH (1988) Transactivation of viral enhancers by the hepatitis B virus X protein. J Virol 62: 427–434

Standring DN, Ou J-H, Rutter WJ (1986) Assembly of viral particles in Xenopus oocytes: pre-surface-antigens regulate secretion of the hepatitis B viral surface envelope particle. Proc Natl Acad Sci USA 83: 9338–9342

Standring DN, Ou J-H, Masiarz FR, Rutter WJ (1989) A signal peptide encoded within the precore region of hepatitis B virus directs the secretion of a heterogeneous population of e antigens in Xenopus oocytes. Proc Natl Acad Sci USA 85: 8405–8409

Summers J (1988) The replication cycle of hepatitis B viruses. Cancer 61: 1957–1962

Summers J, Mason WS (1982) Replication of the genome of a hepatitis B—like virus by reverse transcription of an RNA intermediate. Cell 29: 403–415

Tuttleman J, Pourcel C, Summers J (1986) Formation of the pool of covalently closed circular viral DNA in hepadnavirus infected cells. Cell 47: 451–460

Tuttlemen J, Pugh J, Summers J (1986) In vitro experimental infection of primary duck hepatocyte cultures with duck hepatitis B virus. J Virol 58: 17–25

Twu JS, Schloemer RH (1987) Transcriptional trans-activating function of hepatitis B virus. J virol 61: 3448–3453

Twu JS, Chu K, Robinson WS (1989) Hepatitis B virus X gene activated kB-like enhancer sequences in the long terminal repeat of human immunodeficiency virus 1. Proc Natl Acad Sci USA 86: 5168–5172

Unger T, Shaul Y (1990) The X protein of the hepatitis B virus acts as a transcription factor when targeted to its responsive element. EMBO J 9: 1889–1895

Wang GH, Seeger C (1992) The reverse transcriptase of hepatitis B virus acts as a protein primer for viral DNA synthesis. Cell 71: 663–670

Wang GH, Seeger C (1993) Novel mechanism for reverse transcription in hepatitis B viruses. J Virol 67: 6506–6512

Wang XW, Forrester K, Yeh H, Feitelson MA, Gu J-R, Harris CC (1994) Hepatitis B virus X protein inhibits p53 sequence-specific DNA binding, transcriptional activity and associated with transcription factor ERCC3. Proc Natl Acad Sci USA 91: 2230–2234

Will H (1991) The X-protein of hepatitis B virus. Facts and fiction. J Hepatol 13 Suppl 4: S56–S57

Will H, Reiser W, Weimer T, Pfaff E, Buscher M, Sprengel R, Cattaneo R, Schaller H (1987) Replication strategy of human hepatitis B virus. J Virol 61: 904–911

Wollersheim M, Debelka U, Hofschneider PH (1988) A transactivating function encoded in the hepatitis B virus X gene is conserved in the integrated state. Oncogene 3: 545–552

Yamamura K, Tsurimoto T, Ebihara T, Kamino K, Fujiyama A, Ochiya T, Matsubara K (1987) Methylation of hepatitis B virus DNA and liver-specific suppression of RNA production in transgenic mouse. Jpn J Cancer Res 78: 681–688

Yeh C-T, Liaw Y–F, Ou J-H (1990) The arginine-rich domain of hepatitis B virus precore and core proteins contains a signal for nuclear transport. J Virol 64: 6141–6147

Yokota T, Konno K, Chonan E, Mochizuki S, Kojima K, Shigeta S, De Clercq E (1990) Comparative activates of several nucleoside analogs against duck hepatitis B virus in vitro. Antimicrob Agents Chemother 34: 1326–1330

Zhou S, Standring DN (1991) Production of hepatitis B virus nucleocapsid like core particles in Xenopus oocytes: assembly occurs mainly in the cytoplasm and does not require the nucleus. J Virol 65: 5457–5464

Zhou S, Standring DN (1992) Hepatitis B virus capsid particles are assembled from core-protein dimer precursors. Proc Natl Acad Sci USA 89: 10046–10050

Zhou S, Yang SQ, Standring DN (1992) Characterization of hepatitis B virus capsid particle assembly in Xenopus oocytes. J Virol 66: 3086–3092

Zoulim F, Saputelli J, Seeger C (1994) Woodchuck hepatitis virus X protein is required for viral infection in vivo. J Virol 68: 2026–2030

Transgenic Models of HTLV-I Mediated Disease and Latency

M. Nerenberg[1], X. Xu[1], and D.A. Brown[2]

[1] Department of Neuropharmacology and Molecular and Experimental Medicine, The Scripps Research Institute, La Jolla, CA 92037, USA
[2] *Present address:* Microprobe Corporation, 1725 220th St. SE#104, Bothell, WA 98021, USA

1 Background

1.1 HTLV-I and Oncogenesis

HTLV-I infection is the etiologic agent for a form of adult T cell leukemia (ATLL) (HINUMA et al. 1981; POIESZ et al. 1980; ROBERT et al. 1982; YOSHIDA et al. 1984). A peculiarity of this virus is its long incubation period prior to the development of malignancy and the low percentage of patients who eventually develop disease. Studies in Japan and Jamaica indicate that the latency period may range up to 40 years with a lifetime incidence of ATLL of 2%–4% in seropositive individuals (MURPHY et al. 1989). In vitro, HTLV-I transforms human lymphocytes but carries no clearly identifiable cellular oncogene-like homolog. In addition, no site-specific integration pattern has been identified (SEIKI et al. 1984). Circumstantial data has implicated the HTLV-I tax gene in transformation. This gene encodes a 40 kDa protein with at least partial localization to the nucleus. It is a potent transcriptional activator of its own promoter and those of numerous other oncogenes and cytokines (reviewed SMITH and GREENE 1991; YOSHIDA 1994). Tax is always retained but usually not expressed in ATLL, whereas deletions in other portions of the viral genome are common in transformed cells. Tax alone has been shown to trans-form fibroblasts in vitro (POZZATTI et al. 1990; TANAKA et al. 1990). Nevertheless, most attempts to directly transform human T cells with tax alone have failed. In vitro studies have demonstrated that Tax can activate the promoters of the interleukin-2 (IL-2) receptor (IL-2R) α chain, granulocyte/macrophage colony-stimulating factor (GM-CSF), fos, Platelet-derived growth factor (PDGF), IL-6, vimentin, nerve growth factor (NGF), transforming growth factor-β (TGF-β), HIV long terminal repeat (LTR), as well as its own LTR (reviewed GITLIN et al. 1993). It thus may have pleiomorphic effects on cell growth and inflammation. Mutational analysis of the target sequences of Tax in these promoters have revealed two kinds of sequences, an ATF-like, which occurs in the HTLV-I LTR and in fos, and an NF-κB-like sequence, which occurs in the HIV LTR and in the IL-2R (reviewed in SMITH and GREENE 1991; NERENBERG 1992; GITLIN et al. 1993).

1.2 Analysis of the U3 Region of the HTLV-I LTR

The vast majority of the footprinting and functional analyses of the transcriptional regulation of the HTLV-I LTR have focused on the U3 region, as this is the region which has been best shown to influence transcription in other retroviruses. Deletion and transient transfection analysis of HTLV-I-LTR-CAT (Chloromphenicol acetyl transferase) expression vectors has shown that there are three 21 bp repeats in the HTLV-I LTR that are responsive to trans-activation by Tax. Site-directed mutagenesis of these 21 bp repeats abolishes responsiveness to Tax or cAMP and has demon-strated a core sequence TGACGT to be essential. The first and fifth bases of this sequence appear to be the most crucial for trans-activation (GIAM and XU 1989). Tax does not alter the footprinting patterns on the HTLV-I LTR (JEANG et al. 1988; ALTMAN et al. 1988; NYBORG et al. 1990; NYBORG and DYNAN 1990), and Tax alone does not directly bind to the LTR. However, a number of cellular proteins which bind to this consensus have been described under different conditions and different cell types. The HTLV-I LTR 21 bp repeats, the CRE and the ATF binding sites all contain the TGACG motif and all bind proteins of similar size, leading to the suggestion that the proteins that bind these sites are highly related or identical (TAN et al. 1989). Protein interactions with Tax may lead to coassociation and augmented binding affinity of proteins such as CREB, NF-κB and SRF to their target sequence (WAGNER and GREEN 1993; FRANKLIN et al. 1993; SUZUKI et al. 1993). In some cases, Tax may even alter the preferred binding sequence (PACA-UCCARALERTKUN et al. 1994). Tax may also activate the HTLV-I LTR by influencing cellular signal transduction pathways (reviewed in YOSHIDA 1994). These effects may be very rapid since experiments with inhibitors have demonstrated that new protein synthesis is not necessary for Tax to cause transcriptional augmentation (GIAM et al. 1986; RUBEN et al. 1989). Another study has implicated sequences located between the first and second 21 bp repeats of the HTLV-I LTR as an additional Tax response region, and an oligonucleotide encoding this region removes Tax from solution when precipitated with an insoluble avidin matrix (MARRIOTT et al. 1989). Insertion of this sequence between the first and second 21 bp repeats is said to give 30-fold higher Tax-induced gene expression than a nonspecific DNA of the same length. In this region (also called the Tax responsive region 2) the transcription factors AP-2, Elk-1 and THP-1 have been shown to bind under some conditions (MUCHARD et al. 1992; SEELER et al. 1993; CLARK et al. 1993; TANIMURA et al. 1993).

1.3 Role of the R Region in Transcriptional Regulation

A much more limited range of studies has been performed on the R region of the LTR. This region may be capable of assuming extensive secondary structure. The region is strongly bound by nuclear proteins (LEVINGER and LAUTENBERGER 1987). When placed downstream of heterologous promoters, the R region has variable though sometimes profound effects on augmenting transcription (NAKAMURA et al. 1988). A ten fold *trans*-activation by Tax has also been seen when this region, devoid of the U3 region, is included in CAT constructs (BRADY et al. 1987). Finally, 3' sequential deletions have revealed elements which both positively and negatively regulate transcription (GARTENHAUSE et al. 1991). A region near the R-U5 junction has been proposed to bind proteins involved in basal expression of the virus (KASHANCHI et al. 1994).

2 Significance

Considerable progress has been made in understanding the epidemiology, biology, and molecular biology of HTLV-I in the 15 years since its discovery.

Throughout much of this history, the HTLV-I transcriptional activator Tax has been hypothesized to play a major role, as the major regulator of latency, as an oncogene, and as an immune target. In vitro studies suggest that it is an extremely potent transcriptional *trans*-activator and that its growth promoting potential is both potent and pleiomorphic. A unique aspect of this virus is its extremely long latency period, and the low penetrance level of disease, which is almost uniformly fatal when it occurs. Our current knowledge is inadequate to explain these peculiarities of HTLV-I biology. In particular, it is difficult to understand how latency could be maintained in the presence of Tax, why expression in peripheral tissues of humans is so weak, even after transformation by this virus, and how the lymphoma phenotypes appears so malignant in the absence of detectable virus expression. To answer these questions, it was necessary to establish in vivo models of HTLV-I mediated transformation, and to study the biochemical mechanisms of transformation and latency regulation.

3 LTR-tax Transgenic Mice Reproduce Aspects of HTLV-I Latency and Transformation

In the initial LTR-tax model, the Tax protein was expressed under the direction of the HTLV-I LTR promoter (NERENBERG et al. 1987). This promoter was chosen to allow Tax to autoregulate its own expression in a way analogous to that which occurs in the course of natural infection. These mice showed low to absent levels of tax transcription at birth. However, with aging, mice expressed tax in scattered cells in skeletal muscle, fibroblasts, chondrocytes (Fig. 1) and salivary gland (Fig. 2) (NERENBERG and WILEY 1989; NERENBERG et al. 1987; GREEN et al. 1989). The distribution and frequency of the resulting tumors varied with the genetic background of the mice even though the morphology of these tumors did not change. For example, more tumors occurred on a CD-1 background than on a C57Bl/6 background. In CD-1 mice, they occurred at a much earlier time (3 months as opposed to 7–9 months in the C57 mice) and in atypical locations such as nerve associated, or adrenal medulla (HINRICHS et al. 1987; GREEN et al. 1992). However, even when associated with peripheral nerves, these tumors appeared to be fibroblast outgrowths of the nerve sheaths and had little or no neural or Schwann cell characteristics (NERENBERG and WILEY 1989). Further, no association between neurofibromatosis and HTLV-I was found (NERENBERG et al. 1991a). Thus, disease in these mice did not closely mimic neurofibromatosis, as first proposed (HINRICHS et al. 1987).

The transgenic mice also developed adenomas of salivary ductal epithelium (Fig. 2). These tumors also occurred more frequently in mice bred on a CD-1 background, and though there was frequent hyperplasia of salivary lymph nodes we were unable to detect inflammatory infiltrate in the salivary gland itself. In addition, we did not see an increase in autoreactive antibodies (either to Tax or

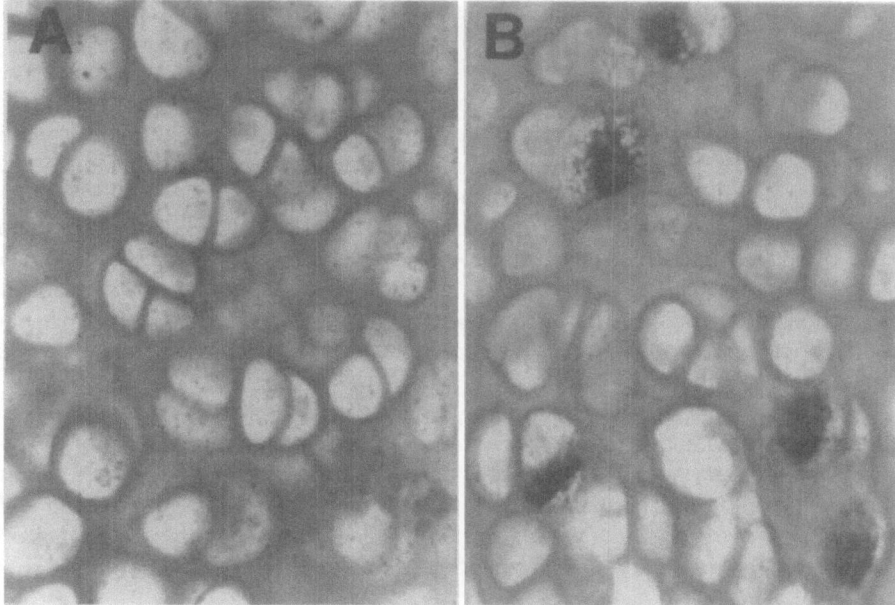

Fig. 1A,B. In situ analysis of tax expression in LTR-tax transgenic mice. **A** Sense and **B** antisense ^{35}S-labeled probes (HIGUCHI et al. 1992). Highly expressing cells are seen next to those with undetectable levels of tax

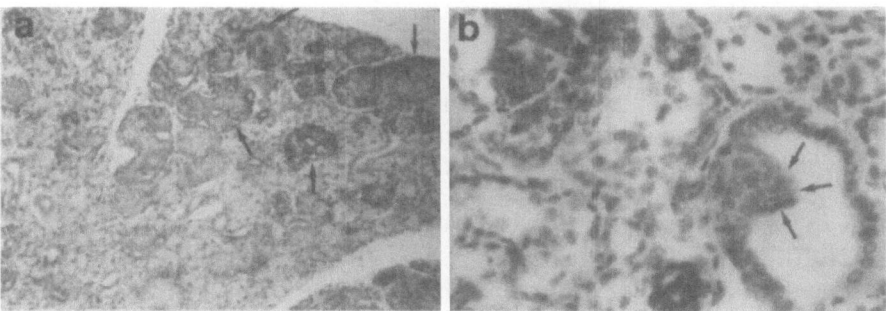

Fig. 2a,b. Histologic analysis of intraductal salivary gland adenomas **a** low and **b** high power micrographs obtained from LTR-tax on CD-1 background. C57Bl/6 background mice develop similar lesions at lower frequency

other proteins). Thus, proposed similarities of salivary lesions in these mice to those of Sjogren's disease (GREEN et al. 1989) remain unclear.

Expression of tax could be seen in parenchymal cells of the thymus in some lines (NERENBERG et al. 1987; NERENBERG and WILEY 1989). Expression was associated with the development of fibroblastic tumors by 6 months, which occurred more frequently in males and clustered around the sites of tissue trauma and biopsies necessary for transgene analysis (NERENBERG et al. 1987; BROWN et al. 1994).

Further in vivo and in vitro studies of LTR-tax lymphocytes, explained why baseline tax expression was not seen in lymphoid organs. Purified resting lymphocytes from both spleen and thymus did not express tax. Stimulation with mitogens such as CD3, or conconavalin A transiently induced low levels of expression. Individually, forskolin, phorbol esters, or calcium ionophores had little effect on LTR directed expression. In contrast, induction of cells with a combination of phorbol ester plus calcium ionophore resulted in cell cycling, and induced tax expression within 1 h. This expression was superinduced by addition of cycloheximide. Thus, the pattern of tax induction was typical of that for a cellular early gene, with the HTLV-I LTR serving as a site of convergence of signal transduction pathways.

In the animal, up-modulation in lymphocytes appears to be overridden by suppression, leading to latency. This pattern of expression mimics that of HTLV-I transformed peripheral leukemic cells obtained from blood of patients in whom virus expression was not detected. However when these cells were blasted with mitogens in vitro or cocultivated with umbilical cord blood, there was rapid up-modulation of virus expression and efficient transmission of infection.

4 Constitutive tax Expression Causes Transformation of Fibroblasts but Not Thymocytes

LTR-tax mice died of fibroblastic tumors by 4–8 months. Unlike HTLV-I infection in humans, none of these mice developed lymphomas. Lack of constitutive expression in lymphoid tissues provided a possible explanation for this. To overcome this block, tax expression was redirected to thymocytes through the use of the murine Thy 1 promoter in a new set of transgenic mice (NERENBERG et al. 1991a). Further enhancement of tax expression was achieved by creating doubly transgenic Thy-tax X LTR-tax mice. Trans-activation of the LTR-tax transgene by Thy-tax led to augmented levels of tax expression in the thymus which were comparable to those seen in the fibroblastic tumors of LTR-tax mice. This also proved that the Tax protein expressed in the thymus was functional, being able to transactivate a silent promoter (NERENBERG et al. 1991b).

Despite high constitutive tax expression in lymphocytes, no increase in proliferation and no discernible structural or functional abnormalities were found in lymphocytes. Given the long period of latency required to develop lymphomas in humans, it was possible that additional cofactors were necessary to accelerate leukemogenesis. To address this, the additional oncogenic effects of γ irradiation was studied in the Thy-tax mice. C57 control mice or transgenic Thy-tax mice on a C57 genetic background were compared. Mice were given 850 rads from a cobalt source in three treatments prior to age of 3 months. This is a regimen which causes thymic lymphomas in virtually 100% of mice with a very characteristic latency of 120 days. Any shift in this latency period would represent an independent oncogenic effect of tax. No changes were seen in latency or in the phenotype of the resultant lymphomas when analyzed by histology, surface markers or ability to grow when transplanted into nontransgenic mice (NERENBERG et al. 1991b). In addition, transformation through radiation was not associated with up-modulation of tax expression.

However, singly transgenic Thy-tax mice developed fibroblastic tumors with similar morphology to those seen in LTR-tax mice. The incubation period for these tumors was shifted from the 6 months seen in the LTR-tax mice to 18 months in Thy-tax mice (NERENBERG et al. 1991b). Fibroblastic tumors from these Thy-tax mice expressed high levels of tax, comparable to that seen in the LTR-tax fibroblastic tumors. High levels of expression and transformation of fibroblasts, even with a mostly lymphocyte-specific promoter, suggested a strong biologic difference in the susceptibility to transformation by tax between fibroblasts and lymphocytes. The spontaneous up-modulation of tax in fibroblasts and not thymocytes most likely reflected a specific growth advantage conferred by tax for fibroblasts during early stages of oncogenesis.

5 Tax Transformed Mouse Fibroblasts Bear Phenotypic Similarities to HTLV-I Transformed Human Lymphocytes

The above studies established tax as a potent oncogene for mouse fibroblasts. This differs from the usual target of HTLV-I transformation in humans which is CD4+ T cells. So far, we have not observed a link between HTLV-I infection and human fibroblastic tumors (NERENBERG 1992; NERENBERG et al. 1991a). This could result from differences in route of delivery or species-specific cofactors.

In order to investigate similarities between HTLV-I transformed lymphocytes and tax transformed fibroblasts, we examined gene pathways in tumor progression using both fresh and cultured fibroblastic tumors. Northern blot analyses were performed using a variety of early and immediate early cellular genes as probes. Both fibroblast tumors and cell lines expressed constitutively high levels of fos, PDGF, GM-CSF, Il-6, and TGF-β (KITAJIMA et al. 1992a). Most of these genes demonstrated levels greater than 20 times higher than those found in unstimulated Balb/3T3 cells. Though fibrosarcomas may constitutively express one or more of these genes, it is highly unusual for them to express all of them. In contrast, none of the other tax expressing tissues in transgenic mice (including thymocytes) showed constitutive elevations of these genes. Thus, long-term tax expression alone was not sufficient to maintain cellular activation, but this required an additional step in transformation.

6 Maintenance of the Transformed State Requires Continued Expression of NF-κB but Not tax

Studies of human cells suggest that HTLV-I may transform lymphocytes through a "hit and run mechanism". The early stage requires infection and replication of the HTLV-I virus. The tax gene is likely to be important early on since this portion of the provirus is preferentially retained despite common deletions of other portions of the provirus. However, fully malignant cells rarely express Tax or other retroviral proteins. Thus tax may become dispensible at later stages of human leukemia. To investigate the need for continued tax expression in maintenance of mouse fibroblast tumor growth, we utilized an antisense approach to inhibit protein expression in transformed cells. Cells were treated with 3' thiol substituted oligonucleotides complementary to the initiation codons (KITAJIMA et al. 1992a) to specifically block tax expression. Immunoprecipitation of tax demonstrated that expression could be decreased by 20-fold when tax antisense treated cells were compared with tax sense treated controls. To demonstrate that this extent of inhibition had functional consequences, the tax responsive plasmid HTLV-I-LTR-Cat was cotransfected and levels of CAT expression were compared between sense and antisense oligonucleotide treated cells. CAT expression decreased up to 95% in the presence of tax antisense oligonucleotides. Parallel northern analysis of treated cells revealed no apparent change in expression of a number of oncogenes or cytokines. In addition, antisense treated cells showed no obvious differences in phenotype or growth rate. This suggested that continued high levels of tax expression were not required for maintenance of the transformed phenotype. Thus, tax may also transform mouse fibroblasts by a hit and run mechanism, and continued growth may be dependent on expression of downstream pathways.

HTLV-I transformed human T cells have a highly activated phenotype, frequently expressing many cytokines and early cellular genes (see background section). In vitro studies have demonstrated that this may occur through a second pathway involving NF-κB, which is an important transcriptional regulator of these genes. We thus wished to evaluate the role of NF-κB on expression of these genes in transformed mouse fibroblasts. As performed before for tax, antisense oligonucleotides were synthesized to block expression of the p50 and p65 subunits of NF-κB. The efficiency of inhibition of NF-κB was accessed by electrophoretic mobility shift analysis (EMSA) assays on nuclear extracts prepared from oligonucleotide treated cells. Specificity was demonstrated by efficient competition with the native but not mutant fragment. Greater than 90% inhibition of binding was achieved with p65 antisense oligonucleotides. To prove that these levels of inhibition were functionally significant, trans-activation assays were performed similar to those described for tax inhibition. In this case the test plasmid was HIV-LTR-CAT, which contains two copies of the NF-κB consensus sequence and has been shown to be strongly NF-κB responsive. A 20-fold inhibition of CAT expression was seen with NF-κB antisense. Northern blot analysis of cells treated with either p50 or p65 antisense oligonucleotides showed no differences in the levels of the NF-κB independent genes fos, tax, or actin. In contrast, profound inhibition of IL- 6 and GM-CSF expression was seen. p65 proved more effective at a given dose than p50, and the effects were additive when combined.

Growth curves performed over a variety of conditions showed similar results. When compared with tax antisense, there was a profound difference in growth dependence on NF-κB over all serum concentrations. In contrast, inhibition of NF-κB had little effect on growth of Balb/3T3 cells (KITAJIMA et al. 1992b).

Though there appeared to be a profound difference in sensitivity to NF-κB between normal and tax transformed cells, it was difficult to know from in vitro assays how this correlated with tumor growth in an intact animal. To test this, antisense manipulations were performed in mice. A total of 10^7 tumor cells were transferred to one hindlimb of syngeneic animals. Tumors were allowed to establish themselves by growth for 7 days. Mice were then treated with varying doses of antisense oligonucleotides administered i.p. At a dose of 40 μg tax antisense per gram body weight, tax expression was fully inhibited. As before, no difference in tumor growth was seen when tax sense or antisense treated animals were compared.

Established tumors in mice were also treated with antisense directed against the p65 subunit of NF-κB. In contrast to tax treatment, NF-κB had profound effects on tumor growth. Two antisense treatments within the first week after tumor establishment were sufficient to completely ablate tumor growth.

These data demonstrated that growth of tumors both in vitro and in vivo is dependent on continued NF-κB but not on tax expression. However the relevance of this mouse model to native viral transformation remained uncertain. To address this, similar in vitro growth curves were performed on the HTLV-I transformed human lymphocyte line MT-2. These cells were treated with both tax and p65 NF-κB oligodeoxynucleotides (ODNs). The tax ODNs were resynthesized to match the sequence differences between the 5′ untranslated regions of the mouse transgene and the human virus. The results were the same as for mouse fibroblasts. No effect on cell growth was seen after tax inhibition while profound growth retardation occurred with antisense treatment against NF-κB.

These data strongly support the hit and run model for transformation of mouse fibroblasts by tax or human T cells by HTLV-I. We hypothesize that tax provides an early selective advantage for cell growth. This could occur by transient up-modulation of nuclear NF-κB directly by tax, leading to transient autocrine growth through one of the NF-κB-dependent cytokines. In vitro studies from other labs have shown that long-term expression of tax alone is not sufficient to maintain constitutive nuclear NF-κB expression. Thus with time, homeostatic mechanisms tend to down-modulate NF-κB expression. Constitutive expression of nuclear NF-κB must then occur through a second genetic event, such as mutation of an NF-κB inhibitor. At this point, continued high level expression of tax would no longer be required. A difference in the number or susceptibility of inhibitors between fibroblasts and lymphocytes could explain the difference in ease of transforming these two cell lines and the profound differences in latency till transformation. These hypotheses might also explain some aspects of the peculiar phenotype of tax transformed cells. If growth of tumors is driven exclusively by autocrine (or intracrine) NF-κB stimulation, these tumors may not need to progress through the cascade of inactivation of tumor suppressors, which has been shown to be an essential part of malignant progression in other systems. In preliminary studies, we have in fact been unable to detect differences in expression of either p53 or retinoblastoma (RB). Further, it might explain the extreme sensitivity of these tumors to antisense inhibition of NF-κB, which would serve as the Achilles heel for these tumors.

7 An In Situ β-gal Marker Reveals Both Transcriptional Activation and Suppression During Fibroblast Tumor Progression

The data obtained from the study of the LTR-tax and Thy-tax transgenic mice allowed us to make considerable progress in characterizing the tax transformed phenotype in fibroblasts. These data suggest an early important role for tax and later events which are dominated by constitutive NF-κB expression. The ultimate phenotype may be the result of many accumulated steps. Unfortunately, it was difficult to study the earlier intervening steps using these models. To aid studies of the relationship between tax mediated transactivation of the HTLV-I LTR and tumor progression, we made an additional set of transgenic mice which express the E. coli Lac Z gene (coding for the sensitive histologic marker β-galactosidase) (MUCKE et al. 1991). Expression was directed by the HTLV-I LTR (LTR-β-gal). Expression of β-gal in singly transgenic mice in the absence of tax was undetectable. Doubly transgenic LTR-tax XLTR-β-gal mice were obtained by breeding the two lines and selecting for both independently segregating transgenes. These mice demonstrated a sporadic pattern of expression in muscle cells and chondrocytes, which correlated with tax expression as demonstrated by the laborious and less sensitive in situ RNA or immunocytochemical methods.

Skin wounding of the doubly transgenic mice led to profound up-modulation of β-gal staining in dermal fibroblasts and keratinocytes surrounding the wound. As wound healing progressed, tax expression was shut off in differentiated keratinocytes, but remained high in dermal fibroblasts, and was associated with local limited proliferation (Fig. 3). After several months, malignant fibroblastic tumors developed at the sites of the wounds. These cells continued to express extremely high levels of tax, but low to undetectable β-gal. Northern blot analysis demonstrated that this difference occurred at the level of transcription. The selective suppression of only the β-gal gene was initially surprising since the two LTR promoters were identical in sequence. Subsequent studies in src transformed cells suggest that this difference between the two promoters is likely to be quantitate and not qualitative, since the LTR driving tax expression in these cells may also be suppressed. This difference likely relates to the differences in transgene copy number between the two constructs (Xu et al. 1994). The LTR-tax construct is present in approximately 40 copies whereas that of LTR-β-gal has five copies and is thus more easily suppressed. Stronger stimuli which are able to suppress both transgenes are discussed below.

In order to study this suppression in greater detail, cell lines were established from the various fibroblastic tumors which arose in doubly (LTR-tax X LTR-β-gal) transgenic mice. Primary cultures showed two distinct populations of fibroblasts when stained with X-gal: a negative and a brightly staining population (Xu et al. 1994). By using a fluorescent β-gal analog on live cells, FACS sorting was able to cleanly separate these two populations. Northern analysis of the β-gal positive and negative cells showed comparable levels of tax expression, with profound diminution of β-gal mRNA in the negative cells.

In each case, the β-gal positive cells were found to be slightly slower growing and to be gradually lost in mixed cultures if not selected for. All of these cells express approximately equal levels of tax. None of the clones derived from β-gal positive cells gave rise to tumors when innoculated into animals, whereas β-gal negative clones did. Thus tumor transfer assays also suggested that β-gal positive cells were premalignant precursors.

8 In Vivo Footprinting of the HTLV-I LTR Suggests Redundant Low to Moderate Affinity Binding Sites

Previously published in vitro footprint analyses of the U3 region of the LTR have revealed considerable variability depending on the tissue type from which the extracts are made. In addition, computer analysis of the sequence reveals few exact consensus sequences for transcription factors. In order to determine the sites of transcription factor contact in the mouse model, in vivo footprint analyses (Mueller et al. 1989; Pfeifer et al. 1991) were performed on transgenic tissues and cell lines (Brown et al. 1994). An important potential advantage of in vivo footprinting over in vitro studies, is that the native conformation of transcription factors, long range protein interactions, and higher order chromatin structure are not disturbed. In addition, ligation mediated PCR allowed for selective amplification of the transcription factor binding pattern from the separate LTRs directing expression of β-gal and tax within the same sample.

Small tissue-specific differences in footprints were seen between lymphocytes and fibroblasts. However, we were surprised to find no significant changes between active and inactive LTRs. The footprint pattern of LTR-β-gal in transgenic thymocytes was identical in singly positive (LTR-βgal) or doubly positive (LTR-β-gal X Thy-tax) mice. It was identical in resting or phorbol ester plus ionomycin treated LTR-tax splenocytes, and there were also no significant footprint differences between blue and white (LTR-tax X LTR-β-gal) fibroblasts, despite profound differences in expression levels. In each case, the chromatin was open and highly

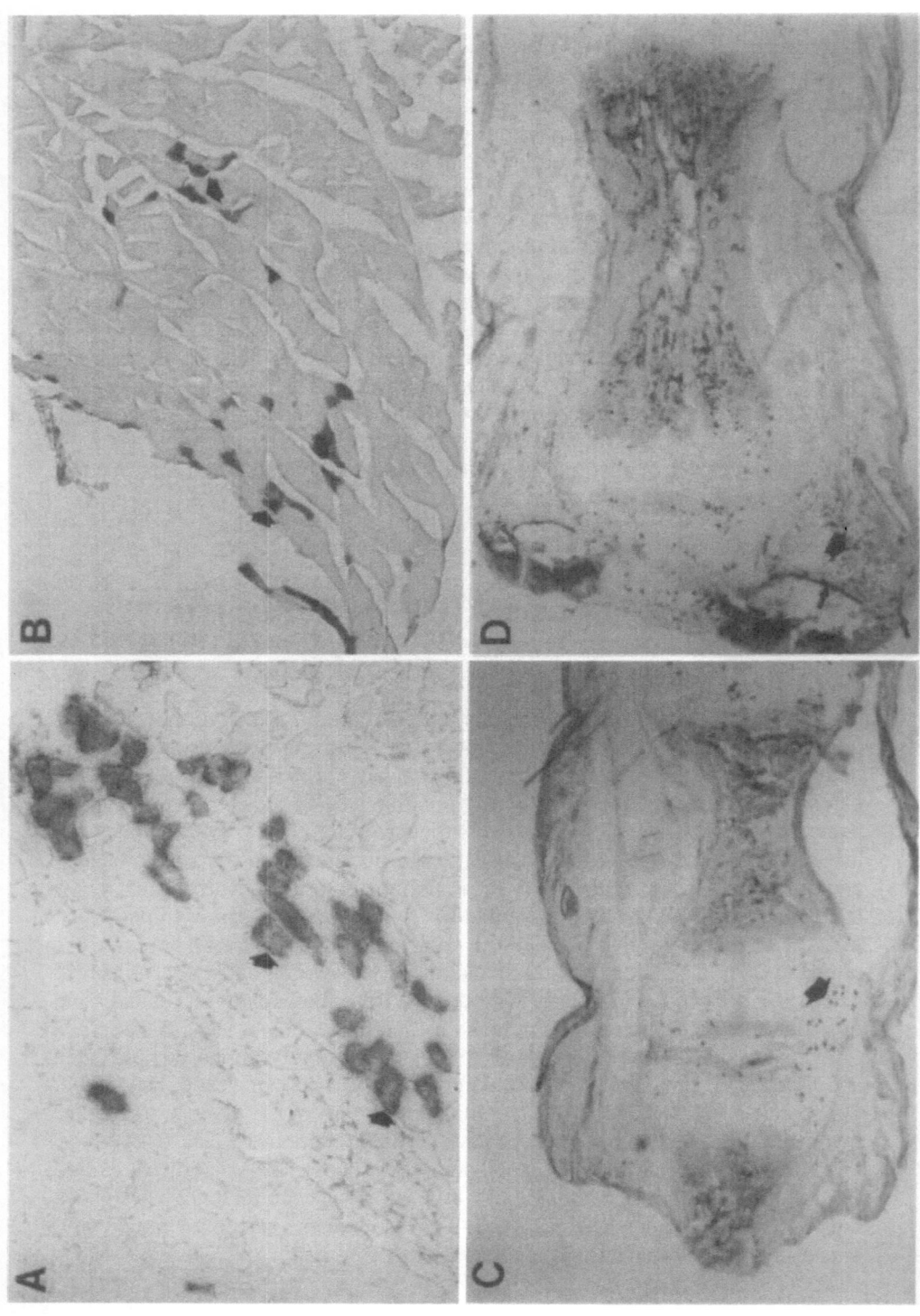

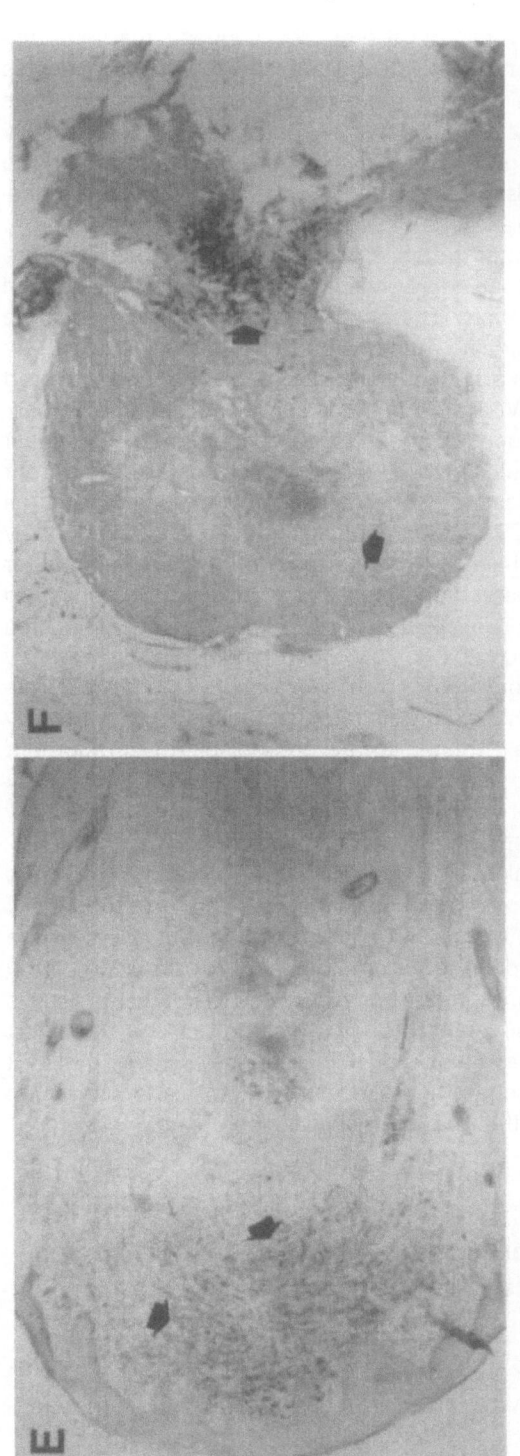

Fig. 3A–F. β-galactosidase expression in LTR-tax X LTR-βgal doubly transgenic mice. **A** Tax expression (*arrows*) in skeletal muscle from a LTR-tax transgenic mouse shown by in situ hybridization with a tax probe. **B** β-gal expression (*arrows*) in skeletal muscle from a LTR-tax X LTR-βgal doubly transgenic mouse shown by X-gal staining. **C–F** β-gal expression in tail. The tails of doubly transgenic mice were wounded by cutting and samples were collected for X-gal staining at **C** 3h, **D** 2 weeks, and **E** 4 weeks after wounding. *Large arrows* point to dermal fibroblasts and *small arrows* to chondrocytes near the wound site. Sections are longitudinal. Scattered chondrocytes and osteocytes are stained. **F** A fibroblastic tumor arising in the tail of a doubly transgenic mouse 4 months after wounding. Nontransformed fibrous capsule cells in the stalk of the tumor stain strongly for β-gal, while cells in the bulk of the tumor do not exhibit β-gal staining

occupied by transcription factors over identical sites. This suggests that transcription factor binding is independent of activation state and that transcription is determined by binding of additional activators through protein-protein interactions or through modification of already bound proteins. This agrees with previously published studies of in vitro binding which show no significant differences in the presence or absence of tax. Our data extends these observations to other *trans*-activating transcription factors, such as those induced by the phorbol plus ionomycin.

In vivo footprinting demonstrated multiple copies of sequences which are related to known transcription factor binding sites. In each case, these contained deviations from the known high affinity consensus sequences which would be predicted to lead to decrease in binding affinity and transcriptional activity. It is possible that these represent high affinity binding sites for as yet undiscovered factors present at low abundance. We feel that this is unlikely for several reasons. First, deviations occur at different sites between repeated elements. This is best exemplified by the three 21 bp repeats in the U3 region. Each contains a different sequence deviation at a different position (6/8 –7/8 match) from the two half-sites which constitute the high affinity CREB site, resulting in the conservation of only one intact half-site per repeat. If these were important new transcription factor binding sites, they should be conserved between repeats. Second, gel shift and in vitro footprinting of the U3 region suggest extremely weak factor binding in a variety of cellular extracts when compared to protein binding on high affinity synthetic regions or similar regulatory regions from a variety of viral and inflammatory factor genes. Third, LTR-β-gal mice demonstrate that the LTR is an extremely weak promoter in the absence of tax. Though tax and phorbol/ionomycin augment transcription, they do not increase factor binding in vivo or in vitro. Thus, weak binding correlates with weak basal transcription, and other modifiers apparently make this weak interaction more effective at stimulating transcription.

We thus hypothesize that weak but redundant transcription factor binding to the HTLV-I LTR causes the chromosomal locus to remain open, independent of transcriptional activity. The locus is poised for transcription, but silent in the absence of Tax. Tax may then induce rapid up-modulation of transcription by serving as a bridge between proteins already present on the LTR, and tax itself is at least partially regulated by transcription through the same LTR and by a short half-life (less than 4 h). This model explains the rapid response of the HTLV-I LTR to environmental stimuli, and perhaps the ease with which this locus is suppressed, but is insufficient by itself to explain maintenance of latency.

9 The HTLV-I LTR is Suppressed by a CREB-Like Pathway Which Interacts with the R Region

Studies in doubly transgenic animals showed that suppression of LTR-βgal correlated with more rapid cell growth, less serum dependence, and tumorigenesis. This suggested that these events may be linked through changes in expression of growth factors or, more directly, through a common signal transduction-like event. We therefore sought to identify agents which would induce malignant-like features in the blue cell precursors (Xu et al. 1994). A simple assay for this was suppression of β-gal expression in blue cells. After extensive trials of growth factors and activating drugs, we found that suppression could partially be reproduced (greater than three fold reduction) with forskolin treatment. More profound suppression (six fold) was induced by okadaic acid at a concentration greater than 10 nM, a range consistent with inhibition of protein phosphatase 1. Neither of these agents independently affected tax expression. Inhibition of protein phosphatase 1 is known to augment the effects of kinase A, and combination of these two treatments was in fact synergistic on suppression of β-gal expression. The combination of drugs slightly suppressed tax expression; however, evaluation of this was limited by toxicity of these drugs.

The above data suggested that the serine/threonine protein kinase A pathway may play a central role in suppression which occurs during tumor progression. However, we also wished to evaluate the role of other protein kinases. Since a large body of literature already exists describing the effects of the tyrosine kinase src on fibroblasts, this kinase was selected for further analysis. Doubly transgenic (LTR-β-gal × LTR-tax) cells were transfected with the avian v-src. Transformation completely shut off both β-gal and tax expression in multiple independent cell lines. Suppression of tax mRNA expression could be partially reversed with optimal doses of herbimycin, a tyrosine kinase inhibitor which has been shown to reverse transformation by src. This demonstrated that suppression was specifically induced by src expression and was not a consequence of clone selection. A similar extent of derepression was achieved with the protein kinase A inhibitor H8. No further derepression was achieved with inhibition of both kinase A and C with optimal concentrations of the more pleiomorphic inhibitor H7.

Thus, despite use of the more pleomorphic upstream src kinase, suppression appeared to occur through the protein kinase A pathway. Since CREB-like factors are well characterized transcriptional targets of protein kinase A, we sought to determine whether they were involved in suppression. To test the involvement of CREB, electrophoretic mobility shift analyses were performed on extracts derived from a variety of cells lines. Since we had no a priori knowledge where within the LTR this interaction would occur, fragments encompassing the complete LTR were used. Gel shift patterns were compared between the tax transformed parent and src transformed extracts. In all of these extracts, binding to the U3 region was weak. There were no significant differences in intensity or mobility between the two extracts. Further, antibodies to CREB failed to supershift U3 binding complexes. These data suggested that the U3 region was not involved in CREB mediated suppression.

Gel shift analysis using a probe encompassing the R region showed a gradation of binding between extracts, which directly correlated with suppression. 3T3 showed the lowest level of binding, followed by doubly transgenic blue fibroblasts, followed by white cells. The src transformed cells showed the highest level of binding with altered mobility of the major binding complex. Phosphatase treatment restored the extent of binding and the mobility to that of the parent blue cell. This suggests the binding of an additional protein to the R region in src transformed cells which is phosphorylation dependent. The binding of the major R region complex was blocked by competition with a high affinity CRE oligonucleotide but not by a high affinity AP-1 oligo which differed by only one base in the core region. Irrelevant control oligos also failed to compete for binding to the R region. In addition, antibody to CREB but not c-jun or jun-D specifically supershifted this complex. These data strongly suggested that CREB binds the R but not U3 regions. Further, this increased CREB binding in the R region correlated with suppression of transcription of the LTR. Though this binding activity was not constitutively present in the nucleus of Balb/3T3 cells, it could rapidly be induced by the protein kinase agonist dibutryl cAMP (dBcAMP). Computer search of the R region failed to identify any sequence which matched a high affinity CREB binding site, i.e., a palindromic repeat of CREB half-sites. In contrast, one match to a perfect single CREB half-site (TGAC) was identified, suggesting that only one subunit of CREB may be involved in suppressing HTLV-I expression.

Indeed, in vitro DNAse footprinting revealed an 18 base region of protection near the R-U5 junction, just 3′ of the TGAC CREB half-site *TGA* CCCTGCTTGCT-CAACTCT. Competition with a synthetic oligonucleotide bracketing the protected region demonstrated specific competition. Thus, specific binding of factors to a discrete region near the R-U5 junction is associated with suppression.

10 The HTLV-I Suppressor Complex Contains a 70 kDa Protein Complexed to CREB

In order to better define the components of the suppressor binding to the HTLV-I R region, partial affinity purification was undertaken. Partially purified fractions were then analyzed by a combination of western, southwestern, and gel shift analyses. The results of these studies demonstrated that the suppressor complex contained a 70 kDa protein with direct DNA binding affinity for the R region target and that the 49 kDa CREB protein was tightly and specifically associated with p70. CREB itself showed no direct DNA binding to the target on southwestern blots. A conventional CREB homodimer interaction with a high affinity binding site would be easily detected by our southwestern blots. Therefore these data suggested that CREB may bind as a monomer. The 70 kDa protein has subsequently been identified as the transcription factor ATF2.

11 Conclusions from Studies of the HTLV-I LTR-tax Transgenic Model

The development of a transgenic tax model has greatly facilitated progress in understanding the mechanisms of tax mediated transformation. It has provided us with reagents to systematically study both early and late steps in tax mediated tumor progression which were not available from clinical materials or in vitro transformation models. The data have led us to a series of conclusions about HTLV-1 mediated transformation. These are summarized in Fig. 4 and below.

11.1 The HTLV-I Promoter Functions Like an Early Cellular Gene

Like an early gene, the LTR is not expressed in most unactivated mouse tissues, including lymphocytes in G_0 phase. The LTR serves as a site of convergence of external transduction signals. Previously published data suggest that this occurs through the U3 region of the virus, though it is possible that downstream regions also play a role. Activation occurs within 2 h by combinations of phorbol esters and

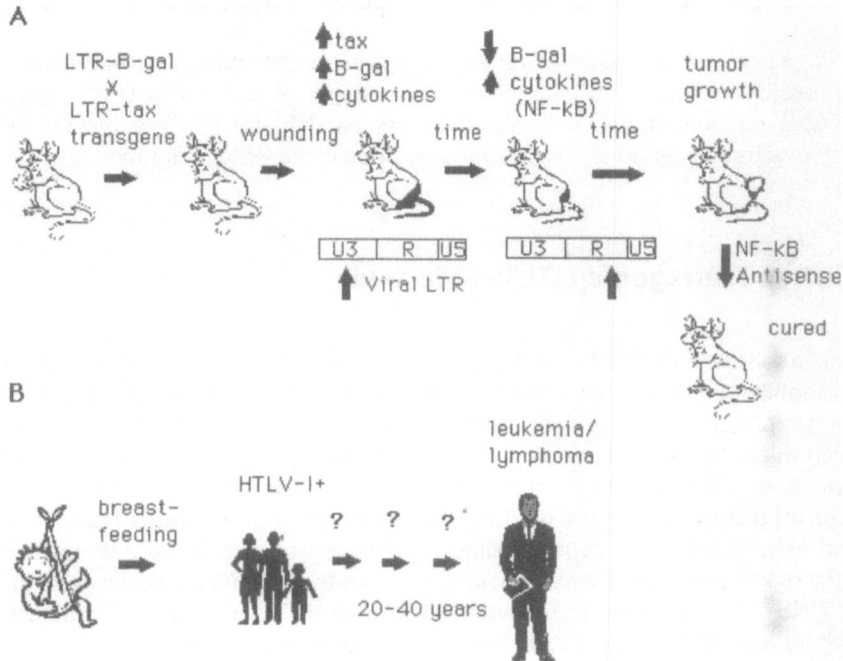

Fig. 4A,B. Steps in HTLV-I mediated oncogenesis. **A** Transgenic mouse model; **B** human

calcium ionophore which can be superinduced by cycloheximide. Wounding serves a similar function in dermal fibroblasts. In vivo footprinting suggests transcriptional poising which is chromosomal site-independent even in the absence of detectable mRNA expression. Like other early genes (e.g., fos, myc) transcription is initially amplified by feedback of the gene product (Tax) on the promoter and suppressed at later stages.

11.2 CREB-Like Factors Play a Crucial Role in Suppression of LTR Expression

In our models, we have been unable to demonstrate direct binding of CREB in U3. In contrast CREB binding occurs at an unconventional site within the R region which correlates with suppression of LTR expression. This may occur through heterodimerization with ATF2. We know from phosphatase studies that binding of this complex is crucially dependent on phosphorylation and that it is specifically increased in kinase A stimulated or src transformed cells. The rapidity of this response in dBcAMP stimulated cells suggests that only posttranslational changes (e.g., phosphorylation) are required for binding.

11.3 Maintenance of the tax Transformed Phenotype Is Dependent on Constitutive Nuclear Expression of NF-κB

Tax expression is clearly required for early steps in transformation. At later stages its expression seems largely dispensable. Later stages of tumor growth are highly dependent on constitutive expression of nuclear NF-κB. The mechanism of constitutive translocation of NF-κB to the nucleus is currently unknown.

12 Other Transgenic HTLV-I Models

Studies in which the HTLV-I or HTLV-2 tax gene were expressed under direction of the metallothionein or Moloney sarcoma virus LTR developed tail tumors of similar morphology to those described above (ROSENBERG et al. 1988). In a model in which the entire HTLV-I genome was introduced, mice developed arthopathy (IWAKURA et al. 1991). However, these results occurred in strains predisposed to immune mediated disease. In a model in which a tropical spastic paraperesis (TSP) derived HTLV-I LTR was used to direct expression of β-galactosidase in the absence of tax, expression was seen in neurons of the CNS (GONZALEZ-DUNIA et al. 1992). CNS expression also occurs when ATL derived LTRs are used as promoters (Shinohara and Nerenberg, unpublished data). However, tissue-specific expression is dramatically altered by the presence of tax (BROWN et al. 1994). In another permutation, doubly transgenic mice were made in which the HTLV-I LTR was used to direct myc and the Ig promoter to direct tax. Pathology was seen only when the myc transgene was present, which was dramatically up-modulated in the presence of Tax (BENVENISTY et al. 1992). Like our studies, these demonstrate that Tax trans-activation works well in a variety of mouse tissues. The presence of brain tumors in some of these animals suggests that the HTLV-I LTR may also have weak basal transcriptional activity in CNS neurons.

Recently, Ratner and colleagues (GROSSMAN 1995), have expressed tax in mice under the granzyme B promoter. These mice develop tumors which the authors believe to be NK cell derived. Recent studies of transgenic rats, demonstrate that tax may also induce mammary tumors in this host (YAMADA 1994).

13 Conclusions and Future Outlook

HTLV-I is a virus which has become very well adapted to its host, and in fact rarely causes disease. It has developed a mode of transmission (sex and breast milk) which insured its survival prior to recent technologic intervention. The virus is almost exclusively transmitted in a cell associated manner and thus may be mostly controlled by cell mediated immunity. It has a strong tendency toward latency, which may allow it to escape immune surveillance in its host. Its expression is rapidly, though transiently, up-modulated when cells are activated, which may occur during transmission of infected cells into a foreign host. Redundancy of weak binding sites may promote transcriptional poising of the virus independent of chromosomal integration site without obligate gene expression. Expression of tax would provide a rapid way to break latency while tax mediated activation of NF-κB may lead to further activation of the host cell and indirect activation of the LTR. Balanced activation and suppression by CREB-like factors may insure rapid reestablishment of latency after a brief period of viral production.

Leukemia may result from a rare but devastating disturbance in this cycle. Leukemia usually progresses rapidly, is refractory to chemotherapy and is uniformly fatal. The tax transgenic models may be a good model for tumor progression and for development of therapeutics. Results from these studies have, in fact, already suggested a rationale for the use of antisense therapeutics in HTLV-I infected patients. Inhibition of tax is only likely to work in preleukemic stages of disease. This may be extremely important for the treatment of the immunopathologic diseases associated with HTLV-I infection such as TSP/HAM (HTLV-I associated myclopathy) or polymyositis. In contrast NF-κB antisense may be a good agent for intervention in advanced stages of leukemia, in which there are currently no satisfactory therapeutic agents.

References

Altman R, Harrich D, Garcia JA, Gaynor RB (1988) Human T-cell leukemia virus types I and II exhibit different D Nase I protection patterns. J Virol 62(4): 1339–1346

Benvenisty N, Ornitz DM, Bennett GL, Sahagan BG, Kuo A, Cardiff RD, Leder P (1992) Brain tumours and lymphomas in transgenic mice that carry HTLV-I LTR/c-myc and Ig/tax genes. Oncogene 7(12): 2399–2405

Brady J, Jeang KT, Duvall J, Khoury G (1987) Identification of p40x-responsive regulatory sequences within the human T-cell leukemia virus type I long terminal repeat. J virol 61: 2175–2181

Brown DA, Xu X, Kitajima I, Shinohara T, Bilakovics J, Fey LW, Shiner T, Nerenberg MI (1994) Genomic footprinting of the HTLV-1 LTR in a transgenic mouse tumorigenesis model. Transgene 1: 297

Clark NM, Smith MJ, Hilfinger JM, Markovitz DM (1993) Activation of the human T-cell leukemia virus type I enhancer is mediated by binding sites for Elf-1 and the pets factor. J Virol 67(9): 5522–5548

Franklin AA, Kubik MF, Uittenbogaard MN, Brauweiler A, Utaisincharoen P, Matthews MA, Dynan WS, Hoeffler JP, Nyborg JK (1993) Transactivation by the human T-cell leukemia virus Tax protein is mediated through enhanced binding of activating transcription factor-2 (ATF-2) ATF-2 response and cAMP element-binding protein (CREB). J Biol Chem 268(28): 21225–21231

Gartenhause RB, Wong-Staal F, Klotman MR (1991) The promoter of human T-cell leukemia virus type 1 is repressed by the immediate early gene region of human cytomegalovirus in primary blood lymphocytes. Blood 78: 2956–2961

Giam CZ, Xu YL (1989) HTLV-I tax gene product activates transcription via pre-existing cellular factors and cAMP responsive element. J Biol Chem 264(26): 15236–15441

Giam CZ, Nerenberg M, Khoury G, Jay G (1986) Expression of the complete human T-cell leukemia virus type I pX coding sequence as a functional protein in Escherichia coli. Proc Natl Acad Sci USA 83(19): 7192–7196

Gitlin SD, Dittmer J, Reid RL, Brady JN (1993) The molecular biology of human T-cell leukemia viruses. In: Cullen B (ed) Frontier in molecular biology. Oxford University Press, Oxford, pp 159–192

Gonzalez-Dunia D, Grimber G, Briand P, Brahic M, Ozden S (1992) Tissue expression pattern directed in transgenic mice by the LTR of an HTLV-I provirus isolated from a case of tropical spastic paraparesis. Virology 187(2): 705–710

Green JE, Hinrichs SH, Vogel J, Jay G (1989) Exocrinopathy resembling Sjogren's syndrome in HTLV-I tax transgenic mice. Nature 341(6237): 72–74

Green JE, Baird AM, Hinrichs SH, Klintworth GK, Jay G (1992) Adrenal medullary tumors and iris proliferation in a transgenic mouse model of neurofibromatosis. Am J Pathol 140(6): 1401–1410

Grossman WJ, Kimata JT, Wong FH, Zutter M, Ley TJ, Ratner L (1995) Development of leukemia in mice transgenic for the tax gene of human T-cell leukemia virus type I. Proc Natl Acad Sci USA 92(4): 1057–1061

Higuchi I, Nerenberg M, Yoshimine K, Osame M, Yoshida M, Yoshimina K, Fukumaga H (1992) Failure to detect HTLV-1 by in situ hybridization in the biopsied muscles of viral carriers with polymyositis. Muscle Nerve 15: 43–47

Hinrichs SH, Nerenberg M, Reynolds RK, Khoury G, Jay G (1987) A transgenic mouse model for human neurofibromatosis. Science 237(4820): 1340–1343

Hinuma Y, Nagata K, Hanaoka M, Nakai M, Matsumoto T, Kinoshita KI, Shirakawa S, Miyoshi I (1981) Adult T-cell leukemia: antigen in an ATL cell line and detection of antibodies to the antigen in human sera. Proc Natl Acad Sci USA 78(10): 6476–6480

Iwakura Y, Tosu M, Yoshida E, Takiguchi M, Sato K, Kitajima I, Nishioka K, Yamamoto K, Takeda T, Hatanaka M et al. (1991) Induction of inflammatory arthropathy resembling rheumatoid arthritis in mice transgenic for HTLV-I. Science 253(5023): 1026–1028

Jeang KT, Boros I, Brady J, Radonovich M, Khoury G (1988) Characterization of cellular factors that interact with the human T-cell leukemia virus type I p40x-responsive 21-base-pair sequence. J Virol 62(12): 4499–4509

Kashanchi F, Duvall JF, Dittmer J, Mireskandari A, Reid RL, Gitlin SD, Brady JN (1994) Involvement of transcription factor YB-1 in human T-cell lymphotropic virus type I basal gene expression. J Virol 68(1): 561–565

Kitajima I, Shinohara T, Minor T, Bibbs L, Bilakovics J, Nerenberg M (1992a) Human T-cell leukemia virus type I tax transformation is associated with increased uptake of oligodeoxynucleotides in vitro and in vivo. J Biol Chem 267: 25881–25888

Kitajima I, Shinohara T, Bilakovics J, Brown DA, Xu X, Nerenberg M (1992b) Ablation of transplanted tax-transformed tumors in mice by antisense inhibition of NF-κB. Science 258: 1792–1795

Levinger L, Lautenberger JA (1987) Human protein binding to DNA sequences surrounding the human T-cell lymphotropic virus type I long terminal repeat polyadenylation site. Eur J Biochem 166: 519–526

Marriott SJ, Boros I, Duvall JF, Brady JN (1989) Indirect binding of human T-cell leukemia virus type I tax 1 to a responsive element in the viral long terminal repeat. Mol Cell Biol 9(10): 4152–4160

Muchardt C, Seeler J-S, Gaynor RB (1992) Regulation of HYLV-I gene expression by tax and AP-2. New Biol 4: 541–550

Mucke L, Oldstone MBA, Morris JC, Nerenberg M (1991) Rapid activation of astrocyte-specific expression of GFAP-lacZ transgene by focal injury. New Biol 3: 465–474

Mueller PR, Wold B (1989) footprinting of a muscle specific enhancer by ligation medi. Science 246: 780–786

Murphy EL, Hanchard B, Figueroa JP, Gibbs WN, Lofters WS, Campbell M, Goedert JJ, Blattner WA (1989) Modelling the risk of adult T-cell leukemia/lymphoma in persons infected with human T-lymphotropic virus type I. Int J Cancer 43(2): 250–253

Nakamura M, Ohtani K, Hinuma Y, Sugamura K (1988) Functional mapping of the activity of the R region in the human T-cell leukemia virus type I long terminal repeat to increase gene expression. Virus Genes 2(2): 147–155

Nerenberg MI (1990) An HTLV-1 transgenic mouse model: role of the tax gene in pathogenesis in multiple organ systems. Curr Top Microbiol Immunol 160(121): 121

Nerenberg M (1992) Biologic and molecular biologic aspects of HTLV-1-associated diseases. In: Roos R (ed) Molecular biologic approaches to the study of CNS Viral Disease. Humana, Clifton, pp 225–247

Nerenberg MI, Wiley CA (1989) Degeneration of oxidative muscle fibers in HTLV-I tax transgenic mice. Am J Pathol 135(6): 1025–1033

Nerenberg M, Hinrichs SH, Reynolds RK, Khoury G, Jay G (1987) The tat gene of human T-lymphotropic virus type 1 induces mesenchymal tumors in transgenic mice. Science 237(4820): 1324–1329

Nerenberg MI, Minor T, Nagashima K, Takebayashi K, Akai K, Wiley CA, Riccardi VM (1991a) Absence of association of HTLV-1 infection with type 1 neurofibromatosis in the United States or Japan. Neurology 41: 1687–1689

Nerenberg MI, Minor T, Price J, Ernst DN, Shinohara T, Schwarz H (1991b) Transgenic thymocytes are refractory to transformation by the human T-cell leukemia virus type I tax gene. J Virol 65: 3349–3353

Nyborg JK, Dynan WS (1990) Interaction of cellular proteins with the human T-cell leukemia virus type I transcriptional control region. Purification of cellular proteins that bind the 21-base pair repeat elements. J Biol Chem 265(14): 8230–8236

Nyborg JK, Mathews MA, Yucel J, Walls L, Golde WT, Dynan WS, Wachsman W (1990) Interaction of host cell proteins with the human T-cell leukemia virus type I transcriptional control region. II. A comprehensive map of protein-binding sites facilitates construction of a simple chimeric promoter responsive to the viral tax2 gene product. J Biol Chem 265(14): 8237–8242

Paca-Uccaralertkun S, Zhao LJ, Adya N, Cross JV, Cullen BR, Boros IM, Giam CZ (1994) In vitro selection of DNA elements highly responsive to the human T-cell lymphotropic virus type I transcriptional activator, Tax. Mol Cell Biol 14(1): 456–462

Pfeifer GP, Riggs AD (1991) Chromatin differences between active and inactive X chromosomes revealed by genomic footprinting of permeabilized cells using DNAse I and ligation mediated PCR. Genes Dev 5: 1102–1113

Pfeifer GP, Tanguay RL, Steiqerwald SD (1991) In vivo footprinting and methylation analysis by PCR-aided genomic sequencing: comparison of active and inactive X chromosomal DNA at the CpG island and promoter of human PGK-1. Genes Dev 4: 1277–1287

Poiesz BJ, Ruscetti FW, Gazdar AF, Bunn PA, Minna JD, Gallo RC (1980) Detection and isolation of type C retroviral particles from fresh and cultured lymphocytes of a patient with cutaneous T-cell lymphoma. Proc Natl Acad Sci USA 77: 7415–7419

Pozzatti R, Vogel J, Jay G (1990) The human T-lymphotropic virus type I tax gene can cooperate with the ras oncogene to induce neoplastic transformation of cells. Mol Cell Biol 10(1): 413–417

Robert GM, Nakao Y, Notake K, Ito Y, Sliski A, Gallo RC (1982) Natural antibodies to human retrovirus HTLV in a cluster of Japanese patients with adult T cell leukemia. Science 215(4535): 975–978

Rosenberg MP, Felber BK, Walton EM, Swing DA, Grammatikakis N, Pavlakis GN, Jenjins NA, Copeland NG (1988) Molecular mechanisms of HTLV transactivator gene in transgenic mice (Abstr). Cold Spring Harbor Mouse Molecular Genetics Meeting, Cold Spring Harbor

Ruben SM, Perkins A, Rosen CA (1989) Activation of NF-kappa B by the HTLV-I trans-activator protein Tax requires an additional factor present in lymphoid cells. New Biol 1(3): 275–284

Seeler JS, Muchardt C, Podar M, Gaynor RB (1993) Regulatory elements involved in tax-mediated transactivation of the HTLV-1 LTR. Virology 196(2): 442–450

Seiki M et al. (1984) Nonspecific integration of the HTLV-1 provirus genome into adult T-cell leukemia cells. Nature 309: 640–642

Smith MR, Greene WC (1991) Molecular biology of the type I human T-cell leukemia virus (HTLV-1) and adult T-cell leukemia. J Clin Invest 87(3): 761–766

Suzuki T, Hirai H, Fujisawa J, Fujita T, Yoshida M (1993) A trans-activator Tax of human T-cell leukemia virus type 1 binds to NF-kappa B p50 and serum response factor (SRF) and associates with enhancer DNAs of the NF-kappa B site and CArG box. Oncogene 8(9): 2391–2397

Tan TH, Horikoshi M, Roeder RG (1989) Purification characterization of multiple nuclear factors that bind to the TAX-inducible enhancer within the human T-cell leukemia virus type 1 long terminal repeat. Mol Cell Biol 9: 1733–1745

Tanaka A et al. (1990) Oncogenic transformation by the tax gene of human T-cell leukemia virus type I in vitro. Proc Natl Acad Sci USA 87(3): 1071–1075

Tanimura K, Tsehima H, Fujisawa JI, Yoshida M (1993) A new element that augments the tax-dependent enhancer of human T-cell leukemia virus type 1 and cloning of cDNAs encoding its binding proteins. J Virol 67: 5375–5382

Wagner S, Green MR (1993) HTLV-1 Tax protein stimulation of DNA binding of bZIP proteins by enhancing dimerization. Science 262(5132): 395–399

Xu YL, Adya N, Siores E, Gao QS, Giam CZ (1990) Cellular factors involved in transcription and Tax-mediated trans-activation directed by the TGACGT motifs in human T-cell leukemia virus type I promoter. J Biol Chem 265(33): 20285–20292

Xu X, Brown DA, Bilakovics J, Kitajima I, Bilakovics J, Fey LW, Nerenberg M (1994) Transcriptional suppression of the HTLV-I LTR occurs by an unconventional interaction of a CREB factor with the R region. Mol Cell Biol 14(8): 5371–5383

Yamada S (1994) Pathological and molecular analyses of mammary tumors induced in HTLV-I pX transgenic rats. Hokkaido Igaku Zasshi 69(3): 479–491, 493, 495–497

Yoshida M (1994) Mechanism of transcriptional activation of viral and cellular genes by oncogenic protein of HTLV-1. Leukemia 8 Suppl 1: S51-S53

Yoshida M, Seiki M, Yamaguchi K, Takatsuki K (1984) Monoclonal integration of human T-cell leukemia provirus in all primary tumors of adult T-cell leukemia suggests causative role of human T-cell leukemia virus in the disease. Proc Natl Acad Sci USA 81(8): 2534–2537

Transgenic Models
of Human Immunodeficiency Virus Type-1

P.E. Klotman[1] and A.L. Notkins[2]

[1] Mount Sinai Medical Center, Box 1243, Division of Nephrology, New York, NY 10029, USA
[2] National Institute of Dental Research, Building 30, NIH, Bethesda, MD 20892, USA

1 Introduction

The use of transgenic technology has provided significant insights into the molecular pathogenesis of HIV-1 and its clinical manifestations. In particular, transgenic mice have suggested a potential role for Tat in Kaposi's sarcoma, and a potential mechanism for the AIDS dementia complex. Furthermore, they have provided the strongest evidence that HIV-associated nephropathy is a direct consequence of viral infection rather than a complication of immune dysfunction or drug abuse. Transgenic mice have also shed light onto the molecular regulation of HIV-1 both at the transcriptional and posttranscriptional levels. In this chapter, we will review HIV-1 transgenic models, focusing on both single (Table 1) and multigenic (Table 2) murine lines that express viral genes and on transgenic mice expressing indicator genes under the control of the LTR.

2 Single Gene Constructs

2.1 LTR-*tat*

Clearly one of the most important transgenic models in furthering our understanding of HIV pathogenesis was that generated by VOGEL et al. (1988) in which *tat* was placed under the control of the HIV-1 LTR and introduced into mice as a transgene. From the six founding animals, three lines manifested epidermal proliferation and dermal tumors (Fig. 1). The development of dermal tumors occurred late in life with epidermal proliferation often the first evidence of disease. After 12–18 months, erythematous, vascular skin tumors were detected in 15% of male mice. Because of the marked dermal infiltration with spindle shaped cells and the evidence of angiogenesis, these tumors were suggested by the authors to have many of the histological features of Kaposi's sarcoma in humans.

A number of additional clinical features of this murine model were similar to Kaposi's sarcoma in humans. Thus, as in Kaposi's lesions in man (DELLI BOVI et al. 1986; WERNER et al. 1989; HUANG et al. 1992), expression of the transgene was detected in the neighboring epidermis but not in the tumor itself. In addition, although the level of *tat* expression appeared to be equivalent in male and female mice, tumor formation was gender-specific. Since only 15% of mice developed the lesions, additional cofactors appeared to be required for the development of Kaposi's lesions, just as they appear necessary in humans. Despite the excitement generated over this initial report suggesting a role for Tat in the pathogenesis of Kaposi's sarcoma, the exact mechanism(s) for the development of this tumor remains enigmatic.

As mentioned previously, epidermal proliferation was also a commonly observed phenotype in the LTR-*tat* transgenic mouse model and actually was

Table 1. Single HIV-1 gene constructs utilized in transgenic mice

Transgene	Viral clone/strain	Promoter-enhancer	Murine strain	Predominant phenotype	Reference
tat	HTL V$_{IIIB}$	αA-Cry	FVB/N	Tat transactivates an LTR-CAT transgenic in vivo (eye)	VHILLAN et al. (1988)
tat	ARV-2	LTR	CD-1	Epidermal proliferation and dermal Kaposi's-like tumors	VOGEL et al. (1988)
tat	ARV-2	LTR	CD-1	High incidence of hepatocellular carcinoma	VOGEL et al. (1991)
tat	LAV-1$_{Bru}$	HSV1 TK-hCD2	C57B1/6 x SJL	Tat transactivates an LTR-α1-antitrypsin reporter gene in vivo	MEHTALI et al. (1992)
tat	ARV-2	LTR	CD-1	Epidermal expression of transgene expression is UV-light inducible	VOGEL et al. (1992)
tat	ARV-2	LTR	CD-1	Systemic liposomes can be delivered to dermal Kaposi's lesions	HUANG et al. (1993)
BKV-tat	HTLV$_{IIIB}$	LTR	BDFI x CD1	Multiple Ca: adeno ca, leiomyosarcoma, squamous papillomas, B cell lymphomas, ±hepatocellular ca, ±Kaposi's sarcoma	CORALLINI et al. (1993)
nef	NL4-3 or HxB3	CD3	C3HeB/FeJ	CD4+ depletion with NL4-3 nef but not with the HxB3 nef	SKOWRONSKI et al. (1993)
nef	NL4-3	LTR or MMTV	FVB/N	Spontaneous and UV light-inducible epidermal papillomas	DICKIE et al. (1993)
nef	HIV-1$_{Bru}$	mVβ8.3 TCRβ	C57B1/6 x DBA/2	Severe immunodeficiency with thymic and peripheral CD4+ depletion	LINDEMANN et al. (1994)
gp120	LAV	GFAP	C57B1/6 x SJL	Neuronal and glial changes consistent with AIDS associated dementia	TOGGAS et al. (1994)

BKV-tat, chimeric gene expressing both the early region of the BK virus T antigen and tat; GFAP, glial fibriallary acidic protein; mVβ8.3 TCRβ, murine Vβ8.3 T cell receptor β chain promoter-enhancer; HSV1 TK, thymidine kinase promoter from the herpes simplex virus; HMGCoA, hydroxy-methylglutaryl-coenzyme A reductase; ca, carcinoma.

Table 2. Multigenic constructs or infectious proviral clones utilized in transgenic mice

Transgene	Viral clone/strain	Promoter-enhancer	Murine strain	Predominant phenotype	Reference
Infectious NL4-3	NL4-3	LTR	FVB/N	Wasting syndrome observed in F1 heterozygotes with epidermal hyperplasia, lymphadenopathy, slenomegaly, interstitial pneumonitis; infectious virus rescued from skin, spleen, and lymph node	LEONARD et al. (1988)
Infectious NL4-3	NL4-3	LTR	FVB/N	Follow-up of LEONARD et al. (1988): skin involvement prerequisite for systemic wasting phenotype	ABRAMCZUK et al. (1992)
Δ gag/pol	NL4-3	LTR	FVB/N	Histologic evidence of HIV-associated nephropathy	DICKIE et al. (1991)
Δ gag/pol	NL4-3	LTR	FVB/N	Functional evidence of HIV-associated nephropathy and glomerulosclerosis with increased collagen IV expression	KOPP et al. (1992)
Δ gag/pol	NL4-3	LTR	FVB/N	Spontaneous and UV-inducible epidermal proliferation with papilloma formation	KOPP et al. (1993)
Δ gaag/pol	NL4-3	LTR	FVB/N	Vaccination with gp160 reduces the incidence of renal disease	SHIRAI and KLINMAN (1993)
Δ gag/pol	NL4-3	LTR	FVB/N	Renal disease manifested by nephrotic syndrome, enlarged kidneys, without evidence of immune complex deposition	KOPP et al. (1994)
Δ gag/pol	NL4-3	LTR	FVB/N	Wasting syndrome in homozygotes, epidermal hyperplasia, lymphadenopathy, splenomegaly, interstitial pneumonitis and reduced peripheral CD4+ T cells	SANTORO et al. (1994)
Δ gag/pol	NL4-3	LTR	FVB/N	Both bFGF and its low affinity receptors participate in the interstitial proliferative disease of HIV-associated nephropathy	RAY et al. (1994)
Δ gag/pol	NL4-3	LTR	FVB/N	Tissue-specific expression and posttranscriptional processing of transgene mRNA in vivo. Splice acceptor site utilization in humans and mouse the same with the exception of vpu/env	BRUGGEMAN et al. (1994)
Δ gag/pol	NL4-3	LTR	FVB/N	Wasting syndrome occurs postnatally and transgene expression increases after delivery	FRANKS et al. (1994)
Δ RT	NL4-3	LTR	C3H/HeN	Cataracts	IWAKURA et al. (1992)
Δ 5'and 3' LTRs	BH10	MMTV	C57BL/6 x C3H	Gag and Env proteins detected in vivo in serum, mammary gland, milk, epididymis, epididymal secretions, spleen, and liver	JOLICOEUR et al. (1992)

MMTV, mouse mammary tumor virus; FGF, fibroblast growth factor.

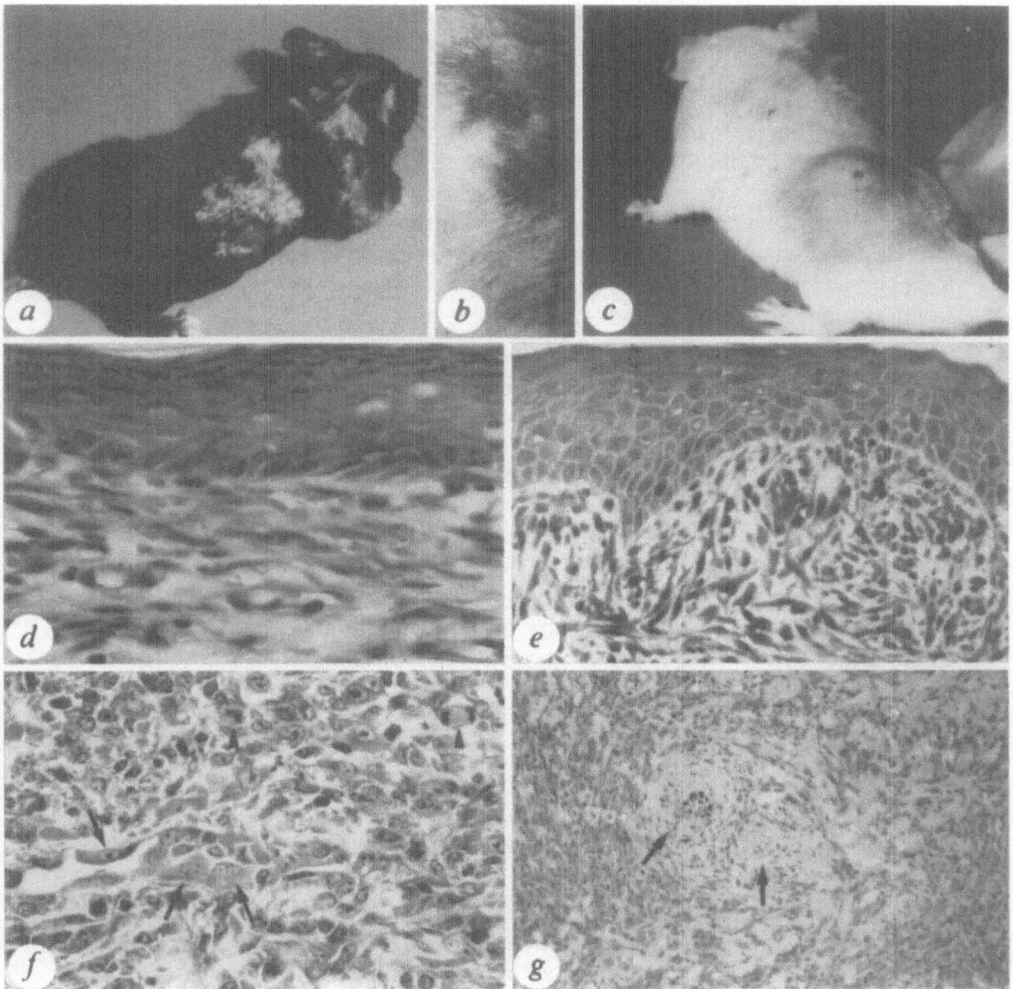

Fig. 1a–g. Skin lesions in the LTR-*tat* transgenic mice. Erythematous lesions apparent in a 14 month old F1 (**a**), an 8 month old F1 (**b**), and a 10 month old F1 (**c**) from three different founder lines. Histology of lesions at 10 months (**d**), 12 months (**e**), and 14 months (**f**) as well as a dermal tumor (**g**). In **f**, *arrows* indicate pleomorphic endothelial cells; in **g**, neoplastic cells surrounding nerve roots. (Adapted from VOGEL et al. 1988)

more common than the development of dermal tumors. The majority of animals experienced an increase in the epidermal cell layer thickness with hyperkeratosis and parakeratosis (VOGEL et al. 1988). These lesions resemble the psoriasiform lesions described in association with HIV-1 infection (DUVIC 1991). Thus, Tat may play a role in both HIV-associated epidermal and dermal lesions.

2.1.1 UV-Inducible Expression of the LTR

The LTR-*tat* transgenic mice have provided additional insights into host pathogenesis, the regulation of the LTR in vivo, and potential therapeutic interventions. Using the LTR-*tat* mice, VOGEL et al. (1992) reported that UV light could induce expression of the LTR in vivo. Stimulated expression was again restricted to the epidermis and dermal expression was not detected either before or after UV irradiation. UV light did not appear to induce the development of Kaposi's sarcoma lesions despite the increase in transgene expression. Whether this was due to the transient nature of the stimulus or evidence that Tat is not specifically required for the development of dermal lesions remains a critical issue.

2.1.2 Liposome Delivery to Dermal Lesions

HUANG et al. (1993) utilized the LTR-*tat* transgenic mice to explore whether changes in vascular permeability surrounding Kaposi's lesions could be employed to direct therapeutic interventions. Liposomes containing colloidal gold were delivered systemically to transgenic mice. Within 24 h, gold particles were detected in dermal lesions and in spindle cells and macrophages surrounding the tumors. These data provided promising evidence that systemic liposomes may have potential as vehicles for drugs or gene therapeutic constructs for the treatment of Kaposi's sarcoma or other vascular tumors.

2.1.3 Hepatocellular Carcinoma

After several years of following these lines, VOGEL et al. (1991) reported that the LTR-*tat* mice were at high risk for the development of hepatocellular dysplasia and liver carcinomas. Like the dermal tumors, the incidence of liver cancer was observed predominantly in male mice (33%–52%) with <3% of female mice developing this lesion. Reminiscent of the findings in the dermal tumors, *tat* mRNA could not be detected even by PCR in two of three lines. Although the authors speculated that liver tumors may be observed increasingly in AIDS patients as survival is prolonged, the exact relationship of these tumors to humans or to patients infected with HIV-1 remains to be established.

2.2 LTR-BK$_{(early region)}$ *tat* Chimeric

Recently, CORALLINI et al. (1993) generated transgenic mice using a chimeric construct in which the BK virus early region including the T antigen and Tat were expressed under the control of the HIV-1 LTR. These authors attempted to achieve a more wide spread expression of Tat by using the BK promoter. This goal was achieved with both Tat- and T antigen-encoding mRNA detected in brain, liver, spleen, kidney, stomach, gonads, thymus, and skin; the lung was the only organ in which transgene message was not detected. These animals also manifested epidermal proliferation with papilloma formation as well as superficial ulcerations. Similar to the findings of Vogel et al. (1988), epidermal lesions were seen in males only and the frequency of skin lesions was increased when the transgene was bred onto a CD1 background. The clinical course, however, was quite different, with skin lesions resolving spontaneously after 2–4 months. Also, unlike the LTR-*tat* mice of Vogel, these mice did not develop lesions resembling Kaposi's sarcoma although some mice developed hepatocellular carcinoma. Other tumors developed after 6 months which probably reflected the expression of the BK virus T antigen. Tumors included adenocarcinomas, fibrosarcomas, squamous cell papillomas, and carcinomas involving the skin. In addition, some animals developed B cell lymphomas that involved abdominal wall, liver, and spleen. These findings are difficult to interpret in view of the known transforming effects of T antigen. That the phenotype was dependent upon genetic background, however, again suggests that genetic cofactors play an important role in the phenotypic expression of certain HIV-1-related diseases, particularly Kaposi's sarcoma.

2.3 αA-Crystallin-*tat*

Khillan et al. (1988) addressed the ability of Tat to transactivate the LTR in vivo in murine tissue by generating transgenic mice in which *tat* was placed under the control of the αA-crystallin promoter. These mice were mated to transgenic mice expressing the chloramphenicol acetyl transferase (CAT) gene linked to the LTR. A six-fold increase in CAT expression was detected in the lens of the eye in mice expressing both *tat* and the LTR-CAT genes when compared to LTR-CAT mice alone. These data suggest that the host factors required for Tat activation of the LTR are present in murine systems, an important issue for those interested in studying physiological regulation of the LTR.

2.4 HSV TK/hCD2-*tat*

In a attempt to develop in vivo systems to test potential therapies to inhibit LTR-Tat interactions, Mehtali et al. (1992) generated transgenic mice expressing *tat* under the control of the herpes simplex virus thymidine kinase promoter and the CD2 enhancer to restrict expression to lymphocytes. As an indicator line, a second line of transgenics was generated in which α_1-antitrypsin with its signal peptide was placed under control of the LTR. When these lines were mated, α_1-antitrypsin mRNA could be detected in the thymus and circulating α_1-antitrypsin was found to be 10- to 20- fold higher in double transgenic mice. Of note, these transgenic *tat* mice manifested no phenotype. These studies provide important corroborating evidence that Tat can transactivate the LTR in murine systems in vivo. In addition, these animals may serve as potential models to test various therapeutic interventions.

2.5 LTR-*nef* and MMTV-*nef*

Epidermal pathology was observed by DICKIE et al. (1993) in transgenic mice expressing *nef* (NL4–3) under the control of the HIV-1 LTR or a heterologous promoter from the mouse mammary tumor virus (MMTV). In the LTR-*nef* lines, proliferative epidermal lesions were noted in four of five lines with variable penetrance (from 10% to 75%). The MMTV-*nef* construct was less efficient in generating papillomatous lesions (i.e. 5%–25%), but the lesions were grossly and histologically identical to those from the LTR-*nef* mice. Expression of the transgene was detectable in diseased skin and immunofluorescences with an anti-Nef antibody revealed staining in the proliferating basal cells of the epidermis. This study suggests that expression of LTR-*nef* in skin leads to epidermal proliferation, but the authors concluded that epidermal proliferation represents one of the in vivo functions of Nef. The latter conclusion, however, is difficult to reconcile with the data generated by VOGEL et al. (1988), in which expression of *tat* without *nef* induces similar epidermal proliferation.

2.6 CD3 Enhancer/Promoter-*nef*$_{NL4-3}$

Transgenic models have provided important insights into the potential pathogenic role for *nef* in vivo. SKOWRONSKI et al. (1993) generated mice transgenic for *nef* (derived from the cloned virus NL4–3) in which expression was restricted to T cells by the murine CD3 enhancer/promoter. Expression of *nef* mRNA and protein in thymus and thymic T cells resulted in a reduction in peripheral CD4$^+$ T cells. As a control, additional transgenics were generated using *nef* derived from the cloned virus HxB3 (HAMMES et al. 1989). In this clone, a premature stop codon in *nef* prevents expression of the complete protein. Transgenics derived from the HxB3 clone had no apparent reduction in CD4$^+$ cells in marked contrast to the transgenics derived from the NL4–3 clone. Thus, these data suggest that Nef plays a critical role in reducing CD4$^+$ number.

2.7 vβ8.3 Promoter/T Cell Receptor β Chain Enhancer-*nef*$_{Bru}$

LINDEMANN et al. (1994) have recently extended these findings by demonstrating that expression of *nef* induces a severe immunodeficiency syndrome in transgenic mice. These investigators placed HIV-1$_{Bru}$ *nef* under the control of the murine Vβ8.3 T cell receptor β chain promoter-enhancer to restrict expression to lymphoid cells. Like the mice of SKOWRONSKI et al. (1993), transgenic mice experienced a fall in CD4$^+$ lymphocyte number. These mice also manifested a syndrome characterized by thymic atrophy, splenomegaly, lymphadenopathy, and increased mortality, with the majority of mice dead by 4–6 months of age. At 8 weeks of age, mice were unable to respond normally to viral challenge with VSV. Thus, in this model, expression of *nef* alters CD4+ cell number but importantly produces a syndrome of immunodysfunction.

2.8 LTR-gp120

Toggas et al. (1994) generated transgenic mice expressing gp120 in the brain to explore potential mechanisms of HIV-induced AIDS dementia complex. The *env* gene encoding gp120 was placed under the control of the glial fibrillary acid protein promoter, restricting expression to astrocytes. Three of five transgenic lines developed histological changes in the brain that are similar to those observed in humans, with dendritic vacuolization and decreased synaptodendritic complexity. All lines expressed mRNA for the transgene but the protein was not detected in brain. These studies suggest that gp120 can induce a pathological response in brain if it gains access to the central nervous system. The mechanisms responsible for gp120 expression in the central nervous system in humans however, remain to be defined.

3 Multigenic Constructs

While single gene constructs have the advantage of producing a more straightforward interpretation of a phenotype, multigenic constructs have been very successful at modeling many of the manifestations of HIV-1 infection.

3.1 The Infectious Proviral Clone NL4–3

In an attempt to develop a model of viral infection in rodents, Leonard and colleagues placed the entire infectious provirus NL4–3 into mice as a transgene (Leonard et al. 1988; Pezen et al. 1991). All seven founders were healthy but one developed antibodies to HIV gp120 and p65 reverse transcriptase. Heterozygous transgenic F1 mice from this founder developed a syndrome characterized by growth retardation and cachexia with death within 1 month of birth. These animals manifested diffuse epidermal proliferation, lymphadenopathy, splenomegaly, thymic atrophy, and interstitial pneumonitis (Fig. 2). They did not develop Kaposi's-like lesions but they may not have been able to survive long enough for the development of this phenotype. Virus was rescued from mice by cocultivation of skin, spleen, and lymph node with a permissive human T cell line. Access to the colony was severely restricted by the required isolation under biosafetly level 4 conditions. Ultimately, this transgenic line of mice was lost in a laboratory accident. While promising, the utility of mice bearing infectious proviral transgenes as models of HIV-1 infection remains to be established.

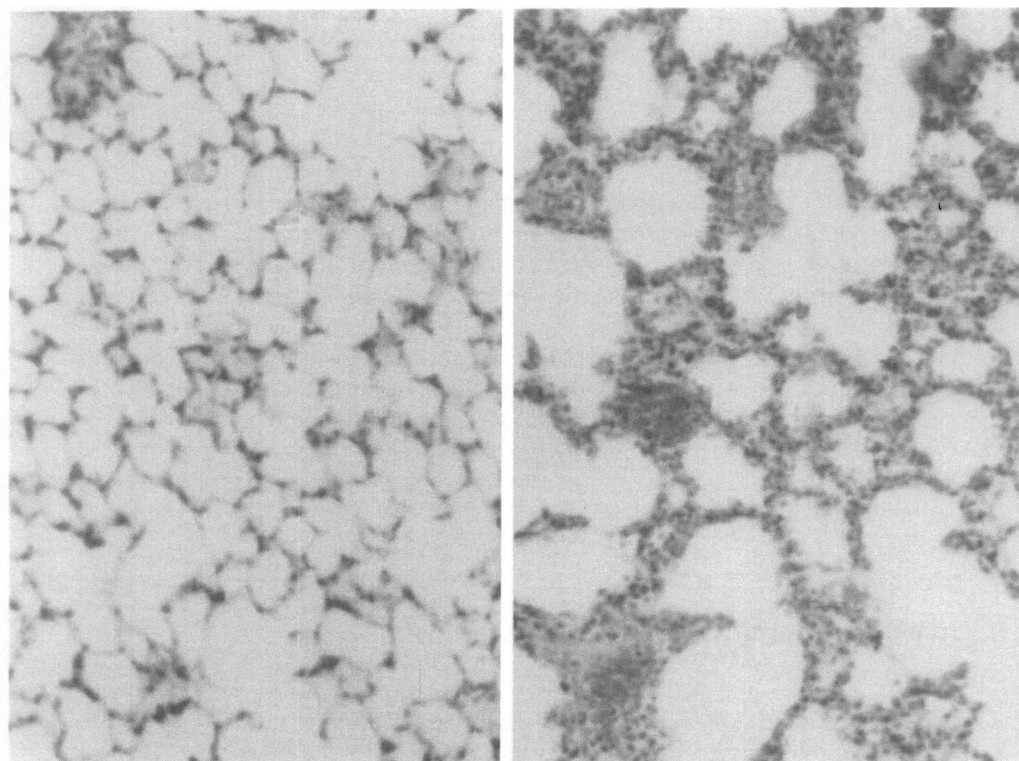

Fig. 2. Interstitial pneumonitis in an affected HIV transgenic mouse heterozygous for the infectious provirus NL4–3. *Left*, lung from a normal littermate; *right*, lung from a transgenic mouse. (From Leonard et al. 1988), × 725

3.2 NL4–3Δ *gag/pol*

As a logical follow-up to the introduction of the infectious provirus, Dickie et al. (1991) introduced a *gag/pol* deletion mutant (Δ) of NL4–3 into the germline of mice as a transgene. This deletion rendered the provirus noninfectious and allowed animals to be housed under normal rodent conditions. These transgenic mice have provided some of the most rewarding findings regarding HIV-1 pathogenesis.

3.2.1 Epidermal Lesions

Epidermal proliferation has been a common finding in LTR-directed constructs; it was observed in the LTR-*tat* mice (Vogel et al. 1988; Corallini et al. 1993), the LTR-*nef* mice (Dickie et al. 1993), and the mice expressing the infectious proviral construct NL4–3 as well (Leonard et al. 1988). In the NL4–3Δ*gag/pol* transgenic line, epidermal proliferation was again a prominent finding (Kopp et al. 1993). In homozygous mice, diffuse involvement of the epidermis was observed and was identical to that in the mice bearing the wild-type infectious provirus (LEONARD et al. 1988). Heterozygotes also manifested epidermal disease but it was much more focal in nature with papilloma formation (Fig. 3). Expression of gp120 was detected by immunohistochemistry and by in situ hybridization in areas of proliferating epithelium. Furthermore, UV stimulation increased the level of transgene expression (Fig. 4) and induced papilloma formation (KOPP et al. 1993). Thus, expression of the transgene appeared to be directly related to the development of epidermal disease.

3.2.2 HIV-Associated Nephropathy

Heterozygous mice from the NL4–3Δ*gag/pol* transgenic line developed a renal disease (DICKIE et al. 1991; KOPP et al. 1992) that is histologically identical to that observed in humans (RAO et al. 1984, 1987; RAO 1991a,b; reviewed in RAPPAPORT et al. 1994) (Fig. 5). The renal disease was characterized by two striking pathological processes, sclerosis and proliferation (KOPP et al. 1994). Sclerosis was evidenced by the glomerular deposition of the basement membrane proteins laminin, collagen IV, and heparan sulfate proteoglycan in a mesangial distribution (Fig. 6) (KOPP et al. 1992). Steady state mRNA for the α1 chain of type IV collagen was increased, suggesting that enhanced production of extracellular matrix proteins contributed, at least in part, to renal pathogenesis (KOPP et al. 1992). The finding that the renal disease in transgenic animals (DICKIE et al. 1991; KOPP et al. 1992, 1994) reproduces the phenotype in humans has provided the strongest evidence supporting a direct role for expression of viral gene products in the pathogenesis of HIV-associated nephropathy.

The transgenic line expressing NL4–3Δ *gag/pol* has also provided an outstanding model to explore mechanisms of renal pathogenesis. RAY et al. (1994) explored the mechanisms responsible for renal epithelial proliferation, a prominent feature of HIV-associated nephropathy. Kidneys from heterozygous mice with renal disease were enlarged, weighed more, and had increased protein and DNA content. Primary tubular epithelial cells isolated from transgenic kidneys proliferated at a faster rate than those isolated from normal nontransgenic littermate controls. Furthermore, when grown on an artificial basement membrane, transgenic epithelial cells tended to form cyst-like structures. Basic fibroblast growth factor (bFGF) appeared to be a likely candidate as a mediator of this pathological process. Thus, transgenic models have provided some of the best evidence for a direct role of HIV-1 in the pathogenesis of HIV-associated nephropathy as well as a useful model to explore mechanisms of renal pathogenesis. This model has also begun to provide an avenue for the exploration of potential therapies for HIV-associated nephropathy (SHIRAI and KLINMAN 1993).

3.2.3 Growth Failure and Cachexia in Transgenic Mice

One of the most striking findings of the NL4–3Δ *gag/pol* line was reported by SANTORO et al. (1994), describing a syndrome characterized by growth failure,

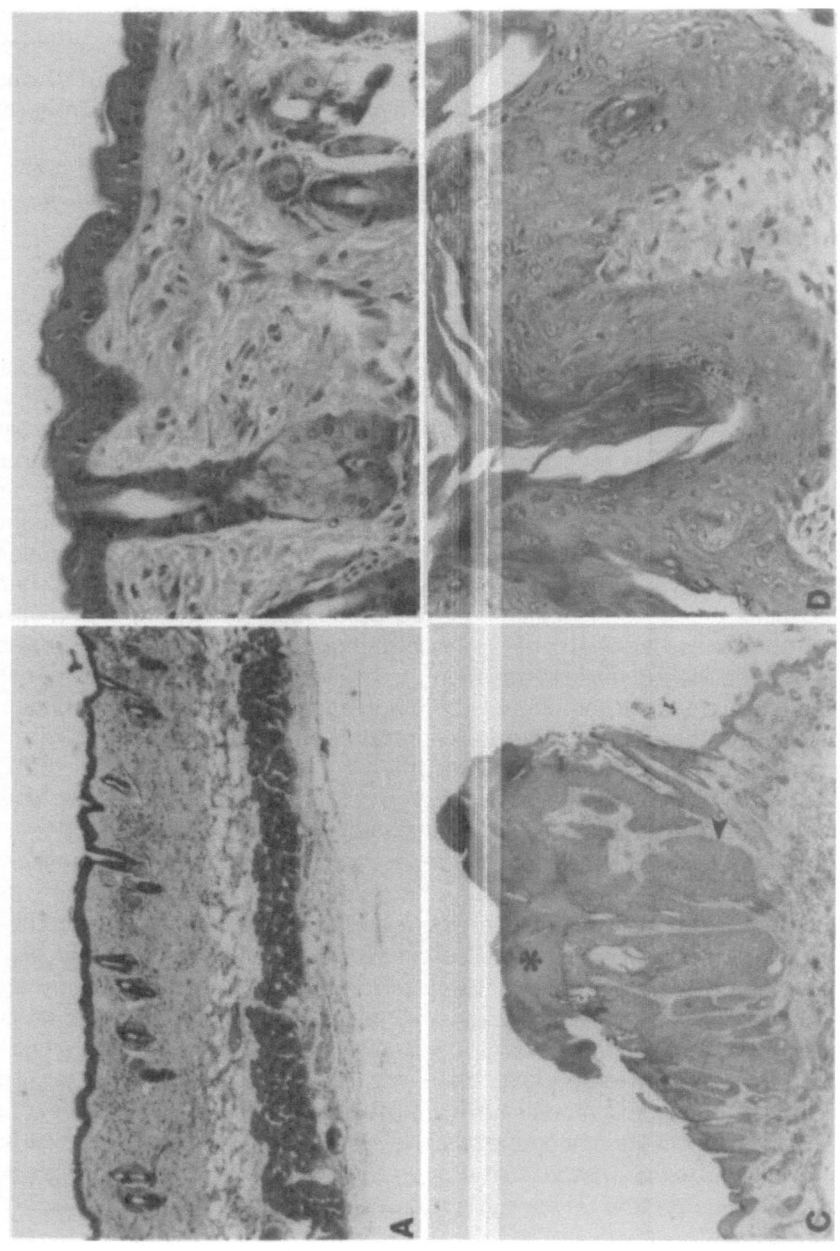

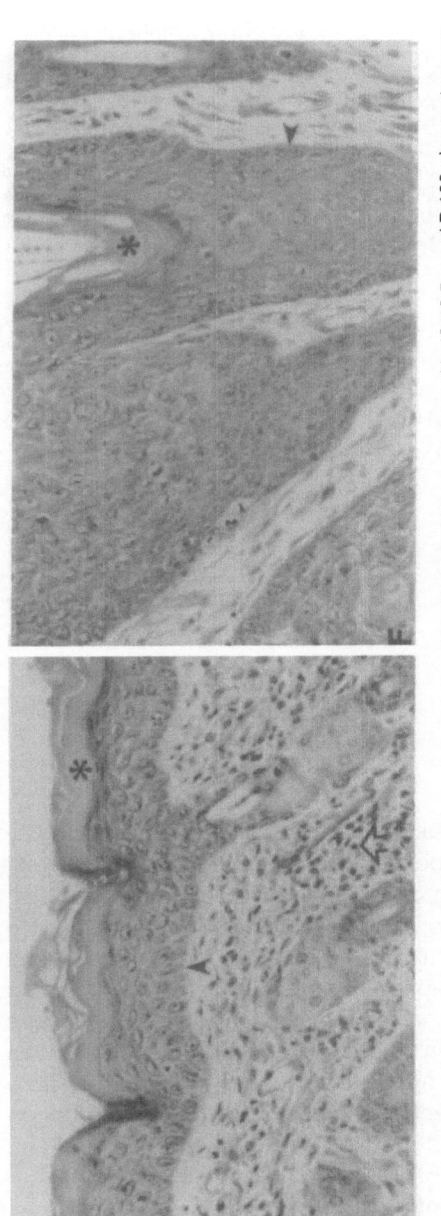

Fig. 3A–F. Histology of epidermis from transgenic mice expressing the NL4–3Δgag/pol. Normal mouse skin (**A** 12.5 × and **B** 100 ×); spontaneous papilloma in heterozygous transgenic mouse (**C** 12.5 × and **D** 100 ×); and diffuse epidermal proliferation with a mild dermal infiltrate in a homozygous trangenic mouse (**E** 100 ×). UV-induced papilloma in heterozygous transgenic mouse (**F** 100 ×). (From KOPP et al. 1993)

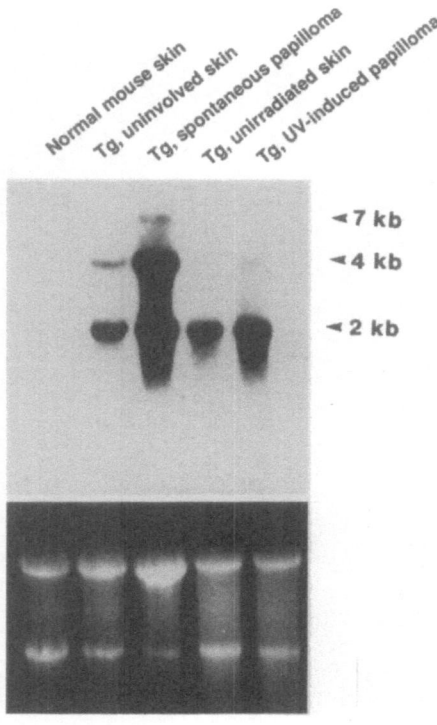

Fig. 4. Expression of HIV-1 mRNA in spontaneous and UV-induced papillomas from heterozygous transgenic mice expressing NL4–3Δ *gag/pol.* Normal skin, *lane 1;* uninvolved skin from heterozygous transgenic mice, *lane 2;* spontaneous papilloma, *lane 3;* unirradiated control skin from heterozygous transgenic mice, *lane 4;* and UV-irradiated skin, *lane 5.* Ethidium bromide staining of the RNA is indicated *below.* (From KOPP et al. 1993)

cachexia, and increased mortality that was identical to the phenotype reported by LEONARD et al. (1988) in transgenic mice bearing the infectious provirus NL4–3. Like the mice of LEONARD et al. (1988) homozygotes manifested lymphadenopathy, splenomegaly, and interstitial pneumonitis. These deletion mutant mice also had reduced numbers of circulating CD4⁺ T cells, thymic hypoplasia (Fig. 7), and early death. Growth retardation was not apparent until the postnatal period and was associated with increased expression of the transgene after delivery (FRANKS et al. 1995). The phenotype persisted even when the transgene was bred onto a nude mouse background (Fig. 8), suggesting that the cachexia and growth failure are T cell-independent. Although the exact mechanisms for this phenomenon has not been established, this transgenic mouse model may provide a useful avenue for exploring the pathogenesis of AIDS-related pediatric growth failure and for developing potential therapeutic interventions.

Fig. 5A-D. Histopathology of HIV-associated nephropathy in humans (**A**, **B**) and the renal disease in transgenic mice (**C, D**) expressing NL4–3Δ *gag/pol.* **A** Kidney from a patient demonstrating glomerular senescence (*arrow head*), dilated tubules with casts (*asterisk*), and modest interstitial infiltrate; × 50. **B** Segmental sclerosis in a glomerulus from the same patient; ×125 **C, D** Identical features are apparent in the HIV transgenic mice × 50, × 200. PAS stain. (From DICKIE et al. 1991; KOPP et al. 1992)

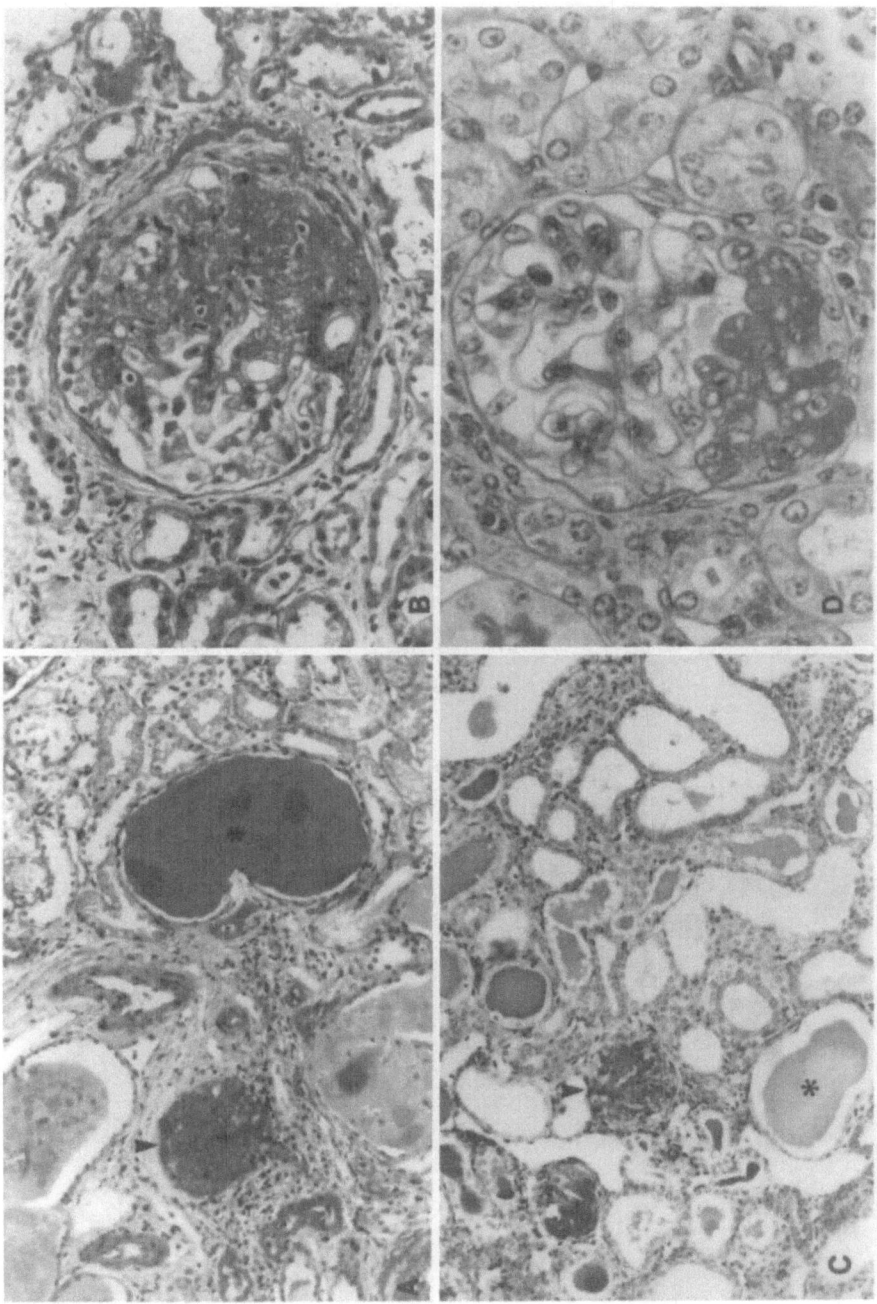

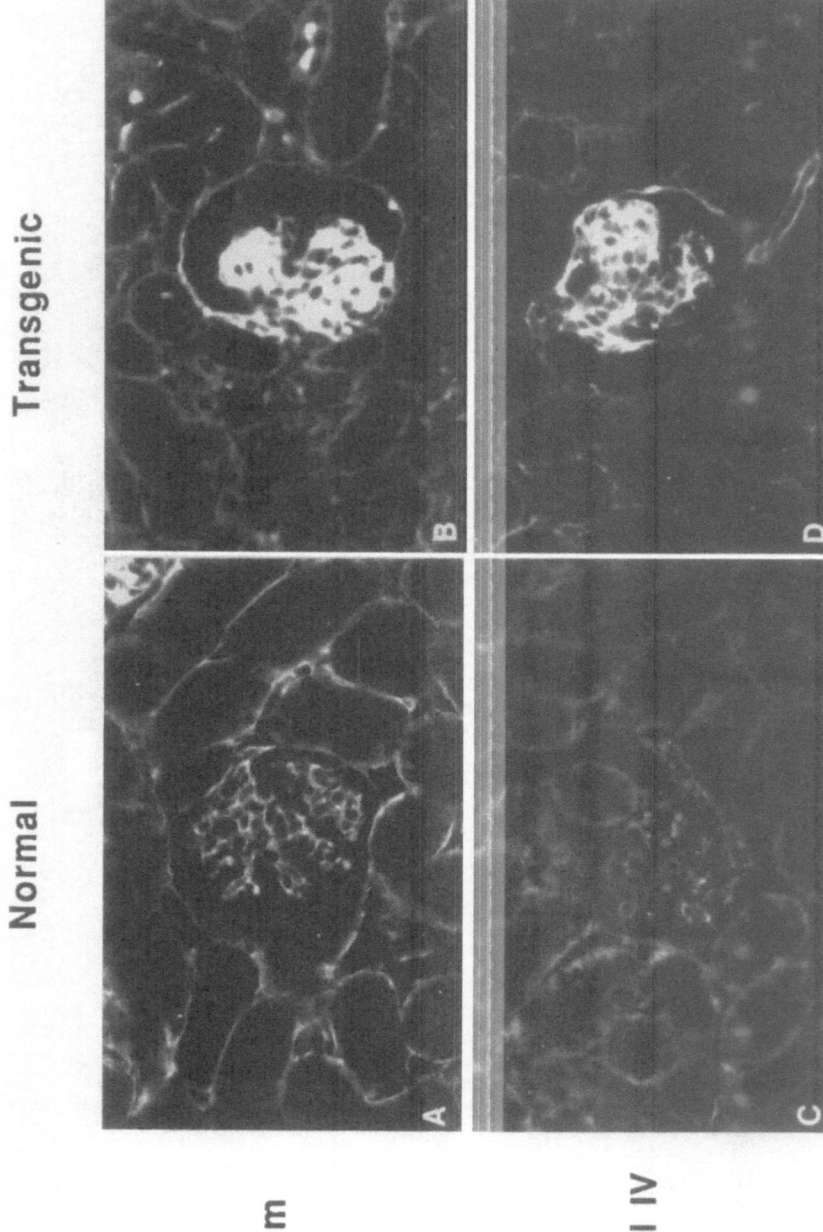

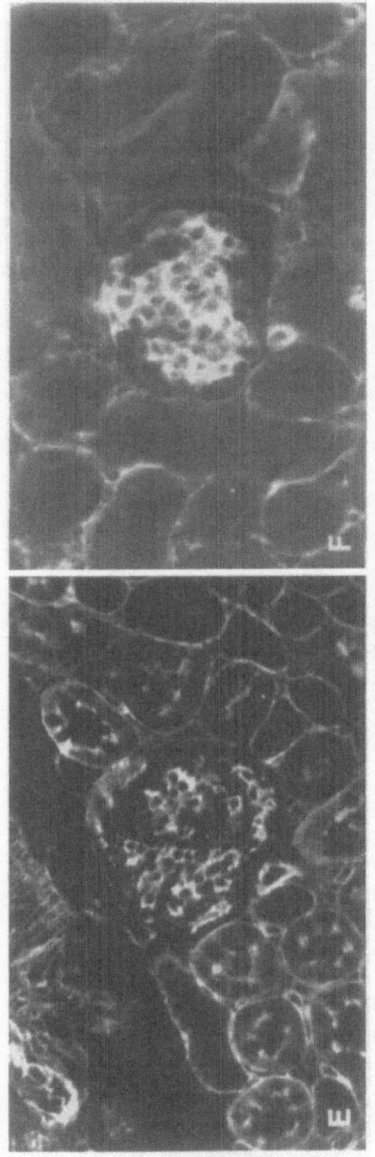

Fig. 6A-F. Immunohistochemistry for basement membrane proteins in kidneys from controls and transgenic mice expressing NL4-3Δ *gag/pol*. Kidneys were isolated from 34 day old animals . **A, B** Control and transgenic sections incubated with antibodies to laminin; **C, D** collagen IV; and **E, F** heparan sulfate proteoglycan. (From KOPP et al. 1992)

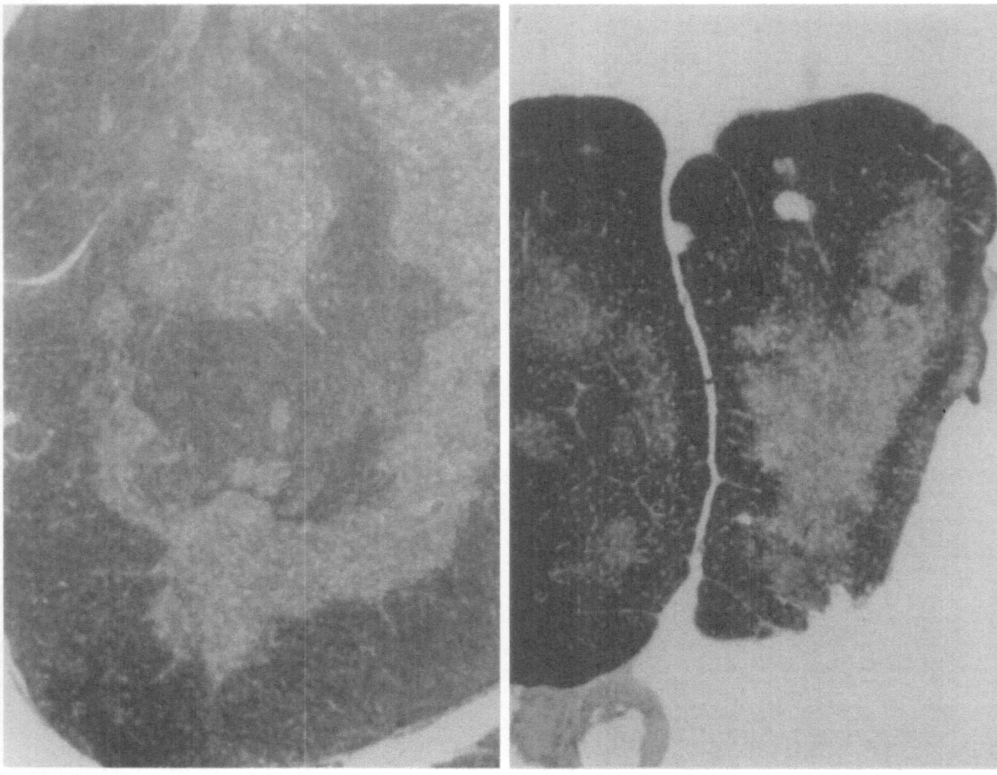

Fig. 7. Thymus from control and homozygous transgenic mice expressing NL4–3Δ *gag/pol. Left,* thymus from normal nontransgenic littermate; *right,* thymus from the homozygous mouse with cachexia. (From SANTORO et al. 1994)

3.2.4 Transcriptional and Posttranscriptional Processing of HIV-1 mRNA In Vivo

The NL4–3Δ *gag/pol* transgenic model has also provided insights into the regulation of the LTR as well as posttranscriptional processing of HIV-1 mRNA in vivo (BRUGGEMAN et al. 1994). Expression of the transgene in heterozygous animals was tissue-specific with high levels of expression in intestine, muscle, and skin, low levels of expression in lymphoid tissue, and virtually no expression in liver and lung (Fig. 9). Renal expression was low and transient with renal disease developing shortly after expression. Patterns of mRNA species were also tissue-specific with predominant expression of the multiply spliced 2 kb family of early regulatory messages in intestine and muscle. In kidney and lymphoid tissue, however, there was balanced expression of mRNAs that were multiply spliced, singly spliced, and full-length (Fig. 10). These data suggest that the kidney and lymphoid tissue posses the cellular machinery necessary to support the expression of late structural genes. The ability of local sites to

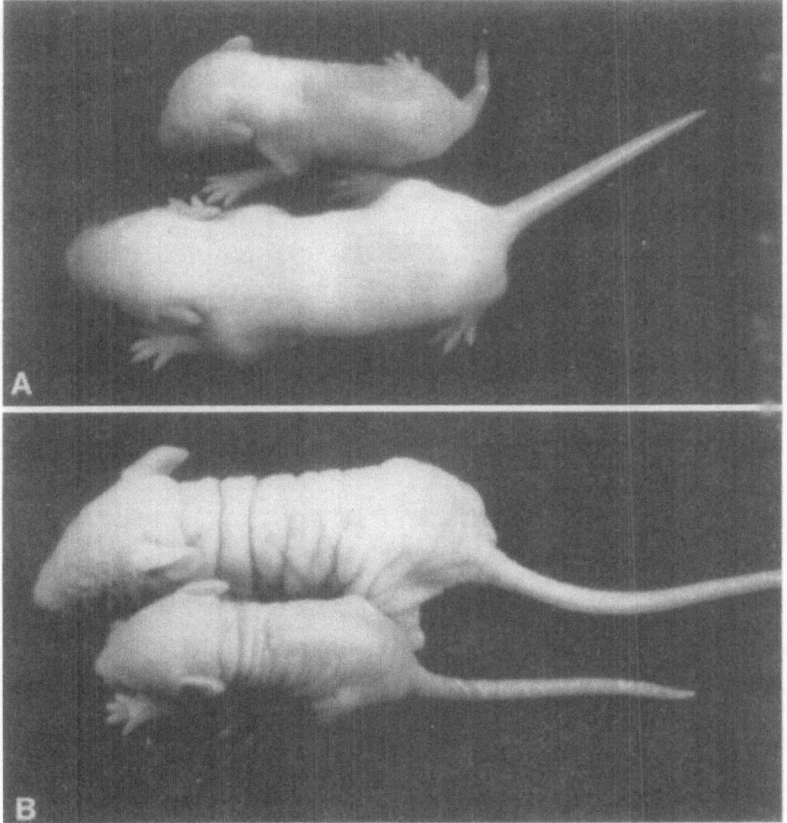

Fig. 8A,B. Homozygous transgenic mice expressing NL4–3Δ *gag/pol* on FVB/N and athymic nude backgrounds. **A** An 8 day old homozygous mouse next to a nontransgenic animal from the same litter. **B** The same age animals are shown from control and homozygous mice on the nude background.

express and possibly translate *env* and *gag/pol* genes may well contribute to the pathogenesis of HIV-1 in these tissues.

3.3 NL4–3Δrt

IWAKURA et al. (1994b) generated four transgenic murine lines using a reverse transcriptase deletion mutant of pNL4–3–2. Like other LTR-directed constructs (FURTH et al. 1990; FRUCHT et al. 1991; LEONARD et al. 1989), expression was apparent in the eyes of transgenic mice. Three of four founding lines developed cortical cataracts. The penetrance of this phenotype varied in the three lines (from 6.5% to 100%). Since p24 Gag was detected in the lens of affected mice, the authors concluded that Gag was detrimental to the lens fiber cells. We have also found cataract formation in 100% of transgenic mice expressing the NL4–3Δ

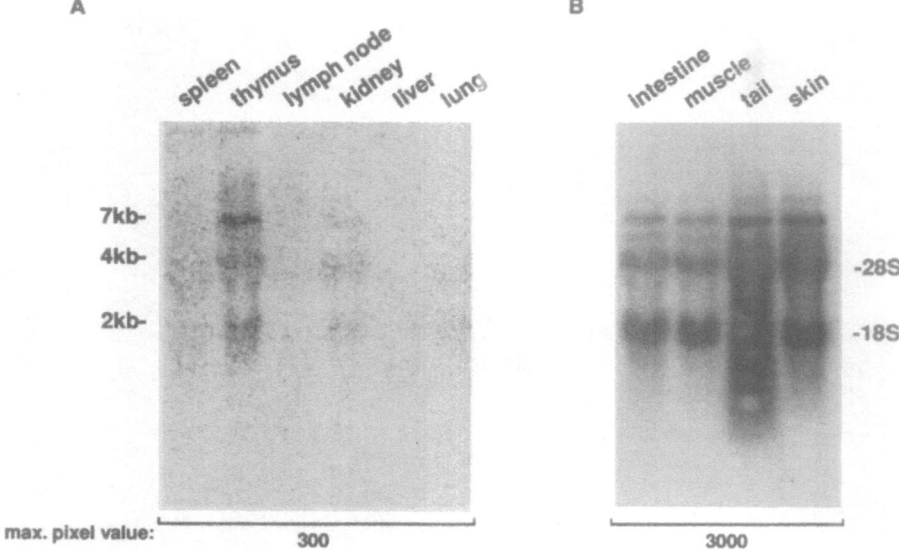

Fig. 9A,B. HIV-1 transgene expression in various organs of a heterozygous animal expressing the NL4–3Δ *gag/pol*. **A** Northern blot of spleen, thymus, lymph node, kidney, liver and lung. The pixel value indicated *below* is proportional to the amount of radioactivity in each band with the greater value indicating a stronger signal. **B** Northern blot of intestine, muscle, tail and skin. (From BRUGGEMAN et al. 1994)

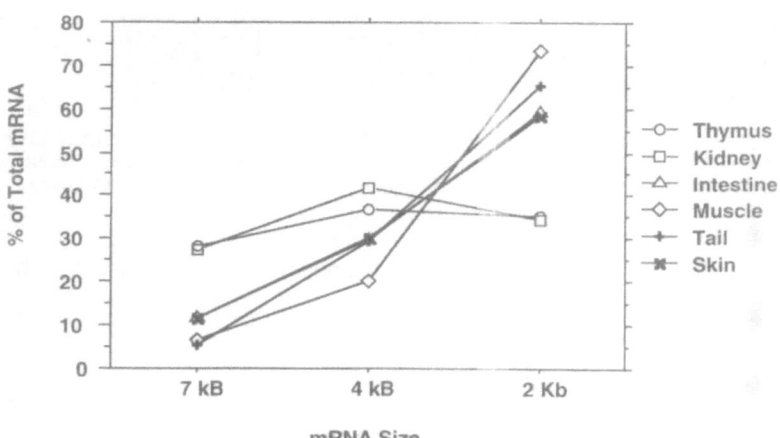

Fig. 10. Interaction line plot to determine a main effect of tissue type on mRNA size vs. percent of total RNA (BRUGGEMAN et al. 1994). This analysis was performed using a repeated measures analysis of variance. Thymus and kidney behaved similarly in supporting the expression of a greater percentage of singly spiced and full-length mRNA

gag/pol construct. Thus, expression of LTR-directed HIV-1 transgenes can induce cataract formation in mice, but the mechanism for epithelial proliferation remains unclear as does the relevance of this finding to patients with AIDS.

3.4 MMTV-BH10Δ 5' and 3' LTRs

To test the ability of the murine system to support translation of HIV-1 proteins in vivo, JOLICOEUR et al. (1992) generated transgenic mice expressing a full length construct in which the 5' LTR had been replaced by the mouse mammary tumor virus (MMTV) promoter. Gag and Env were detected in serum and in mammary glands, salivary glands, harderian glands, liver, seminal vesicles, epididymis, and testes. These findings provide further evidence that the murine system provides the cellular machinery for the expression and translation of HIV-1 genes.

4 LTR-Indicator Constructs

Numerous transgenic lines expressing LTR-indicator constructs have been generated for the purpose of evaluating the regulation of the LTR in vivo. These are discussed briefly below and are listed in Table 3.

4.1 LTR$_{CNS}$-β-Galactosidase

To explore the mechanisms for HIV-induced dementia, CORBOY et al. (1992) cloned and sequenced the LTRs from strains of HIV-1 isolated from the central nervous system of two patients with AIDS dementia complex. Multiple base pair differences were found in the U3 region of the LTR, 5' to the NF-κB and SP-1 sites. When these LTRs were linked to β-galactosidase and introduced as transgenes, expression was restricted to brain. Control transgenic lines were generated using the LTR of cloned strains of non-neuropathic HIV-1 linked to β-galactosidase. LTRs from non-neuropathic strains were unable to direct expression to the brain. These studies have provided important insight into the potential molecular basis for HIV expression in the central nervous system.

4.2 LTR-CAT, LTR-Luciferase, LTR-β-Galactosidase, and LTR-*lacZ*

4.2.1 Environmental Stimuli to LTR Activity

Several transgenic lines have been generated in which the HIV-1 LTR has directed expression of an indicator gene. These have been generated to explore the regulation of LTR in vivo and most have provided evidence for epithelial

Table 3. LTR-indicator constructs utilized in transgenic mice

Transgene	Viral clone/strain	Promoter-enhancer	Murine strain	Predominant phenotype	Reference
CAT	LAV	LTR	FVB/N	Constitutive expression in eye, heart, thymus, and tail with augmented expression in circulating lymphocytes and macrophages in response to activation stimuli	LEONARD et al. (1989)
CAT	LAV	LTR	FVB/N	UV-C and UV-B light activates the LTR in vivo	FRUCHT et al. (1991)
SV40 T-ag	HxB2	LTR	C33HeB/FeJ	Expression of transgene in lymphoid tissue and skin but greatest in B cells and thymic stromal cells	SKOWRONSKI (1991)
HSV1 TK	HxB2	LTR	B6D2F2	With ganciclovir, dendritic cells in spleen and thymus selectively ablated, thymic atrophy, and loss of $CD4^+$ $CD8^+$ thymocytes	SALOMON et al. (1994)
β-galactosidase Luciferase	HxB2	LTR	FVB/N	UV-A (with psoralen) and UV-B light activation of the LTR in vivo	MORREY et al. (1991)
β-galactosidase	HIV-1$_{JR-CSF}$ HIV-1$_{JR-FL}$	LTR	CD1	The LTR of primary isolates from the CNS contain cis elements responsible for CNS-restricted expression	CORBOY et al. (1992)
Luciferase	HxB2	LTR	FVB/N	Topical DMSO, retinoic acid, phorbol ester, UV light, hydrogen peroxide, and skin wounds, increase LTR activation in vivo	MORREY et al. (1992)
Luciferase	HxB2	LTR	FVB/N	Topically administered dinitrochlorobenzene can induce LTR activity	MORREY et al. (1993)
lacZ	HIV-1$_{Bru}$	LTR	B6D2F1	Spontaneous activation of the eye and epidermis. UV-induced increase in expression in the epidermis	CAVARD et al. (1990)
lacZ	HIV-1$_{Bru}$	LTR Δ NFκB LTR	B6D2F1	NFκB is not required for UV-induced activation of the LTR in vivo	ZIDER et al. (1993)

DMSO, dimethylsulfoxide; HSV1 TK, herpes virus type-1 thymidine kinase; CAT, chloramphenicol acetyl transferase.

expression and UV light sensitivity. LEONARD et al. (1989) introduced an HIV LTR-chloramphenicol acetyltransferase (CAT) construct into transgenic mice. Eye, heart, spleen, thymus, and tail skin expressed the CAT gene. Of the marrow-derived cells, Langerhans cells from skin expressed the most CAT activity, a finding substantiated by SALOMON et al. (1994) using an LTR-directed dendritic cell ablation strategy. In skin, keratinocytes were second only to Langerhans cells in the level of expression. None of the animals were reported to develop epidermal pathology, despite expression of the indicator gene in epidermis.

UV light has been shown to activate the HIV-1 LTR in vitro (VALERIE et al. 1988; IWAKURA et al. 1992; STEIN et al. 1989) by a Tat-independent mechanism (SADAIE et al. 1990) and transgenic animals have been utilized extensively to explore this issue in vivo. UV-induced expression has been demonstrated in transgenic mice expressing CAT (FRUCHT et al. 1991), LTR-luciferase or LTR-β-galactosidase (MORREY et al. 1991, 1992; ZIDER et al. 1993), and LTR-*lacZ* (CAVARD et al. 1990). LTR activity can be induced by a number of stimuli that produce either activation of cells or cellular injury (MORREY et al. 1992, 1993). The exact mechanism by which UV light or other environmental factors induce LTR-directed transcription is complex but appears to be independent of NF-κB in vivo (ZIDER et al. 1993). Furthermore, other factors such as cytokines that are induced by injury may activate the LTR, and DNA damage and repair (STEIN et al. 1989) may also participate in the process of LTR activation in vivo.

4.2.2 Pregnancy-Induced LTR Activity and Fetal Development

Transgenic mouse models have also been used to explore issues related to maternal transmission of HIV-1 and expression during fetal development. In mice transgenic for the CAT gene under the control of the HIV-1 LTR, FURTH et al. (1990) demonstrated increased CAT activity in the uterus and placenta of pregnant mice. In studies by FRANKS et al. (1995) utilizing the multigenic construct NL4–3Δ *gag/pol*, increased expression of the transgene was also detected in pregnant uterus and fetus. Fetal expression of the HIV-1 transgene, however, did not alter development or birth weight. Since children who acquire the virus through maternal-fetal transmission are clearly at risk for in utero growth impairment, it appears that other maternal and environmental factors must play a critical role in affecting intrauterine growth and development.

4.3 LTR-SV40 T Antigen

To explore tissue-specific regulation of the LTR, SKOWRONSKI (1991) generated three lines of transgenic mice in which the SV40 T antigen was placed under the control of the HIV-1 LTR. The LTR (-450 to +82) included the majority of *cis*-acting elements defined in the promoter including the negative regulatory region, binding sites for NFAT1, NF-κB, and SP-1, and the TATA box as well as the Tat activation region TAR. Three transgenic lines were generated and expression was noted consistently in lymphoid tissues including thymus, lymph node, and spleen and in the skin. Of note, expression was greater in B cells, thymic stromal cells, and monocytes and macrophages than in quiescent T cells. Other tissues expressed variably but the exact pattern was both organ and line-dependent. Tissues that expressed the transgene in one or more lines included small intestine, kidney, mesenteric lymph node, skeletal muscle, and pancreas. These data suggest that the expression of the LTR is tissue-specific and that expression is regulated by critical interactions between host transcriptional machinery and *cis* elements in the viral promoter. Furthermore, these data confirm the importance of in vivo analysis of transcriptional regulation of HIV-1.

5 Summary

Transgenic models have provided significant insights into HIV-1 pathogenesis, particularly with regards to Kaposi's sarcoma, HIV-associated nephropathy, the tissue-restricted expression of CNS strains of HIV-1, and the function of Nef in vivo. Both multigenic and single gene constructs have contributed to our understanding of HIV-1-induced diseases. While failing to provide models suitable for vaccine development, these transgenic models have provided great insight into HIV pathogenesis and may yet provide a means for the development and testing of molecular based therapies for AIDS.

References

Abramczuk JW, Pezen DS, Leonard JM, Monell Torrens E, Belcher JH, Martin MA, Notkins AL (1992) Transgenic mice carrying intact HIV provirus: biological effects and organization of a transgene. J Acquir Immune Defic Syndr 5: 196–203

Bruggeman LA, Thomson MM, Nelson PJ, Kopp JB, Rappaport J, Klotman PE, Klotman ME (1994) Patterns of HIV-1 mRNA expression in transgenic mice are tissue-dependent. Virology 202: 940–948

Cavard C, Zider A, Vernet M, Bennoun M, Saragosti S, Grimber G, Briand P (1990) In vivo activation by ultraviolet rays of the human immunodeficiency virus type 1 long terminal repeat. J Clin Invest 86: 1369–1374

Corallini A, Altavilla G, Pozzi L, Bignozzi F, Negrini M, Rimessi P, Gualandi F, Barbanti-Brodano G (1993) Systemic expression of HIV-1 tat gene in transgenic mice induces endothelial proliferation and tumors of different histotypes. Cancer Res 53: 5569–5575

Corboy JR, Buzy JM, Zink MC, Clements JE (1992) Expression directed from HIV long terminal repeats in the central nervous system of transgenic mice. Science 258: 1804–1808

Delli Bovi P, Donti E, Knowles DM, Friedman Kien A, Luciw PA, Dina D, Dalla Favera R, Basilico C (1986) Presence of chromosomal abnormalities and lack of AIDS retrovirus DNA sequences in AIDS-associated Kaposi's sarcoma. Cancer Res 46: 6333–6338

Dickie P, Felser J, Eckhaus M, Bryant J, Silver J, Marinos N, Notkins AL (1991) HIV-associated nephropathy in transgenic mice expressing HIV-1 genes. Virology 185: 109–119

Dickie P, Ramsdell F, Notkins AL, Venkatesan S (1993) Spontaneous and inducible epidermal hyperplasia in transgenic mice expressing HIV-1 Nef. Virology 197: 431–448

Duvic M (1991) Papulosquamous disorders associated with human immunodeficiency virus infection. Dermatol Clin 9: 523–530

Franks RF, Ray PE, Babbot CC, Bryant JL, Notkins AL, Santoro TJ, Klotman PE (1995) Maternal-fetal interactions affect growth of HIV-1 transgenic mice. Pediatr. Res. 37: 56–63

Frucht DM, Lamperth L, Vicenzi E, Belcher JH, Martin MA (1991) Ultraviolet radiation increases HIV-long terminal repeat-directed expression in transgenic mice. AIDS Res. Hum. Retroviruses 7: 729–733

Furth PA, Westphal H, Hennighausen L (1990) Expression from the HIV-LTR is stimulated by glucocorticoids and pregnancy. AIDS Res. Hum. Retroviruses 6: 533–560

Hammes SR, Dixon EP, Malim MH, Cullen BR, Greene WC (1989) Nef protein of human immunodeficiency virus type 1: evidence against its role as a transcriptional inhibitor. Proc Natl Acad Sci USA 86: 9549–9553

Huang SK, Martin FJ, Jay G, Vogel J, Papahadjopoulos D, Friend DS (1993) Extravasation and transcytosis of liposomes in Kaposi's sarcoma-like dermal lesions of transgenic mice bearing the HIV tat gene. AM J Pathol 143: 10–14

Huang YQ, Buchbinder A, Li JJ, Nicolaides A, Zhang WG, Friedman Kien AE (1992) The absence of Tat sequences in tissues of HIV-negative patients with epidermic Kaposi's sarcoma AIDS 6: 1139–1142

Iwakura Y, Shioda T, Tosu M, Yoshida E, Hayashi M, Nagata T, Shibuta H (1992) The induction of cataracts by HIV-1 in transgenic mice. AIDS 6: 1069–1075

Jolicoeur P, Laperriere A, Beaulieu N (1992) Efficient production of human immunodeficiency virus proteins in transgenic mice. J Virol 66: 3904–3908

Khillan JS, Deen KC, Yu SH, Sweet RW, Rosenberg M, Westphal H (1988) Gene transactivation mediated by the TAT gene of human immunodeficiency virus in transgenic mice. Nucleic Acids Res 16: 1423–1430

Kopp JB, Klotman ME, Adler SH, Bruggeman LA, Eckhaus M, Dickie P, Marinos NJ, Bryant JL, Notkins AL, Klotman PE (1992) Progressive glomerulosclerosis and enhanced renal accumulation of basement membrane components in mice transgenic for HIV-1 genes. Proc Natl Acad Sci USA 89: 1577–1581

Kopp JB, Rooney JF, Wohlenberg C, Dorfman N, Marinos NJ, Bryant JL, Katz SI, Notkins AL, Klotman PE (1993) Cutaneous disorders and viral gene expression in HIV-1 transgenic mice. AIDS Res Hum Retroviruses 9: 267–275

Kopp JB, Ray PE, Adler SH, Bruggeman LA, Mangurian CV, Owens JW, Eckhaus MA, Bryant JL, Klotman PE (1994) Nephropathy in HIV-transgenic mice. In: Koide H, Hayashi T (eds) Extracellular matrix in the kidney; Karger, Basel, pp 194–204

Leonard JM, Abramczuk JW, Pezen DS, Rutledge R, Belcher JH, Hakim F, Shearer G, Lamperth L, Travis W, Fredrickson T, Notkins AL, Martin MA (1988) Development of disease and virus recovery in transgenic mice containing HIV proviral DNA. Science 242: 1665–1670

Leonard JM, Khillan JS, Gendelman HE, Adachi A, Lorenzo S, Westphal H, Martin MA, Meltzer MS (1989) The human immunodeficiency virus long terminal repeat is preferentially expressed in Langerhans cells in transgenic mice. AIDS Res Hum Retroviruses 5: 421–430

Lindemann D, Wilhelm R, Renard P, Althage A, Zinkernagel R, Mous J (1994) Severe immunodeficiency associated with a human immunodeficiency virus 1 NEF/3'-long terminal repeat transgene. J Exp Med 179: 797–807

Mehtali M, Munschy M, Ali Hadji D, Kieny MP (1992) A novel transgenic mouse model for the in vivo evaluation of anti-human immunodeficiency virus type 1 drugs. AIDS Res Hum Retroviruses 8: 1959–1965

Morrey JD, Bourn SM, Bunch TD, Jackson MK, Sidwell RW, Barrows LR, Daynes RA, Rosen CA (1991) In vivo activation of human immunodeficiency virus type 1 long terminal repeat by UV type A (UV-A) light plus psoralen and UV-B light in the skin of transgenic mice. J Virol 65: 5045–5051

Morrey JD, Bourn SM, Bunch TD, Sidwell RW, Rosen CA (1992) HIV-1 LTR activation model: evaluation of various agents in skin of transgenic mice. J Acquir Immune Defic Syndr 5: 1195–1203

Morrey JD, Jackson MK, Bunch TD, Sidwell RW (1993) Activation of the human immunodeficiency virus type 1 long terminal repeat by skin-sensitizing chemicals in transgenic mice. Intervirology 36: 65–71

Pezen DS, Leonard JM, Abramczuk JW, Martin MA (1991) Transgenic mice carrying HIV proviral DNA. Biotechnology 16: 213–226

Rao TK (1991a) Human immunodeficiency virus (HIV) associated nephropathy. Annu Rev Med 42: 391–401

Rao TK (1991b) Clinical features of human immunodeficiency virus associated nephropathy. Kidney. Int Suppl 35: S13–S18

Rao TK, Filippone EJ, Nicastri AD, Landesman SH, Frank E, Chen CK, Friedman EA (1984) Associated focal and segmental glomerulosclerosis in the acquired immunodeficiency syndrome. N Engl J Med 310: 669–673

Rao TK, Friedman EA, Nicastri AD (1987) The types of renal disease in the acquired immunodeficiency syndrome. N Engl J Med 316: 1062–1068

Rappaport J, Kopp JB, Klotman PE (1994) Host virus interactions and the molecular regulation of HIV-1: Role in the pathogenesis of HIV-associated nephropathy. Kidney Int 46: 16–27

Ray PE, Bruggeman LA, Weeks BS, Kopp JB, Bryant J, Owens JW, Notkins AL, Klotman PE (1994) bFGF and its low affinity receptors in the pathogenesis of HIV-associated nephropathy. Kidney Int 46: 759–772

Sadaie MR, Tschacler E, Valerie K, Rosenberg M, Felber BK, Pavlakis GN, Klotman ME, Wong-Staal F (1990) Activation of tat defective human immunodeficiency virus by ultraviolet light. New Biol 2: 479–486

Salomon B, Lores P, Pioche C, Racz P, Jami J, Klatzmann D (1994) Conditional ablation of dendritic cells in transgenic mice. J Immunol 152: 537–548

Santoro TJ, Bryant JL, Pellicoro J, Klotman ME, Kopp JB, Bruggeman LA, Franks RR, Notkins AL, Klotman PE (1994) Growth failure and AIDS-like cachexia syndrome in HIV-1 transgenic mice. Virology 201: 147–151

Shirai A, Klinman DM (1993) Immunization with recombinant gp160 prolongs the survival of HIV-1 transgenic mice. AIDS Res Hum Retroviruses 9: 979–983

Skowronski J (1991) Expression of a human immunodeficiency virus type 1 long terminal repeat/ simian virus 40 early region fusion gene in transgenic mice. J Virol 65: 754–762

Skowronski J, Parks D, Mariani R (1993) Altered T cell activation and development in transgenic mice expressing the HIV-1 nef gene. EMBO J 12: 703–713

Stein B, Rahmsdorf HJ, Steffen A, Litfin M, Herrlich P (1989) UV induced DNA damage is an intermediate step in UV induced expression of human immunodeficiency virus type I, collagenase, c-fos and metallothionein. Mol Cell Biol 9: 5169–5181

Toggas SM, Masliah E, Rockenstein EM, Rall GF, Abraham CR, Mucke L (1994) Central nervous system damage produced by expression of the HIV-1 coat protein gp120 in transgenic mice. Nature 367: 188–193

Valerie K, Delers A, Bruck C, Thiriart C, Rosenberg H, Debouck C, Rosenberg M (1988) Activation of human immunodeficiency virus type I by DNA damage in human cells. Nature 333: 78–81

Vogel J, Hinrichs SH, Reynolds RK, Luciw PA, Jay G (1988) The HIV tat gene induces dermal lesions resembling Kaposi's sarcoma in transgenic mice. Nature 335: 606–611

Vogel J, Hinrichs SH, Napolitano LA, Ngo L, Jay G (1991) Liver cancer in transgenic mice carrying the human immunodeficiency virus tat gene. Cancer Res. 51: 6686–6690

Vogel J, Cepeda M, Tschachler E, Napolitano LA, Jay G (1992) UV activation of human immuno-deficiency virus gene expression in transgenic mice. J Virol 66: 1–5

Werner S, Hofschneider PH, Roth WK (1989) Cells derived from sporadic and AIDS-related Kaposi's sarcoma reveal identical cytochemical and molecular properties in vitro. Int J Cancer 43: 1137–1144

Zider A, Mashhour B, Fergelot P, Grimber G, Vernet M, Hazan U, Couton D, Briand P, Cavard C (1993) Dispensable role of the NF-kappa B sites in the UV-induction of the HIV-1 LTR in transgenic mice. Nucleic Acids Res 21: 79–86

Transgenic Models
in the Study of AIDS Dementia Complex

S.M. Toggas and L. Mucke

1 HIV-1 Associated Neurologic Disease

1.1 Introduction

A significant number of HIV-1 infected individuals develop central nervous system (CNS) complications which can include motor difficulties, behavioral abnormalities, and cognitive impairments. These symptoms can progress to nearly vegetative end stage clinical presentations of paralysis and severe dementia (see PRICE and PERRY 1994, for detailed review). Variably referred to as "AIDS dementia complex," "HIV encephalopathy," and "HIV-1 associated cognitive/motor complex," this syndrome has been observed most often in the late stages of systemic HIV-1 infection. Nevertheless, the CNS is an early target of infection (Ho et al. 1985; GOUDSMIT et al. 1986; MARSHALL et al. 1988; RESNICK et al. 1988; APPLEMAN et al. 1988; MCARTHUR et al. 1988; ELOVAARA et al. 1990). In pediatric patients, CNS disease occurs more frequently and its accompanying clinical syndrome has been

Department of Neuropharmacology, The Scripps Research Institute, 10666 North Torrey Pines Road, La Jolla, CA 92037, USA

termed "progressive encephalopathy" (for review, see PRICE and PERRY 1994). The transgenic models discussed in this chapter are pertinent to both pediatric and adult forms of HIV-1-related brain damage.

At the structural level, CNS damage associated with HIV-1 infection in humans is manifested by loss of specific neuronal subpopulations, synaptic dropout, widespread degeneration of neuronal dendrites, reactive gliosis, microglial nodule and multinucleated giant cell formation, myelin pallor, and blood-brain barrier (BBB) permeability changes (for reviews see SHARER 1992; BUDKA 1992; SPENCER and PRICE 1992; PRICE and PERRY 1994). Despite extensive analysis and characterization of the clinical and neuropathological features of the AIDS dementia complex, its pathogenesis remains unclear. Direct infection of neurons by HIV-1 has been difficult to demonstrate; however, a role of HIV-1 in the production of CNS pathology is suggested strongly by the correlation between viral burden and synapto-dendritic damage in brains of HIV-1-infected patients (MASLIAH et al. 1992).

HIV-1 is thought to enter the CNS via infected peripheral blood mononuclear cells, and neuroinvasive strains have been shown to replicate primarily within macrophages/microglia. How can infection of nonneuronal cells bring about such extensive impairments of the CNS? It is possible that HIV-1 affects neuronal function indirectly, for example, via disturbance of glial support cells, as well as through the generation of soluble, diffusible viral and/or host-derived neurotoxic substances which could produce widespread detrimental effects (for review see MUCKE and EDDLESTON 1993; EPSTEIN and GENDELMAN 1993; PRICE and PERRY 1994; LIPTON 1994).

Before discussing transgenic approaches designed to test these hypotheses, we will briefly outline other approaches that are used, along with their advantages and disadvantages. By doing so, we intend to: (1) illustrate the relative contributions these analyses make toward the study of AIDS dementia complex; (2) demonstrate the utility of the transgenic approach in assessing the neuropathogenic potential of individual factors in vivo, and (3) emphasize that the combination of data obtained from diverse approaches has the highest chance to unravel the complex etiology of HIV-1 associated CNS damage.

1.2 Clinical and Neuropathological Studies

Conventional clinical and neuropathological approaches have revealed changes in cellular structure and in specific protein and metabolite levels in the CNS of patients with neurologic complications of HIV-1. For example, immunocytochemical staining of autopsy tissue and assays of cerebrospinal fluid (CSF) have established cytokine alterations to be a characteristic feature of HIV-1 infection of the CNS (MERRILL and CHEN 1991; GELEZIUNAS et al. 1992; EPSTEIN and GENDELMAN 1993). In addition, factors such as β_2-microglobulin (BREW et al. 1989), quinolinic acid (HEYES et al. 1991), neopterin (BREW et al. 1990), and platelet-activating factor (PAF) (GELBARD et al. 1994) have been shown to be elevated in the CSF of AIDS patients with CNS dysfunction.

Clinical and neuropathological approaches offer the distinct advantage of providing information from what is clearly the most relevant source—tissue from

HIV-1-infected human patients. However, the picture provided by both postmortem material and clinical specimens represents a dynamic of causative and compensatory processes; therefore, such analyses often cannot conclusively establish critical cause-effect relationships. Further, while levels of factors may correlate directly with the severity of dementia (HEYES et al. 1991; GELBARD et al. 1994), this is not always the case (TYOR et al. 1992; WESSELINGH et al. 1993). Finally, because of their immune-compromised state, patients with advanced HIV-1 infection are susceptible to opportunistic infections which result in the introduction of additional CNS pathogens, including viruses, bacteria, parasites, and fungi. Many of these microbes may also generate factors that could contribute directly or indirectly to the neurological impairment. Consequently, correlative information gained from such studies, while provocative, should be interpreted with caution until tested in experimental paradigms.

1.3 In Vitro Studies

In vitro studies have identified a growing number of peptides, proteins, and other biochemicals with neurotoxic properties. For example, in vitro studies were the first to demonstrate that the HIV-1 envelope glycoprotein gp120 has toxic effects on neurons (see LIPTON 1994 for review). It is unlikely that this molecule would have been identified as a candidate neurotoxin by clinical and neuropathological approaches, since it has been impossible to identify in the CNS by conventional immunochemical techniques. Nonetheless, earlier in vitro work had shown this protein to be shed from the virus and from virally infected cells (SCHNEIDER et al. 1986), making it a potentially toxic diffusible factor.

In vitro studies have the advantage that conditions can be manipulated effectively by the investigator to achieve a specific, controlled environment, allowing for relative ease of dissection of molecular mechanisms. For instance, in vitro studies have provided evidence that macrophage/microglia-derived factors may be essential mediators of HIV-1-associated neuronal damage (LIPTON 1992b; GIULIAN et al. 1993). Further, agonists and antagonists of the N-methyl-D-aspartate (NMDA) subtype of glutamate receptors have been added to culture systems along with gp120 and shown to influence the degree of damage produced (reviewed by LIPTON 1994). This NMDA-mediated excitotoxicity has been suggested to have several upstream triggers and downstream activators, including generation of heat-stable, protease-resistant, macrophage-derived toxins (GIULIAN et al. 1990, 1993), and nitric oxide (DAWSON et al. 1993; PIETRAFORTE et al. 1994), respectively. Direct effects of gp120 on astrocyte ion channel fluxes (BENOS et al. 1994) and increased production of lipid biomediators following interactions of infected macrophages with cultured astrocytes (GENIS et al. 1992) have also been implicated in HIV-1-induced neurotoxicity. Therefore, results from in vitro experiments support the hypothesis that HIV-1-related neuronal injury can be mediated by effects of the virus or viral products on nonneuronal cells.

Unfortunately, it is often difficult to extrapolate from in vitro studies to the in vivo situation because of the extensive interactions between different types of predominantly nondividing cells that comprise the adult CNS. Thus, in vitro studies can help identify and characterize the effects of specific molecules; however, conclusions should be followed up in vivo to determine whether such effects are physiologically or pathologically relevant.

1.4 Nontransgenic In Vivo Studies

The development of a practical animal model of HIV-1 infection has been a long-time goal which, so far, has not been realized. The experimental systems currently available which approximate HIV-1 infection include simian immuno-deficiency virus (SIV) in monkeys (for review, see PRICE and PERRY 1994), feline immunodeficiency virus (FIV) in cats (LACKNER et al. 1991; HURTREL et al. 1992), and the severe combined immunodeficiency disease (scid) mouse reconstituted with human immune cells and infected with HIV-1 (NAMIKAWA et al. 1988; MOSIER et al. 1991). CNS damage is associated with both SIV and FIV infection, and similarities to that seen in HIV-1 infection of humans include presence of multinucleated giant cells, restriction of brain infection primarily to monocytes/macrophages, and development of neurobehavioral abnormalities. (See LACKNER et al. (1991) for a review and comparison of HIV-, SIV-, and FIV-associated neurologic disease). Recently, intracerebral injection of human peripheral blood mononuclear cells and HIV-1 into scid mice has been shown to result in CNS tumor necrosis factor-α (TNF-α) and interleukin-1 (IL-1) secretion and in reactive astrocytosis in a proportion of animals (TYOR et al. 1993). While enabling the study of issues related to specific phases of the viral life cycle and stages of systemic infection, these models produce an end result which reflects the potential contributions of, and host responses to, all viral products. Therefore, in such systems, it is often difficult to compare the relative pathogenetic importance of different viral and host-derived factors.

Other nontransgenic in vivo studies have been designed to address such questions. In particular, systemic injection of gp120 in neonatal rats has been shown to damage nerve cells in the CNS and to retard acquisition of behavioral developmental milestones (HILL et al. 1993). Injection of gp120 into the cerebral ventricles of adult rats decreased cerebral glucose utilization (KIMES et al. 1991) and impaired learning ability (GLOWA et al. 1992). Peptides representing portions of the HIV-1 proteins Tat and Rev have been found to be lethal when injected intracerebroventricularly into rodents (SABATIER et al. 1991; MABROUK et al. 1991), although exact cause of death and degree of CNS involvement were not determined. A key advantage of this type of in vivo approach is that isolated factors can be studied for their effects on the fully developed, adult nervous system. Disadvantages are that direct injection of substances into the CNS can result in gross variations in amount delivered, and both direct injection and osmotic minipump infusion methods can induce vigorous secondary responses to mechanical injury. Experiments using peripheral administration of compounds are frequently limited by the BBB.

2 Transgenic Models to Study HIV-1 Associated Neurologic Disease

2.1 Introduction

The transgenic approach involves the permanent introduction of an exogenous DNA sequence into the genome of a host animal. In many cases, the sequence encodes a specific factor to be analyzed. Alternatively, gene regulatory sequences ligated to reporter genes can serve as transgenes in the study of determinants of transcription. To produce transgenic animals, the DNA of interest is microinjected into pronuclei of fertilized oocytes, where it integrates apparently into random sites within the genome. The embryos are transferred into pseudopregnant females; the resulting FO generation of transgenic progeny, termed "founders," usually transmit the transgene(s) to approximately 50% of their offspring. Further characterization and breeding of transgenic F1 and F2 generations derived from individual founders allows for the establishment of distinct, genetically (and often also phenotypically) homogenous strains/lines of transgenic mice. Transgene constructs can use the native promoter of the gene of interest to direct gene expression to host cells, or can be designed such that cell-type specific promoters will target the in vivo expression of a heterologous protein to specific populations of CNS cells.

The goals of transgenic expression of potential neurotoxic factors include the following: (1) determination of their relative pathogenetic importance, (2) identification of mechanisms by which they produce damage and (3) development and evaluation of therapeutic strategies aimed at amelioration of such damage. Information from clinical and neuropathological studies, in vitro studies, and nontransgenic in vivo studies may aid in selection of candidate agents to test in the transgenic models.

Because genes encoding individual factors are introduced, comparison of transgenic animals with nontransgenic controls yields information that is not confounded by the presence of other viral products or pathogens. Further, while allowing for the assessment of specific factors selected by the investigator (as in in vitro studies), transgenic models enable the assessment of these factors in an environment where all cell types are represented in the appropriate ratio and can be expected to respond in a physiologic manner. Synthesis of transgene products within the CNS parenchyma eliminates the problems associated with direct injection into the CNS and with peripheral injection/BBB permeability issues. Replacement of viral regulatory sequences by eukaryotic regulatory sequences allows for expression of products to be controlled by the investigator. In addition, constructs expressing HIV-1 sequences can be assessed in mice, thus overcoming species barriers and providing experimental animals which are easy and inexpensive to work with relative to monkeys and cats.

Despite their numerous advantages, transgenic approaches are not without disadvantages. A major consideration is that expression of most conventional transgenes begins early in development. Therefore, a potentially toxic transgene-derived factor may produce damage in the developing CNS that is beyond its capabilities in the adult nervous system. Conversely, it may show a reduced effect if the plasticity of the developing nervous system can compensate for its detrimental actions. Although this feature can be problematic in the analysis of effects of some molecules on the adult nervous system, it can actually be very useful in studying questions relating to neurologic diseases of infants and children. Furthermore, regulatable promoter systems, such as the one developed by GOSSEN and BUJARD (1992), allows for controlled delays of transgene expression in vivo, thereby enabling assessment of effects of the transgene product on the CNS at defined developmental stages (GOSSEN et al. 1993; FURTH et al. 1994; PASSMAN and FISHMAN 1994).

While the chronic expression of transgenes may yield an outcome not representative of more acute forms of exposure, many CNS diseases, including those associated with HIV-1 infection, are characterized by a subacute or chronic time course. Hence, in these situations, the chronic expression of pathogenic factors provided by transgenic approaches approximates the clinical picture more closely. Notably, for chronic conditions, transgenic animals also provide suitable models in which to analyze both early and late events in pathogenesis. This represents a significant advantage, as CNS tissue from patients in the early stages of chronic neurologic disease is rarely available. Because transgenic models enable the mapping of CNS alterations along a continuum, cause/effect relationships can be more effectively revealed and dissected .

Genetic perturbation analyses can lead to important findings, some predictable, others unexpected and, sometimes, serendipitous. Three types of models which have been used to study the pathogenesis of AIDS dementia complex are described below. These models have addressed the following general questions: (1) What cells possess the transcription factors necessary to induce expression from the HIV-1 long terminal repeat (LTR)? (2) What roles do specific viral products play in the production of HIV-1-associated neuropathology? (3) What are the contributions of host factors to the damage produced? In addition to providing information to answer such questions, these models constitute important tools in the assessment of therapeutic strategies targeted against specific aspects of HIV-1 toxicity.

2.2 Reporter Gene Transgenic Models

Information about the potential tropism of and host factor interactions with specific strains of HIV-1 in the CNS can be gained using reporter gene constructs. Typically, these constructs have consisted of the HIV-1 LTR, which contains transcriptional regulatory regions, linked to a reporter gene such as chloramphenicol acetyl transferase (CAT) or lac Z. Thus, LTR-driven reporter gene expression in transgenic mice identifies host cells capable of activating transcription of HIV-1 genes. Several HIV-1 LTR/reporter gene transgenic models have been established (see KOCH and RUPRECHT 1992, for review); however, in those studies which analyzed the CNS, the majority of models did not show expression of their respective reporter genes in brain tissue.

Since strains of HIV-1 vary in tropism for macrophages and T cells, CORBOY et al. (1992) developed transgenic mice which contained the LTRs of either of two CNS-derived strains, or of a T cell-derived strain, fused to the lac Z reporter gene to compare the host cell specificity directed by these different viral LTRs. The CNS-derived strains had been isolated from either the CSF or the frontal lobe of an individual patient and shown to differ in nucleotide sequence and in ability to infect macrophage/monocytes and brain glioma explant cultures (KOYANAGI et al. 1987). Although the HIV-1 transcriptional regulatory sites NF-κB and SP-1 were conserved between the two strains, differences were found upstream of these sites in the U3 regions of the LTRs. Both of these constructs were expressed in the CNS, whereas no β-galactosidase expression was seen in the CNS of mice transgenic for the T cell-derived LTR.

Surprisingly, CNS expression of both CNS-derived LTR constructs was found predominantly in neurons, cells in which replication of HIV-1 has not been demonstrated conclusively in vivo. This result invites both speculation and further experimentation. For instance, replication of HIV-1 in neurons may be an early, transient event in the CNS which has thus far escaped detection. Early neuronal infection could result in cell death, which may induce, either alone or in concert with invading infected peripheral blood mononuclear cells, activation and subsequent productive infection of the CNS resident macrophages.

It has been shown for visna virus, another lentivirus which affects the CNS and the immune system, that macrophages are required to be in an activated state in order to achieve expression from the LTR (SMALL et al. 1989). Since the transgenic model of CORBOY et al. (1992) monitored viral LTR-directed expression in the absence of neurotoxic viral factors, it is likely that microglia were not activated in these mice. It would be interesting to analyze the expression patterns of these constructs in a model containing activated microglia to determine if, in this state, these cells are capable of driving β-galactosidase expression from the LTR.

Although puzzling, the picture provided by the CNS-derived LTR transgenic model demonstrates that neurons are capable of activating transcription from the HIV-1 LTR in vivo. Thus, the lack or paucity of detectable HIV-1 within neurons in vivo may not be due to a block in neuronal transcription of the viral genome. This model could be used to study the effects of neuron-specific transcription factors and the potential stimulatory or inhibitory effects of pharmacologic agents on the activity of the CNS-derived HIV-1 LTR.

2.3 Transgenic Models Expressing Viral Proteins

In addition to those discussed above, many transgenic models have been generated to analyze the role of viral proteins in the pathology produced by HIV-1 infection. Mice transgenic for HIV-1 proviral DNA have been developed and, in one study, were reported to produce infectious virus (LEONARD et al. 1988). However, in this study, HIV-1-specific nucleotide sequences were not detected in CNS tissue, and no pathologic effects on the CNS were reported. The toxic effects of peripheral expression of HIV-1 sequences, which include epidermal hyperplasia, lymphadenopathy, splenomegaly, pulmonary lymphoid infiltration, growth retardation, cachexia, wasting, and death, are reviewed elsewhere (KOCH and RUPRECHT 1992; KLOTMAN and NOTKINS, this volume).

Studies performed in our laboratory have demonstrated that expression of the HIV-1 envelope glycoprotein gp120 within the otherwise unmanipulated murine CNS is sufficient to induce pathologic effects on neurons, astrocytes, and microglia, that closely resemble those seen in brains of patients with AIDS dementia complex (TOGGAS et al. 1994). While monocytoid cells are the primary source of HIV-1 proteins in the human CNS, there was no promoter available at the time that enabled effective constitutive expression of transgenes in brain microglia/macrophages. Instead, we used regulatory sequences from the glial fibrillary acidic protein (GFAP) gene to target the expression of gp120 to astrocytes. Astroglial expression of GFAP-driven fusion genes has been shown to be an effective means of delivering diverse proteins to the CNS parenchyma (MUCKE et al. 1992; MUCKE and ROCKENSTEIN 1993; CAMPBELL et al. 1993a,b; JOHNSON et al. 1995; WYSS-CORAY et al. 1995). Furthermore, increasing evidence suggests that astrocytes are involved in the pathogenesis of HIV-1 associated neurologic disease (LIPTON 1992c; GENIS et al. 1992; LEVI et al. 1993; PULLIAM et al. 1993; MUCKE and EDDLESTON 1993; TORNATORE et al. 1994; SAITO et al. 1994; BENOS et al. 1994). As new regulatory sequences become available, it will be important to assess the effects of microglial expression of gp120 on the CNS and to compare these effects with those seen in the GFAP-gp120 model.

To construct the GFAP-gp120 transgene, the *env* gene of HIV-1$_{LAV}$ was truncated to encode a secretable form of gp120 (without the gp41 transmembrane domain) and inserted into exon 1 of a modified murine GFAP gene (Fig. 1). In vitro studies confirmed that astroglial cell lines expressing this transgene secrete gp120 into the culture media (TOGGAS et al. 1994). By in situ hybridization, widespread astroglial expression of gp120 mRNA was seen in GFAP-gp120 transgenic mice, with highest levels in neocortex, olfactory bulb, hippocampus, tectum, selected white matter tracts, and along the glia limitans. Three distinct GFAP-gp120 transgenic lines with low, intermediate, and high levels of gp120 expression were analyzed in detail (TOGGAS et al. 1994). Transgenic mice were compared with nontransgenic littermates and with transgenic mice expressing other fusion gene products from the same GFAP cassette. Inclusion of these controls ensured that the pathologic effects observed in GFAP-gp120 mice were not due to an insertional mutation or to nonspecific effects resulting from the astroglial expression of any foreign protein.

Expression of gp120 induced a widespread astrocytosis, microglial activation (Fig. 2), vacuolization of neuronal dendrites (Fig. 3), and loss of synapses and neuronal subpopulations. Quantitative analysis determined that the extent of CNS pathology occurred in direct correlation to levels of gp120 transgene expression,

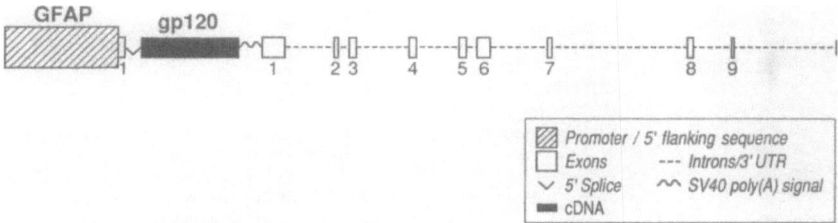

Fig. 1. GFAP-gp120 transgene construct. The region of HIV-1 *env* gene encoding gp120 (derived from clone pHenv; FREED et al. 1989) was inserted into the modified exon 1 of the murine glial fibrillary acidic protein (GFAP) gene, targeting production of gp120 to astrocytes. See TOGGAS et al. (1994) for a detailed description of the cloning scheme. This construct contains all of the exons and introns of the GFAP gene. A simian virus 40 (SV40) late gene splice and polyadenylation signal were incorporated to enhance in vivo expression and prevent expression of GFAP coding sequences, respectively. *UTR*, 3' untranslated region

both across different brain regions and across different lines of mice, with the greatest damage found in regions of highest expression and in mice of the highest expressing line. Notably, similar structural changes are found in brains of AIDS patients with CNS complications (reviewed in PRICE and PERRY 1994).

Since HIV-1-infected brains frequently show microglial abnormalities and activated microglial cells are thought to produce factors which contribute to HIV-1-associated CNS pathology (see above), we were particularly interested in the microglial alterations induced in the GFAP-gp120 transgenic model. As shown in Fig. 2, transgenic mice of the highest expressing line displayed a significant increase in the number of cells positive for the mononuclear phagocyte marker F4/80. Changes in morphology consisted of increased branching and cytoplasmic distension, suggesting that these cells, both in the hippocampus (Fig. 2) and the fronto-parietal cortex (TOGGAS et al. 1994), were in an activated state.

A provocative connection between activated microglia, which could produce neurotoxins, and NMDA-like neurotoxicity may exist in this model and is under further investigation (see below). While high expressor mice showed both damage to and loss of neurons, the only form of structural neuronal damage detected in low expressor mice was dendritic vacuolization (Fig. 3). Thus, the neuropathologic analysis of transgenic lines expressing different levels of gp120 revealed a dose effect which established dendritic vacuolization as one of the earliest forms of gp120-induced neuronal damage in the CNS. It is interesting in this context that dystrophic neurites and dendritic vacuolizations are also typically seen in other models of excitotoxicity (LIPTON 1994).

Vacuolar degeneration of neuronal dendrites was evident in gp120 transgenic mice but not in nontransgenic littermate controls by postnatal day 7 (Toggas et al. 1995). This early development of brain damage in GFAP-gp120 mice is consistent with the postnatal increase in expression directed by the GFAP promoter (TARDY et al. 1989). Notably, it also correlates well with the developmental expression of NMDA receptors (MACDONALD et al. 1993). This model should allow assessment

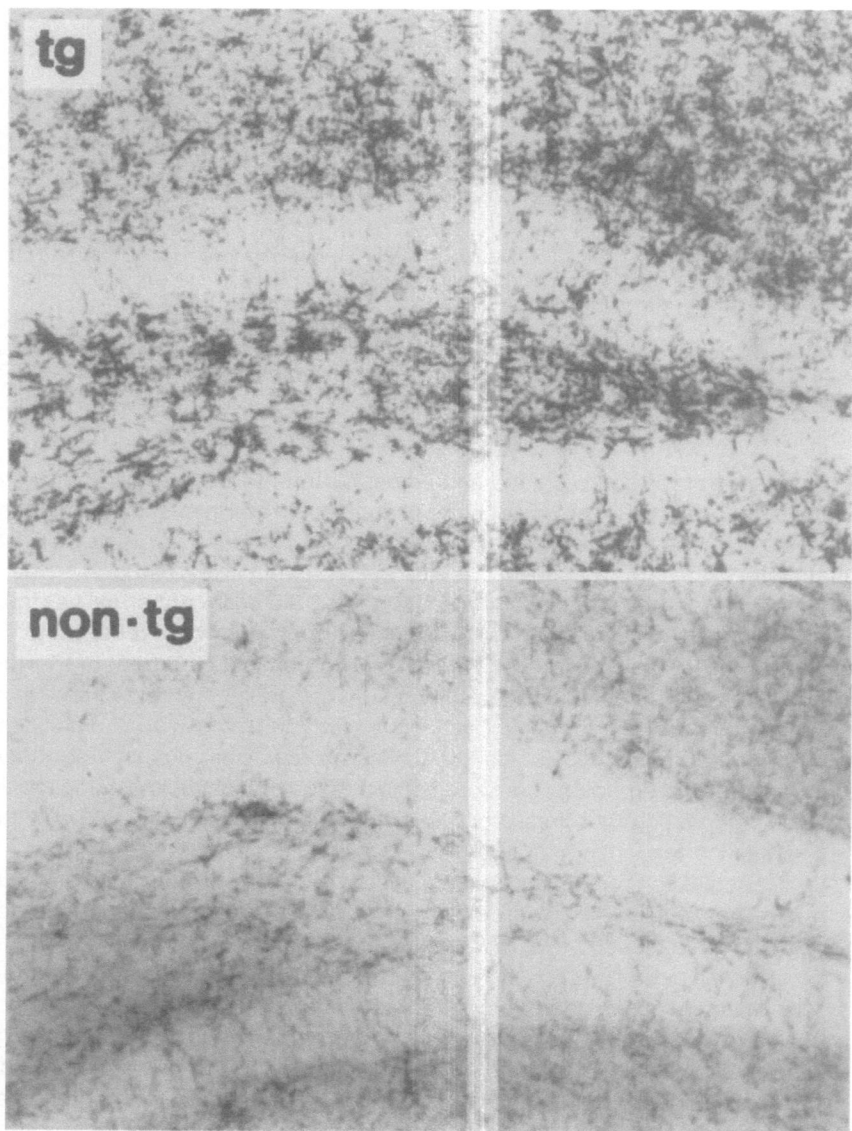

Fig. 2. Microglial alterations in GFAP-gp120 transgenic mice. F4/80 immunostaining of transgenic (tg) versus nontransgenic (non-tg) hippocampi revealed greatly increased numbers of F4/80-reactive microglia in the hippocampus of transgenic GFAP-gp120 mice. The hypertrophied, ramified morphology of the stained cells in the section from the transgenic animal is suggestive of an activated state. Vibratome sections (40 μm) were immunolabeled with a monoclonal antibody against F4/80 as described previously. (TOGGAS et al. 1994)

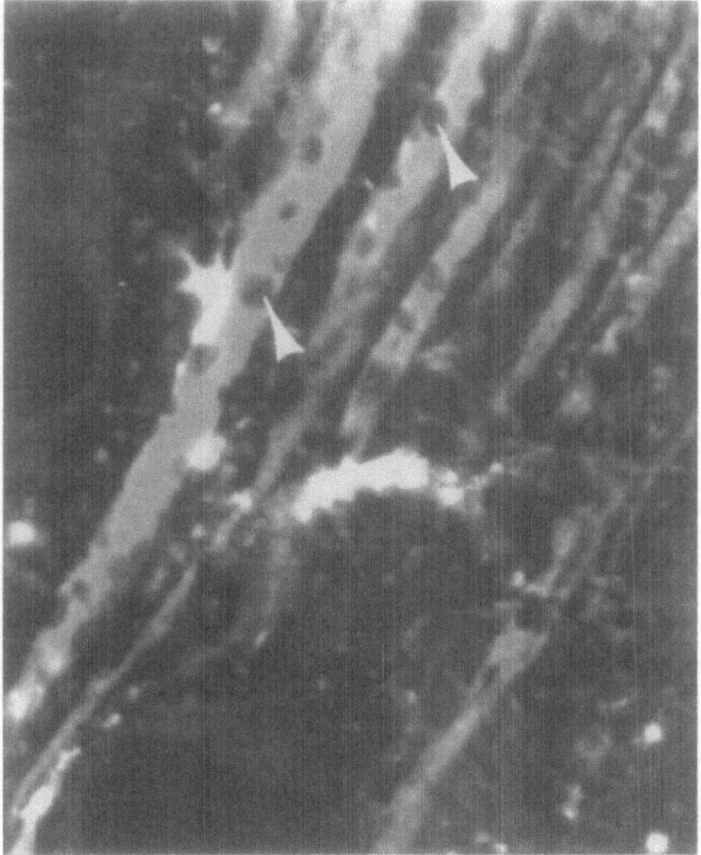

Fig. 3. Vacuolar degeneration of neuronal dendrites in the neocortex of low expressor gp120 transgenic mice. This laser scanning confocal micrograph of cells immunolabeled with a monoclonal antibody against microtubule-associated protein-2 (MAP-2) illustrates neuronal damage in the form of dendritic vacuolizations (*arrows*). Because low expressor mice did not show other forms of neuronal damage detected in high expressor gp120 transgenic mice (TOGGAS et al. 1994), e.g., loss of neurons or decreases in presynaptic terminal density, this dendritic vacuolization probably represents an early form of gp120-induced neuronal damage. Double immunolabeling of vibratome sections (40 μm) for MAP-2 and GFAP and subsequent confocal imaging were performed as described (TOGGAS et al. 1994)

of the sensitivity of the developing nervous system to gp120 and may identify more clearly the role of this factor in HIV-1-infected human infants.

Hence, expression of gp120 within the CNS of transgenic mice has provided in vivo evidence that this molecule may play an important part in the production of CNS damage associated with HIV-1 infection. Since this transgenic model simulates many of the alterations seen in the CNS of AIDS patients, it should help determine relevant mechanisms of gp120 neurotoxicity and facilitate the evaluation of pharmacologic interventions prior to advancing drugs into human trials (see below). Currently, CNS tissue of GFAP-gp120 transgenic mice is being subjected to numerous analyses to identify biochemical alterations implicated in other studies of HIV-1 CNS damage as well as those seen in other CNS injury states but not yet investigated for HIV-1.

HIV-1 infection of the CNS is associated not only with structural and biochemical alterations, but also with functional impairments. Obvious behavioral abnormalities have not been observed in GFAP-gp120 transgenic mice on the B6xSJL background, and preliminary studies using Y-maze avoidance paradigms have not identified deficits in learning in these mice. However, as demonstrated for FMRI gene knockout mice (BAKKER et al. 1994), these basic studies may require much finer dissection to reveal alterations in CNS function. Notably, neurological impairments in HIV-1-infected individuals are also often subtle until the later stages of disease, when patients have developed frank AIDS dementia complex (reviewed in PRICE and PERRY 1994). Identification of a reliable behavioral or other clinically obvious phenotypic alteration in GFAP-gp120 transgenic mice would constitute an informative endpoint for pharmacologic analyses, both from a clinical and a neuropathogenetic point of view. Preliminary studies indicate that a distinctive phenotype can be achieved when the GFAP-gp120 transgene is expressed in the CNS of Balb/c mice (Toggas and Mucke, unpublished observations).

Transgenic models of infection by other retroviruses may also provide insights into neurologic disease associated with HIV-1. A transgenic model in which the envelope glycoprotein gp70 of Cas-Br-E murine leukemia virus was expressed in the CNS showed that this envelope protein is sufficient to produce neurotoxic effects resembling those seen in infection with this virus (KAY et al. 1993). Two different cassettes were used to establish distinct transgenic lines. Expression was directed by the viral LTRs and gp70 RNA was detected in the spinal cord and brainstem. Although no behavioral effects were reported, histologic damage, similar for both lines, consisted of spongiform changes in brainstem nuclei and anterior horns of the spinal cord. Gliosis was also present in these regions and in the cortex of more severely diseased animals. Consequently, mechanisms of CNS damage for different retroviruses may involve neurotoxic properties of their envelope glycoproteins.

2.4 Transgenic Models Expressing Host Factors

Host factors expressed in response to HIV-1 infection may contribute to the production of CNS pathology. As mentioned above, clinical and neuropathological studies, along with in vitro studies, suggest that HIV-1 may induce the production of cytokines such as IL-1, IL-6, transforming growth factor (TGF)-β and TNF-α from T cells, B cells, macrophages, microglia, and/or astrocytes (ROSENBERG and FAUCI 1990; EMILIE et al. 1990; VITKOVIC et al. 1990; WAHL et al. 1991; SUNDAR et al. 1991; MERRILL et al. 1992; GENIS et al. 1992). Although these soluble factors may help coordinate the antiviral host response and contribute to the repair of tissue injury, they may also possess considerable toxicity. Furthermore, cytokines have been shown to influence HIV-1 gene expression, with IL- 6 and TNF-α having primarily enhancing (ROSENBERG and FAUCI 1990) and TGF-β suppressing (KEKOW et al. 1990) effects. Thus, important regulatory controls exist between HIV-1 and the host cytokine network that may affect strongly the development of damage in the HIV-1-infected CNS.

To study the role of increased cytokine expression in the production of CNS injury, the modified GFAP gene was used to overexpress IL- 6, a cytokine implicated in the pathogenesis of HIV-1-associated neurologic impairment (GALLO et al. 1989; TYOR et al. 1992), in astrocytes of transgenic mice (CAMPBELL et al. 1993a). Expression of transgene-derived murine IL-6 was found predominantly in subcortical regions such as the thalamus, cerebellum and brainstem and was associated with a slow, progressive, neurologic syndrome in which both the clinical severity and neuropathology correlated positively with the level and distribution of transgene expression (CAMPBELL et al. 1993a; CHIANG et al. 1994). Increased IL-1α, IL-1β, and TNF-α mRNA levels, extensive neurodegeneration, and gliosis were also detected in the CNS of these mice. These transgenic studies demonstrated that cerebral expression of IL-6 can induce many of the characteristic molecular, cellular, and functional alterations (see MUCKE et al. 1995b, for review) seen in HIV-1-associated neurologic disease.

Recently, the GFAP-driven expression of several other cytokines has been evaluated. In particular, expression of interferon (IFN)-α, IL-3 (Campbell et al., unpublished), and TGF-β (WYSS-CORAY et al. 1995) in the CNS resulted in severe neurologic complications, which in high expressors culminated in death. The initial characterization of these GFAP-cytokine transgenic mice revealed both overlapping (e.g., gliosis) and model-specific forms of damage (e.g., angiogenesis, up-modulation of acute phase vs extracellular matrix proteins, prominent inflammatory infiltrates). Differences between distinct transgenic models likely reflect a combination of the specific properties of the transgene products expressed and the selective vulnerability of the various cell populations in the brain. Pharmacologic approaches to define mechanisms of damage operating in these models should reveal if the diverse factors expressed ultimately converge upon a final common pathogenetic pathway or employ distinct mechanisms to produce neural injury.

3 Development and Evaluation of Therapeutic Strategies Using Transgenic Models

3.1 Pharmacologic Studies

Mice expressing a transgene product that produces a biologic effect can be used to determine the efficacy of exogenously administered compounds at blocking the effect. Dose-response experiments can determine optimum dose ranges for inhibition of damage induced by expression of the transgene product at different levels. While these analyses may suffer from the same limitations as those described in Sect. 1.4, they are particularly useful for compounds which can cross the BBB. Two such compounds, the clinically tolerated NMDA receptor antagonist memantine and the free radical spin-trapping agent phenylbutylnitrone, have been shown to be protective in vivo against neuronal injury induced by excitotoxicity (Seif el Nasr et al. 1990; Keilhoff and Wolf 1992) and ischemia (Oliver et al. 1990), respectively. In addition, memantine can block the effects of gp120 in vitro (Lipton 1992a; Müller et al. 1992). Trials involving systemic administration of these drugs to GFAP-gp120 transgenic mice to inhibit the toxic effects of gp120 expression in vivo are now in progress. These trials indicate that NMDA receptors may mediate a component of gp120-induced CNS damage in vivo (Toggas et al. 1995).

3.2 Generation of Bigenic Models

In cases of early postnatal transgene expression, therapeutic interventions may be required shortly after birth, which can complicate the design of effective treatment paradigms. The increased use of transgenic approaches to study a wide variety of neurobiologic questions has resulted in the availability of numerous transgenic models which may be useful in the evaluation of specific aspects of AIDS dementia complex, as well as other diseases. For example, the GFAP-gp120 transgenic mice generated in our laboratory can be bred with transgenic mice expressing putative neurotrophic and neuroprotective factors to determine potential mechanisms of gp120-mediated CNS damage and to assess the potential therapeutic usefulness of these factors. Our studies indicate that neuronal overexpression of the amyloid precursor protein (APP) effectively protects bigenic gp120/APP mice against gp 120 -induced neuronal damage (Mucke et al. 1995a). Since APP has been implicated in the stabilization of neuronal calcium homeostasis (Mattson et al.1993), these results provide further in vivo support for a role of calcium in gp120-induced damage. This example illustrates that establishment of bigenic mice can be an effective alternative to drug administration studies in singly transgenic mice.

As mentioned above, damage due to free radical generation is also a proposed mechanism of HIV-1 CNS toxicity. Transgenic mice which express the human copper/zinc superoxide dismutase enzyme, a free radical scavenger, have been generated and shown to be protected from neurotoxic damage in several models of CNS injury (Przedborski et al. 1992; Chan et al. 1994; Cadet et al. 1994; Yang et al. 1994). Currently, brains of mice bigenic for this enzyme and gp120 are being evaluated to determine if free radicals play a role in the injury produced by expression of gp120 in vivo. Careful study of transgenic model databases should enable selection of additional lines of mice expressing factors which may counteract or synergize with HIV-1-derived/induced neurotoxins, both situations allowing for further interesting bigenic analyses. Finally, the study of offspring from HIV-1 transgenic mice and mice engineered (homologous knockout, anti-sense, ribozymes, etc.) to lack specific mediators may determine the relative contributions, helpful or harmful, of these molecules to the overall CNS pathology. In addition to helping dissect mechanisms of neuropathogenicity, this approach should establish fruitful collaborations among scientists from diverse backgrounds.

4 Conclusions

As outlined above, the combination of clinical studies, in vitro studies, and non-transgenic in vivo studies can act in concert with transgenic investigations to provide the most complete picture possible of the etiology of AIDS dementia. Notably, these approaches are complementary; that is, specific molecules may be identified in patient specimens, analyzed for activity acutely in vitro and in vivo and, if promising, can be used for the assessment of more chronic effects in transgenic models. Transgenic models can help establish or confirm specific, pathologically relevant cause/effect relationships, and may identify novel alterations which can then be assayed in patient samples. Ultimately, key pathogenetic targets may be identified and pharmacologic interventions tested using transgenic models. Transgenic models which can reveal a correlation between peripheral indices and neurologic damage would be of particular value in monitoring CNS disease onset and progression, and in the efficacy assessment of therapeutic strategies.

Acknowledgement. Supported by NIH grants MH-47680(LM), NS-33056(LM), NS-34602(LM) and AG-00080 (SMT).

References

Appleman ME, Marshall DW, Brey RL, Houk RW, Beatty DC, Winn RE, Melcher GP, Wise MG, Sumaya CV, Boswell RN (1988) Cerebrospinal fluid abnormalities in patients without AIDS who are seropositive for the human immunodeficiency virus. J Infect Dis 158: 193–199

Bakker CE, Verheij C, Willemsen R, van der Helm R, Oerlemans F, Vermey M, Bygrave A, Hoogeveen AT, Oostra BA, Reyniers E, De Boulle K, D'Hooge R, Cras P, van Velzen D, Nagels G, Martin J-J, De Deyn PP, Darby JK, Willems PJ (1994) Fmr 1 knockout mice: a model to study fragile X mental retardation. Cell 78: 23–33

Benos DJ, Hahn BH, Bubien JK, Ghosh SK, Mashburn NA, Chaikin MA, Shaw GM, Benveniste EN (1994) Envelope glycoprotein gp120 of human immunodeficiency virus type 1 alters ion transport in astrocytes: implications for AIDS dementia complex. Proc Natl Acad Sci USA 91: 494–498

Brew BJ, Bhalla RB, Fleisher M, Paul M, Khan A, Schwartz MK, Price RW (1989) Cerebrospinal β_2 microglobulin in patients infected with human immunodeficiency virus. Neurology 39: 830–834

Brew BJ, Bhalla RB, Paul M, Gallardo H, McArthur JC, Schwartz MK, Price RW (1990) Cerebrospinal fluid neopterin in human immunodeficiency virus type 1 infection. Ann Neurol 28: 556–560

Budka H (1992) Cerebral pathology in AIDS: a new nomenclature and pathogenetic concepts. Curr Opinion Neurol Neurosurg 5: 917–923

Cadet JL, Sheng P, Ali S, Rothman R, Carlson E, Epstein C (1994) Attenuation of methamphetamine-induced neurotoxicity in copper/zinc superoxide dismutase transgenic mice. J Neurochem 62: 380–383

Campbell IL, Abraham CR, Masliah E, Kemper P, Inglis JD, Oldstone MBA, Mucke L (1993a) Neurologic disease induced in transgenic mice by cerebral overexpression of interleukin 6. Proc Natl Acad Sci USA 90: 10061–10065

Campbell IL, Mucke L, Sandberg K (1993b) Cerebral overexpression of interleukin-6 or interferon-α1 induces distinct neuropathology in transgenic mice (Abstr). Soc Neurosci Abstr 19: 98.2

Chan PH, Chu L, Chen SF, Carlson EJ, Epstein CJ (1994) Attenuation of glutamate-induced neuronal swelling and toxicity in transgenic mice overexpressing human CuZn-superoxide dismutase. Acta Neurochir Suppl (Wiar) 51: 245–247

Chiang C-S, Stalder A, Samimi A, Campbell IL (1994) Reactive gliosis as a consequence of interleukin-6 expression in the brain. Studies in transgenic mice. Dev Neurosci 16: 212–221

Corboy JR, Buzy JM, Zink MC, Clements JE (1992) Expression directed from HIV long terminal repeats in the central nervous system of transgenic mice. Science 258: 1804–1808

Dawson VL, Dawson TM, Uhl GR, Snyder SH (1993) Human immunodeficiency virus type 1 coat protein neurotoxicity mediated by nitric oxide in primary cortical cultures. Proc Natl Acad Sci USA 90: 3256–3259

Elovaara I, Poutiainen E, Raininko R, Valanne L, Virta A, Valle SL, Lahdevirta J, Iivanainen M (1990) Mild brain atrophy in early HIV infection: the lack of association with cognitive deficits and HIV-specific intrathecal immune response. J Neurol Sci 99: 121–136

Emilie D, Peuchmaur M, Maillot MC, Crevon MC, Brousse N, Delfraissy JF, Dormont J, Galanaud P (1990) Production of interleukins in human immunodeficiency virus-1-replicating lymph nodes. J Clin Invest 86: 148–159

Epstein LG, Gendelman HE (1993) Human immunodeficiency virus type 1 infection of the nervous system: pathogenetic mechanisms. Ann Neurol 33: 429–436

Freed EO, Myers DJ, Risser R (1989) Mutational analysis of the cleavage sequence of the human immunodeficiency virus type 1 envelope glycoprotein precursor gp160. J Virol 63: 4670–4675

Furth PA, St Onge L, Böger H, Gruss P, Gossen M, Kistner A, Bujard H, Henninghausen L (1994) Temporal control of gene expression in transgenic mice by a tetracycline-responsive promoter. Proc Natl Acad Sci 91: 9302–9306

Gallo P, Frei K, Rordorf C, Lazdins J, Tavolato B, Fontana A (1989) Human immunodeficiency virus type 1 (HIV-1) infection of the central nervous system: an evaluation of cytokines in cerebrospinal fluid. J Neuroimmunol 23: 109–116

Gelbard HA, Nottet HSLM, Swindells S, Jett M, Dzenko KA, Genis P, White R, Wang L, Choi Y-B, Zhang D, Lipton SA, Tourtellotte VW, Epstein LG, Gendelman HE (1994) Platelet-activating factor: a candidate human immunodeficiency virus type 1-induced neurotoxin. J Virol 68: 4628–4635

Geleziunas R, Schipper HM, Wainberg MA (1992) Pathogenesis and therapy of HIV-1 infection of the central nervous system. AIDS 6: 1411–1426

Genis P, Jett M, Bernton EW, Boyle T, Gelbard HA, Dzenko K, Keane RW, Resnick L, Mizrachi Y, Volsky DJ, Epstein LG, Gendelman HE (1992) Cytokines and arachidonic acid metabolites produced during human immunodeficiency virus (HIV)-infected macrophage-astroglia interactions: Implications for the neuropathogenesis of HIV disease. J Exp Med 176: 1703–1718

Giulian D, Vaca K, Noonan CA (1990) Secretion of neurotoxin by mononuclear phagocytes infected with HIV-1. Science 250: 1593–1596

Giulian D, Wendt E, Vaca K, Noonan CA (1993) The envelope glycoprotein of human immunodeficiency virus type 1 stimulates release of neurotoxins from monocytes. Proc Natl Acad Sci USA 90: 2769–2773

Glowa JR, Panlilio LV, Brenneman DE, Gozes I, Fridkin M, Hill JM (1992) Learning impairment following intracerebral administration of the HIV envelope protein gp120 or a VIP antagonist. Brain Res 570: 49–53

Gossen M, Bujard H (1992) Tight control of gene expression in mammalian cells by tetracycline-responsive promoters. Proc Natl Acad Sci USA 89: 5547–5551

Gossen M, Bonin AL, Bujard H (1993) Control of gene activity in higher eukaryotic cells by prokaryotic regulatory elements. Trends Biochem Sci 18: 471–475

Goudsmit J, Paul DA, Lange JMA, Speelman H, van der Noordaa J, van der Helm HJ, de Wolf F, Epstein LG, Krone WJA, Oleske JM, Coutinho RA (1986) Expression of human immunodeficiency virus antigen (HIV-Ag) in serum and cerebrospinal fluid during acute and chronic infection. Lancet 2: 177–180

Heyes MP, Brew BJ, Martin A, Price RW, Salazar AM, Sidtis JJ, Yergey JA, Mouradian MM, Sadler AE, Keilp J (1991) Quinolinic acid in cerebrospinal fluid and serum in HIV-1 infection: relationship to clinical and neurological status. Ann Neurol 29: 202–209

Hill JM, Mervis RF, Avidor R, Moody TW, Brenneman DE (1993) HIV envelope protein-induced neuronal damage and retardation of behavioral development in rat neonates. Brain Res 603: 222–233

Ho DD, Rota TR, Schooley RT, Kaplan JC, Allan JD, Groopman JE, Resnick L, Felsenstein D, Andrews CA, Hirsch MS (1985) Isolation of HTLV-III from cerebrospinal fluid and neural tissues of patients with neurologic syndromes related to the acquired immunodeficiency syndrome. N Engl J Med 313: 1493–1497

Hurtrel M, Ganiere JP, Guelfi JF, Chakrabarti L, Maire MA, Gray F, Montagnier L, Hurtrel B (1992) Comparison of early and late feline immunodeficiency virus encephalopathies. AIDS 6: 399–406

Johnson WB, Ruppe MD, Rockenstein EM, Price J, Sarthy VP, Verderber LC, Mucke L (1995) Indicator expression directed by regulatory sequences of the glial fibrillary acidic protein (GFAP) gene: in vivo comparison of distinct GFAP-lacZ transgenes. Glia 13: 174–184

Kay DG, Gravel C, Pothier F, Laperriere A, Robitaille Y, Jolicoeur P (1993) Neurological disease induced in transgenic mice expressing the env gene of the Cas-Br-E murine retrovirus. Proc Natl Acad Sci USA 90: 4538–4542

Keilhoff G, Wolf G (1992) Memantine prevents quinolinic acid-induced hippocampal damage. Eur J Pharmacol 219: 451–454

Kekow J, Wachsman W, McCutchan JA, Cronin M, Carson D, Lotz M (1990) Transforming growth factor β and noncytopathic mechanisms of immunodeficiency in human immunodeficiency virus infection. Proc Natl Acad Sci USA 87: 8321–8325

Kimes AS, London ED, Szabo G, Raymon L, Tabakoff B (1991) Reduction of cerebral glucose utilization by HIV envelope glycoprotein gp-120. Exp Neurol 112: 224–228

Koch JA, Ruprecht RM (1992) Animal models for anti-AIDS therapy. Antiviral Res 19: 81–109

Koyanagi Y, Miles S, Mitsuyasu RT, Merrill JE, Vinters HV, Chen ISY (1987) Dual infection of the central nervous system by AIDS viruses with distinct cellular tropisms. Science 236: 819–822

Lackner AA, Dandekar S, Gardner MB (1991) Neurobiology of simian and feline immunodeficiency virus infections. Brain Pathol 1: 201–212

Leonard JM, Abramczuk JW, Pezen DS, Rutledge R, Belcher JH, Hakim F, Shearer G, Lamperth L, Travis W, Fredrickson T (1988) Development of disease and virus recovery in transgenic mice containing HIV proviral DNA. Science 242: 1665–1670

Levi G, Patrizio M, Bernardo A, Petrucci TC, Agresti C (1993) Human immunodeficiency virus coat protein gp120 inhibits the β-adrenergic regulation of astroglial and microglial functions. Proc Natl Acad Sci USA 90: 1541–1545

Lipton SA (1992a) Memantine prevents HIV coat protein-induced neuronal injury in vitro. Neurology 42: 1403–1405

Lipton SA (1992b) Requirement for macrophages in neuronal injury induced by HIV envelope protein gp120. NeuroReport 3: 913–915

Lipton SA (1992c) Models of neuronal injury in AIDS: another role for the NMDA receptor? Trends Neurosci 15: 75–79

Lipton SA (1994) HIV-related neuronal injury: potential therapeutic intervention with calcium channel antagonists and NMDA antagonists. Mol Neurobiol 8: 181–196

Mabrouk K, Van Rietschoten J, Vives E, Darbon H, Rochat H, Sabatier J-M (1991) Lethal neurotoxicity in mice of the basic domains of HIV and SIV rev proteins. FEBS Lett 289: 13–17

MacDonald MC, Robertson HA, Wilkinson M (1993) Age- and dose-related NMDA induction of Fos-like immunoreactivity and c-fos mRNA in the arcuate nucleus of immature female rats. Dev Brain Res 73: 193–198

Marshall DW, Brey RL, Cahill WT, Houk RW, Zajac RA, Boswell RN (1988) Spectrum of cerebrospinal fluid findings in various stages of human immunodeficiency virus infection. Arch Neurol 45: 954–958

Masliah E, Achim CL, Ge N, DeTeresa R, Terry RD, Wiley CA (1992) Spectrum of human immunodeficiency virus-associated neocortical damage. Ann Neurol 32: 321–329

Mattson MP, Cheng B, Culwell AR, Esch FS, Lieberburg I, Rydel RE (1993) Evidence for excito-protective and intraneuronal calcium- regulating roles for secreted forms of the β-amyloid precursor protein. Neuron 10: 243–254

McArthur JC, Cohen BA, Farzedegan H, Cornblath DR, Selnes OA, Ostrow D, Johnson RT, Phair J, Polk BF (1988) Cerebrospinal fluid abnormalities in homosexual men with and without neuropsychiatric findings. Ann Neurol 23 suppl: S34-S37

Merrill JE, Chen IS (1991) HIV-1, macrophages, glial cells, and cytokines in AIDS nervous system disease. FASEB J 5: 2391–2397

Merrill JE, Koyanagi Y, Zack J, Thomas L, Martin F, Chen IS (1992) Induction of interleukin-1 and tumor necrosis factor alpha in brain cultures by human immunodeficiency virus type 1. J Virol 66: 2217–2225

Mosier DE, Gulizia RJ, Baird SM, Wilson DB, Spector DH, Spector SA (1991) Human immunodeficiency virus infection of human-PBL-SCID mice. Science 251: 791–794

Mucke L, Eddleston M (1993) Astrocytes in infectious and immune-mediated diseases of the central nervous system. FASEB J 7: 1226–1232

Mucke L, Rockenstein EM (1993) Prolonged delivery of transgene products to specific brain regions by migratory astrocyte grafts. Transgenics 1: 3–9

Mucke L, Forss-Petter S, Goldgaber D, Johnson W, Picard E, Rockenstein E, Abraham C (1992) Transgenic models to study the pathogenic role of mutated and non-mutated forms of human amyloid proteins in the development of Alzheimer's disease (AD). Neurobiol Aging 13 Suppl 1: S101

Mucke L, Abraham CR, Ruppe MD, Rockenstein EM, Toggas SM, Alford M, Masliah E (1995a) Protection against HIV-1 gp120-induced brain damage by neuronal expression of human amyloid precursor protein. J Exp Med 181: 1551–1556

Mucke L, Masliah E, Campbell IL (1995b) Transgenic models to assess the neuropathogenic potential of HIV-1 proteins and cytokines. Springer, Berlin Heidelberg New York (Current Topics in Microbiology and Immunology, Vol. 202: 187–205)

Müller WEG, Schröder HC, Ushijima H, Dapper J, Bormann J (1992) gp120 of HIV-1 induced apoptosis in rat cortical cell cultures: prevention by memantine. Eur J Pharmacol 226: 209–214

Namikawa R, Kaneshima H, Lieberman M, Weissman IL, McCune JM (1988) Infection of SCID-hu mouse by HIV-1. Science 242: 1684–1686

Oliver CN, Starke-Reed PE, Stadtman ER, Liu GJ, Carney JM, Floyd RA (1990) Oxidative damage to brain proteins, loss of glutamine synthetase activity, and production of free radicals during ischemia/reperfusion-induced injury to gerbil brain. Proc Natl Acad Sci USA 87: 5144–5147

Passman RS, Fishman GI (1994) Regulated expression of foreign genes in vivo after germline transfer. J Clin Invest 94: 2421–2425

Pietraforte D, Tritarelli E, Testa U, Minetti M (1994) gp120 HIV envelope glycoprotein increases the production of nitric oxide in human monocyte-derived macrophages. J Leukocyte Biol 55: 175–182

Price RW, Perry SW (eds) (1994) HIV, AIDS, and the brain. Raven, New York

Przedborski S, Kostic V, Jackson-Lewis V, Naini AB, Simonetti S, Fahn S, Carlson E, Epstein CJ, Cadet JL (1992) Transgenic mice with increased Cu/Zn-superoxide dismutase activity are resistant to N-methyl-4-phenyl-1,2,3,6-tetrahydropyridine-induced neurotoxicity. J Neurosci 12: 1658–1667

Pulliam L, West D, Haigwood N, Swanson RA (1993) HIV-1 envelope gp120 alters astrocytes in human brain cultures. AIDS Res Hum Retroviruses 9: 439–444

Resnick L, Berger JR, Shapshak P, Tourtellote W W (1988) Early penetration of the blood-brain barrier by HIV. Neurology 38: 9–14

Rosenberg ZF, Fauci AS (1990) Immunopathogenetic mechanisms of HIV infection: cytokine induction of HIV expression. Immunol Today 11: 176–180

Sabatier J-M, Vives E, Mabrouk K, Benjouad A, Rochat H, Duval A, Hue B, Bahraoui E (1991) Evidence for neurotoxic activity of tat from human immunodeficiency virus type 1. J Virol 65: 961–967

Saito Y, Sharer LR, Epstein LG, Michaels J, Mintz M, Louder M, Golding K, Cvetkovich TA, Blumberg BM (1994) Overexpression of nef as a marker for restricted HIV-1 infection of astrocytes in postmortem pediatric central nervous tissues. Neurology 44: 474–481

Schneider J, Kaaden O, Copeland TD, Oroszlan S, Hunsmann G (1986) Shedding and interspecies type sero-reactivity of the envelope glycopolypeptide gp120 of the human immunodeficiency virus. J Gen Virol 67: 2533–2538

Seif el Nasr M, Peruche B, Rossberg C, Mennel HD, Krieglstain J (1990) Neuroprotective effect of memantine demonstrated in vivo and in vitro. Eur J Pharmacol 185: 19–24

Sharer LR (1992) Pathology of HIV-1 infection of the central nervous system. A review. J Neuropathol Exp Neurol 51: 3–11

Small JA, Bieberich C, Ghotbi Z, Hess J, Scangos GA, Clements JE (1989) The visna virus long terminal repeat directs expression of a reporter gene in activated macrophages, lymphocytes, and the central nervous system of transgenic mice. J Virol 63: 1891–1896

Spencer DC, Price RW (1992) Human immunodeficiency virus and the central nervous system. Annu Rev Microbiol 46: 655–693

Sundar SK, Cierpial MA, Kamaraju LS, Long S, Hsieh S, Lorenz C, Aaron M, Ritchie JC, Weiss JM (1991) Human immunodeficiency virus glycoprotein (gp120) infused into rat brain induces interleukin 1 to elevate pituitary-adrenal activity and decrease peripheral cellular immune responses. Proc Natl Acad Sci USA 88: 11246–11250

Tardy M, Fages C, Riol H, LePrince G, Rataboul P, Charriere-Bertrand C, Nunez J (1989) Developmental expression of the glial fibrillary acidic protein mRNA in the central nervous system and in cultured astrocytes. J Neurochem 52: 162–167

Toggas SM, Masliah E, Rockenstein EM, Rall GF, Abraham CR, Mucke L (1994) Central nervous system damage produced by expréssion of the HIV-1 coat protein gp120 in transgenic mice. Nature 367: 188–193

Toggas SM, Masliah E, Mucke L (1995) Prevention of HIV-1 gp120-induced neuronal damage in the central nervous system of transgenic mice by the NMDA receptor antagonist memantine. Brain Res (in press)

Tornatore C, Chandra R, Berger JR, Major EO (1994) HIV-1 infection of subcortical astrocytes in the pediatric central nervous system. Neurology 44: 481–487

Tyor WR, Glass JD, Griffin JW, Becker PS, McArthur JC, Bezman L, Griffin DE (1992) Cytokine expression in the brain during the acquired immunodeficiency syndrome. Ann Neurol 31: 349–360

Tyor WR, Power C, Gendelman HE, Markham RB (1993) A model of human immunodeficiency virus encephalitis in *scid* mice. Proc Natl Acad Sci USA 90: 8658–8662

Vitkovic L, Kalebic T, da Cunha A, Fauci AS (1990) Astrocyte-conditioned medium stimulates HIV-1 expression in a chronically infected promonocyte clone. J Neuroimmunol 30: 153–160

Wahl SM, Allen JB, McCartney-Francis N, Morganti-Kossmann MC, Kossmann T, Ellingsworth L, Mai UE, Mergenhagen SE, Orenstein JM (1991) Macrophage- and astrocyte-derived transforming growth factor beta as a mediator of central nervous system dysfunction in acquired immune deficiency syndrome. J Exp Med 173: 981–991

Wesselingh SL, Power C, Glass JD, Tyor WR, McArthur JC, Farber JM, Griffin JW, Griffin DE (1993) Intracerebral cytokine messenger RNA expression in acquired immunodeficiency syndrome dementia. Ann Neurol 33: 576–582

Wyss-Coray T, Feng L, Masliah E, Ruppe MD, Lee HS, Toggas SM, Rockenstein EM, Mucke L (1995) Increased central nervous system production of extracellular matrix components and development of hydrocephalus in transgenic mice overexpressing transforming growth factor-β1. Am J Pathol 147: 53–67

Yang G, Chan PH, Chen J, Carlson E, Chen SF, Weinstein P, Epstein CJ, Kamii H (1994) Human copper-zinc superoxide dismutase transgenic mice are highly resistant to reperfusion injury after focal cerebral ischemia. Stroke 25: 165–170

Regulation of Expression and Pathogenic Potential of Human Foamy Virus In Vitro and in Transgenic Mice

A. Aguzzi[1], S. Marino[1], R. Tschopp[1], and A. Rethwilm[2]

1 Introduction

Increasing scientific interest in the foamy virus subgroup of retroviruses during the last few years has been due to two discoveries: (1) the molecular biology of foamy viruses shows some close similarities to that of lentiviruses and of the HTLV/BLV group of retroviruses, in addition to some features which set the foamy viruses apart from all other retroviruses. Moreover (2), transgenic mouse systems

---

[1] Institute of Neuropathology, Department of Pathology, University of Zürich, Schmelzbergstrasse 12,
  8091 Zürich, Switzerland
[2] Institute of Virology, University of Würzburg, Versbacherstrasse 7, 97078 Würzburg, Germany

demonstrated for the first time a pathogenic potential of foamy viruses, i.e., induction of a fatal encephalopathy and myopathy in mice expressing various combinations of foamy virus genes.

This review will mainly focus on pathogenetic aspects of human foamy virus (HFV) transgenic mice. However, since the foamy virus group has been characterized in some detail only recently, and several speculations have been put forward on the role of HFV as a significant human pathogen, we will first summarize some more general virological, epidemiological and molecular aspects. For a more detailed discussion of those topics only touched here we refer to more extensive reviews (HOOKS and GIBBS 1975; AGUZZI 1993; RETHWILM 1995).

2 Virological Properties of Foamy Viruses

The foamy viruses (latin: Spumavirinae) are a distinct subgroup of the retrovirus family (TEICH 1984). Foamy virions are 100–140 nm in diameter and characteristically decorated by radially arranged spikes measuring 5–15 nm (WERNER and GELDERBLOM 1979). Mature condensed cores are rarely observed by electron microscopy, reflecting the biochemical finding of predominantly uncleaved *gag* precursor protein in extracellular virions (NETZER et al. 1990; HAHN et al. 1994). Foamy viruses induce multinucleated and vacuolated syncytia during lytic replication in cultured cells and viruses preferentially bud from intracellular membranes in specific virus host cell systems (WERNER and GELDERBLOM 1979). For this reason they often remain cell-associated (NEUMANN-HAEFELIN et al. 1983; LOH 1993). Massive vacuolation of infected cells may eventually result in a "foamy" appearance of the culture, which has led to the name of this viral family. The molecular cloning and nucleotide sequencing of several primate foamy virus isolates have allowed comparison of the reverse transcriptase (RT) amino acid sequences of foamy viruses and of other retroviruses (FLÜGEL et al. 1987; MAURER et al. 1988; KUPIEC et al. 1991; RENNE et al. 1992; HERCHENRÖDER et al. 1994). This has led to the incorporation of the foamy viruses as a separate subgroup into the phylogenetic tree of retroviruses (MAURER et al. 1988; DOOLITTLE et al. 1990). These analyses also revealed that foamy virus RT sequences are more closely related to MuLV than to lentivirus or HTLV/BLV sequences (DOOLITTLE et al. 1990; HERCHENRÖDER et al. 1994).

Foamy viruses have been isolated from almost every organ and body fluid of naturally infected hosts (HOOKS and GIBBS 1975). It is not yet known whether the isolates derived in all cases from various cell types infected in vivo or might also have stemmed from infected cells present in every tissue, e.g. lymphoid cells. Foamy virus isolates, obtained by throat swabs, from infected monkeys, and numerous isolates from brain and lymphocytes indicate that the virus naturally resides in these tissues (HOOKS and GIBBS 1975). These isolation studies have more recently been confirmed by polymerase chain reaction (PCR) mediated amplification of foamy virus DNA (NEUMANN-HAEFELIN et al. 1993). However, in general, foamy virus research has been handicapped by a deficiency of in vivo experiments applying modern molecular biology techniques. Appropriate studies could reveal the nature of the infected target cells, the levels of viral gene expression, the state of the viral DNA, the sequence variation of the virus over time, and the requirement of ancillary genes for efficient virus replication in vivo (see Sect. 4.2). In tissue culture foamy viruses show an amphotropic host range and replicate in various kinds of primary cells or cell lines derived from lymphoid, epithelial or fibroblastoid lineages, and they induce characteristic cytopathological changes in all cases.

3 Natural Foamy Viruses Infections

3.1 Prevalence of Nonhuman Foamy Viruses

Foamy Viruses have been isolated from the same species as the immunodeficiency viruses of primates, bovines, and felines (HOOKS and GIBBS 1975). In addition, anecdotal isolations have been reported from various species, such as sea lion and bison (KENNEDY-STOSKOPF et al. 1986; AMBORSKI et al. 1987). Simian foamy viruses (SFV), which have been widely studied in the past, have been isolated from chimpanzee, gorilla, orangutan, numerous Old and New World monkey species, and prosimians (HOOKS and GIBBS 1975; HAHN et al. 1994; McCLURE et al. 1994). More than ten SFV serotypes have been distinguished on the basis of neutralization (HOOKS and GIBBS 1975; McCLURE et al. 1994); however, not all available isolates have been classified using this assay. The natural prevalence of foamy viruses among nonhuman primates is high. When 16 chimpanzees caught in the African bush were tested for foamy virus serological markers, all contained neutralizing antibodies against one or both of the two chimpanzee foamy viruses (SFV-6 and -7) (HOOKS et al. 1972). Moreover, 70% of 57 captured wild cynomolgus macaques were reported to have neutralizing antibodies against SFV-1 (HOOKS and GIBBS 1975). Among monkeys the virus spreads horizontally and vertically (HOOKS and GIBBS 1975). It is believed that horizontal spreading occurs predominantly by biting or the respiratory route (HOOKS and GIBBS 1975). Foamy virus negative captive monkeys are rare exceptions and probably depend on whether the animals are housed in groups or single cages. The natural prevalence of foamy viruses in bovines and felines (BFV and FFV, respectively) is approximately as high as in primates. In individual studies seroprevalences between 30% and 90% have been reported. The presence of BFV in milk points to a possible route of transmission in cattle.

Although foamy viruses have been isolated on some occasions from diseased animals, it has never been possible to establish a clear causality between infection and disease (HOOKS and GIBBS 1975; McCLURE et al. 1994). Moreover, experimental infections of monkeys with SFVs did not reveal any specific pathogenicity of foamy viruses (Hooks and Gibbs 1975; Neumann-Haefelin et al. 1983). It is believed that the majority of the proviruses reside in a latent state in the infected host (Neumann-Haefelin et al. 1993). Immunosuppression of seven SFV-3 infected African green monkeys by the administration of cyclosporin A for 2–3 weeks led to a reactivation of foamy virus as determined by the average time for the appearance of a positive culture upon virus isolation (Neumann-Haefelin, personal communication). No pathogenic effects due to this reactivation were observed during this short-term study. This experiment points to the role of the immune system in controlling foamy virus infection. Our current knowledge of natural foamy virus infections in animals suggests that these viruses may be apathogenic passengers of their hosts and, as is typical for many other retroviruses, may persist in the presence of high antibody titers (Hooks and Gibbs 1975; Hooks et al. 1976).

3.2 Search for Human Foamy Viruses

In 1971 a foamy virus was isolated from the tumor tissue of a Kenyan patient suffering from nasopharyngeal carcinoma (Achong et al. 1971). The HFV isolate has so far been most widely used for epidemiological, molecular and functional studies. Some early sero-epidemiological reports indicating that HFV may be naturally prevalent in East-African and Pacific island populations (Achong and Epstein 1978; Muller et al. 1980; Loh et al. 1980) had previously been challenged (Brown et al. 1978). Since the advent of recombinant DNA technology, foamy virus proteins have been expressed in bacteria and in insect cells and have been used as virus antigen in serological tests (Mahnke et al. 1990; Bartholoma et al. 1992; Hahn et al. 1994). Furthermore, PCR has been applied in addition to classical methods such as neutralization assay, immunoblot and immunofluorescence, to search for human foamy virus infections (Schweizer et al. 1994a,b). Immunofluorescence was shown to be of particular value in the detection of foamy virus antibodies (Neumann-Haefelin et al. 1993; Schweizer et al. 1994a,b). The combination of several of these methods allows a clear-cut virological diagnosis of foamy virus infection (Schweizer et al. 1994a,b). The populations studied included people expected to be at an increased risk for HFV infection, i.e., Africans, tumor patients, i.v. drug users, patients with various neurological and autoimmune diseases, HIV infected and AIDS patients, as well as normal persons. The examination of several thousands of human samples by using tests validated on probes derived from floamy virus-infected monkeys and apes did not reveal any positivity for HFV (Schweizer et al. 1994a,b). In addition, reports suggesting a natural HFV prevalence in the human population (Mahnke et al. 1992) or an association with Graves' disease and amyotrophic lateral sclerosis (Lagaye et al. 1992; Westarp et al. 1992) could not be confirmed by independent investigators examining similar material (Schweizer et al. 1994a,b; Hahn and Rethwilm, unpublished results). Unfortunately, none of the reported HFV isolates, other than the 1971 isolate (Young et al. 1973; Stancek et al. 1975; Cameron et al. 1978; Werner and Gelderblom 1979), is available to the research community for closer analysis.

While no human reservoir of naturally spreading HFV has been detected thus far, some cases of accidental inoculations clearly indicate that humans are susceptible to primate foamy virus infection (HOOKS and GIBBS 1975; NEUMANN-HAEFELIN et al. 1983, 1993; SCHWEIZER et al. 1994a,b). Some of these cases have been followed-up for 10 years or more, but neither health problems attributable to foamy virus infection nor human-to-human transmission could be demonstrated yet (Neumann-Haefelin and Rethwilm, unpublished results).

The recent molecular characterization of a foamy virus isolated from a chimpanzee (SFVcpz) revealed a close relationship to HFV (HERCHENRÖDER et al. 1994). However, on the basis of these data it is still unclear whether HFV may have resulted from trans-species transmission, since the lessons learned from the lentivirus and HTLV retroviral groups do not clarify this point and may not apply to foamy viruses at all. While even the closest pair of HIV-1/SIV is less homologous than the HFV/SFVcpz pair, the homology of HTLV-I/STLV-I pairs is in the foamy virus range (WATANABE et al. 1985, 1986). In addition, there are not enough sequence data available yet for the construction of a reliable primate foamy virus phylogenetic tree (HERCHENRÖDER et al. 1994). Thus, the true origin of HFV remains an open question.

It is intriguing that foamy viruses are so prevalent in higher primates but apparently not present in humans. However, nature provides examples for this. For instance, D-type retroviruses have been found in several monkey species but not in humans. By contrast, humans harbour endemically the hepatitis B virus, which is not naturally present in simians although it can infect them. From the current knowledge of foamy virus epidemiology and biology two scenarios may be drawn. First, foamy viruses may be transmitted from time to time from simians to humans. The human, however, may act as a dead end for the virus, which therefore does not spread further in the human population. From what we have learned so far, these infections may be benign if they occur in otherwise healthy hosts. In the second scenario, viruses adapted to the human host may evolve from such trans-species transmissions and from then on may emerge as "novel" human viruses. Whether or not adaptation may be followed by pathogenicity is, to date, a matter of speculation.

4 Molecular Biology of HFV

4.1 General Features of the HFV Genome

The foamy viruses have the longest genomes of all known retroviruses; the HFV provirus is 13.25 kb long. The 1.42 kb LTR U3 region of HFV contributes mainly to this extraordinary genome length (Fig. 1). There are some specific structural and functional features of foamy virus genomes which are potentially involved in the pathogenesis of HFV transgenic mice.

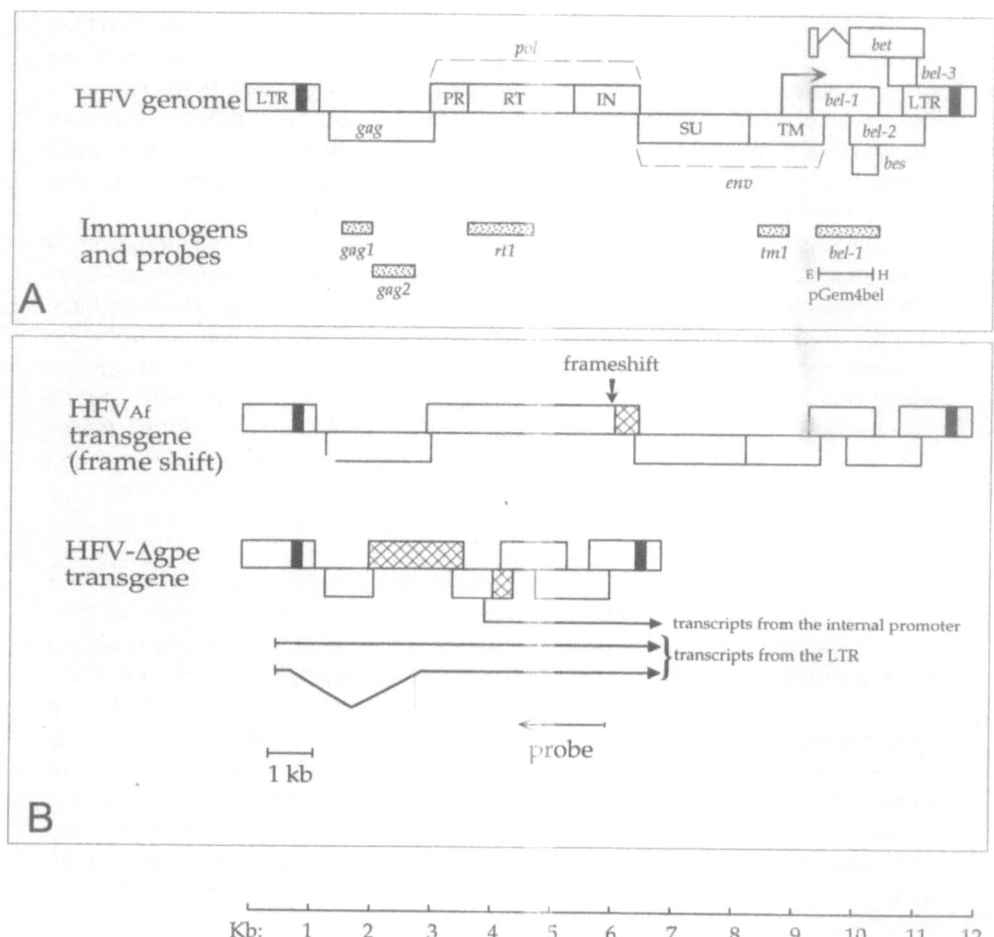

Fig. 1A,B. Genomic structure of A human foamy virus (HFV) and **B** structure of the pHFV$_{Af}$ and pΔgpe DNA constructs used for generation of transgenic mice. The *stippled* boxes in the lower part of **A** indicate the portions of HFV reading frames cloned into E. coli expression vectors. The E-H bar in **A** (*lower right*) indicates the cDNA *Eco*RI-*Hind*III fragment cloned into pGem4bel, an in vitro transcription vector utilized to generate the ribonucleic acid probes for in situ hybridization. Abbreviations for the gene products of HFV are: *gag*, group-specific antigen; *PR*, protease; *RT*, reverse transcriptase; *IN*, integrase (endonuclease); *SU*, surface protein; *TM*, transmembrane protein. Vertical black bars, R region of LTR; *hatched boxes*, regions of HFV that cannot be translated due to frameshifts or start codon deletions

Maturation of retroviral particles, which is initiated by binding of the *gag* precursor proteins to the encapsidation signal on the genomic RNA, is a cytoplasmic event (TEICH 1984). Therefore uncleaved *gag* molecules are mainly if not exclusively found in the cytoplasm of retrovirus-infected cells. In this respect, foamy viruses are an exception because they transiently and quantitatively translocate their *gag* precursor protein into the nuclei of infected cells (SCHLIEPHAKE and RETHWILM 1994). The reason for this unusual behavior is not understood. However, the massive loading of the nuclei with *gag* protein during lytic infection may contribute to the cytopathology of foamy viruses or may perform some unknown regulatory function.

In addition to the canonical retrovirus genes *gag*, *pol* and *env*, HFV contains three reading frames located between *env* and LTR which have been named *bel* (FLÜGEL et al. 1987). The *bel-1* gene is the *tat* or *tax* homologue of foamy viruses; it encodes for a transcriptional *trans*-activator (RETHWILM et al. 1991; KELLER et al. 1991; VENKATESH et al. 1991).*Bel-1* is the only regulatory HFV protein required for in vitro replication (BAUNACH et al. 1993; RETHWILM 1995); its functional details will be discussed in the next section. There is no evidence for a foamy virus post-transcriptional regulator of expression of the structural proteins, as is present in HIV and HTLV (CULLEN 1991).

The *bel-2* frame is predominantly expressed from a multispliced mRNA which consists of a 5' *bel-1* exon and the *bel-2* exon giving rise to a fusion protein named *Bet* (Fig. 1) (MURANYI and FLÜGEL 1991). The 60 kDa cytoplasmic and membrane associated *Bet* phosphoprotein is the second virus protein (after the *bel-1 trans*-activator) that becomes detectable by immunological methods in infected cells (SCHLIEPHAKE and RETHWILM 1994; HAHN et al. 1994; Hahn and Rethwilm, unpublished results). It is the most abundant viral protein, and it has been reported that up to 50% of the virus mRNA may encode for *Bet* (HAHN et al. 1994). However, *Bet* is not required for HFV in vitro replication, since stop codons can be introduced into the *bet* coding sequence or the whole genomic region can be deleted from the virus without greatly affecting its growth characteristics (BAUNACH et al. 1993; YU and LINIAL 1993; RETHWILM 1995). However, the *Bet* protein is believed to play an essential role for foamy virus in vivo replication. The *bel-3* gene is also not essential for HFV replication in tissue culture (BAUNACH et al. 1993; YU and LINIAL 1993; RETHWILM 1995). A *bel-3* gene is missing in the genomes of the lower primate foamy viruses (SFV-1 and SFV-3) and in BFV (KUPIEC et al. 1991; RENNE et al. 1992; RENSHAW and CASEY 1994). In chimpanzee foamy viruses a *bel-3* reading frame has been found, but only with in-frame stop codons (HERCHENRÖDER et al. 1994). In HFV it has not been possible yet to identify a *bel-3*-specific mRNA, although some characteristics of the potential *Bel-3* protein have been published recently (WEISSENBERGER and FLÜGEL 1994).

Another interesting but poorly understood phenomenon has been described for several foamy viruses: the presence of virus DNA derived from reverse transcription of a pregenomic RNA from which the *bet* intron has been "spliced out" (SAIB et al. 1993; HERCHENRÖDER et al. 1994). This genome is defective and depends on a *bel-1* expressing helper virus. It was estimated that the ratio of deleted to undeleted forms may vary from 1: 10 to 20: 1 (SAIB et al. 1993; HERCHENRÖDER et al. 1994). The accumulation of the defective viruses may alter cellular functions, as has been described for other retroviruses (SHARPE et al. 1990). In addition to this finding, high numbers of unintegrated foamy virus DNA copies, in the range of several thousands per cell, have been detected in lytically infected cells (SCHWEIZER et al. 1989; MERGIA and LUCIW 1991).

4.2 Regulation of HFV Gene Expression

In contrast to most other retroviruses, foamy viruses use more than one promoter to direct their own gene expression. In addition to the LTR promoter, an internal start of transcription is located in the *env* gene, approximately 230 bp upstream of the *bel* genes (LÖCHELT et al. 1993; MERGIA 1994). This internal promoter (IP) seems to be active at least in the initial phase of HFV gene expression, and mutagenesis of the TATA elements of the IP slows down virus replication (LÖCHELT et al. 1994; LÖCHELT, personal communication). As described for HFV and for SFV-1, the basal activity of the IP is higher than the basal activity of the LTR promoter, which is beyond the level of detection (LÖCHELT et al. 1994; KELLER et al. 1991; CAMPBELL et al. 1994). Basal expression from the IP might give rise to an initial burst of *bel-1* which in turn switches on LTR directed expression of structural proteins (LÖCHELT et al. 1993, 1994). It is believed that IP is also responsible for the extraordinary high levels of *Bet* protein synthesized by the virus (HAHN et al. 1994; RETHWILM 1995). It is noteworthy that such additional promoters have also been identified in HTLV and MMTV (ELLIOTT et al. 1988; MILLER et al. 1992; GUNZBURG et al. 1993; NOSAKA et al. 1993).

Both foamy virus promoters, the LTR and the IP, require *bel-1* for their full basal and inducible activity (RETHWILM et al. 1991; LÖCHELT et al. 1993). However, in all sequenced foamy virus genomes no other regions of significant homology between the LTRs and IPs are found aside from the regions between the TATA box and the transcriptional start (FLÜGEL et al. 1987; MAURER et al. 1988; KUPIEC et al. 1991; RENNE et al. 1992; HERCHENRÖDER et al. 1994). The *bel-1* response elements of the LTR and the IP are located upstream of the TATA boxes and behave like classical inducible enhancers (ERLWEIN and RETHWILM 1993; LEE et al. 1993; LÖCHELT et al. 1993). Several of these elements have been mapped in the HFV LTR which again show no homology to each other (LEE et al. 1993; ERLWEIN and RETHWILM 1993). Curiously, the *Bel-1* proteins of the individual foamy viruses display a high degree of specificity for their own promoters. For instance, no cross *trans*-activation is observed between HFV and SFV-1 onto either the LTR or the IP (MERGIA et al. 1992; ERLWEIN and RETHWILM 1993). Since a direct interaction of *Bel-1* protein with these diverse target elements could not be demonstrated (VENKATESH and CHINNADURAI 1993; ERLWEIN and RETHWILM 1993), it had been suggested that *Bel-1* requires cellular proteins to recognize specific DNA sequences. This might explain its virus specificity as well as its promiscuity for different target elements (RETHWILM 1995). *Bel-1* is a nuclear phosphoprotein (VENKATESH and CHINNADURAI 1993); however, the mechanism by which it enhances transcription is unknown. Primary protein structure did not disclose significant homologies between *Bel-1* and known transcriptional activators. A minimal activation domain (MAD) conserved among the known foamy virus *trans*-activator proteins has been mapped to a region at the COOH-terminal of the protein (HE et al. 1993; GARRETT et al. 1993; MERGIA et al. 1993; LEE et al. 1994) and functions in lower and higher eukaryotic cells (GARRETT et al. 1993). Exploiting a genetic approach in yeast it was shown that: (1) *bel-1* activation is dependent on the ADA2 transcriptional adapter molecule and (2) mutations of hydrophobic or aromatic residues around the MAD abolish *Bel-1* activity (BLAIR et al. 1994). Both features are shared with the VP16 type of the acidic class of transcriptional activators. Therefore, the main mechanism of *Bel-1* action is likely to consist in attracting cellular transcriptional helper molecules to the foamy virus promoters.

It should be mentioned that so far no oncogenic potential of *bel-1* has been identified. However, the *Bel-1* proteins are able to augment transcription from the HIV/SIV LTR promoters (KELLER et al. 1992; LEE et al. 1992; MERGIA et al. 1992; VENKATESH and CHINNADURAI 1993). The mechanism of this phenomenon is poorly understood.

5 Animal Models for Foamy Virus Infection

5.1 Direct Infection of Animals with Foamy Viruses

The infectibility of small laboratory animals with foamy viruses has been addressed only in a few, relatively old, studies. HOOKS et al. (1972) inoculated rabbits with SFV-7, which was isolated in 1972 from chimpanzees and had been adapted to primary cultures of rabbit kidney. Inoculation of rabbits produced a persistent foamy virus infection, and virus was consistently isolated from peripheral blood mononuclear cells by cocultivation up to 6 months after inoculation. The lymphocytic response of host animals to phytohemaggolutinin stimulation in vitro was depressed during the first 2 weeks after infection. When blood was drawn from experimental animals at later time points, These previously abnormal reactions returned to normal. The authors noted that one out of six inoculated rabbits developed a herpes virus infection, suggesting that these observations may indeed bear clinical relevance. However, while this phenomenon may play a role in determining persistent, chronic infection, the absence of progression to a frank immunodeficient state in the experimental animals questions the clinical relevance of these transiently altered in vitro parameters.

The group of D.C. Gajdusek has also sought to establish a model for foamy virus infection in the mouse (BROWN et al. 1982). SFV type 6, which had originally been isolated from the kidney of a kuru-inoculated chimpanzee, was adapted to produce an asymptomatic infection in Swiss-Webster white mice. Virus was detected in the kidney and in the spleen for up to 10 months after intraperitoneal inoculation, as were serum antibodies to SFV-6. Although this first successful experimental infection of the laboratory mouse by a member of the foamy virus family is undoubtedly relevant as a model for studying the control of viral latency, no pathologies could be detected in the infected animals. Similar experiments have been performed more recently with HFV (SANTILLANA HAYAT et al. 1993). As in the early experiments with SFV-7, infected rabbits experienced a transient immunosuppression mainly affecting cellular immune functions.

5.2 Transgenic Mouse Models of HFV

It is still possible that infection of permissive animals with viable wild-type virus may prove feasible as a strategy for studying the biology of HFV in vivo. However, the suitability of nonprimate species in direct infection systems appears to be limited, and the use of primates requires a specialized infrastructure and adds qualitative and quantitative constraints to the experimental design. Instead, introduction of retroviral genes into mice by the transgenic route represents a flexible approach and offers several distinct advantages over direct retroviral infection, since it bypasses species- and tissue-specific infectibility barriers. Expression of various combinations of native and mutated HFV genes is proving invaluable for detailed molecular dissection of HFV-induced disease. Moreover, transgenic mice have a distinct advantage over the use of infectious virus in terms of biosafety, since introduction of appropriate mutations makes spread of replication-competent viral particles unlikely (AGUZZI et al. 1992a, 1994). Transgenic mice have also been useful for studying human retroviruses, and several studies have shown that expression of retroviral genes in mice may represent a useful complementation to in vitro studies and clinical investigations and have helped defining the pathogenic potential of the *trans*-activating proteins of HTLV-I and HIV-I (HINRICHS et al. 1987; LEONARD et al. 1988; NERENBERG and WILEY 1989; GREEN et al. 1989). Anticipating that this strategy might prove useful for obtaining relevant information on the behaviour of HFV genes in vivo, we have chosen to explore the introduction of HFV genes into the germ line of mice.

5.2.1 Properties of HFV Constructs Used to Generate Transgenic Mice

As shown in Fig. 1, four HFV constructs have been introduced into fertilized mouse eggs following standard procedures (HOGAN et al. 1986). pHFV$_{Af}$ is identical to the replication competent molecular clone pHSRV (RETHWILM et al. 1990) except for a mutation in the COOH-terminal of the endonuclease (RETHWILM et al. 1991). This mutation disrupts a gene essential for proviral integration and was designed to help reducing the likelihood that infectious retroviral particles will be assembled in cells expressing this construct. Although pHFV$_{Af}$ was initially reported to be replication defective (RETHWILM et al. 1991), a closer analysis revealed that the plasmid gives rise to infectious virus, although with retarded kinetics (Baunach and Rethwilm, unpublished results).

In order to differentiate effects due to expression of the structural genes from those elicited by the ancillary *bel* genes, we have constructed an additional mutated form of HFV called pΔgpe (RETHWILM et al. 1991), which contains two large deletions in the *gag-pol* and in the *env* regions of HFV. The transcripts generated by this construct allow only for generation of truncated NH$_2$-terminal fragments of *gag* and *env*, while no *pol* gene products can be expressed. In contrast, the genes encoded by the *bel* region are not affected by the introduced deletions and can be normally expressed. The construct pΔgpe$_H$ has a further in-frame deletion in the *bel-2* and *bel-3* genes (RETHWILM et al. 1991). Finally, the pL*bel-1* plasmid was made by inserting the *bel-1 trans*-activator gene into a vector under control of the HFV U3 region (Rethwilm, unpublished results).

The U3 region of the HFV LTR was initially believed to be 640 bp shorter than the homologous region of SFVs (RETHWILM et al. 1990; LÖCHELT et al. 1991). All constructs used for the generation of transgenic mice had this "short" LTR (BOTHE et al. 1991). It was not until the molecular characterization of the chimpanzee foamy virus that the short LTR was discovered to be an artifact, probably due to a deletion in the virus upon replication in tissue culture (HERCHENRÖDER et al. 1994). Subsequently, an undeleted provirus in an early HFV passage was detected and a full-length molecular clone of the HFV genome was constructed (HERCHENRÖDER et al. 1994; Rappold et al., unpublished results). The consequences of using "short" HFV LTRs for the tissue-specific expression pattern and for the development of pathologies in transgenic mice have not been investigated so far.

DNA analysis of tail tissue identified several pups containing intact integrated transgenes with each construct. Several mice from the pΔgpe, pΔgpe$_H$ and *L-bel*1 group passed the transgene to their progeny, thus allowing establishment of transgenic lines. In these mice, hybridization to a DNA probe encoding the 3' region of HFV demonstrated head-to-tail integration of several intact copies of the HFV transgene in all somatic cells. Transgenic mice from independent lines contained varying numbers of integrated copies of the constructs, ranging from 3 to 30 per haploid genome. However, gene dosage was not found to significantly influence the phenotype observed.

5.2.2 Expression of HFV *bel* Genes
Leads to a Rapidly Progressing Encephalopathy and to Myopathy

Expression of HFV genes in transgenic mice yielded the first demonstration that foamy virus genes, besides causing cytopathic effects in cultured cells, can indeed provoke pathologies in vivo. Mice harboring either the pHFV$_{Af}$ or the pΔgpe construct developed a severe neurological syndrome at 6–8 weeks of age (BOTHE et al. 1991; AGUZZI et al. 1992a). The most prominent neurological symptoms were ataxia, spastic tetraparesis and, with lower penetrance, blindness. The symptoms rapidly progressed and led to death within 4–6 weeks from onset. The clinical features of this syndrome were similar in all animals, but transgenic mice harboring the pΔgpe construct displayed a later onset and slower progression than mice expressing pHFV$_{Af}$. Heterozygous and homozygous transgenic progeny of these founder mice developed a similar disease with 100% penetrance.

Expression analysis by northern blotting highlighted very sustained levels of HFV transgene expression in the brain of adult animals. Expression levels were not dependent on the number of integrated copies of the transgenic DNA and were comparable in heterozygous and homozygous transgenic siblings. Interestingly, expression levels decreased after the eighth postnatal week, suggesting that cells expressing HFV may eventually undergo degeneration. Indeed, this suspicion was confirmed by further studies. Histopathological analysis of HFV associated disease involved use of various morphological techniques, including conventional histology, immunohistochemistry, in situ hybridization and electron microscopy (AGUZZI et al. 1990, 1991). Over 60 HFV transgenic mice from several independent families were analyzed at various ages. In all animals the pathological findings were restricted to the CNS and to striated muscle. By the age of 7 weeks most of the mice analyzed had developed some of the clinical symptoms described above and showed variable degrees of nerve cell loss in the forebrain. Areas affected most frequently were the CA3 layer of the hippocampus (Fig. 2a) and the telencephalic cortex (Fig. 2b). The lesions consisted of selective nerve cell degeneration with tissue atrophy and prominent reactive astrogliosis. A few lesions revealed a more destructive pattern with incomplete cortical and subcortical necroses. At the borders between the lesions and the surrounding CNS tissue, degenerated nerve cells were present with condensed, highly eosinophilic cytoplasm. Although occasional macrophages and minor lymphocytic perivascular infiltrates were observed, their sporadic and mild character argues against a primary inflammatory pathogenesis of the observed lesions. The extent and time of onset of the described lesions were remarkably reproducible: they varied only slightly between individual animals and between the families derived from distinct founders.

Several mice analyzed exhibited foci of degeneration in striated muscle, especially in the spinal musculature. These changes ranged from degeneration of single myotubes to large areas with extensive necroses and focal lymphocytic infiltrates (Fig. 3).

In addition to the findings described above, two founder animals and all their progeny developed a complete elective degeneration of the cerebellar granule cell layer by the age of 2 months in addition to the findings described (Fig. 4). Clinically, these mice suffered from an obvious cerebellar phenotype, consisting mainly of ataxia (A. Aguzzi and J. Lampe, unpublished observations). It remains to be established whether this additional phenotype is caused by certain characteristics of the integration pattern of the transgene (e.g., the number of copies of the integrated array, or deletion/duplication of specific regions of the HFV genome) or by the particular genomic site of integration of the transgene in these two lines of mice.

Four founder mice had integrated HFV sequences but did not develop clinical symptoms. Analysis of the integration sites by Southern hybridization and of expression by northern blotting and in situ hybridization revealed that these mice either harbored rearranged transgenes whose structure was not intact or did not express transgenic RNA, possibly because of integration at an inactive chromosomal location.

Fig. 2a,b. Expression pattern of HFV in transgenic mice. **a** In situ hybridization of the hippocampus of a transgenic mouse at 5 weeks of age. High levels of expression restricted to the pyramidal neurons of the CA3 and CA4 hippocampal sublayer. Neurons in the dentate gyrus and in the CA1–2 sublayers are devoid of transgenic mRNA. **b** Expression of the *gag* protein of HFV in cortical neurons of a transgenic mouse. Immunohistochemical stain with an antibody to a bacterial *lacZ-gag* fusion protein

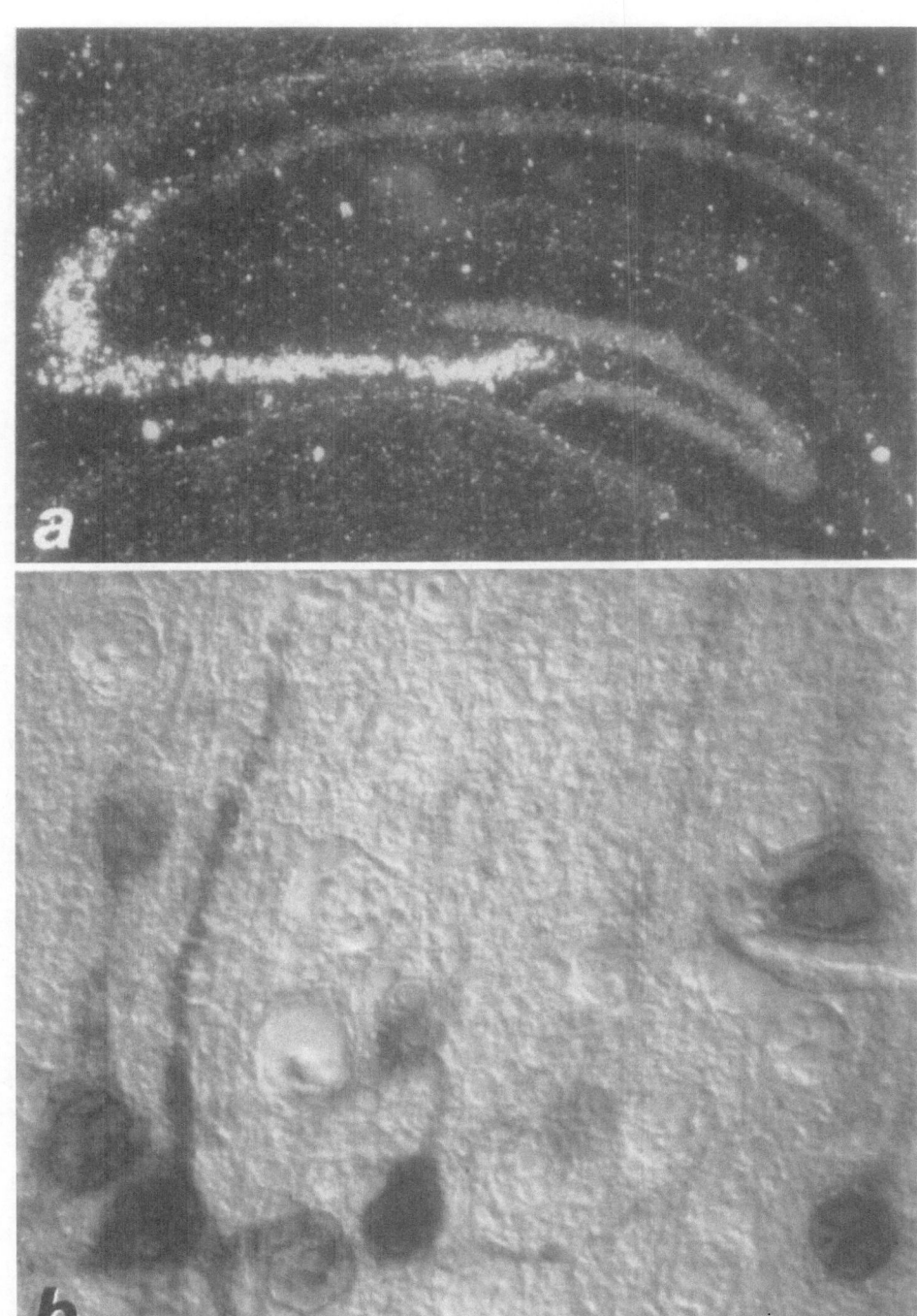

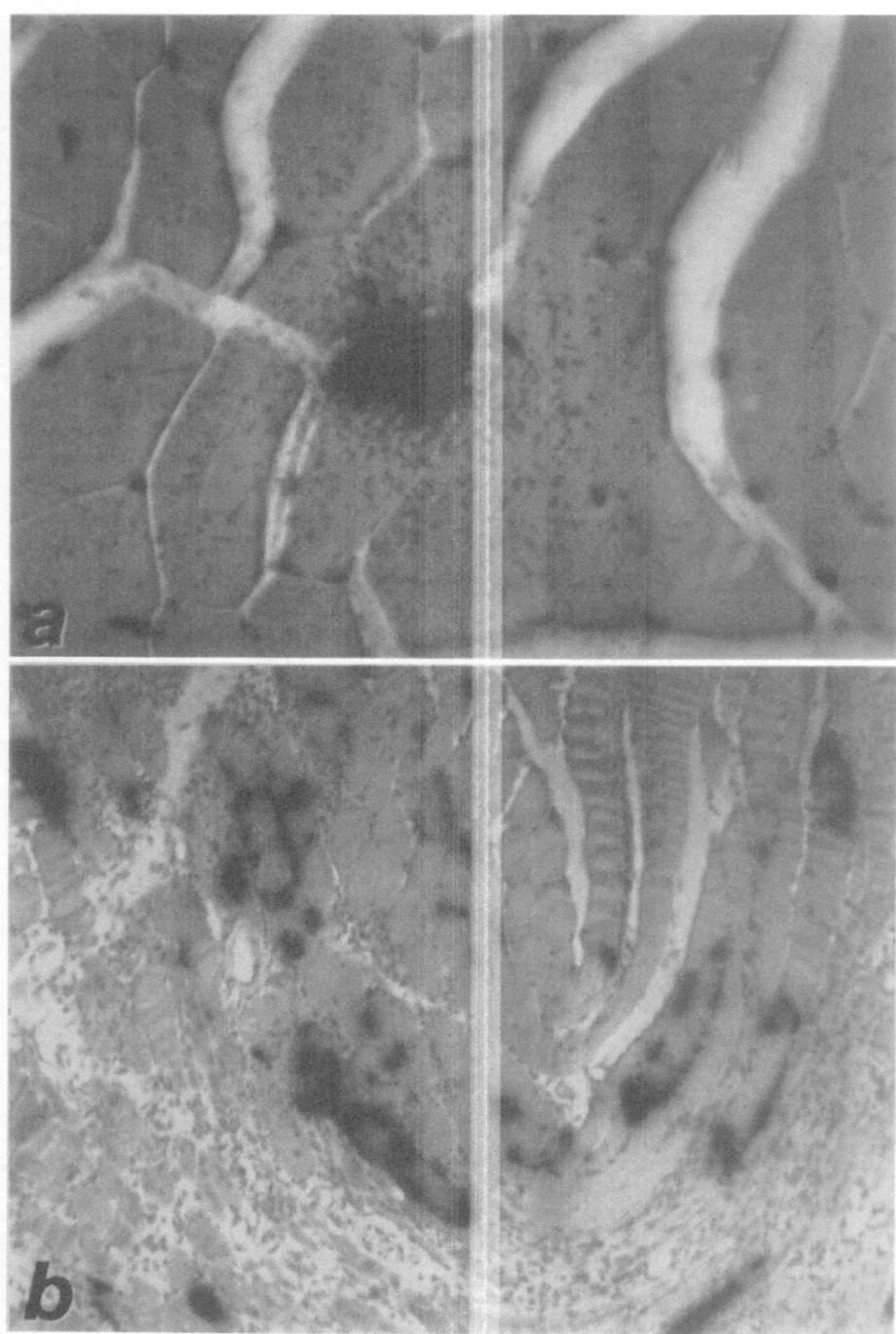

5.2.3 Myelin Pathology and Multinucleated Giant Cells in Mice Expressing HFV Structural Genes

In addition to the neurodegenerative disease described above, further pathological changes were seen in transgenic founder mice carrying the pHFV$_{Af}$ construct. This transgenic construct encodes the entire HFV genome and allows for expression of all retroviral genes except for the disrupted endonuclease domain of the *pol* gene. Macroscopic examination of the brain in these mice revealed a peculiar appearance of the cerebellar white matter with spotty symmetric areas of grayish color, suggestive of demyelination plaques. Indeed, histopathological analysis confirmed severe bilateral damage of the myelinated tracts in forebrain, brain stem, cerebellum and, in a milder form, in the spinal cord. The most dramatic phenotype was observed in the anterior commissure and the corpus callosum, the optic nerves and the optic chiasm, and the cerebellar white matter. These lesions were invariably associated with numerous nonneuronal cells expressing the HFV transgene in the white matter as established by in situ hybridization (Fig. 5).

The microscopic appearance of these white matter lesions revealed a spongy myelinopathy with microcystic changes of variable diameter (Fig. 5). However, areas of true demyelination, such as those seen in experimental allergic encephalitis, were not present, and the staining intensity of myelin between the vacuoles was roughly normal, as evaluated by conventional myelin stains (Luxol) and by immunocytochemistry for myelin basic protein (Fig. 5). Furthermore, virtually no macrophages with myelin degradation products or other inflammatory cells were found in the lesions. These findings confirmed that a structural loss of myelin did not take place. Bodian stains and electron microscopic studies of selected animals did not disclose evidence of a primay axonal lesion. Although immunocytochemistry for glial fibrillary acidic protein (GFAP) revealed pronounced astrogliosis in the lesions, the number and perikaryal morphology of the oligodendrocytes in the above mentioned tracts were almost normal. The additional pathologies in the white matter led to a syndrome much severer than the phenotype observed in pΔgpe transgenic mice and to early death of carrier mice, thereby suggesting that expression of yet unidentified components of the structural genes *gag*, *pol*, or *env* may be specifically toxic to the myelination system.

A typical multinucleated giant cells (MGCs) containing five to ten nuclei with a diameter exceeding 50 µm were often found in the lesioned brains of transgenic mice expressing all structural genes of HFV. The multiple nuclei were arranged in proximity of the plasma membrane in a horseshoe-like fashion reminiscent of Langhans giant cells. Since mitotic figures were not observed in these cells, we believe that MGCs originated through syncytial fusion. Although such cells were

Fig. 3a,b. Degeneration of striated muscle in HFV transgenic mice. Accumulation of transgenic transcripts is present in single muscle fibers before pathologies can be identified, as shown by in situ hybridization (**a**). At later time points, a frank destructive myopathy can be seen (**b**)

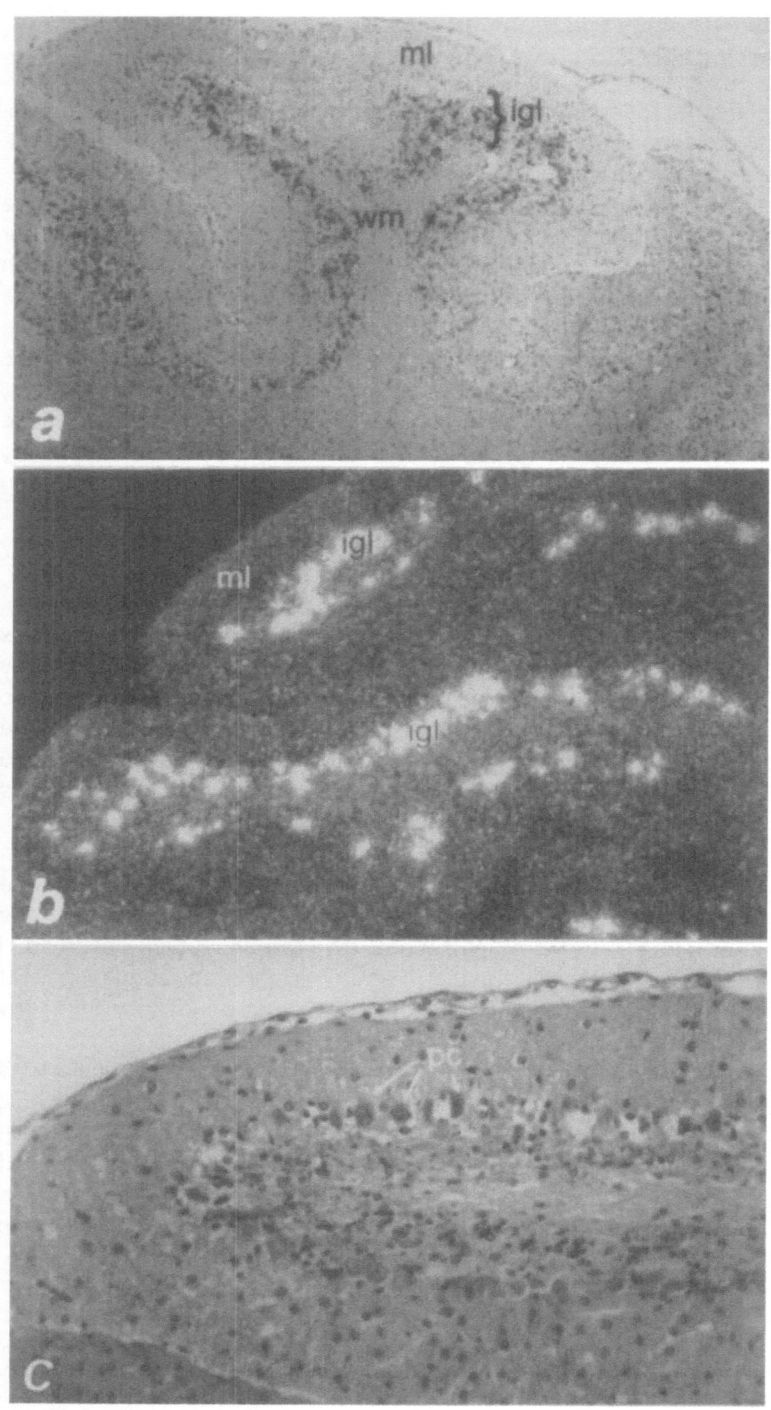

vaguely reminiscent of the giant cells seen in HIV encephalopathy (MAIER et al. 1989; BUDKA 1986), their morphology was similar to that of neighboring activated astrocytes. In fact, the cytoplasm of MGCs displayed strong immunoreactivity for GFAP, betraying their astrocytic origin. MGCs were strictly confined to the areas of telencephalon displaying histological lesions; they were not found in the brain stem or in extracranial locations. We also identified atypical, large, GFAP-positive cells devoid of processes in brain sections of these mice. These cells may represent MGC precursors.

Immunostains of MGCs with antisera to HFV revealed that they contained *gag*, *env*, and *bel-1*, whereas none of these gene products were detectable in the reactive astroglial compartment. *gag* proteins tended to accumulate in the central part of the cytoplasm of MGCs, whereas *env* and *bel-1* were more homogeneously distributed.

5.2.4 Profile of HFV Transgene Expression During Development and Adulthood

In order to gain a better understanding of the relationship between expression of HFV and the development of the disease, we have studied the time course of expression of the HFV transgene during development using in situ hybridization. We found that HFV was widely expressed at low levels and well tolerated during development. The pattern of HFV expression differed radically from that of known retroviruses (AGUZZI et al. 1992a,b).

Transcription of HFV occurred in two distinct phases. At midgestation, widespread expression was first detected in cells of the extra-embryonic membranes and in various tissues originating from mesoderm, neuroectoderm and neural crest. Expression decreased dramatically during late gestation. Surprisingly, the highest levels of expression achieved during embryonic life were not found in the neural tube, but in neural crest-derived tissues such as the dorsal root ganglia. No permanent morphological damage was detected during this time and expression was suppressed in most tissues shortly after birth. However, several weeks later transcription of HFV resumed in a small fraction of single cells distributed irregularly in the central nervous system and in skeletal muscle. At the age of 6–8 weeks expression reached extremely high levels in an increasing number of cells in these tissues and was followed by severe degenerative changes. These findings indicate that the regulatory elements of HFV allow for expression in a broad range of tissues at midgestation and that tissue-specific expression of HFV is differentially regulated later in development. Detailed molecular analysis of the events responsible for these observations may shed

---

Fig. 4a–c. Cerebellar degeneration in two independent strains of transgenic mice. Also in this case, accumulation of transgenic mRNA is present in the neurons of the internal granule cell layer (*igl*) before any pathological changes occur (**a**). These can be sensitively visualized by dark field illumination (**b**). The disease then progresses to subtotal atrophy of the granule cell layer (**c**). Note that the Purkinje cell layer (*pc*) and the molecular layer (*ml*) are not affected. *wm*, white matter

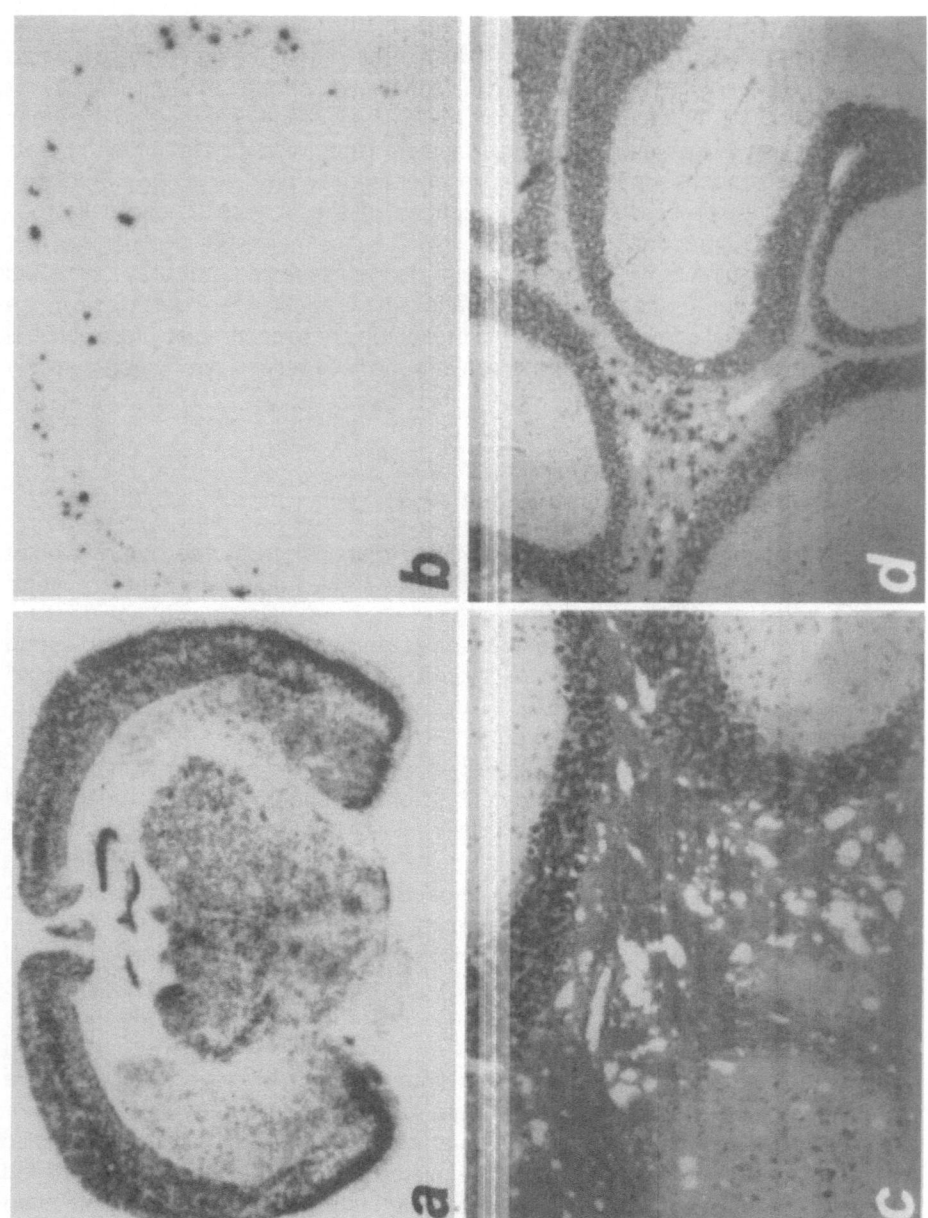

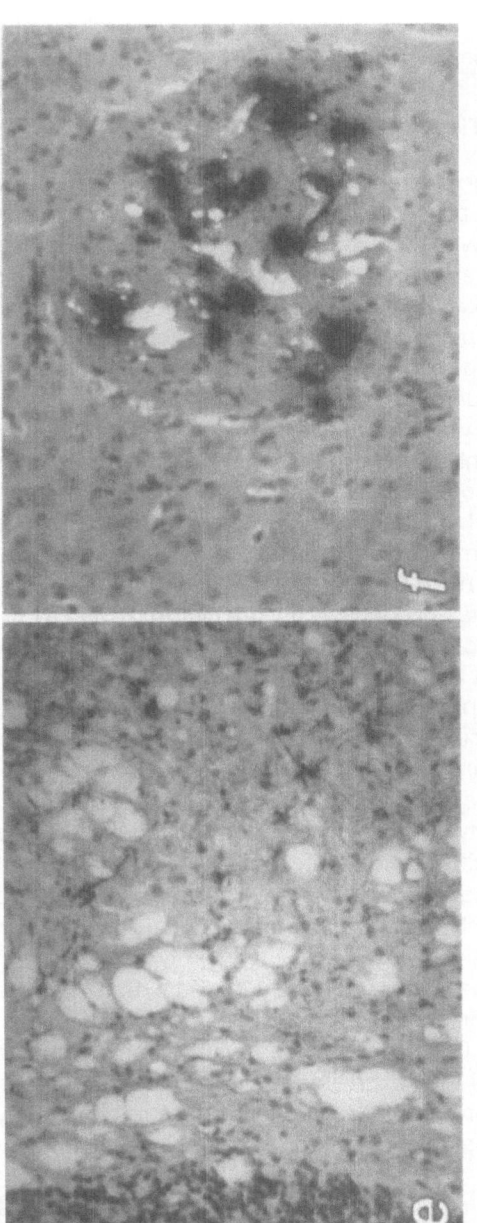

Fig. 5. Film autoradiography of sections through the cerebral cortex of a HFV Δgpe transgenic mouse. Consecutive paraffin sections were hybridized in situ with probes specific for the neuronal protein synaptophysin (**a**) and for the transgene (**b**). Note the highly irregular distribution of neurons expressing the transgene. **c** Pathology of the white matter in a transgenic mouse expressing the ancillary and structural genes of HFV (construct pHFV$_{Al}$). In this histochemical myelin stain, the white matter appears bluish. Large vacuoles are apparent in the myelinized areas. **d** Expression of HFV genes in the white matter of the cerebellum. The distribution of expressing cells correlates with the pathological changes shown in **c**. **e** Occurrence of reactive astrocytes in the white matter of the cerebellum. Glial fibrillary acidic protein (GFAP) immunohisochemistry. **f** Co-localization of vacuolar degeneration and transgene expression in the anterior commissure, a myelinated tract within the forebrain (*bel-1* in situ hybridization, bright field)

light on the *cis*-acting mechanisms controlling retroviral latency and perhaps also on some aspects of vertical transmission of retroviruses.

5.2.5 How Does HFV Accomplish Tissue Damage?

From what has been discussed in the preceding sections a variety of effects in transgenic mice were to be expected. pHFV$_{Af}$ transgenic mice may express all viral proteins in a functionally active form with the possible exception of the integrase. Thus, the accumulation of nuclear located *gag* precursor protein and unintegrated virus DNA, with and without a deleted *bet* intron, and the syncytiogenic property of the *env* protein may contribute to pathogenicity, in addition to toxic effects of the *Bel-1* and *Bet* proteins. While pΔgpe allows expression of both these proteins, pΔgpe$_H$ can only express *bel-1* in an unmutated form.

The developmental profile of HFV gene expression in transgenic mice suggests that cytotoxicity is achieved only when a threshold level of expression is reached, and low levels of transcription, such as those observed during prenatal life, seem to be tolerated. The actual mechanism by which HFV induces tissue degeneration in the CNS and striated muscle is still unclear, and at present we can only provide some suggestions. The most basic question with respect to pathogenesis is: which of the gene products encoded by HFV are cytotoxic? For a long time we have speculated that of all potential gene products encoded by the HFV genome, *Bel-1* represented the most likely candidate for neurotoxicity. This is in agreement with the neurotoxic properties of functionally analogous lentiviral *trans*-activators such as the *Tat* molecules of HIV and visna (GOURDOU et al. 1990). It is conceivable that *Bel-1* exerts *trans*-activating functions on cellular genes in addition to the LTR of HFV, and that the resulting perturbations of gene expression play a role in the observed neurotoxicity. In addition, the spotty distribution of cells expressing *bel* genes observed in in situ hybridization studies suggests that *bel-1* is capable of initiating a positive feedback loop of retroviral *trans*-activation to single cells. This may lead to extremely high levels of transcriptional activity and to cytopathic effects.

More recent data, however, indicate that *Bet* is produced in extremely large amounts in the brain of transgenic mice developing neuropathological findings, and that transgenic mice in which the COOH-terminal of *Bet* has been deleted do not develop any brain disease, even if they express *bel-1*. Taken together, these observations suggest that *Bet*, a protein whose function is still completely unclear, may in fact be one of the most relevant factors for pathogenicity of HFV in vivo.

Bel-1 toxicity may result from "stealing" cellular transcription factors off cellular promoters, for instance of housekeeping genes, while a possible *Bet* toxicity can not be explained so easily. It had been speculated that the *Bet* molecules of foamy viruses might be equivalent to *Nef* of HIV (BAUNACH et al. 1993; RETHWILM 1995). If this holds true, *Bet* might interfere with cellular signal transduction like the *Nef* protein. Finally, the use of different HFV promoters should be investigated to explain the restricted tropism of HFV gene expression in neuronal and muscular tissues. Among other DNA constructs, the pL*bel*-1 construct, which lacked the IP, should be used to this end.

Although much circumstantial evidence incriminates *Bel-1* or *Bet* as a possible neurotoxic protein, we cannot exclude from the studies carried out so far that the NH$_2$-terminal portions of *gag* and/or *env* contribute to the phenotype. Truncated forms of these gene products may be translated from the microinjected constructs and paly a role in at least a part of the spectrum of diseases observed.

5.2.6 Syncytiogenic Activity of HFV In Vivo

Induction of syncytia in infected cells is a property shared by disparate enveloped viruses. HIV is by far the best characterized fusogenic retrovirus: in vitro induction of syncytia in lymphoid cell lines by HIV closely correlates with the spreading of retroviral infection and its cytotoxic potential and has been shown to require the action of viral *env*-derived proteins (BOSCH et al. 1989; FREED et al. 1990). Specifically, a conserved undecapeptide has been identified to be shared by HIV-1 and HIV-2, para-myxoviruses, maedi-visna virus, and several simian retroviruses (GALLAHER 1987). Theoretical calculations predict that this peptide perturbs the cellular membrane bilayer by inserting obliquely into it, and mutations affecting the angle of its insertion into the membrane may abolish fusogenic activity (HORTH et al. 1991).

The HFV *env* gene product is a likely candidate responsible for fusogenic activity. When protein alignment studies were performed in order to identify a fusogenic peptide within the reading frames of HFV, regions of similarity to the fusogenic determinants of human and simian retroviruses were identified in the *env* polypeptide. We have compared the products of the *env* reading frame of HFV to partially conserved peptide sequences characteristic of fusogenic retroviruses and paramyxoviruses which have been shown to be involved in the fusion process (GALLAHER 1987; HORTH et al. 1991). While no significant homologies to the fusion peptides of respiratory syncytial virus and measles virus could be identified, two regions of similarity to retroviral fusogenic sequences are present. The first region is situated at residue 521–536 of the *env* polypeptide and shows homologies to the fusion peptides of three different HIV-1 isolates. The second region extends over 20 amino acid residues (948–967) at the COOH-terminal of the transmembrane protein. Though unrelated to the HIV-1 peptides, this region is significantly similar to the fusion peptides of HIV-2 and SIV$_{mac}$. Limited homology to the putative fusion peptide of SRV was also detected, albeit not at a statistically significant level.

The physiological importance of these amino acid motifs for induction of syncytia is made more likely by our observation that giant cells develop only in pHFV$_{Af}$ mice, which express these sequences, while they are not detectable in pΔgpe transgenic mice, in which these sequences are deleted.

A critical role in HIV-mediated cell fusion is played by host proteins whose presence is required in addition to HIV-*env*. Domains have been identified within CD4 (CAMERINI and SEED 1990; TRUNEH et al. 1991), ICAM-1 (HILDRETH and ORENTAS 1989), and LFA-1 (PANTALEO et al. 1991) whose deletion abrogates the fusogenic potential without affecting infectivity. Since syncytia in HFV transgenic mice were invariably composed of glial cells expressing GFAP, whereas neuronal fusions were not observed despite the much higher proportion of neurons expressing HFV, presumptive cellular partner molecules may be present in vivo only in astrocytes. It will be of interest to search for giant cells in the brain of experimental animals infected with HFV, since in vitro studies have revealed that HFV preferentially replicates in cell lines of astroglial origin (RETHWILM et al. 1990).

5.2.7 Possible Neuropathogenic Relevance of Multinucleated Giant Cells

Although the prominent fusogenic properties of HFV in cultured cells have been recognized for a long time, and indeed foamy viruses have been referred to in the past as "syncytial viruses" (MULLER et al. 1980; LOH et al. 1977, 1980), formation of giant cells in vivo has not been observed in infected patients or in transgenic mice expressing genes of HIV (LEONARD et al. 1988) and HTLV-I (VOGEL et al. 1988; HINRICHS et al. 1987; NERENBERG et al. 1987). Therefore, this feature makes HFV transgenic mice an attractive in vivo model system in which syncytiogenic activity of human retroviruses can be studied.

MGCs of monocytic origin are an important diagnostic hallmark of HIV–associated giant cell encephalitis (GCE) (MAIER et al. 1989; BUDKA 1986). The finding that antibodies inhibiting syncytia formation are of prognostic significance in HIV-seropositive children lends support to the hypothesis that in vivo fusogenic activity may play an important role in HIV pathogenicity (BRENNER et al. 1991). Although at first glance our findings bear striking similarities to those regarding the histopathology of GCE, the resemblance is rather superficial. While MGCs in GCE essentially originate from hematopoietic cells expressing the CD4 receptor, MGCs in HFV transgenic mice are derived from neuroectodermal cells belonging to the astrocytic lineage. The relevance of MGCs to the disease observed in HFV transgenic mice is not yet clear since we observe neurodegenerative pathologies also in the absence of MGCs. However, mice expressing full-length structural genes display MGCs within the brain lesions and have a markedly reduced life expectancy as compared to pΔgpe mice which do not develop MGCs.

5.2.8 Interactions Between HFV and Other Viruses

It is conceivable that foamy viruses may cooperate with other retroviral pathogens in various ways. Double infection of cells may result in formation of pseudotyped virions with altered tissue specificity (LANDAU et al. 1991; LE GUERN and LEVY 1992). It appears indeed that foamy viruses may undergo formation of pseudotypes even with viruses belonging to different families. For example, it has been possible to produce pseudotypes of SFV with vesicular stomatitis virus (SCHNITZER 1982). It is not known whether HFV can form pseudotypes with human retroviruses in vivo, but in this context it is interesting to note that a SFV was obtained when reisolation of the virus was attempted in chimpanzees which had been inoculated with HIV (NARA et al. 1987).

It will be interesting to test whether foamy viruses are capable of exploiting alternative strategies for retroviral cooperativity, such as up-regulation of the viral receptor (Lusso et al. 1991). Another mechanism which may be biologically and clinically relevant is heterologous transcriptional trans-activation, a relatively common phenomenon among viruses encoding trans-acting factors (Siekevitz et al. 1987; Mosca et al. 1987; Davis et al. 1987; Gendelman et al. 1986). This phenomenon may possibly bear a considerable clinical relevance in the case of human herpes virus type 6 (Lusso et al. 1989).

Bel-1 can interact with responsive elements located on the LTR of HIV and trans-activate its transcription in vitro (Keller et al. 1992; Lee et al. 1992). The molecular mechanisms underlying heterologous trans-activation have been studied in great detail. A region between −158 and −118 upstream of the transcription initiation site of HIV, immediately in front of the core enhancer element, was identified as responsible for the trans-activation by Bel-1. This region does not coincide with any previously described factor-binding site. Gene expression directed by an HIV-1 LTR lacking functional sites for the inducible cellular transcription factor NFκ-B was activated over 100-fold by coexpression of Bel-1. Interestingly, Bel-1 trans-activated a heterologous promoter when this region was positioned upstream of it in both sense and antisense orientations. Optimal trans-activation of the HIV-1 LTR by Bel-1 did not require an intact HIV-tat responsive element, suggesting a fundamentally different mode of action for Bel-1 and Tat (Keller et al. 1992; Lee et al. 1992). Indeed, sequence comparisons between the region of the HIV-1 LTR with that of the HFV LTR necessary for the Bel-1 mediated trans-activation have identified a nonamer sequence which is well conserved between HIV-1 (position −124 to −116 relative to the transcriptional start) and HFV (position −134 to −126). These results may suggest Bel-1 exerts its action, at least in part, via a specific DNA sequence which is shared by the LTR of both HIV-1 and HFV, although other authors have found that the region of homology does not play an important role for Bel-1-dependent trans-activation. Moreover, Bel-1 is likely to enhance the level of HIV-1 gene expression in cells dually infected with HIV-1 and HFV. The basal transcriptional activity of the LTR of HTLV-I, by contrast, does not seem to be affected by co-expression of bel-1 (Keller et al. 1991).

In order to assess the trans-activating potential of the Bel-1 protein of HFV on the cis-regulatory element of HIV in vivo, we used double transgenic mice carrying the bel-1 trans-activator of HFV under transcriptional control of its own regulatory elements, and the full-length U_3R region of the HIV-1 LTR directing the expression of the β-galactosidase gene fused to a heterologous poly-A addition motif.

Two lines of HFV Δgpe transgenic mice (see above and Fig. 1) carrying a linearized form of the plasmid Δgpe (Rethwilm et al. 1991) have been used. Two strains of HIV transgenic mice carrying the lacZ gene under control of the HIV-1 LTR (U_3R region) have been used (Cavard et al. 1990). The HFV and the HIV-1 LTR transgenic strains were intercrossed to obtain double transgenics. The expression of bel-1 and of the reporter gene were examined at various time points by in situ hybridization, immunohistochemistry and X-gal in situ enzymatic reaction (Aguzzi and Theuring 1994).

. In situ hybridization and immunohistochemistry for *Bel-1* showed positive cells in the hippocampus and in the telencephalic cortex. This pattern could only be detected in 6, 7 and 8 week old animals. In younger and in older animals neither transcription of the HFV transgene nor expression of *Bel-1* protein was detectable. In older mice, degeneration mainly in CA$_3$ and CA$_4$ sublayers was present in those areas previously expressing *Bel-1* protein. Prominent reactive gliosis in the affected areas has been detected by immunohistochemistry for GFAP.

Not all the hippocampal neurons showing *Bel-1* positivity in CA$_3$ and CA$_4$ were found to express also *trans*-activational activity. In the cerebral cortex, no *lacZ* positive cells have been found even where *Bel-1* positive cells were present in adjacent sections. In the two strains carrying only the HIV-1 LTR-lacZ transgene, no lacZ staining was observed in the hippocampus whatever the age tested. These findings may suggest that full *trans*-activating property of *Bel-1* is obtained in cooperation with a cellular transcription factor only present in specific neuronal subpopulations.

We analyzed two regions of the HIV-1 LTR with modulating activity on transcription. The NRE region maps between nucleotides −340 and −185 and is known to be a negative regulatory element in vitro. A deletion of this sequence increases LTR directed gene expression (ROSEN 1991). The NRE region down-regulates HIV-1 LTR directed gene expression. In situ hybridization and immuno-histochemistry for *Bel-1* protein as well as lacZ staining revealed the same distribution of positive cells as observed in the previous group. The temporal bel-1 expression was detected from 6 to 11 weeks of age and therefore covered a broader range. We assume that the negative regulatory activity of the NRE region is an early action and its removal allows the expression of *Bel-1* protein as well as its *trans*-activational activity until these neurons degenerate. The NF-κB motifs are two conserved 11 bases pair sequences located between position −108 and −83. It is known that NF-κB participates in the induction of HIV expression by UV light in vitro. To investigate the role of the NF-κB binding site as target of *Bel-1* action in vivo we used two lines of transgenic mice carrying a 29 bp deletion in the HIV-1 LTR that removes the two NF-κB sites without insertion of additional DNA sequences (ΔNF-κB-HIV-1 LTR) (ZIDER et al. 1993).

However, we found that the deletion of both NF-κB sites completely abol-ishes the *trans*-activational activity of *Bel-1* on HIV-1 expression in vivo. These mice displayed a UV light-inducible epidermal expression of HIV-1. Our results are therefore in apparent contrast to previous studies which have shown that the NF-κB site is not involved as a target of the *Bel-1 trans*-activation on HIV-1 in vitro (KELLER et al. 1992; LEE et al. 1992).

6 Conclusions and Future Perspectives

The HFV transgenic mice discussed in the present article represent a novel development in human retrovirology, since this model system has provided the first clear-cut evidence of pathogenicity of HFV. However, caution has to be exercised when a transgenic animal model is used to draw conclusions valid for the natural route of infection: transgenic biology may differ in important aspects from horizontal spread of infectious retroviral particles. In particular, all somatic and germ cells of a transgenic mouse contain the transgenic DNA in equal amounts, in striking contrast to horizontal transmission through the natural route, in which only cells permissive to infection will integrate into the retroviral genome and experience the consequences of its expression. It is possible that presumptive target cells in the CNS are not accessible to the virus in healthy individuals, e.g., because of their protected location beyond the blood-brain barrier.

Despite this limitation, the transgenic mouse model has enabled us to address questions not easily accessible by other means, such as the consequences of expression of the single genetic elements of the virus separately and in various combinations in transgenic mice. These experiments represent an important priority in our approach and will allow straightforward identification of the genes involved in the causal pathogenesis of the disease. Further, deletion of portions of the genome may create replication defective mutants and mimic abortive retroviral infection, which seems to play a role in some forms of neuronal degeneration (SHARPE et al. 1990). Finally, we have shown that, in the case of the HFV, expression in mice could be achieved in early embryonic stages, thus simulating some aspects of transplacental infection.

Although for more than 20 years virologists and epidemiologists have been searching without success for a disease caused by HFV, the issue is not yet resolved. Certain intriguing similarities of the phenotype of HFV transgenic mice to known human retroviral diseases have prompted us to speculate about the possible role of HFV in human medicine. The human retroviruses HTLV-I and HIV are frequently associated with neurological syndromes. While HTLV-I infection often results in inflammatory degeneration of the spinal cord, a majority of AIDS patients develop complex and variable pathologies of the CNS during their illness (PETITO 1988; GONZALES and DAVIS 1988). CNS involvement in adult AIDS patients is most often characterized by a microglial nodular encephalitis, opportunistic infections, and a progressive diffuse encephalopathy of the white matter (KLEIHUES et al. 1985). While these features were not observed in HFV mice, AIDS encephalopathy following congenital and early childhood HIV infection often presents with distinct features, including prominent nerve cell loss and subcortical necroses (GIANGASPERO et al. 1989; LEWIS et al. 1990; EUROPEAN COLLABORATIVE STUDY 1991), which are highly reminiscent of the neuropathological changes in HFV mice.

In addition, the microcystic changes seen in the $pHFV_{AF}$ mice resemble a condition called vacuolar myelopathy, which occurs in the spinal cord of 10%–20% of AIDS patients (PETITO et al. 1985; MAIER et al. 1989). Vacuolar myelopathy affects the myelin sheaths of the long tracts of the cord while sparing the axons and closely resembles subacute combined degeneration. As in HFV myelinopathy, vacuolar myelopathy often occurs in the absence of a significant cellular reaction and is probably due to a direct toxic action of retroviral gene products. The morphological analogies between the HFV phenotype and this condition, together with the reported difficulties in insolating HIV from vacuolar myelopathic lesions, suggests that it will be important to exclude coinfection or superinfection with HFV in the subset of AIDS patients developing spinal cord pathologies. However, since the prevalence of HFV in human populations seems to be very low, it is also possible that the morphological similarities result from HFV and HIV utilizing convergent pathogenetic pathways for damaging the white matter.

Our studies with transgenic mice have uncovered some significant consequences of expression of the genes of HFV in an intact mammalian host and have enabled us to advance some speculations on the biological properties of the virus. However, the relevance of HFV for human pathology is still an unresolved matter, and predictions are difficult since the prevalence of the virus in human populations appears to be very low and possibly only confined to zoonotic situations. In the case of lower primates, it is possible that foamy viruses do not damage the central nervous system of otherwise healthy hosts, and it is unclear whether they may do so in situations of immune depression (Neumann-Haefelin and Luciw, personal communications). Even so, it can be expected that the molecular dissection of the disease seen in the transgenic mouse strains described above will further our understanding of the pathogenic properties of retroviruses and may also have an impact on current concepts of how neurodegenerative diseases arise.

Acknowledgments. We are indebted to Dieter Neumann-Haefelin for communicating unpublished results and for discussion and comments on the manuscript and to Lee Dunster for reviewing the manuscript. We thank Ivan Horak, Martin Löchelt, Myra McClure, Erwin F. Wagner and Brian Cullen for communicating results prior to publication. Our work is supported by grants from the DFG (SFB165), the Bundesministerium für Forschung und Technologie, the Bayerische Forschungsstiftung, the "Nationales AIDS-Forschungsprogramm der Schweiz" and Swiss National Foundation (no. 31–36059.92), and the European Union (BMHI-CT93–1142 and Concerted Action on the Neuropathology of AIDS).

References

Achong BG, Epstein MA (1978) Preliminary seroepidemiological studies on the human syncytial virus. J Gen Virol 40: 175–181

Achong BG, Mansell PW, Epstein MA, Clifford P (1971) An unusual virus in cultures from a human nasopharyngeal carcinoma. J Natl Cancer Inst 46: 299–307

Aguzzi A (1993) The foamy virus family: molecular biology, epidemiology and neuropathology. Biochim Biophys Acta 1155: 1–24

Aguzzi A, Theuring F (1994) An improved in situ beta-galactosidase staining for histological analysis of transgenic mice. Histochemistry 102: 477–481

Aguzzi A, Wagner EF, Williams RL, Courtneidge SA (1990) Sympathetic hyperplasia and neuroblastomas in transgenic mice expressing polyoma middle T antigen. New Biol 2: 533–543

Aguzzi A, Kleihues P, Heckl K, Wiestler OD (1991) Cell-type specific tumor induction by oncogenes in fetal forebrains transplants. Oncogene 6: 113–118

Aguzzi A, Bothe K, Anhauser I, Horak I, Rethwilm A, Wagner EF (1992a) Expression of human foamy virus is differentially regulated during development in transgenic mice. New Biol 4: 225–237

Aguzzi A, Ellmeier W, Weith A (1992b) Dominant and recessive molecular changes in neuroblastoma. Brain Pathol 2: 195–208

Aguzzi A, Brandner S, Sure U, Rüedi D, Isenmann S (1994) Transgenic and knock-out mice: models of neurological disease. Brain Pathol 4: 3–20

Amborski GF, Storz J, Keney D, Lo J, McChesney AE (1987) Isolation of a retrovirus from the American bison and its relation to bovine retroviruses. J Wildl Dis 23: 7–11

Bartholoma A, Muranyi W, Flügel RM (1992) Bacterial expression of the capsid antigen domain and identification of native gag proteins in spumavirus-infected cells. Virus Res 23: 27–38

Baunach G, Maurer B, Hahn H, Kranz M, Rethwilm A (1993) Functional analysis of human foamy virus accessory reading frames. J Virol 67: 5411–5418

Blair WS, Bogerd H, Cullen BR (1994) Genetic analysis indicates that the human foamy virus Bel-1 protein contains a transcription activation domain of the acidic class. J Virol 68: 3803–3808

Bosch MI, Earl PL, Fargnoli K, Picciafuoco S, Giombini F, Wong Staal F, Franchini G (1989) Identification of the fusion peptide of primate immunodeficiency viruses. Science 244: 694–697

Bothe K, Aguzzi A, Lassmann H, Rethwilm A, Horak I (1991) Progressive encephalopathy and myopathy in transgenic mice expressing human foamy virus genes. Science 253: 555–557

Brenner TJ, Dahl KE, Olson B, Miller G, Andiman WA (1991) Relation between HIV-1 syncytium inhibition antibodies and clinical outcome in children. Lancet 337: 1001–1005

Brown P, Nemo G, Gajdusek DC (1978) Human foamy virus: further characterization, seroepidemiology, and relationship to chimpanzee foamy viruses. J Infect Dis 137: 421–427

Brown P, Moreau-Dubois MC, Gajdusek DC (1982) Persistent asymptomatic infection of the laboratory mouse by simian foamy virus type 6: a new model of retrovirus latency. Arch Virol 71: 229–234

Budka H (1986) Multinucleated giant cells in brain: a hallmark of the acquired immune deficiency syndrome (AIDS). Acta Neuropathol (Berl) 69: 253–258

Camerini D, Seed B (1990) A CD4 domain important for HIV-mediated syncytium formation lies outside the virus binding site. Cell 60: 747–754

Cameron KR, Birchall SM, Moses MA (1978) Isolation of foamy virus from patient with dialysis encephalopathy. Lancet 2: 79666

Campbell M, Renshaw Gegg L, Renne R, Luciw PA (1994) Characterization of the internal promoter of simian foamy viruses. J Virol 68: 4811– 4820

Cavard C, Zider A, Vernet M, Bennoun M, Saragosti S, Grimber G, Briand P (1990) In vivo activation by ultraviolet rays of the human immunodeficiency virus type 1 long terminal repeat. J Clin Invest 86: 1369–1374

Cullen BR (1991) Human immunodeficiency virus as a prototypic complex retrovirus. J Virol 65: 1053–1056

Davis MG, Kenney SC, Kamine J, Pagano JS, Huang ES (1987) Immediate-early gene region of human cytomegalovirus trans-activates the promoter of human immunodeficiency virus. Proc Natl Acad Sci USA 84: 8642–8646

Doolittle RF, Feng DF, McClure MA, Johnson MS (1990) Retrovirus phylogeny and evolution. Curr Top Microbiol Immunol 157: 1–18

Elliott JF, Pohajdak B, Talbot DJ, Shaw J, Paetkau V (1988) Phorbol diester-inducible, cyclosporine-suppressible transcription from a novel promoter within the mouse mammary tumor virus env gene. J Virol 62: 1373–1380

Erlwein O, Rethwilm A (1993) Bel-1 transactivator responsive sequences in the long terminal repeat of human foamy virus. Virology 196: 256–268

European Collaborative Study (1991) Children born to women with HIV-1 infection: natural history and risk of transmission. Lancet 337: 253–260

Flügel RM, Rethwilm A, Maurer B, Darai G (1987) Nucleotide sequence analysis of the env gene and its flanking regions of the human spumaretrovirus reveals two novel genes. EMBO J 6: 2077–2084

Freed EO, Myers DJ, Risser R (1990) Characterization of the fusion domain of the human immunodeficiency virus type 1 envelope glycoprotein gp41. Proc Natl Acad Sci USA 87: 4650–4654

Gallaher WR (1987) Detection of a fusion peptide sequence in the transmembrane protein of human immunodeficiency virus. Cell 50: 327–328

Garrett ED, He F, Bogerd HP, Cullen BR (1993) Transcriptional trans activators of human and simian foamy viruses contain a small, highly conserved activation domain. J Virol 67: 6824–6827

Gendelman HE, Phelps W, Feigenbaum L, Ostrove JM, Adachi A, Howley PM, Khoury G, Ginsberg HS, Martin MA (1986) Trans-activation of the huan immunodeficiency virus long terminal repeat sequence by DNA viruses. Proc Natl Acad Sci USA 83: 9759–9763

Giangaspero F, Scanabissi E, Baldacci MC, Betts CM (1989) Massive neuronal destructiona in human immunodeficiency virus (HIV) encephalitis. A clinico-pathological study of a pediatric case. Acta Neuropathol (Berl) 78: 662–665

Gonzales MF, Davis RL (1988) Neuropathology of acquired immunodeficiency syndrome. Neuropathol Appl Neurobiol 14: 345–363

Gourdou I, Mabrouk K, Harkiss G, Marchot P, Watt N, Hery F, Vigne R (1990) Neurotoxicity in mice due to cysteine-rich parts of visna virus and HIV-1 Tat proteins. C R Acad Sci [III] 311: 149–155

Green JE, Begley CG, Wagner DK, Waldmann TA, Jay G (1989) trans activation of granulocyte-macrophage colony-stimulating factor and the interleukin-2 receptor in transgenic mice carrying the human T-lymphotropic virus type 1 tax gene. Mol Cell Biol 9: 4731–4737

Gunzburg WH, Heinemann F, Wintersperger S, Miethke T, Wagner H, Erfle V, Salmons B (1993) Endogenous superantigen expression controlled by a novel promoter in the MMTV long terminal repeat. Nature 364: 154–158

Hahn H, Baunach G, Bräutigam S, Mergia A, Neumann-Haefelin D, Daniel MD, McClure MO, Rethwilm A (1994) Reactivity of primate sera to foamy virus gag and bet proteins. J Gen Virol 75: 2635–2644

He F, Sun JD, Garrett ED, Cullen BR (1993) Functional organization of the Bel-1 trans activator of human foamy virus. J Virol 67: 1896–1904

Herchenröder O, Renne R, Loncar D, Cobb EK, Murthy KK, Schneider J, Mergia A, Luciw PA (1994) Isolation, cloning, and sequencing of simian foamy viruses from chimpanzees (SFVcpz): high homology to human foamy virus (HFV). Virology 201: 187–199

Hildreth JE, Orentas RJ (1989) Involvement of a leukocyte adhesion receptor (LFA-1) in HIV-induced syncytium formation. Science 244: 1075–1078

Hinrichs SH, Nerenberg M, Reynolds RK, Khoury G, Jay G (1987) A transgenic mouse model for human neurofibromatosis. Science 237: 1340–1343

Hogan B, Constantini F, Lacy E (1986) Manipulating the mouse embryo. A laboratory manual. Cold Spring Harbor Laboratory, New York

Hooks JJ, Gibbs CJ Jr (1975) The foamy viruses. Bacteriol Rev 39: 169–185

Hook JJ, Gibbs CJ Jr, Cutchins EC, Rogers NG, Lampert P, Gajdusek DC (1972) Characterization and distribution of two new foamy viruses isolated from chimpanzees. Arch Gesamte Virusforsch 38: 38–55

Hooks JJ, Burns W, Hayashi K, Geis S, Notkins AL (1976) Viral spread in the presence of neutralizing antibody: mechanisms of persistence in foamy virus infection. Infect Immun 14: 1172–1181

Horth M, Lambrecht B, Chuah Lay Khim M, Bex F, Thiriart C, Ruysschaert JM, Burny A, Brasseur R (1991) Theoretical and functional analysis of the SIV fusion peptide. EMBO J 10: 2747–2755

Keller A, Partin KM, Löchelt M, Bannert H, Flügel RM, Cullen BR (1991) Characterization of the transcriptional trans activator of human foamy retrovirus. J Virol 65: 2589–2594

Keller A, Garrett ED, Cullen BR (1992) The Bel-1 protein of human foamy virus activates human immunodeficiency virus type 1 gene expression via a novel DNA target site. J Virol 66: 3946–3949

Kennedy-Stoskopf S, Stoskopf MK, Eckhaus MA, Strandberg JD (1986) Isolation of a retrovirus and a herpesvirus from a captive California sea lion. J Wildl Dis 22: 156–164

Kleihues P, Lang W, Burger PC, Budka H, Vogt M, Maurer R, Lüthy R, Siegenthaler W (1985) Progressive diffuse leukoencephalopathy in patients with acquired immunodeficiency syndrome (AIDS). Acta Neuropathol (Berl) 333: 339

Kupiec JJ, Kay A, Hayat M, Ravier R, Peries J, Galibert F (1991) Sequence analysis of the simian foamy virus type 1 genome. Gene 101: 185–194

Lagaye S, Vexiau P, Morozov V, Guenebaut-Claudet V, Tobaly-Tapiero J, Canivet M, Cathelineau G, Peries J, Emanoil-Ravier R (1992) Human spumaretrovirus-related sequences in the DNA of leukocytes from patients with Graves' disease. Proc Natl Acad Sci USA 89: 10070–10074

Landau NR, Page KA, Littman DR (1991) Pseudotyping with human T-cell leukemia virus type I broadens the human immunodeficiency virus host range. J Virol 65: 162–169

Le Guern M, Levy JA (1992) Human immunodeficiency virus (HIV) type 1 can superinfect HIV-2 infected cells. Pseudotypes virions produced with expanded cellular host range. Proc Natl Acad USA 89: 363–367

Lee AH, Lee KJ, Kim S, Sung YC (1992) Transactivation of human immunodeficiency virus type 1 long terminal repeat-directed gene expression by the human foamy virus bell protein requires a specific DNA sequence. J Virol 66: 3236–3240

Lee CW, Chang J, Lee KJ, Sung YC (1994) The Bell protein of human foamy virus contains one positive and two negative control regions which regulate a distinct activation domain of 30 amino acids. J Virol 68: 2708–2719

Lee KJ, Lee AH, Sung YC (1993) Multiple positive and negative cis-acting elements that mediate transactivation by bell in the long terminal repeat of human foamy virus. J Virol 67: 2317–2326

Leonard JM, Abramczuk JW, Pezen DS, Rutledge R, Belcher JH, Hakim F, Shearer G, Lamperth L, Travis W, Fredrickson T, Notkins AL, Martin MA (1988) Development of disease and virus recovery in transgenic mice containing HIV proviral DNA. Science 242: 1665–1670

Lewis SH, Reynolds-Kohler C, For HE, Nelson JA (1990) HIV-1 in trophoblastic and villous Hofbauer cells, and haematological precursors in eight-week fetuses. Lancet 335: 565–568

Löchelt M, Zentgraf H, Flügel RM (1991) Construction of an infectious DNA clone of the full length human spumaretrovirus genome and mutagenesis of the bel-1 gene. Virology 184: 34–54

Löchelt M, Aboud M, Flügel RM (1993) Increase in the basal transcriptional activity of the human foamy virus internal promoter by the homologous long terminal repeat promoter in cis. Nucleic Acids Res 21: 4226–4230

Löchelt M, Flügel RM, Aboud M (1994) The human foamy virus internal promoter directs the expression of the functional Bel 1 transactivator and Bet protein early after infection. J Virol 68: 638–645

Loh P (1993) Spumavirinae. In: Levy J (ed) The retroviridae. Plenum, New York, pp 361–397

Loh PC, Achong BG, Epstein MA (1977) Further biological properties of the human syncytial virus. Intervirology 8: 204–217

Loh PC, Matsuura F, Mizumoto C (1980) Seroepidemiology of human syncytial virus: antibody prevalence in the Pacific. Intervirology 13: 87–90

Lusso P, Ensoli B, Markham PD, Ablashi DV, Salahuddin SZ, Tschachler E, Wong-Staal F, Gallo RC (1989) Productive dual infection of human CD4+ T lymphocytes by HIV-1 and HHV-6. Nature 337: 370–373

Lusso P, De Maria A, Malnati M, Lori F, DeRocco SE, Baseler M, Gallo RC (1991) Induction of CD4 and susceptibility to HIV-1 infection in human CD8+ T lymphocytes by human herpesvirus 6. Nature 349: 533–535

Mahnke C, Löchelt M, Bannert H, Flügel RM (1990) Specific enzyme-linked immunosorbent assay for the detection of antibodies to the human spumavirus. J Virol Methods 29: 13–22

Mahnke C, Kashaiya P, Rössler J, Bannert H, Levin A, Blattner WA, Dietrich M, Luande J, Löchelt M, Friedman-Kien A, Komaroff AL, Loh PC, Westarp ME, Flügel RM (1992) Human spumavirus antibodies in sera from african patients. Arch Virol 123: 243–253

Maier H, Budka H, Lassmann H, Pohl P (1989) Vacuolar myelopathy with multinucleated giant cells in the acquired immune deficiency syndrome (AIDS). Light and electron microscopic distribution of human immunodeficiency virus (HIV) antigens. Acta Neuropathol (Berl) 78: 497–503

Maurer B, Bannert H, Darai G, Flügel RM (1988) Analysis of the primary structure of the long terminal repeat and the gag and pol genes of the human spumaretrovirus. J Virol 62: 1590–1597

McClure MO, Bienasz PD, Schulz TF, Chrystie IL, Simpson G, Aguzzi A, Hoad JG, Cunningham A, Kirkwood J, Weiss RA (1994) Isolation of a new foamy virus from orangutans. J Virol 68: 7124–7130

Mergia A (1994) Simian foamy virus type 1 contains a second promoter located at the 3' end of the env gene. Virology 199: 219–222

Mergia A, Luciw PA (1991) Replication and regulation of primate foamy viruses. Virology 184: 475–482

Mergia A, Pratt Lowe E, Shaw KE, Renshaw Gegg LW, Luciw PA (1992) cis-acting regulatory regions in the long terminal repeat of simian foamy virus type 1. J Virol 66: 251–257

Mergia A, Renshaw Gegg LW, Stout MW, Renne R, Herchenröder O (1993) Functional domains of the simian foamy virus type 1 transcriptional transactivator (Taf). J Virol 67: 4598–4604

Miller CL, Garner R, Paetkau V (1992) An activation-dependent, T-lymphocyte-specific transcriptional activator in the mouse mammary tumor virus env gene. Mol Cell Biol 12: 3262–3272

Mosca JD, Bednarik DP, Raj NBK, Rosen CA, Sodroski JG, Haseltine WA, Pitha PM (1987) Herpes simplex virus type-1 can reactivate transcription of latent human immunodeficiency virus. Nature 325: 67–70

Muller HK, Ball G, Epstein MA, Achong BG, Lenoir G, Levin A (1980) The prevalence of naturally occurring antibodies to human syncytial virus in East African populations. J Gen Virol 47: 399–406

Muranyi W, Flügel RM (1991) Analysis of splicing patterns of human spumaretrovirus by polymerase chain reaction reveals complex RNA structures. J Virol 65: 727–735

Nara PL, Robey WG, Arthur LO, Gonda MA, Asher DM, Yanagihara R, Gibbs CJ Jr, Gajdusek DC, Fischinger PJ (1987) Simultaneous isolation of simian foamy virus and HTLV-III/LAV from chimpanzee lymphocytes following HTLV-III or LAV inoculation. Arch Virol 92: 183–186

Nerenberg MI, Wiley CA (1989) Degeneration of oxidative muscle fibers in HTLV-1 tax transgenic mice. Am J Pathol 135: 1025–1033

Nerenberg M, Hinrichs SH, Reynolds RK, Khoury G, Jay G (1987) The tat gene of human T-lymphotropic virus type 1 induces mesenchymal tumors in transgenic mice. Science 237: 1324–1329

Netzer KO, Rethwilm A, Maurer B, Ter Meulen V (1990) Identification of the major immunogenic structural proteins of human foamy virus. J Gen Virol 71: 1237–1241

Neumann-Haefelin D, Rethwilm A, Bauer G, Gudat F, Zur Hausen H (1983) Characterization of a foamy virus isolated from Cercopithecus aethiops lymphoblastoid cells. Med Microbiol Immunol (Berl) 172: 75–86

Neumann-Haefelin D, Fleps U, Renne R, Schweizer M (1993) Foamy viruses. Intervirology 35: 196–207

Nosaka T, Ariumi Y, Sakurai M, Takeuchi R, Hatanaka M (1993) Novel internal promoter/enhancer of HTLV-I for Tax expression. Nucleic Acids Res 21: 5124–5129

Pantaleo G, Poli G, Butini L, Fox C, Dayton AI, Fauci AS (1991) Dissociation between syncytia formation and HIV spreading. Suppression of syncytia formation does not necessarily reflect inhibition of HIV infection. Eur J Immunol 21: 1771–1774

Petito CK (1988) Review of central nervous system pathology in human immunodeficiency virus infection. Ann Neurol 23 Suppl: S54-S57

Petito CK, Navia BA, Cho ES, Jordan BD, George DC, Price RW (1985) Vacuolar myelopathy pathologically resembling subacute combined degeneration in patients with the acquired immunodeficiency syndrome. N Engl J Med 312: 874–879

Renne R, Friedl E, Schweizer M, Fleps U, Turek R, Neumann-Haefelin D (1992) Genomic organization and expression of simian foamy virus type 3 (SFV-3). Virology 186: 597–608

Renshaw RW, Casey JW (1994) Analysis of the 5' long terminal repeat of bovine syncytial virus. Gene 141: 221–224

Rethwilm A (1995) Regulation of foamy virus gene expression. Curr Top Microbiol Immunol 193 (in press)

Rethwilm A, Baunach G, Netzer KO, Maurer B, Borisch B, Ter Meulen V (1990) Infectious DNA of the human spumaretrovirus. Nucleic Acids Res 18: 733–738

Rethwilm A, Erlwein O, Baunach G, Maurer B, Ter Meulen V (1991) The transcriptional transactivator of human foamy virus maps to the bel 1 genomic region. Proc Natl Acad Sci USA 88: 941–945

Rosen CA (1991) Regulation of HIV gene expression by RNA-protein interactions. Trends Genet 7: 9–14

Saib A, Peries J, de The H (1993) A defective human foamy provirus generated by pregenome splicing. EMBO J 12: 4439–4444

Santillana Hayat M, Rozain F, Bittoun P, Chopin Robert C, Lasneret J, Peries J, Canivet M (1993) Transient immunosuppressive effect induced in rabbits and mice by the human spumaretrovirus prototype HFV (human foamy virus). Res Virol 144: 389–396

Schliephake A, Rethwilm A (1994) Nuclear localization of foamy virus gag precursor protein. J Virol 68: 4946–4954

Schnitzer TJ (1982) Simian foamy virus pseudotypes of vesicular stomatitis virus: production and use in sero-epidemiological investigations. J Gen Virol 59: 203–206

Schweizer M, Renne R, Neumann-Haefelin D (1989) Structural analysis of proviral DNA in simian foamy virus (LK-3)-infected cells. Arch Virol 109: 103–104

Schweizer M, Turek R, Hahn H, Schliephake A, Netzer KO, Eder G, Reinhardt M, Rethwilm A, Neumann-Haefelin D (1994a) Markers of foamy virus (FV) infection in monkeys, apes, and accidentally infected humans: appropriate testing fails to confirm suspected FV prevalence in man. AIDS Res Hum Retrovir uses (in press)

Schweizer M, Turek R, Reinhardt M, Neumann-Haefelin D (1994b) Absence of foamy virus DNA in Graves' disease. AIDS Res Hum Retroviruses 10: 601–605

Sharpe AH, Hunter JJ, Chassler P, Jänisch R (1990) Role of abortive retroviral infection of neurons in spongiform CNS degeneration. Nature 346: 181–183

Siekevitz M, Joseph SF, Dukovich M, Peffer N, Wong-Staal F, Greene WC (1987) Activation of the HIV-1 LTR by T cell mitogens and the trans-activator protein of HTLV-I. Science 238: 1575–1578

Stancek D, Stancekova-Gressnerova M, Janotka M, Hnilica P, Oravec D (1975) Isolation and some serological and epidemiological data on the viruses recovered from patients with subacute thyroiditis de Quervain. Med Microbiol Immunol (Berl) 161: 133–144

Teich N (1984) Taxonomy of retroviruses. In: Weiss RA, Teich N, Varmus HE, Coffin J (eds) RNA tumor viruses. Cold Spring Harbor Laboratory, New York, pp 25–208

Truneh A, Buck D, Cassatt DR, Juszczak R, Kassis S, Ryu SE, Healey D, Sweet R, Sattentau Q (1991) A region in domain 1 of CD4 distinct from the primary gp120 binding site is involved in HIV infection and virus-mediated fusion. J Biol Chem 266: 5942–5948

Venkatesh LK, Chinnadurai G (1993) The carboxy-terminal transcription enhancement region of the human spumaretrovirus transactivator contains discrete determinants of the activator function. J Virol 67: 3868–3876

Venkatesh LK, Theodorakis PA, Chinnadurai G (1991) Distinct cis-acting regions in U3 regulate trans-activation of the human spumaretrovirus long terminal repeat by the viral bell gene product. Nucleic Acids Res 19: 3661–3666

Vogel J, Hinrichs SH, Reynolds RK, Luciw PA, Jay G (1988) The HIV tat gene induces dermal lesions resembling Kaposi's sarcoma in transgenic mice. Nature 335: 606–611

Watanabe T, Seiki M, Tsujimoto H, Miyoshi I, Hayami M, Yoshida M (1985) Sequence homology of the simian retrovirus genome with human T-cell leukemia virus type I. Virology 144: 59–65

Watanabe T, Seiki M, Hirayama Y, Yoshida M (1986) Human T-cell leukemia virus type I is a member of the African subtype of simian viruses (STLV). Virology 148: 385–388

Weissenberger J, Flügel R (1994) Identification and characterization of the bel-3 protein of human foamy virus. AIDS Res Hum Retroviruses 10: 595–600

Werner J, Gelderblom HR (1979) Isolation of foamy virus from patients with de Quervain thyroiditis. Lancet 2: 258–259

Westarp ME, Kornhuber HH, Rössler J, Flügel RM (1992) Human spuma virus antibodies in amyotrophic lateral sclerosis. Neurol Psychiatry Brain Res 1: 1–4

Young D, Samuels J, Clarke JK (1973) A foamy virus of possible human origin isolated in BHK-21 cells. Arch Gesamte Virusforsch 42: 228–234

Yu SF, Linial ML (1993) Analysis of the role of the bel and bet open reading frames of human foamy virus by using a new quantitative assay. J Virol 67: 6618–6624

Zider A, Mashhour B, Fergelot P, Grimber G, Vernet M, Hazan U, Couton D, Briand P, Cavard C (1993) Dispensable role of the NF-kappa B sites in the UV-induction of the HIV-1 LTR in transgenic mice. Nucleic Acids Res 21: 79–86

Transgenetics of Prion Diseases*

S.B. Prusiner

Introduction

Prions cause a group of human and animal neurodegenerative diseases which are now classified together because their etiology and pathogenesis involve modification of the prion protein (PrP) (PRUSINER 1991). Prion diseases are manifest as infectious, genetic and sporadic disorders (Table 1). These diseases can be transmitted among mammals by the infectious particle designated "prion" (PRUSINER 1982). Despite intensive searches over the past three decades, no nucleic acid has been found within prions (ALPER et al. 1966, 1967; HUNTER 1972; RIESNER et al. 1992); yet, a modified isoform of the host-encoded PrP, designated PrP^{Sc}, is essential for infectivity (BÜELER et al. 1993; PRUSINER 1991; PRUSINER et al.

Department of Neurology, HSE-781, University of California, San Francisco, CA 94143-0518, USA
*This Chapter also appears in Vol. 207: Prions Prions Prions (S.B. PRUSINER)

Table 1. Human prion diseases

Disease	Etiology
Kuru	Infection
Creutzfeldt-Jakob disease	
Iatrogenic	Infection
Sporadic	Unknown
Familial	PrP mutation
Gerstmann-Sträussler-Scheinker disease	PrP mutation
Fatal familial insomnia	PrP mutation

1983, 1993b,c). In fact, considerable experimental data argue that prions are composed exclusively of PrP^{Sc}. Earlier terms used to describe the prion diseases include: transmissible encephalopathies, spongiform encephalopathies and slow virus diseases (GAJDUSEK 1977, 1985; SIGURDSSON 1954).

The quartet of human (Hu) prion diseases are frequently referred to as kuru, Creutzfeldt-Jakob disease (CJD), Gerstmann-Sträussler-Scheinker (GSS) disease and fatal familial insomnia (FFI). Kuru was the first of the human prion diseases to be transmitted to experimental animals, and it has often been suggested that kuru spread among the Fore people of Papua New Guinea by ritualistic cannibalism (GAJDUSEK 1977; GAJDUSEK et al. 1966). The experimental and presumed human to human transmission of kuru led to the belief that prion diseases are infectious disorders caused by unusual viruses similar to those causing scrapie in sheep and goats. Yet, the occurrence of CJD in families, first reported almost 70 years ago (KIRSCHBAUM 1924; MEGGENDORFER 1930), was perplexing to say the least. The significance of familial CJD remained unappreciated until mutations in the protein coding region of the PrP gene on the short arm of chromosome 20 were discovered (HSIAO et al. 1989; PRUSINER 1994; SPARKES et al. 1986). The earlier finding that brain extracts from patients who had died of familial prion diseases inoculated into experimental animals often transmit disease posed a conundrum that was resolved with the genetic linkage of these diseases to mutations of the PrP gene (MASTERS et al. 1981; PRUSINER 1989; TATEISHI et al. 1992).

The most common form of prion disease is sporadic CJD. Many attempts to show that the sporadic prion diseases are caused by infection have been unsuccessful (BROWN et al. 1987; COUSENS et al. 1990; HARRIES-JONES et al. 1988; MALMGREN et al. 1979). The discovery that inherited prion diseases are caused by germline mutation of the PrP gene raised the possibility that sporadic forms of these diseases might result from a somatic mutation (PRUSINER 1989). The discovery that PrP^{Sc} is formed from PrP^{C}, the cellular isoform of prion protein, by a posttranslational process (BORCHELT et al. 1990) and that overexpression of wild-type (wt) PrP transgenes produces spongiform degeneration and infectivity de novo (WESTAWAY et al. 1994b) has raised the possibility that sporadic prion diseases result from the spontaneous conversion of PrP^{C} into PrP^{Sc}.

CJD has a worldwide incidence of approximately 1 case per 10^{6} population (MASTERS et al. 1978). Less than 1% of CJD cases are infectious and all of those appear to be iatrogenic. Between 10% and 15% of prion disease cases are inherited while the remaining cases are sporadic. Kuru was once the most common cause of death among New Guinea women in the Fore region of the Highlands (GAJDUSEK et al. 1966; GAJDUSEK and ZIGAS 1957, 1959) but has virtually disappeared with the cessation of ritualistic cannibalism (ALPERS 1987). Patients with CJD frequently present with dementia but about 10% of patients exhibit cerebellar dysfunction initially. Patients with either kuru or GSS usually present with ataxia while those with FFI manifest insomnia and autonomic dysfunction (BROWN 1992; HSIAO and PRUSINER 1990; MEDORI et al. 1992b).

PrP^{CJD} has been found in the brains of most patients who died of prion disease. The term PrP^{CJD} is preferred by some investigators when referring to the abnormal isoform of HuPrP in human brain. Here, PrP^{Sc} is used interchangeably with PrP^{CJD}. PrP^{Sc} is always used after human CJD prions have been passaged into an experimental animal since the nascent PrP^{Sc} molecules are produced from host PrP^{C} and the PrP^{CJD} in the inoculum only serves to initiate the process. In the brains of some patients with inherited prion diseases and in transgenic (Tg) mice expressing mouse (Mo) PrP with the human GSS point mutation (Pro → Leu), detection of PrP^{Sc} has been problematic despite clinical and neuropathologic hallmarks of neurodegeneration (HSIAO et al. 1990, 1995). Presumably, neuro-degeneration is due, at least in part, to the abnormal metabolism of mutant PrP (HSIAO et al. 1990). Of note, horizontal transmission of neurodegeneration from the brains of patients with inherited prion diseases to inoculated rodents has been less frequent than with sporadic cases (TATEISHI et al. 1992). Whether this distinc-tion between transmissible and nontransmissible inherited prion diseases will persist is unclear. Tg mice expressing a chimeric Hu/Mo PrP gene have been found to be highly susceptible to Hu prions from sporadic and iatrogenic CJD cases (TELLING et al. 1995). These Tg (MHu2M) mice should make the use of apes and monkeys for the study of human prion diseases unnecessary and allow for tailoring the PrP^{C} translated from the transgene to match the sequence of the PrP^{CJD} in the inoculum. Other Tg mouse studies have demonstrated that PrP^{Sc} in the inoculum interacts preferentially with homotypic PrP^{C} during the propagation of prions (PRUSINER et al. 1990; SCOTT et al. 1993). PrP^{C} is the cellular isoform of the prion protein which has been identified in all mammals and birds examined to date; PrP^{C} is anchored to the external surface of cells by a glycolipid moiety and its function is unknown (STAHL et al. 1987).

Scrapie is the most common natural prion disease of animals. An investiga-tion into the etiology of scrapie followed the vaccination of sheep for looping ill virus with formalin-treated extracts of ovine lymphoid tissue unknowingly con-taminated with scrapie prions (GORDON 1946). Two years later, more than 1500 sheep developed scrapie from this vaccine. While the transmissibility of experi-mental scrapie became well established, the spread of natural scrapie within and among flocks of sheep remained puzzling. PARRY (1962, 1983) argued that host genes were responsible for the development of scrapie in sheep. He was convinced that natural scrapie is a genetic disease which could be eradicated by proper breeding protocols. He considered its transmission by inoculation of importance primarily for laboratory studies and communicable infection of little consequence in nature. Other investigators viewed natural scrapie as an infectious disease and argued that host genetics only modulates susceptibility to an endemic infectious agent (DICKINSON et al. 1965).

The offal of scrapied sheep in Great Britain is thought to be responsible for the current epidemic of bovine spongiform encephalopathy (BSE) or mad cow disease (WILESMITH et al. 1992b). Prions in the offal from scrapie-infected sheep appear to have survived rendering which produced meat and bone meal (MBM). After BSE was recognized, MBM produced from domestic animals was banned from further use. Since 1986 when BSE was first recognized, >130,000 cattle have died of BSE. Whether humans will develop CJD after consuming beef from cattle with BSE prions is of considerable concern.

The fundamental event in prion diseases seems to be a conformational change in PrP. All attempts to identify a posttranslational chemical modification that distinguishes PrP^{Sc} from PrP^{C} have been unsuccessful to date (STAHL et al. 1993). PrP^{C} contains ~45% α-helix and is virtually devoid of β-sheet (PAN et al. 1993). Conversion to PrP^{Sc} creates a protein which contains ~30% α-helix and 45% β-sheet. The mechanism by which PrP^{C} is converted into PrP^{Sc} remains unknown but PrP^{C} appears to bind to PrP^{Sc} to form an intermediate complex during the formation of nascent PrP^{Sc}.

Prions differ from all other known infectious pathogens in several respects. First, prions do not contain a nucleic acid genome which codes for their progeny. Viruses, viroids, bacteria, fungi and parasites all have nucleic acid genomes that code for their progeny. Second, the only known component of the prion is a modified protein that is encoded.by a cellular gene. Third, the major, and possibly only, component of the prion is PrP^{Sc} which is a pathogenic conformer of PrP^{C}.

As our knowledge of the prion diseases increases and more is learned about the molecular and genetic characteristics of prion proteins, these disorders will undoubtedly undergo modification with respect to their classification. Indeed, the discovery of PrP and the identification of pathogenic PrP gene mutations have already forced us to view these illnesses from perspectives not previously imagined.

1 Transgenetics and Gene Targeting

While transgenetic studies have yielded a wealth of new knowledge about infectious, genetic and sporadic prion diseases, the laborious production of Tg mice limits the number of studies that can be performed. The relatively long gestation period of mice coupled with the need to do microinjections of fertilized embryos prevents the creation of the very large numbers of different Tg mice that would yield the greatest amount of new information. Assays that permit screening of a multitude of possible phenotypes in genetic experiments are generally the most informative. While the limited number of mice expressing different transgenes is definitely a liability, experiments with Tg mice expressing foreign and mutant PrP molecules have been extraordinarily useful in advancing our understanding of prion biology. It is important to stress that transgenetic studies can readily yield an incomplete, and sometimes erroneous, interpretation of the data if the number of lines of mice examined expressing a particular construct is inadequate. Defining an adequate number of lines is difficult but comparisons of lines expressing high and low levels of a given PrP transgene have proved to be quite helpful (HSIAO et al. 1995; PRUSINER et al. 1990).

1.1 Species Barriers for Transmission of Prion Diseases

The passage of prions between species is a stochastic process characterized by prolonged incubation times (Pattison 1965, 1966; PATTISON and JONES 1967). Prions synthesized de novo reflect the sequence of the host PrP gene and not that of the PrP^{Sc} molecules in the inoculum (BOCKMAN et al. 1987). On subsequent passage in a homologous host, the incubation time shortens to that recorded for all subsequent passages and it becomes a nonstochastic process. The species barrier concept is of practical importance in assessing the risk for humans of developing CJD after consumption of scrapie-infected lamb or BSE-infected beef (GOLDMANN et al. 1991; HOPE et al. 1988; PRUSINER et al. 1993a; WILESMITH et al. 1992a,b).

   To test the hypothesis that differences in PrP gene sequences might be responsible for the species barrier, Tg mice expressing Syrian hamster PrP (SHaPrP) were constructed (PRUSINER et al. 1990; SCOTT et al. 1989). The PrP genes of Syrian hamsters and mice encode proteins differing at 16 positions. Incubation times in four lines of Tg(SHaPrP) mice inoculated with Mo prions were prolonged compared to those observed for non-Tg, control mice (Fig. 1A). Inoculation of Tg(SHaPrP) mice with SHa prions demonstrated abrogation of the species barrier resulting in abbreviated incubation times due to a nonstochastic process (Fig. 1B) (PRUSINER et al. 1990; SCOTT et al. 1989). The length of the incubation time after inoculation with SHa prions was inversely proportional to the level of SHaPrP^C in the brains of Tg(SHaPrP) mice (Fig. 1B,C) (PRUSINER et al. 1990). SHaPrP^{Sc} levels in the brains of clinically ill mice were similar in all four Tg(SHaPrP) lines inoculated with SHa prions (Fig. 1D). Bioassays of brain extracts from clinically ill Tg(SHaPrP) mice inoculated with Mo prions revealed that only Mo prions but no SHa prions were produced (Fig. 1E). Conversely, inoculation of Tg(SHaPrP) mice with SHa prions led to only the synthesis of SHa prions (Fig. 1F). Thus, the de novo synthesis of prions is species-specific and reflects the genetic origin of the inoculated prions. Similarly, the neuropathology of Tg(SHaPrP) mice is determined by the genetic origin of prion inoculum. Mo prions injected into Tg(SHaPrP) mice produced a neuropathology characteristic of mice with scrapie. A moderate degree of vacuolation in both the gray and white matter was found while amyloid plaques were rarely detected (Fig. 1G) (Table 2). Inoculation of Tg(SHaPrP) mice with SHa prions produced intense vacuolation of the gray matter, sparing of the white matter, and numerous SHaPrP amyloid plaques characteristic of Syrian hamsters with scrapie (Fig. 1H).

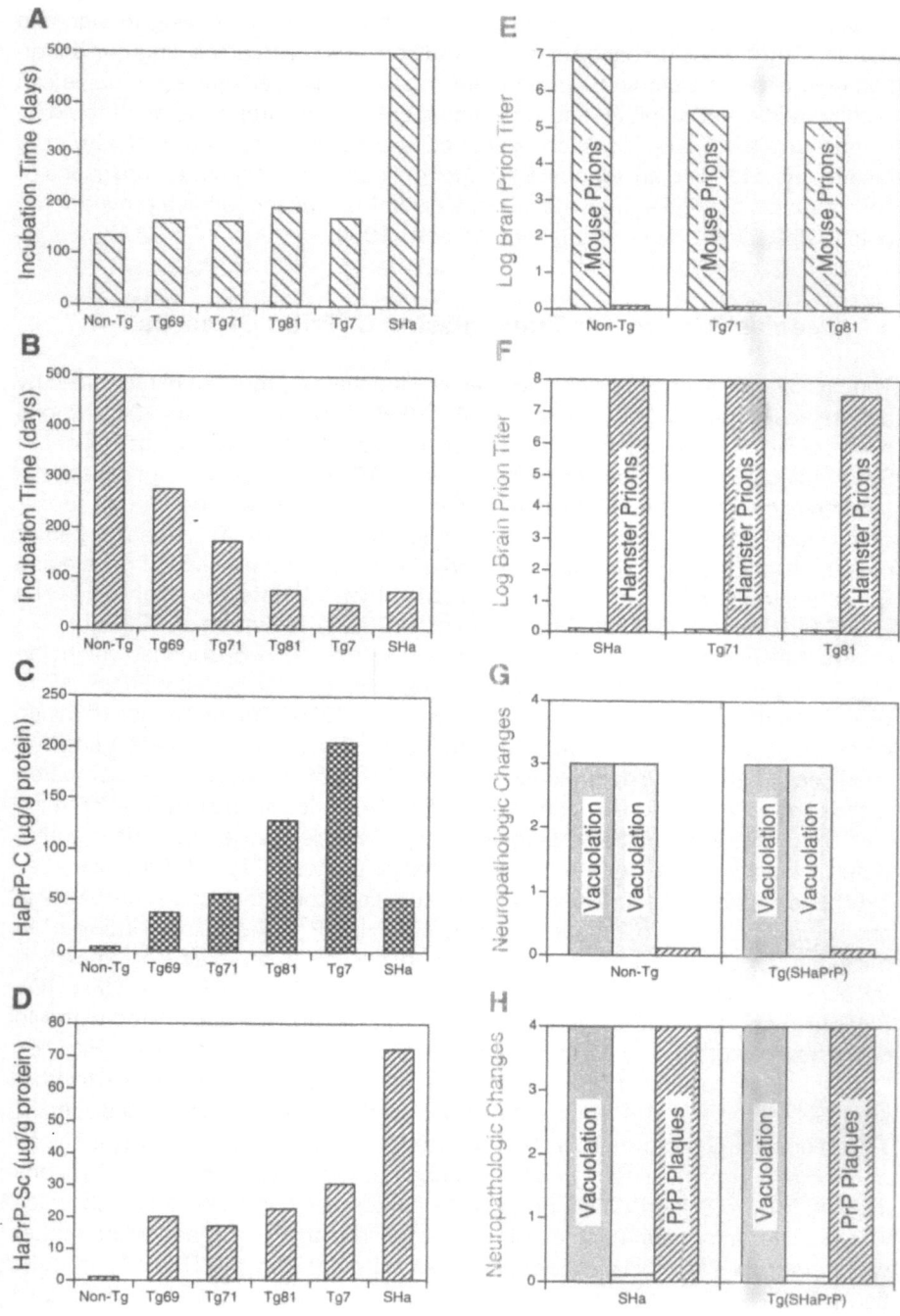

Table 2. Species-specific prion inocula determine the distribution of spongiform change and deposition of PrP amyloid plaques in transgenic mice (from PRUSINER 1992)

Animal	SHa prions					Mo prions			
		Spongiform change[a]		PrP plaques[b]			Spongiform change[a]		PrP plaques[b]
	n^a	Gray	White	Frequency	Diameter[d]	n^c	Gray	White	Frequency
Non-Tg		N.D.		N.D		10	+	+	−
Tg-69	6	+	−	Numerous	6.5 ± 3.1 (389)	2	+	+	−
Tg-71	5	+	−	Numerous	8.1 ± 3.6 (345)	2	+	+	−
Tg-81	7	+	−	Numerous	8.3 ± 3.0 (439)	3	+	+	Few
Tg-7	3	+[e]	−	Numerous	14.0 ± 8.3 (19)	4	+	+	−
SHa	3	+	−	Numerous	5.7 ± 2.7 (247)		N.D.		N.D.

SHa, Syrian hamster; Mo, mouse; N.D., not determined.

[a] Spongiform change evaluated in hippocampus, thalamus, cerebral cortex and brainstem for gray matter and the deep cerebellum for white matter.

[b] Plaques in the subcallosal region were stained with SHaPrP monoclonal antibody 13A5, anti-PrP rabbit antisera R073 and trichrome stain.

[c] Number of brains examined.

[d] Mean diameter of PrP plaques given in microns ± standard error with the number of observations in parentheses.

[e] Focal: Confirmed to the dorsal nucleus of the raphe.

1.2 Overexpression of wtPrP Transgenes

During transgenetic studies, we discovered that uninoculated older mice harboring high copy-numbers of wtPrP transgenes derived from Syrian hamsters, sheep, and PrP-B mice spontaneously developed truncal ataxia, hind-limb paralysis and tremors (WESTAWAY et al. 1994a). These Tg mice exhibited a profound necrotizing myopathy involving skeletal muscle, a demyelinating polyneuropathy

Fig. 1A–H. Transgenic (Tg) mice expressing Syrian hamster (SHa) prion protein exhibit species-specific scrapie incubation times, infectious prion synthesis and neuropathology. **A** Scrapie incubation times in nontransgenic mice (Non-Tg) and four lines of Tg mice expressing SHaPrP and Syrian hamsters inoculated intracerebrally with ~10^6 ID$_{50}$ units of Chandler Mo prions serially passaged in Swiss mice. The four lines of Tg mice have different numbers of transgene copies: Tg69 and 71 mice have two to four copies of the SHaPrP transgene, whereas Tg81 have 30 to 50 and Tg7 mice have >60. Incubation times are number of days from inoculation to onset of neurologic dysfunction. **B** Scrapie incubation times in mice and hamsters inoculated with ~10^7 ID$_{50}$ units of Sc237 prions serially passaged in Syrian hamsters and as described in **A**. **C** Brain SHaPrP^C in Tg mice and hamsters. SHaPrP^C levels were quantitated by an enzyme-linked immunoassay. **D** Brain SHaPrP^{Sc} in Tg mice and hamsters. Animals were killed after exhibiting clinical signs of scrapie. SHaPrP^{Sc} levels were determined by immunoassay. **E** Prion titers in brains of clinically ill animals after inoculation with Mo prions. Brain extracts from Non-Tg, Tg71, and Tg81 mice were bioassayed for prions in mice (left Columns) and hamsters (right Columns). **F** Prion titers in brains of clinically ill animals after inoculation with SHa prions. Brain extracts from Syrian Hamsters as well as Tg71 and Tg81 mice were bioassayed for prions in mice (left columns) and hamsters (right columns). **G** Neuropathology in Non-Tg mice and Tg(SHaPrP) mice with clinical signs of scrapie after inoculation with Mo prions. Vacuolation in gray (left columns) and white matter (center columns); PrP amyloid plaques (right columns). Vacuolation score: 0, none; 1, rare; 2, modest; 3, moderate; 4, intense. **H** Neuropathology in Syrian hamsters and transgenic mice inoculated with SHa prions. Degree of vacuolation and frequency of PrP amyloid plaques as described in **G**. (Adapted from PRUSINER 1991)

and focal vacuolation of the CNS. Development of disease was dependent on transgene dosage. For example, Tg(SHaPrP$^{+/+}$)7 mice homozygous for the SHaPrP transgene array regularly developed disease between 400 and 600 days of age, while hemizygous Tg(SHaPrP$^{+/0}$)7 mice also developed disease, but after >650 days.

Attempts to demonstrate PrP^{Sc} in either muscle or brain were unsuccessful but transmission of disease with brain extracts from Tg(SHaPrP$^{+/+}$)7 mice inoculated into Syrian hamsters did occur. These Syrian hamsters had PrP^{Sc} as detected by immunoblotting and spongiform degeneration (Groth and Prusiner, unpublished data). Serial passage with brain extracts from these animals to recipients was observed. De novo synthesis of prions in Tg(SHaPrP$^{+/+}$)7 mice overexpressing wtSHaPrP^C provides support for the hypothesis that sporadic CJD does not result from infection but rather is a consequence of the spontaneous, although rare, conversion of PrP^C into PrP^{Sc}. Alternatively, a somatic mutation in which mutant SHaPrP^C is spontaneously converted into PrP^{Sc} as in the inherited prion diseases could also explain sporadic CJD. These findings as well as those described below for Tg(MoPrP-P101L) mice argue that prions are devoid of foreign nucleic acid, in accord with many earlier studies that used other experimental approaches as described above.

1.3 Ablation of the PrP Gene

Ablation of the PrP gene in Tg(Prn-p$^{0/0}$) mice has, unexpectedly, not affected the development of these animals (Büeler et al. 1992). In fact, they are healthy at almost 2 years of age. Prn-p$^{0/0}$ mice are resistant to prions (Fig. 2) and do not propagate scrapie infectivity (Büeler et al. 1993; Prusiner et al. 1993b).

Prn-p$^{0/0}$ mice crossed with Tg(SHaPrP) mice were rendered susceptible to SHa prions but remained resistant to Mo prions (Büeler et al. 1993; Prusiner et al. 1993b). Since the absence of PrP^c expression does not provoke disease, it is likely that scrapie and other prion diseases are a consequence of PrP^{Sc} accumulation rather than an inhibition of PrP^C function (Büeler et al. 1992).

Mice heterozygous (Prn-p$^{0/+}$) for ablation of the PrP gene had prolonged incubation times when inoculated with Mo prions (Fig. 2) (Prusiner et al. 1993b). The Prn-p$^{0/+}$ mice developed signs of neurologic dysfunction at 400–460 days after inoculation. These findings are in accord with studies on Tg(SHaPrP) mice in which increased SHaPrP expression was accompanied by diminished incubation times (Fig. 1B) (Prusiner et al. 1990).

Since Prn-p$^{0/0}$ mice do not express PrP^C, we reasoned that they might more readily produce α-PrP antibodies. Prn-p$^{0/0}$ mice immunized with Mo or SHa prion rods produced α-PrP antisera which bound Mo, SHa and Hu PrP (Prusiner et al. 1993b). These findings contrast with earlier studies in which α-MoPrP antibodies could not be produced in mice presumably because the mice had been rendered tolerant by the presence of MoPrP^C (Barry and Prusiner 1986; Kascsak et al. 1987; Rogers et al. 1991). That Prn-p$^{0/0}$ mice readily produce α-PrP antibodies is

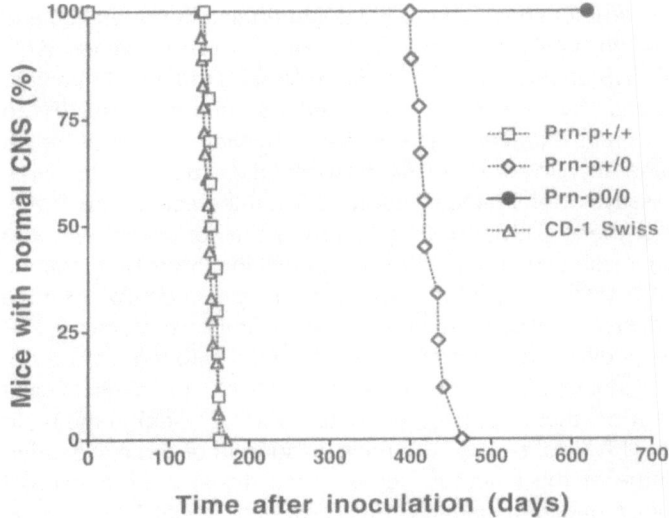

Fig. 2. Incubation times in PrP gene ablated Prn-p$^{0/+}$ and Prn-p$^{0/0}$ mice and wt Prn-p$^{+/+}$ and CD-1 mice inoculated with (RML) Rocky Mountain Laboratory strain mouse prions. The RML prions were heated and irradiated at 254 nm prior to intracerebral inoculation into CD-1 Swiss mice (*open triangles*), Prn-p$^{+/+}$ mice (*open squares*), Prn-p$^{0/+}$ mice (*open diamonds*) or Prn-p$^{0/0}$ mice (*filled circle*)

consistent with the hypothesis that the lack of an immune response in prion diseases is due to the fact that PrP^C and PrP^{Sc} share many epitopes. Whether Prn-p$^{0/0}$ mice produce α-PrP antibodies that specifically recognize conformational-dependent epitopes present on PrP^{Sc} but absent from PrP^C remains to be determined.

1.4 Modeling of GSS in Tg(MoPrP-P101L) Mice

The codon 102 point mutation found in GSS patients was introduced into the MoPrP gene and Tg(MoPrP-P101L)H mice were created expressing high (H) levels of the mutant transgene product. The two lines of Tg(MoPrP-P101L)H mice designated 174 and 87 spontaneously developed CNS degeneration, characterized by clinical signs indistinguishable from experimental murine scrapie and neuropathology consisting of widespread spongiform morphology and astrocytic gliosis (HSIAO et al. 1990) and PrP amyloid plaques (Fig. 3) (HSIAO et al. 1995). By inference, these results contend that PrP gene mutations cause GSS, familial CJD and FFI.

Brain extracts prepared from spontaneously ill Tg(MoPrP-P101L)H mice transmitted CNS degeneration to Tg196 mice expressing low levels of the mutant transgene product and some Syrian hamsters (HSIAO et al. 1995). Many Tg196 mice and some Syrian hamsters developed CNS degeneration between 200 and 700 days after inoculation, while inoculated CD-1 Swiss mice remained well. Serial transmission of CNS degeneration in Tg196 mice required about one year while serial transmission in Syrian hamsters occurred after about 75 days (HSIAO et al. 1995). Although brain extracts prepared from Tg(MoPrP-P101L)H mice transmitted CNS degeneration to some inoculated recipients, little or no PrP^{Sc} was detected by immunoassays after limited proteolysis. Undetectable or low levels of PrP^{Sc} in the brains of these Tg(MoPrP-P101L)H mice are consistent with the results of these transmission experiments which suggest low titers of infectious prions. Though no PrP^{Sc} was detected in the brains of inoculated Tg196 mice exhibiting neurologic dysfunction by immunoassays after limited proteolysis, PrP amyloid plaques and spongiform degeneration were frequently found. The neuro-degeneration found in inoculated Tg196 mice seems likely to result from a modification of mutant PrP^C that is initiated by mutant PrP^{Sc} present in the brain extracts prepared from ill Tg(MoPrP-P101L)H mice. In support of this explanation are the findings in some of the inherited human prion diseases as described above, in which neither protease-resistant PrP (BROWN et al. 1992; MEDORI et al. 1992a) nor transmission to experimental rodents could be demonstrated (TATEISHI et al. 1992). Furthermore, transmission of disease from Tg(MoPrP-P101L)H mice to Tg196 mice but not to Swiss mice is consistent with earlier findings which demonstrate that homotypic interactions between PrP^C and PrP^{Sc} feature in the formation of PrP^{Sc}.

In other studies, modifying the expression of mutant and wtPrP genes in Tg mice permitted experimental manipulation of the pathogenesis of both inherited and infectious prion diseases. Although overexpression of the wtPrP-A transgene by approximately eight fold was not deleterious to the mice, it did shorten scrapie incubation times from 145 days to 45 days after inoculation with Mo scrapie prions (TELLING et al., in preparation). In contrast, overexpression at the same level of a PrP-A transgene mutated at codon 101 produced spontaneous, fatal neuro-degeneration between 150 and 300 days of age in two new lines of Tg(MoPrP-P101L) mice designated 2866 and 2247. Genetic crosses of Tg(MoPrP-P101L) 2866 mice with gene targeted mice lacking both PrP alleles (Prn-p$^{0/0}$) produced animals with a highly synchronous onset of illness between 150 and 160 days of age. The Tg(MoPrP-P101L) 2866/Prn-p$^{0/0}$ mice had numerous PrP plaques and widespread spongiform degeneration in contrast to the Tg2866 and 2247 mice that exhibited spongiform degeneration but only a few PrP amyloid plaques. Another line of mice designated Tg2862 overexpress the mutant transgene 32-fold and develop fatal neurodegeneration between 200 and 400 days of age. Tg2862 mice exhibited the most severe spongiform degeneration and had numerous, large PrP amyloid plaques. While mutant PrP^C (P101L) clearly produces neurodegeneration, wtPrP^C profoundly modifies both the age of onset of illness and the neuropathology for a given level of transgene expression. These findings and those from other studies (TELLING et al. 1995) suggest that mutant and wtPrP interact, perhaps through a chaperone-like protein, to modify the pathogenesis of the dominantly inherited prion diseases.

2 Transmission of Prions

For three decades, the transmission of human prion diseases was studied largely with apes and monkeys where >90% of cases are thought to be transmissible (Brown et al. 1994; Gajdusek et al. 1966). Inoculations of mice, rats and hamsters produced variable results (Manuelidis et al. 1978; Tateishi and Kitamoto 1995; Tateishi et al. 1983). In our experience, only ~10% of intracerebrally inoculated mice developed CNS dysfunction with incubation times of >500 days (Prusiner 1987; Telling et al. 1994). Since previous investigations had shown that the "species barrier" between mice and Syrian hamsters for the transmission of prions can be abrogated by expression of a SHaPrP transgene in mice (Scott et al. 1989), Tg mice expressing HuPrP were constructed. These Tg(HuPrP) mice expressed levels of HuPrP^C that were 4–8 fold higher than those of endogenous MoPrP^C; yet, upon inoculation with Hu prions, they failed to develop CNS dysfunction more frequently than non-Tg controls (Telling et al. 1994).

Because of the resistance of Tg(HuPrP) mice to Hu prions, we constructed mice expressing a chimeric Hu/Mo PrP transgene designated MHu2M. Earlier studies had shown that chimeric SHa/Mo PrP transgenes supported transmission of either Mo or SHa prions (Scott et al. 1993; Scott et al. 1992). The Tg(MHu2M) mice expressing the chimeric transgene at a level slightly below that of endogenous MoPrP^C were found to be highly susceptible to Hu prions suggesting that Tg(HuPrP) mice have considerable difficulty converting HuPrP^C into PrP^{Sc} (Telling et al. 1994). Although MoPrP and HuPrP differ at 28 residues, only nine or perhaps fewer amino acids in the region between codons 96 and 167 feature in the species barrier in the transmission of Hu prions into mice, as demonstrated by the susceptibility of Tg(MHu2M) mice to Hu prions.

To explore why Hu prions transmit disease to Tg(MHu2M) mice expressing chimeric PrP but not to Tg(HuPrP) mice, we crossed the Tg(HuPrP)FVB mice with those in which the MoPrP gene had been ablated, designated Prnp$^{0/0}$ (Büeler et al. 1992). The resulting Tg(HuPrP) Prnp$^{0/0}$ mice were found to be susceptible to Hu prions, whereas Tg(MHu2M) Prnp$^{0/0}$ mice were rendered only slightly more susceptible (Telling et al. In press). These observations contend that MoPrP^C inhibited the conversion of HuPrP^C into PrP^{Sc}; once MoPrP^C was removed by gene ablation, inhibition was abolished.

2.1 Protein X and Prion Propagation

The results of our studies suggest that two separate domains of HuPrP^C participate in the formation of PrP^{Sc}: (1) the central domain delimited by codons 96 and 167 as defined by the Hu sequence in chimeric MHu2MPrP^C that binds to PrP^{Sc} and (2) an additional domain through which HuPrP^C binds to a macromolecule other than PrP^{Sc} (Telling et al. In press). We assume that this macromolecule is a protein and have provisionally designated it "protein X". From our chimeric transgene studies, the second domain of PrP^C must be at the N- or C-terminus,

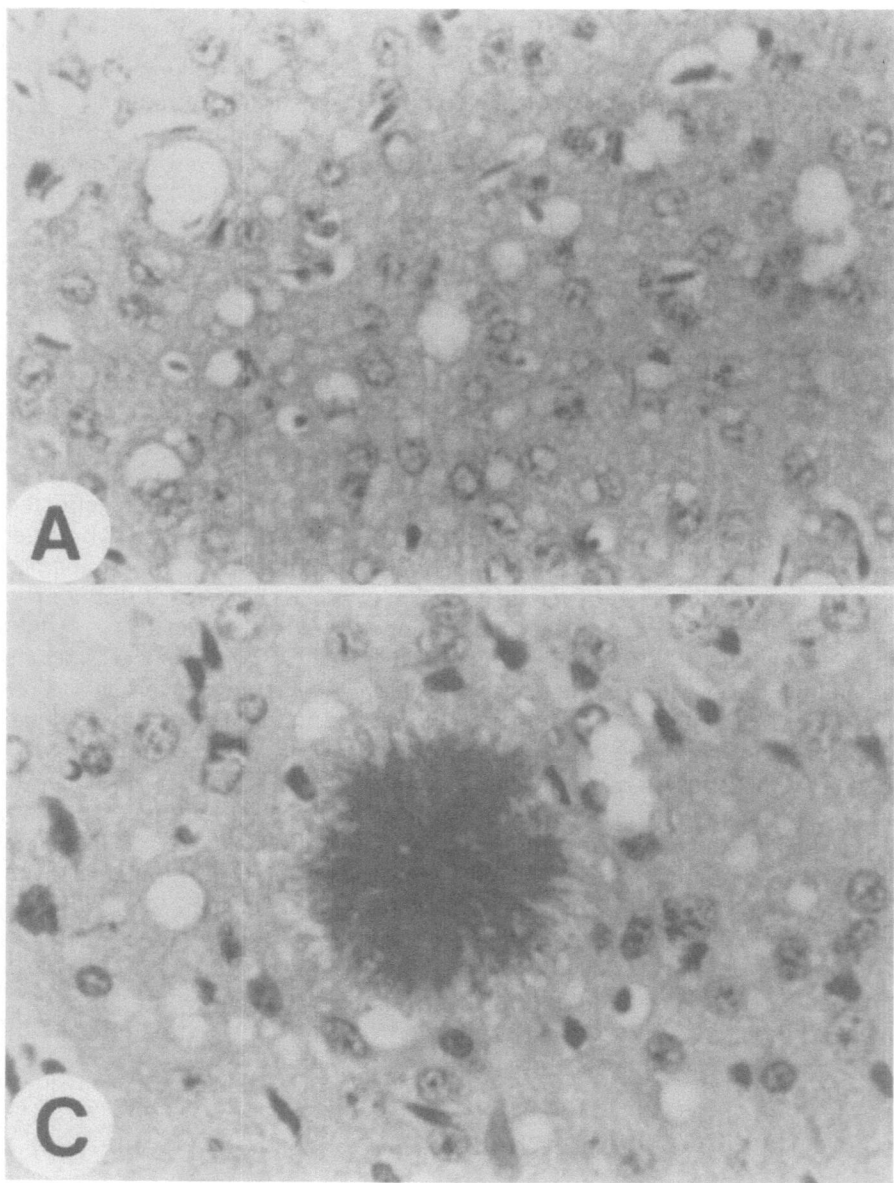

Fig. 3A–D. Neuropathology of Tg(MoPrP-P101L) mice developing neurodegeneration spontaneously. **A** Vacuolation in cerebral cortex of a Swiss CD-1 mouse that exhibited signs of neurologic dysfunction at 138 days after intracerebral inoculation with ~10^6 Id$_{50}$ units of Rocky Mountain Laboratory (RML) strain scrapie prions. **B** Vacuolation in cerebral cortex of a Tg(MoPrP-P101L) mouse that exhibited signs of neurologic dysfunction at 252 days of age. **C** Kuru-type PrP amyloid plaque stained with periodic acid-Schiff in the caudate nucleus of a Tg(MoPrP-P101L) mouse that exhibited signs of neurologic dysfunction. **D** PrP amyloid plaques stained with α-PrP antiserum (R073) in the caudate nucleus of a Tg(MoPrP-P101L) mouse that exhibited signs of neurologic dysfunction. Bar in **A** and **B** = 50 μm; bar in **C** and **D** = 25 μm. (From Prusiner 1993)

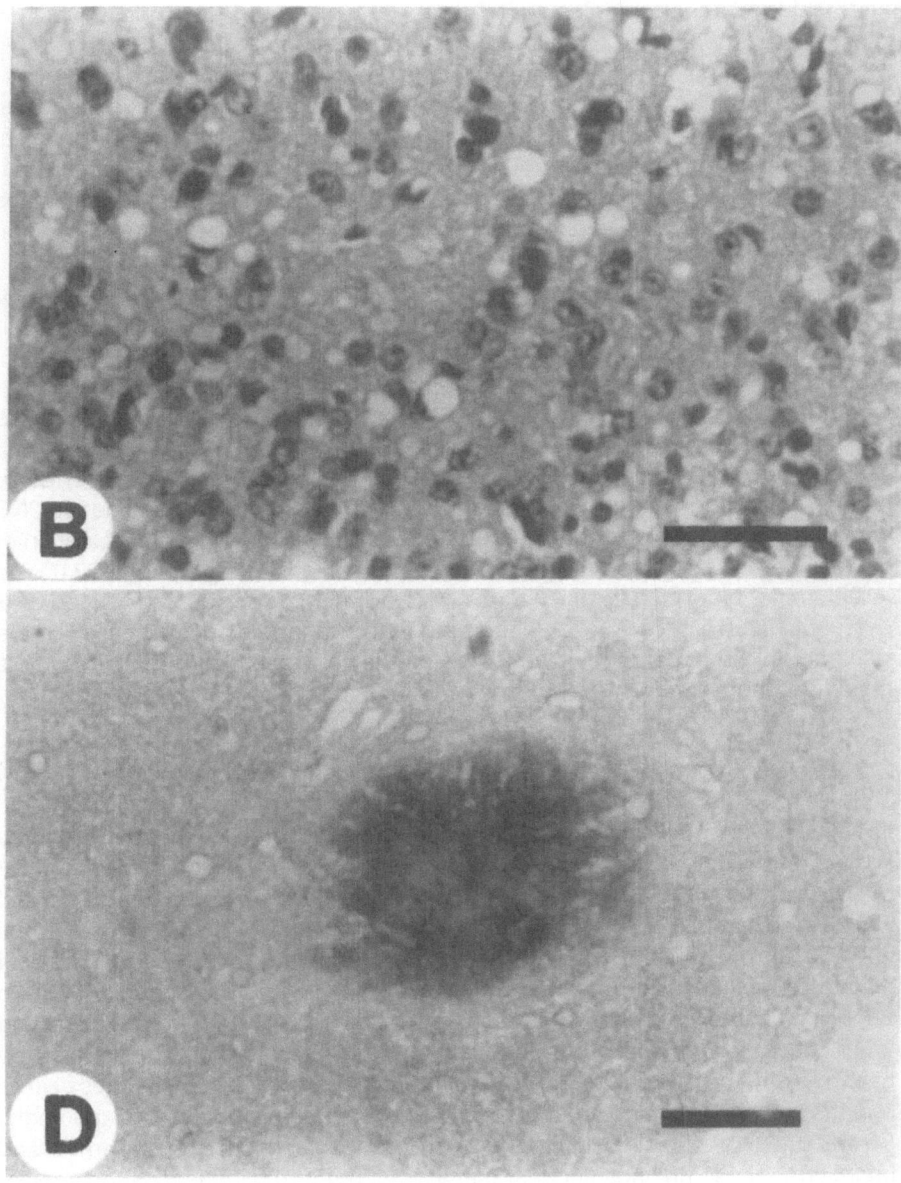

Fig. 3B,D

i.e. outside the central region of PrP. Like the binding of PrP^C to PrP^{Sc} which is most efficient when the two isoforms have the same sequence (PRUSINER et al. 1990), the binding of PrP^C to protein X seems to exhibit the highest affinity when these two proteins are from the same species. Although the level of MoPrP^C is only 10–20% of the transgene product HuPrP^C in the brains of the Tg(HuPrP) mice, it prevented the conversion of HuPrP^C into PrP^{Sc}. These findings suggest that MoPrP^C binds to Mo protein X with a considerably higher affinity than does HuPrP^C, which provides an explanation for why MoPrP^C inhibits the transmission of Hu prions in Tg(HuPrP) mice.

Since truncation of the N-terminus of recombinant PrP expressed in cultured cells still permitted the formation of PrP^{Sc}-like molecules (ROGERS et al. 1993), it seems likely that the site at which PrP^C binds to another protein is at the C-terminal end of PrP^C. Mature HuPrP differs from MoPrP at only 5 positions at the C-terminus which lie between residues 215 and 230, some of which are likely to form the protein X binding site for PrP^C (TELLING et al. In press).

The proposed model for prion propagation involving protein X is supported by studies on an inherited form of prion disease modeled in mice. Spontaneous CNS disease was found in uninoculated mice expressing the P102L point mutation of GSS when this substitution was introduced into MoPrP (HSIAO et al. 1994). The P102L mutation expressed in chimeric MHu2MPrP but not HuPrP produced CNS dysfunction in Tg mice. These findings argue that inherited prion disease like the transmissible disorder requires protein X for the conversion of mutant PrP^C into a pathologic isoform.

2.2 Protein Y and the Neuropathology of Prion Disease

Four lines of congenic mice were produced by crossing the PrP gene of the ILn/J mouse onto C57BL. The four lines of congenic mice are designated: B6.1-4 for B6.1-*B2m^a*, B6.1-1 for B6.1-*Prnp^b*, B6.1-2 for B61-*Il-1ad Prnp^b*. B6-3 for B6.1-*B2m^a Prnp^b* (CARLSON et al. 1993). Neuropathologic examination of B6.1-1, B6.1-2, 1/LnJ and VM/Dk mice inoculated with 87 V prions showed numerous PrP amyloid plaques in accord with an earlier report on VM/Dk mice (BRUCE et al. 1976). In B6.1-1 mice intense spongiform degeneration, gliosis, and PrP immunostaining were found in the ventral posterior lateral (VPL) nucleus of the thalamus, the habenula and the raphe nuclei of the brainstem (CARLSON et al. 1994). These same regions showed intense immunoreactivity for PrP^{Sc} on histoblast. Unexpectedly, B6.1-2 and ILn/J mice exhibited only mild vacuolation of the thalamus and brainstem. These findings suggest that a locus near *Prnp* influences the deposition of PrP^{Sc}, and thus vacuolation, in the thalamus, the habenula and raphe nuclei. We have provisionally designated the product of this gene protein Y.

Identification of the gene that encodes protein Y that is distinct from but near *Prnp* will be important. The gene Y product appears to control, at least in part, neuronal vacuolation and presumably PrP^{Sc} deposition in mice inoculated with scrapie prions. Isolation of protein Y should be helpful in dissecting the molecular events that feature in the pathogenesis of the prion diseases.

3 Prion Diversity

3.1 Prion Strains and Variations in Patterns of Disease

For many years, studies of experimental scrapie were performed exclusively with sheep and goats. The disease was first transmitted by intraocular inoculation (CUILLÉ and CHELLE 1939) and later by intracerebral, oral, subcutaneous, intramuscular and intravenous injections of brain extracts from sheep developing scrapie. Incubation periods of 1–3 years were common and often many of the inoculated animals failed to develop disease (DICKINSON and STAMP 1969; HADLOW et al. 1980, 1982). Different breeds of sheep exhibited markedly different susceptibilities to scrapie prions inoculated subcutaneously, suggesting that the genetic background might influence host permissiveness (GORDON 1966).

The diversity of scrapie prions was first appreciated in goats inoculated with "hyper" and "drowsy" isolates (PATTISON and MILLSON 1961). Subsequently, studies in mice demonstrated the existence of many scrapie "strains" (BRUCE and DICKINSON 1987; DICKINSON and FRASER 1979; DICKINSON and OUTRAM 1988; KIMBERLIN et al. 1987) which continues to pose a fascinating conundrum. What is the macromolecule that carries the information required for each strain to manifest a unique set of biological properties if it is not a nucleic acid?

There is good evidence for multiple "strains" or distinct isolates of prions as defined by specific incubation times, distribution of vacuolar lesions and patterns of PrP^{Sc} accumulation (BRUCE et al. 1989; DICKINSON et al. 1968; FRASER and DICKINSON 1973; HECKER et al. 1992). The lengths of the incubation times have been used to distinguish prion strains inoculated into sheep, goats, mice and hamsters. DICKINSON and his colleagues developed a system for "strain typing" by which mice with genetically determined short and long incubation times were used in combination with the F1 cross (DICKINSON et al. 1968, 1984; DICKINSON and MEIKLE 1971). For example, C57BL mice exhibited short incubation times of ~150 days when inoculated with either the Me7 or Chandler isolates; VM mice inoculated with these same isolates had prolonged incubation times of ~300 days. The mouse gene controlling incubation times was labelled *Sinc* and long incubation times were said to be a dominant trait because of prolonged incubation times in F1 mice. Prion strains were categorized into two groups based upon their incubation times: (1) those causing disease more rapidly in "short" incubation time C57BL mice and (2) those causing disease more rapidly in "long" incubation time VM mice.

3.2 PrP Gene Dosage Controls the Length of the Scrapie Incubation Time

More than a decade of study was required to unravel the mechanism responsible for the "dominance" of long incubation times; not unexpectedly, long incubation times were found not to be dominant traits. Instead, the apparent dominance of long incubation times is due to a gene dosage effect (CARLSON et al. 1994).

Our own studies began with the identification of a widely available mouse strain with long incubation times. ILn/J mice inoculated with Rocky Mountain Laboratory Strain (RML) prions were found to have incubation times exceeding 200 days (KINGSBURY et al. 1983), a finding that was confirmed by others (CARP et al. 1987). Once molecular clones of the PrP gene were available, we asked whether or not the PrP genes of short and long mice segregate with incubation times. A restriction fragment length polymorphism (RFLP) of the PrP gene was used to follow the segregation of MoPrP genes (*Prnp*) from short NZW or C57BL mice with long ILn/J mice in F1 and F2 crosses. This approach permitted the demonstration of genetic linkage between a *Prnp* and a gene modulating incubation times (*Prn-i*) (CARLSON et al. 1986). Other investigators have confirmed the genetic linkage, and one group has shown that the incubation time gene *Sinc* is also linked to PrP (HUNTER et al. 1987; RACE et al. 1990). It now seems likely that the genes for PrP, *Prn-i* and *Sinc* are all congruent; the term *Sinc* is no longer used (ZIEGLER 1993). The PrP sequences of NZW with short and long scrapie incubation times, respectively, differ at codons 108 (L → F) and 189 (T → V) (WESTAWAY et al. 1987).

Although the amino acid substitutions in PrP that distinguish *Prnp^a* from *Prnp^b* mice argued for the congruency of *Prnp* and *Prn-i*, experiments with *Prnp^a* mice expressing *Prnp^b* transgenes demonstrated a "paradoxical" shortening of incubation times (WESTAWAY et al. 1991). We had predicted that these Tg mice would exhibit a prolongation of the incubation time after inoculation with RML prions based on (*Prnp^a* × *Prnp^b*) F1 mice which do exhibit long incubation times. We described those findings as "paradoxical shortening" because we and others had believed for many years that long incubation times are dominant traits (CARLSON et al. 1986; DICKINSON et al. 1968). From studies of congenic and transgenic mice expressing different numbers of the *a* and *b* alleles of *Prnp* (Table 3), we now realize that these findings were not paradoxical; indeed, they result from increased PrP gene dosage (CARLSON et al. 1994). When the RML isolate was inoculated into congenic and transgenic mice, increasing the number of copies of the *a* allele was found to be the major determinant in reducing the incubation time; however, increasing the number of copies of the *b* allele also reduced the incubation time, but not to the same extent as that seen with the *a* allele (Table 3).

The discovery that incubation times are controlled by the relative dosage of *Prnp^a* and *Prnp^b* alleles was foreshadowed by studies of Tg(SHaPrP) mice in which the length of the incubation time after inculation with SHa prions was inversely proportional to the transgene product, SHaPrP^C (PRUSINER et al. 1990). Not only does the PrP gene dose determine the length of the incubation time, but also the passage history of the inoculum, particularly in *Prnp^b* mice (Table 4). The PrP^Sc allotype in the inoculum produced the shortest incubation times when it was the same as that of PrP^C in the host (CARLSON et al. 1989). The term "allotype" is used to describe allelic variants of PrP. To address the issue of whether gene products other than PrP might be responsible for these findings, we inoculated B6

Table 3. MoPrP-A expression is a major determinant of incubation times in mice inoculated with the RML strain scrapie prions

Mice	Prnp genotype (copies)	Prnp transgenes (copies)	Alleles a	b	Incubation time[a] (days ± SEM)	n
Prn-p[0/0]	0/0		0	0	>600	4
Prn-p[+/0]	a/0		1	0	426 ± 18	9[a]
B6.1-1	b/b		0	2	360 ± 16	7[b]
B6.1-2	b/b		0	2	379 ± 8	10[b]
B6.1-3	b/b		0	2	404 ± 10	20
(B6 × B6.1-1)F1	a/b		1	1	268 ± 4	7
B6.1-1 × Tg(MoPrP-B[0/0])15	a/b		1	1	255 ± 7	11[c]
B6.1-1 × Tg(MoPrP-B[0/0])15	a/b		1	1	274 ± 3	9[d]
B6.1-1 × Tg(MoPrP-B[+/0])15	a/b	bbb/0	1	4	166 ± 2	11[c]
B6.1-1 × Tg(MoPrP-B[+/0])15	a/b	bbb/0	1	4	162 ± 3	8[d]
C57BL/6J (B6)	a/a		2	0	143 ± 4	8
B6.1-4	a/a		2	0	144 ± 5	8
non-Tg(MoPrP-B[0/0])15	a/a		2	0	130 ± 3	10
Tg(MoPrP-B[+/0])15	a/a	bbb/0	2	3	115 ± 2	18
Tg(MoPrP-B[+/+])15	a/a	bbb/bbb	2	6	111 ± 5	5
Tg(MoPrP-B[+/0])94	a/a	>30b	2	>30	75 ± 2	15[e]
Tg(MoPrP-A[+/0])B4053	a/a	>30a	>30	0	50 ±	16

RML, Rocky Mountain Laboratory.
[a] Data from PRUSINER et al. (1993b).
[b] Data from CARLSON et al. (1993).
[c] The homozygous Tg(MoPrP-B[+/+])15 mice were maintained as a distinct subline selected for transgene homozygosity two generations removed from the (B6 x LT/Sv)F2 founder.
Hemizygous Tg(MoPrP-B[+/0])15 mice were produced by crossing the Tg(MoPrP-B[+/+])15 line with B6 mice.
[d] Tg(MoPrP-B[+/0])15 mice were maintained by repeated backcrossing to B6 mice.
[e] Data from WESTAWAY et al. (1991).

Table 4. Mismatching of PrP allotypes between PrP[Sc] in the inoculum and PrP[C] in the inoculated host extends prion incubation times in congenic mice

Mice	Host Genotype	Donor	Inoculum Genotype	Donor Incubation Time	(n)
C57BL/6J (B6)	a/a	CD-1	a/a	143 ± 4	8
B6.1-4	a/a	NZW	a/a	144 ± 5	8
B6.1-4	a/a	I/Ln	b/b	150 ± 6	6
B6.1-2	b/b	CD-1	a/a	360 ± 16	8
B6.1-2	b/b	NZW	a/a	404 ± 4	20
B6.1-2	b/b	I/Ln	b/b	237 ± 8	17
I/LnJ[a]	b/b	CD-1	a/a	314 ± 13	11
I/LnJ	b/b	NZW	a/a	283 ± 21	8
I/LnJ	b/b	I/Ln	b/b	193 ± 6	16

[a]I/LnJ results previously reported (CARLSON et al. 1994).

and B6.1-4 mice carrying *Prnp[a/a]* as well as I/Ln, and B6.1-2 mice (CARLSON et al. 1993, 1994), with RML prions passaged in mice homozygous for either the *a* or *b* allele of *Prnp* (Table 4). CD-1 and NZW/LacJ mice produce prions containing PrP[Sc]-A encoded by *Prnp[a]* while I/LnJ mice produce PrP[Sc]-B prions. The incubation times in the congenic mice reflected the PrP allotype, rather than other factors acquired during prion passage. The effect of the allotype barrier was small when measured in $Prnp^{a/a}$ mice but was clearly demonstrable in $Prnp^{b/b}$ mice. B6.1-2 congenic mice inoculated with prions from I/Ln mice had an incubation time of 237 ± 8 days compared to times of 360 ± 16 days and 404 ± 4 days for mice inoculated with prions passaged in CD-1 and NZW mice, respectively. Thus, previous passage of prions in $Prnp^b$ mice shortened the incubation time by ~40% when assayed in $Prnp^b$ mice, compared to those inoculated with prions passaged in $Prnp^a$ mice (CARLSON et al. 1989).

3.3 Overdominance

The phenomenon of "overdominance" in which incubation times in F1 hybrids are longer than those of either parent (DICKINSON and MEIKLE 1969) contributed to the confusion surrounding control of scrapie incubation times. When the 22A scrapie isolate was inoculated into B6, B6.1-1 and (B6 × B6.1-1) F1, overdominance was observed: the scrapie incubation time in B6 mice was 405 ± 2 days, in B6.1 mice 194 ± 10 days and in (B6 × B6.1-1)F1 mice 508 ± 14 days (Table 5). Shorter incubation times were observed in Tg(MoPrP-B)15 mice which were either homozygous or hemizygous for the $Prnp^b$ transgene. Hemizygous Tg(MoPrP-B$^{+/0}$) 15 mice exhibited a scrapie incubation time of 395 ± 12 days while the homozygous mice had an incubation time of 286 ± 15 days.

As with the results with the RML isolate (Table 3), the findings with the 22A isolate can be explained on the basis of gene dosage; however, the relative effects of the a and b alleles differ in two respects. First, the b allele is the major determinant of the scrapie incubation time with the 22A isolate, not the a allele. Second, increasing the number of copies of the a allele does not diminish the incubation but prolongs it: the a allele is inhibitory with the 22A isolate (Table 5). With the 87V prion isolate the inhibitory effect of the $Prnp^a$ allele is even more pronounced since only a few $Prnp^a$ and ($Prnp^a$ × $Prnp^b$)F1 mice develop scrapie after >600 days postinoculation (CARLSON et al. 1994).

Table 5. MoPrP^C-A inhibits the synthesis of 22A scrapie prions

Mice	Prnp genotype	Prnp transgenes (copies)	Alleles (copies)		Incubation time	n
			a	b	(days ± SEM)	
B6.1-1	b/b		0	2	194 ± 10	7
(B6 × B6.1-1)F1	a/b		1	1	508 ± 14	7
C57BL/6J (B6)	a/a		2	0	405 ± 2	8
non-Tg(MoPrP-B$^{0/0}$)15	a/a		2	0	378 ± 8	3[a]
Tg(MoPrP-B$^{+/0}$)15	a/a	bbb/0	2	3	318 ± 14	15[a]
Tg(MoPrP-B$^{+/0}$)15	a/a	bbb/0	2	3	395 ± 12	6[b]
Tg(MoPrP-B$^{+/+}$)15	a/a	bbb/bbb	2	6	266 ± 1	6[a]
Tg(MoPrP-B$^{+/+}$)15	a/a	bbb/bbb	2	6	286 ± 15	5[b]

[a] The homozygous Tg(MoPrP-B$^{+/+}$)15 mice were maintained as a distinct subline selected for transgene homozygosity two generations removed from the (B6 × LT/Sv) F2 founder
Hemizygous Tg(MoPrP-B$^{+/0}$)15 mice were produced by crossing the Tg(MoPrP-B$^{+/+}$)15 line with B6 mice.
[b] Tg(MoPrP-B$^{+/0}$)15 mice were maintained by repeated backcrossing to B6 mice.

The most interesting feature of the incubation time profile for 22A is the overdominance of the a allele of *Prnp* in prolonging incubation period. On the basis of overdominance, DICKINSON and OUTRAM (1979) put forth the replication site hypothesis, postulating that dimers of the *Sinc* gene product feature in the replication of the scrapie agent. The results in Table 5 are compatible with the interpretation that the target for PrP^{Sc} may be a PrP^C dimer or multimer. The assumptions under this model are that PrP^C-B dimers are more readily converted to PrP^{Sc} than are PrP^C-A dimers and that PrP^c-A: PrP^c-B heterodimers are even more resistant to conversion to PrP^{Sc} than PrP^C -A dimers. Increasing the ratio of PrP-B to PrP-A would lead to shorter incubation times by favoring the formation of PrP^C-B homodimers (Table 5). A similar mechanism may account for the relative paucity of individuals heterozygous for the Met/Val polymorphism at codon 129 of the human PrP gene in spontaneous and iatrogenic CJD (PALMER et al. 1991). Alternatively, PrP^C-PrP^{Sc} interaction can be broken down to two distinct aspects: binding affinity and efficacy of conversion to PrP^{Sc}. If PrP-A has a higher affinity for 22A PrP^{Sc} than does PrP^C-B, but is inefficiently converted to PrP^{Sc}, the exceptionally long incubation time of *Prnp$^{a/b}$* heterozygotes might reflect reduction in the supply of 22A prions available for interaction with the PrP^C-B product of the single *Prnp^b* allele. Additionally, PrP^C-A may inhibit the interaction of 22A PrP^{Sc} with PrP^C-B leading to prolongation of the incubation time. This interpretation is supported by prolonged incubation times in Tg(SHaPrP) mice inoculated with mouse prions in which SHaPrP^C is thought to inhibit the binding of MoPrP^{Sc} to the substrate MoPrP^C (PRUSINER et al. 1990).

3.4 Patterns of PrP^{Sc} Deposition

Besides measurements of the length of the incubation time, profiles of spongiform degeneration have also been used to characterize different prion strains (FRASER 1979; FRASER and DICKINSON 1973). With the development of a new procedure for in situ detection of PrP^{Sc}, designated histoblotting (TARABOULOS et al. 1992), it became possible to localize and quantify PrP^{Sc} and to determine whether or not strains produce different, reproducible patterns of PrP^{Sc} accumulation (Fig. 4) (DEARMOND et al. 1993; HECKER et al. 1992).

Histoblotting overcame two obstacles that plagued PrP^{Sc} detection in brain by standard immunohistochemical techniques: the presence of PrP^C and weak antigenicity of PrP^{Sc} (DEARMOND et al. 1987). The histoblot is made by pressing 10 μm thick cryostat sections of fresh frozen brain tissue to nitrocellulose paper. To localize protease-resistant PrP^{Sc} in brain, the histoblot is digested with proteinase K to eliminate PrP^C, followed by denaturation of the undigested PrP^{Sc} to enhance binding of PrP antibodies. Immunohistochemical staining yields a far more intense, specific and reproducible PrP signal than can be achieved by immunohistochemistry on standard tissue sections. The intensity of immunostaining correlates well with neurochemical estimates of PrP^{Sc} concentration in homogenates of dissected brain regions. PrP^C can be localized in histoblots of normal brains by eliminating the proteinase K digestion step.

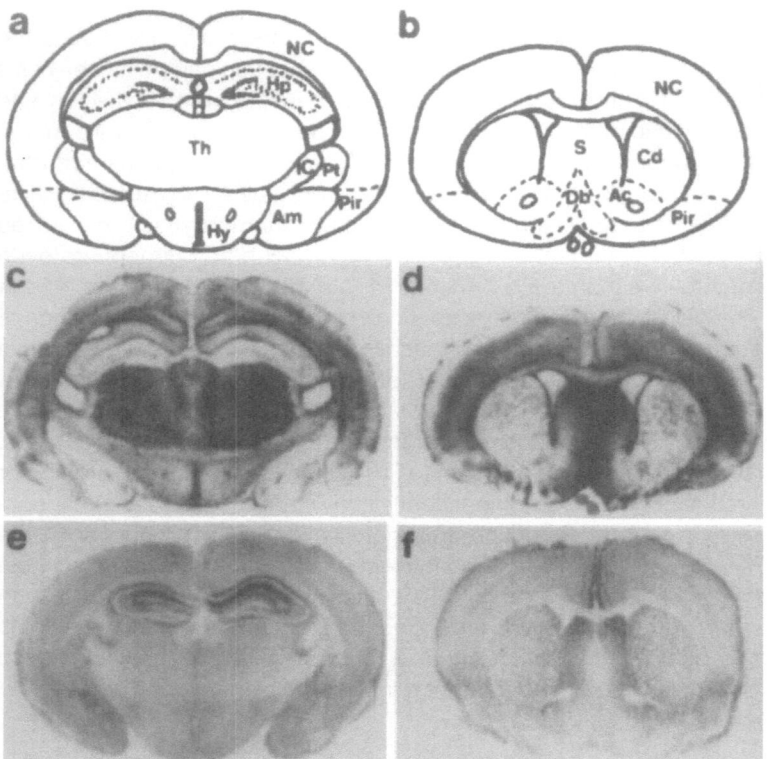

Fig. 4a–f. Histoblots of Syrian hamster brain immunostained for PrP^C or PrP^Sc. Coronal sections through the hippocampus-thalamus (**a, c** and **e**) and the septum-caudate (**b, d** and **f**). Brain sections of a Syrian hamster clinically ill after inoculation with Sc237 prions (**c** and **d**) and an uninfected control animal (**e** and **f**). Immunostaining for PrP^Sc shown in **c** and **d**; for PrP^C in **e** and **f**. *Ac*, nucleus accumbens; *Am*, amygdala; *Cd*, caudate nucleus; *Db*, diagonal band of Broca; *H*, habenula; *Hp*, hippocampus; *Hy* hypothalamus; *IC*, internal capsule; *NC*, neocortex. (From TARABOULOS et al. 1992)

Comparisons of PrP^Sc accumulation of histoblots with histologic sections showed that PrP^Sc deposition preceded vacuolation and only those regions with PrP^Sc underwent degeneration. Microdissection of individual brain regions confirmed the conclusions of the histoblot studies: those regions with high levels of PrP 27–30 had intense vacuolation (CASACCIA-BONNEFIL et al. 1993). Thus, we concluded that the deposition of PrP^Sc is responsible for the neuropathologic changes found in the prion diseases.

While studies with both mice and Syrian hamsters established that each isolate has a specific signature, as defined by a specific pattern of PrP^Sc accumulation in the brain (CARLSON et al. 1994; DEARMOND et al. 1993; HECKER et al. 1992), comparisons must be done on an isogenic background (HSIAO et al. 1995; SCOTT et al. 1993). Variations in the patterns of PrP^Sc accumulation were found to be equally as great as those seen between two strains when a single strain is inoculated in mice expressing different PrP genes. Based upon the initial studies which were performed in animals of a single genotye, we suggested that PrP^{Sc} synthesis occurs in specific populations of cells for a given distinct prion isolate.

3.5 Are Prion Strains Different PrP^{Sc} Conformers?

Explaining the problem of multiple distinct prion isolates might be accommodated by multiple PrP^{Sc} conformers that act as templates for the folding of de novo synthesized PrP^{Sc} molecules during prion "replication" (Fig. 5). Although it is clear that passage history can be responsible for the prolongation of incubation time when prions are passed between mice expressing different PrP allotypes (CARLSON et al. 1989) or between species (PRUSINER et al. 1990), many scrapie strains show distinct incubation times in the same inbred host (BRUCE et al. 1991).

In recent studies we inoculated three strains of prions into congenic and Tg mice harboring various numbers of the a and b alleles of Prn-P (CARLSON et al. 1994). The number of Prnp^a genes was the major determinant of incubation times in mice inoculated with the RML prion isolate and was inversely related to the length of the incubation time (Table 3). In contrast, the Prnp^a allele prevented scrapie in mice inoculated with 87V prions. Prnp^b genes were permissive for 87V prions and shortened incubation times in most mice inoculated with 22A prions (Table 5). Experiments with the 87V isolate suggest that a genetic locus encoding protein Y, distinct from Prnp, controls the deposition of PrP^{Sc} and the attendant neuropathology. While each prion isolate produced distinguishable patterns of PrP^{Sc} accumulation in brain, a comparison of these patterns showed that those patterns found with RML and 22A prions in congenic Prnp^b mice were more similar than those with RML prions in Prnp^a and Prnp^b congenic mice. Thus, both the PrP genotype and prion isolate modify the distribution of PrP^{Sc} and the length of the incubation time. These findings suggest that prion strain specified properties result from different affinities of PrP^{Sc} in the inocula for PrP^{C}-A and PrP^{C}-B allotypes encoded by the host.

Although the proposal for multiple PrP^{Sc} conformers is rather unorthodox, we already know that PrP can assume at least two profoundly different conformations: PrP^{C} and PrP^{Sc} (PAN et al. 1993). Of note, two different isolates from mink dying of transmissible mink encephalopathy exhibit different sensitivities of PrP^{Sc} to proteolytic digestion, supporting the suggestion that isolate-specific information might be carried by PrP^{Sc} (BESSEN and MARSH 1992a,b; MARSH et al. 1991). How many confor-mations PrP^{Sc} can assume is unknown. The molecular weight of a PrP^{Sc} homodimer is consistent with the ionizing radiation target size of $55,000 \pm 9000$ daltons, as determined for infectious prion particles independent of their polymeric form (BELLINGER-KAWAHARA et al. 1988). If prions are oligomers of PrP^{Sc}, which seems likely, then this offers another level of complexity which in turn generates additional diversity.

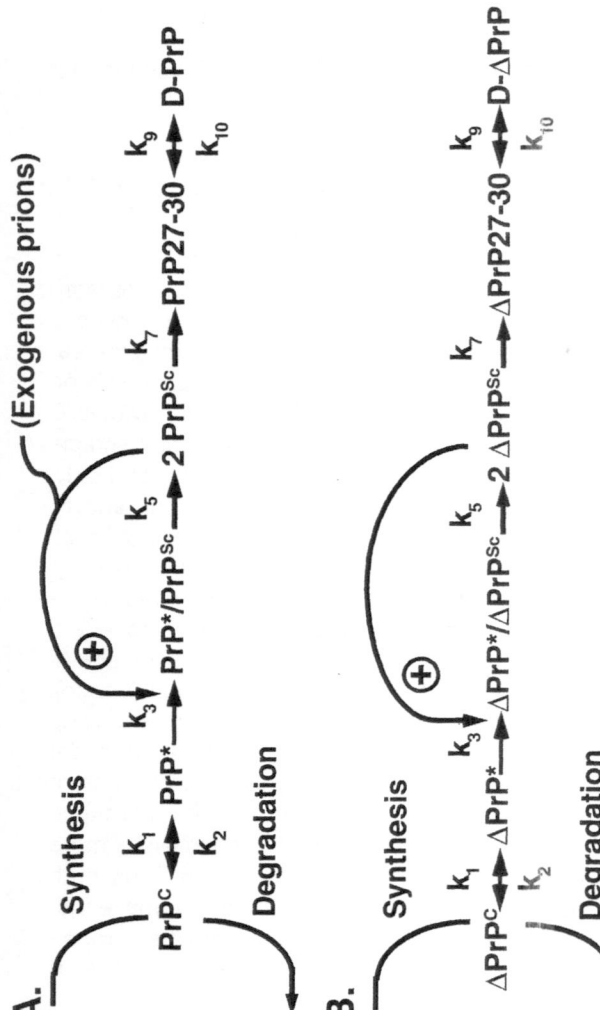

Fig. 5A,B. Models for the replication of prions. **A** Proposed scheme for the replication of prions in sporadic and infectious prion diseases. wtPrP^C is synthesized and degraded as part of the normal metabolism of many cells. Stochastic fluctuations in the structure of PrP^C can create (k_1) a rare, partially unfolded, monomeric structure, PrP*, that is an intermediate in the formation of PrP^{Sc}, but can revert (k_2) to PrP^C or be degraded prior to its conversion (k_3) into PrP^{Sc}. Normally, the concentration of PrP* is small and PrP^{Sc} formation is insignificant. In infectious prion diseases, exogenous prions enter the cell and stimulate conversion of PrP* into PrP^{Sc}. In the absence of exogenous prions, the concentration of PrP^{Sc} may eventually reach a threshold level in sporadic prion diseases after which a positive feedback loop would stimulate the formation of PrP^{Sc}. Limited proteolysis of the NH$_2$-terminal of PrP^{Sc} produces (k_7) PrP27-30 which can also be generated in scrapie-infected cells from a recombinant vector encoding PrP truncated at the NH$_2$-terminal (ROGERS et al. 1993). Denaturation (k_9) of PrP^{Sc} or PrP27-30 renders these molecules protease sensitive and abolishes scrapie infectivity; attempts to renature (k_{10}) these PrP^{Sc} or PrP27-30 have been unsuccessful to date (PRUSINER et al. 1983, 1993c). **B** Scheme for the replication of prions in genetic prion diseases. Mutant (Δ) PrP^C is synthesized and degraded as part of the normal metabolism of many cells. Stochastic fluctuations in the structure of ΔPrP^C are increased compared to wtPrP^C, which creates (k_1) a partially unfolded, monomeric structure, PrP*, that is an intermediate in the formation of ΔPrP^{Sc}, but can revert (k_2) to ΔPrP^C or be degraded prior to its conversion (k_3) into ΔPrP^{Sc}. Limited proteolysis of the NH$_2$-terminal of ΔPrP^{Sc} produces (k_7) ΔPrP27-30 which in some cases may be less protease resistant than wtPrP27-30 (MONARI et al. 1994; PRUSINER and HSIAO 1994). (Adapted from COHEN et al. 1994)

4 Concluding Remarks

4.1 Prions are not Viruses

The study of prions has taken several unexpected directions over the past few years. The discovery that prion diseases in humans are uniquely both genetic and infectious has greatly strengthened and extended the prion concept. To date, 18 different mutations in the human PrP gene all resulting in nonconservative substitutions have been found to either be linked genetically to or segregate with the inherited prion diseases. Yet, the transmissible prion particle is composed largely, if not entirely, of an abnormal isoform of the prion protein designated PrP^{Sc} (PRUSINER 1991). These findings argue that prion diseases should be considered pseudoinfections, since the particles transmitting disease appear to be devoid of a foreign nucleic acid and thus differ from all known microorganisms as well as viruses and viroids. Because much information, especially about scrapie of rodents, has been derived using experimental protocols adapted from virology, we continue to use terms such as infection, incubation period, transmissibility and endpoint titration in studies of prion diseases.

4.2 Do Prions Exist in Lower Organisms?

In *S. cerevisiae*, ure2 and [URA3] mutants were described that can grow on ureidosuccinate under conditions of nitrogen repression such as glutamic acid and ammonia (LACROUTE 1971). Mutants of URE2 exhibit Mendelian inheritance, whereas [URE3] is cytoplasmically inherited (WICKNER 1995). The [URE3] phenotype can be induced by UV irradiation and by overexpression of ure2p, the gene product of ure2; deletion of ure2 abolishes [URE3]. The function of ure2p is unknown but it has substantial homology with glutathione-S-transferase; attempts to demonstrate this enzymic activity with purified ure2p have not been successful (COSCHIGANO and MAGASANIK 1991). Whether the [URE3] protein is a posttranslationally modified form of ure2p which acts upon unmodified ure2p to produce more of itself remains to be established.

Another possible yeast prion is the [PSI] phenotype (WICKNER 1995). [PSI] is a non-Mendelian inherited trait that can be induced by expression of the PNM2 gene (COX et al. 1988). Both [PSI] and [URE3] can be cured by exposure of the yeast to 3 mM GdnHCl. The mechanism responsible for abolishing [PSI] and [URE3] with a low concentration of GdnHCl is unknown. In the filamentous fungus *Podospora anserina*, the het-s locus controls the vegetative incompatibility; conversion from the S^s to the s state seems to be a posttranslational, autocatalytic process (DELEU et al. 1993).

If any of the above cited examples can be shown to function in a manner similar to prions in animals, then many new, more rapid and economical approaches to prion diseases should be forthcoming.

4.3 Common Neurodegenerative Diseases

The knowledge accrued from the study of prion diseases may provide an effective strategy for defining the etiologies and dissecting the molecular pathogenesis of the more common neurodegenerative disorders such as Alzheimer's disease, Parkinson's disease and amyotrophic lateral sclerosis (ALS). Advances in the molecular genetics of Alzheimer's disease and ALS suggest that, like the prion diseases, an important subset is caused by mutations that result in nonconservative amino acid substitutions in proteins expressed in the CNS (GOATE et al. 1991; LEVY et al. 1990; MULLAN et al. 1992; ROSEN et al. 1993; SCHELLENBERG et al. 1992; ST. GEORGE-HYSLOP et al. 1992; VAN BROECKHOVEN et al. 1990, 1992). Since people at risk for inherited prion diseases can now be identified decades before neurologic dysfunction is evident, the development of an effective therapy is imperative.

4.4 Future Studies

Transgenic mice expressing foreign or mutant PrP genes now permit virtually all facets of prion diseases to be studied and have created a framework for future investigations. Furthermore, the structure and organization of the PrP gene suggested that PrP^{Sc} is derived from PrP^C or a precursor by a posttranslational process. Studies with scrapie-infected cultured cells have provided much evidence that the conversion of PrP^C to PrP^{Sc} is a posttranslational process that probably occurs within a subcellular compartment bounded by cholesterol-rich membranes. The molecular mechanism of PrP^{Sc} formation remains to be elucidated, but chemical and physical studies have shown that the conformations of PrP^C and PrP^{Sc} are profoundly different.

The study of prion biology and diseases seems to be a new and emerging area of biomedical investigation. While prion biology has its roots in virology, neurology and neuropathology, its relationships to the disciplines of molecular and cell biology and protein chemistry have become evident only recently. Certainly, the possibility that learning how prions multiply and cause disease will open up new vistas in biochemistry and genetics seems likely.

Acknowledgments. I thank M. Baldwin, D. Borchelt, G. Carlson, F. Cohen, C. Cooper, S. DeArmond, R. Fletterick, M. Gasset, R. Gabizon, D. Groth, R. Koehler, L. Hood, K. Hsiao, Z. Huang, V. Lingappa, K-M. Pan, D. Riesner, M. Scott, A. Serban, N. Stahl, A. Taraboulos, M. Torchia, and D. Westaway for their help in these studies. Special thanks to C. Rogers who assembled this manuscript. Supported by grants from the National Institutes of Health (NS14069, AG08967, AGO2132, NS22786 and AG10770) and the American Health Assistance Foundation, as well as by gifts from Sherman Fairchild Foundation, Bernard Osher Foundation and National Medical Enterprises.

References

Alper T, Haig DA, Clarke MC (1966) The exceptionally small size of the scrapie agent. Biochem Biophys Res Commun 22: 278–284
Alper T, Cramp WA, Haig DA, Clarke MC (1967) Does the agent of scrapie replicate without nucleic acid? Nature 214: 764–766

Alpers M (1987) Epidemiology and clinical aspects of kuru. In: Prusiner SB, McKinley MP (eds) Prions - novel infectious pathogens causing scrapie and Creutzfeldt-Jakob disease. Academic, Orlando, pp 451–465

Barry RA, Prusiner SB (1986) Monoclonal antibodies to the cellular and scrapie prion proteins. J Infect Dis 154: 518–521

Bellinger-Kawahara CG, Kempner E, Groth DF, Gabizon R, Prusiner SB (1988) Scrapie prion liposomes and rods exhibit target sizes of 55, 000 Da. Virology 164: 537–541

Bessen RA, Marsh RF (1992a) Biochemical and physical properties of the prion protein from two strains of the transmissible mink encephalopathy agent. J Virol 66: 2096–2101

Bessen RA, Marsh RF (1992b) Identification of two biologically distinct strains of transmissible mink encephalopathy in hamsters. J Gen Virol 73: 329–334

Bockman JM, Prusiner SB, Tateishi J, Kingsbury DT (1987) Immunoblotting of Creutzfeldt-Jakob disease prion proteins: host species-specific epitopes. Ann Neurol 21: 589–595

Borchelt DR, Scott M, Taraboulos A, Stahl N, Prusiner SB (1990) Scrapie and cellular prion proteins differ in their kinetics of synthesis and topology in cultured cells. J Cell Biol 110: 743–752

Brown P (1992) The phenotypic expression of different mutations in transmissible human spongiform encephalopathy. Rev Neurol (Paris) 148: 317–327

Brown P, Cathala F, Raubertas RF, Gajdusek DC, Castaigne P (1987) The epidemiology of Creutzfeldt-Jakob disease: conclusion of 15-year investigation in France and review of the world literature. Neurology 37: 895–904

Brown P, Goldfarb LG, Kovanen J, Haltia M, Cathala F, Sulima M, Gibbs CJ Jr, Gajdusek DC (1992) Phenotypic characteristics of familial Creutzfeldt-Jakob disease associated with the codon 178^{Asn} PRNP mutation. Ann Neurol 31: 282–285

Brown P, Gibbs CJ Jr, Rodgers-Johnson P, Asher DM, Sulima MP, Bacote A, Goldfarb LG, Gajdusek DC (1994) Human spongiform encephalopathy: the National Institutes of Health series of 300 cases of experimentally transmitted disease. Ann Neurol 35: 513–529

Bruce ME, Dickinson AG (1987) Biological evidence that the scrapie agent has an independent genome. J Gen Virol 68: 79–89

Bruce ME, Dickinson AG, Fraser H (1976) Cerebral amyloidosis in scrapie in the mouse: effect of agent strain and mouse genotype. Neuropathol Appl Neurobiol 2: 471–478

Bruce ME, McBride PA, Farquhar CF (1989) Precise targeting of the pathology of the sialoglycoprotein, PrP, and vacuolar degeneration in mouse scrapie. Neurosci Lett 102: 1–6

Bruce ME, McConnell I, Fraser H, Dickinson AG (1991) The disease characteristics of different strains of scrapie in Sinc congenic mouse lines: implications for the nature of the agent and host control of pathogenesis. J Gen Virol 72: 595–603

Büeler H, Fischer M, Lang Y, Bluethmann H, Lipp H-P, DeArmond SJ, Prusiner SB, Aguet M, Weissmann C (1992) Normal developement and behaviour of mice lacking the neuronal cell-surface PrP protein. Nature 356: 577–582

Büeler H, Aguzzi A, Sailer A, Greiner R-A, Autenried P, Aguet M, Weissmann C (1993) Mice devoid of PrP are resistant to scrapie. Cell 73: 1339–1347

Carlson GA, Kingsbury DT, Goodman PA, Coleman S, Marshall ST, DeArmond SJ, Westaway D, Prusiner SB (1986) Linkage of prion protein and scrapie incubation time genes. Cell 46: 503–511

Carlson GA, Westaway D, DeArmond SJ, Peterson-Torchia M, Prusiner SB (1989) Primary structure of prion protein may modify scrapie isolate properties. Proc Natl Acad Sci USA 86: 7475–7479

Carlson GA, Ebeling C, Torchia M, Westaway D, Prusiner SB (1993) Delimiting the location of the scrapie prion incubation time gene on chromosome 2 of the mouse. Genetics 133: 979–988

Carlson GA, Ebeling C, Yang S-L, Telling G, Torchia M, Groth D, Westaway D, DeArmond SJ, Prusiner SB (1994) Prion isolate specified allotypic interactions between the cellular and scrapie prion proteins in congenic and transgenic mice. Proc Natl Acad Sci USA 91: 5690–5694

Carp RI, Moretz RC, Natelli M, Dickinson AG (1987) Genetic control of scrapie: incubation period and plaque formation in mice. J Gen Virol 68: 401–407

Casaccia-Bonnefil P, Kascsak RJ, Fersko R, Callahan S, Carp RI (1993) Brain regional distribution of prion protein PrP27-30 in mice stereotaxically microinjected with different strains of scrapie. J Infect Dis 167: 7–12

Cohen FE, Pan K-M, Huang Z, Baldwin M, Fletterick RJ, Prusiner SB (1994) Structural clues to prion replication. Science 264: 530–531

Coschigano PW, Magasanik B (1991) The URE2 gene product of Saccharomyces cerevisiae plays an important role in the cellular response to the nitrogen source and has homology to glutathione S-transferases. Mol Cell Biol 11: 822–832

Cousens SN, Harries-Jones R, Knight R, Will RG, Smith PG, Matthews WB (1990) Geographical distribution of cases of Creutzfeldt-Jakob disease in England and Wales 1970–84. J Neurol Neurosurg Psychiatry 53: 459–465

Cox BS, Tuite MF, McLaughlin CS (1988) The psi factor of yeast: a problem in inheritance. Yeast 4: 159–178

Cuillé J, Chelle PL (1939) Experimental transmission of trembling to the goat. C R Acad Sci 208: 1058–1060

DeArmond SJ, Mobley WC, DeMott DL, Barry RA, Beckstead JH, Prusiner SB (1987) Changes in the localization of brain prion proteins during scrapie infection. Neurology 37: 1271–1280

DeArmond SJ, Yang S-L, Lee A, Bowler R, Taraboulos A, Groth D, Prusiner SB (1993) Three scrapie prion isolates exhibit different accumulation patterns of the prion protein scrapie isoform. Proc Natl Acad Sci USA 90: 6449–6453

Deleu C, Clavé C, Bégueret J (1993) A single amino acid difference is sufficient to elicit vegetative incompatibility in the fungus Podospora anserina. Genetics 135: 45–52

Dickinson AG, Fraser H (1979) An assessment of the genetics of scrapie in sheep and mice. In: Prusiner SB, Hadlow WJ (eds) Slow transmissible diseases of the nervous system, vol 1. Academic, New York, pp 367–386

Dickinson AG, Meikle VM (1969) A comparison of some biological characteristics of the mouse-passaged scrapie agents, 22A and ME7. Genet Res 13: 213–225

Dickinson AG, Meikle VMH (1971) Host-genotype and agent effects in scrapie incubation: change in allelic interaction with different strains of agent. Mol Gen Genet 112: 73–79

Dickinson AG, Outram GW (1979) The scrapie replication-site hypothesis and its implications for pathogenesis. In: Prusiner SB, Hadlow WJ (eds) Slow transmissible diseases of the nervous system, vol 2. Academic, New York, pp 13–31

Dickinson AG, Outram GW (1988) Genetic aspects of unconventional virus infections: the basis of the virino hypothesis. Ciba Found Symp 135: 63–83

Dickinson AG, Stamp JT (1969) Experimental scrapie in Cheviot and Suffolk sheep. J Comp Pathol 79: 23–26

Dickinson AG, Young GB, Stamp JT, Renwick CC (1965) An analysis of natural scrapie in Suffolk sheep. Heredity 20: 485–503

Dickinson AG, Meikle VMH, Fraser H (1968) Identification of a gene which controls the incubation period of some strains of scrapie agent in mice. J Comp Pathol 78: 293–299

Dickinson AG, Bruce ME, Outram GW, Kimberlin RH (1984) Scrapie strain differences: the implications of stability and mutation. In: Tateishi J (ed) Proceedings of workshop on slow transmissible diseases. Japanese Ministry of Health and Welfare, Tokyo, pp 105–118

Fraser H (1979) Neuropathology of scrapie: the precision of the lesions and their diversity. In: Prusiner SB, Hadlow WJ (eds) Slow transmissible diseases of the nervous system, vol 1. Academic, New York, pp 387–406

Fraser H, Dickinson AG (1973) Scrapie in mice. Agent-strain differences in the distribution and intensity of grey matter vacuolation. J Comp Pathol 83: 29–40

Gajdusek DC (1977) Unconventional viruses and the origin and disappearance of kuru. Science 197: 943–960

Gajdusek DC (1985) Subacute spongiform virus encephalopathies caused by unconventional viruses. In: Maramorosch K, McKelvey JJ Jr (eds) Subviral pathogens of plants and animals: viroids and prions. Academic, Orlando, pp 483–544

Gajdusek DC, Zigas V (1957) Degenerative disease of the central nervous system in New Guinea - The endemic occurrence of "kuru" in the native population. N Engl J Med 257: 974–978

Gajdusek DC, Zigas V (1959) Clinical, pathological and epidemiological study of an acute progressive degenerative disease of the central nervous system among natives of the eastern highlands of New Guinea. Am J Med 26: 442–469

Gajdusek DC, Gibbs CJ Jr, Alpers M (1966) Experimental transmission of a kuru-like syndrome to chimpanzees. Nature 209: 794–796

Goate A, Chartier-Harlin M-C, Mullan M, Brown J, Crawford F, Fidani L, Giuffra L, Haynes A, Irving N, James L, Mant R, Newton P, Rooke K, Roques P, Talbot C, Pericak-Vance M, Roses A, Williamson R, Rossor M, Owen M, Hardy J (1991) Segregation of a missense mutation in the amyloid precursor protein gene with familial Alzheimer's disease. Nature 349: 704–706

Goldmann W, Hunter N, Martin T, Dawson M, Hope J (1991) Different forms of the bovine PrP gene have five or six copies of a short, G-C-rich element within the protein-coding exon. J Gen Virol 72: 201–204

Gordon WS (1946) Advances in veterinary research. Vet Res 58: 516–520

Gordon WS (1966) Variation in susceptibility of sheep to scrapie and genetic implications. USA Department of Agriculture, Washington, pp 53–67 (Report of scrapie seminar, ARS 91–53)

Hadlow WJ, Kennedy RC, Race RE, Eklund CM (1980) Virologic and neurohistologic findings in dairy goats affected with natural scrapie. Vet Pathol 17: 187–199

Hadlow WJ, Kennedy RC, Race RE (1982) Natural infection of Suffolk sheep with scrapie virus. J Infect Dis 146: 657–664

Harries-Jones R, Knight R, Will RG, Cousens S, Smith PG, Matthews WB (1988) Creutzfeldt-Jakob disease in England and Wales, 1980–1984: a case-control study of potential risk factors. J Neurol Neurosurg Psychiatry 51: 1113–1119

Hecker R, Taraboulos A, Scott M, Pan K-M, Torchia M, Jendroska K, DeArmond SJ, Prusiner SB (1992) Replication of distinct prion isolates is region specific in brains of transgenic mice and hamsters. Genes Dev 6: 1213–1228

Hope J, Reekie LJD, Hunter N, Multhaup G, Beyreuther K, White H, Scott AC, Stack MJ, Dawson M, Wells GAH (1988) Fibrils from brains of cows with new cattle disease contain scrapie-associated protein. Nature 336: 390–392

Hsiao K, Prusiner SB (1990) Inherited human prion diseases. Neurology 40: 1820–1827

Hsiao KK, Scott M, Foster D, Groth DF, DeArmond SJ, Prusiner SB (1990) Spontaneous neuro-degeneration in transgenic mice with mutant prion protein. Science 250: 1587–1590

Hsiao KK, Baker HF, Crow TJ, Poulter M, Owen F, Terwilliger JD, Westaway D, Ott J, Prusiner SB (1989) Linkage of a prion protein missense variant to Gerstmann-Sträussler syndrome. Nature 338: 342–345

Hsiao KK, Groth D, Scott M, Yang S-L, Serban H, Rapp D, Foster D, Torchia M, DeArmond SJ, Prusiner SB (1995) Serial transmission in rodents of neurodegeneration from transgenic mice expressing mutant prion protein. Proc Natl Acad Sci USA (in press)

Hunter GD (1972) Scrapie: a prototype slow infection. J Infect Dis 125: 427–440

Hunter N, Hope J, McConnell I, Dickinson AG (1987) Linkage of the scrapie-associated fibril protein (PrP) gene and Sinc using congenic mice and restriction fragment length polymorphism analysis. J Gen Virol 68: 2711–2716

Kascsak RJ, Rubenstein R, Merz PA, Tonna-DeMasi M, Fersko R, Carp RI, Wisniewski HM, Diringer H (1987) Mouse polyclonal and monoclonal antibody to scrapie-associated fibril proteins. J Virol 61: 3688–3693

Kimberlin RH, Cole S, Walker CA (1987) Temporary and permanent modifications to a single strain of mouse scrapie on transmission to rats and hamsters. J Gen Virol 68: 1875–1881

Kingsbury DT, Kasper KC, Stites DP, Watson JD, Hogan RN, Prusiner SB (1983) Genetic control of scrapie and Creutzfeldt-Jakob disease in mice. J Immunol 131: 491–496

Kirschbaum WR (1924) Zwei eigenartige Erkrankungen des Zentralnervensystems nach Art der spastischen Pseudosklerose (Jakob). Z Ges amte Neurol Psychiatr 92: 175–220

Kretzschmar HA, Stowring LE, Westaway D, Stubblebine WH, Prusiner SB, DeArmond SJ (1986) Molecular cloning of a human prion protein cDNA. DNA 5: 315–324

Lacroute F (1971) Non-Mendelain mutation allowing ureidosuccinic acid uptake in yeast. J Bacteriol 106: 519–522

Levy E, Carman MD, Fernandez-Madrid IJ, Power MD, Lieberburg I, van Duinen SG, Bots GTAM, Luyendijk W, Frangione B (1990) Mutation of the Alzheimer's disease amyloid gene in hereditary cerebral hemorrhage, Dutch type. Science 248: 1124–1126

Malmgren R, Kurland L, Mokri B, Kurtzke J (1979) The epidemiology of Creutzfeldt-Jakob disease. In: Prusiner SB, Hadlow WJ (eds) Slow transmissible diseases of the nervous system, vol 1. Academic, New York, pp 93–112

Manuelidis E, Gorgacz EJ, Manuelidis L (1978) Interspecies transmission of Creutzfeldt-Jakob disease to Syrian hamsters with reference to clinical syndromes and strains of agent. Proc Natl Acad Sci USA 75: 3422–3436

Marsh RF, Bessen RA, Lehmann S, Hartsough GR (1991) Epidemiological and experimental studies on a new incident of transmissible mink encephalopathy. J Gen Virol 72: 589–594

Masters CL, Harris JO, Gajdusek DC, Gibbs CJ Jr, Bernouilli C, Asher DM (1978) Creutzfeldt-Jakob disease: patterns of worldwide occurrence and the significance of familial and sporadic clustering. Ann Neurol 5: 177–188

Masters CL, Gajdusek DC, Gibbs CJ Jr (1981) Creutzfeldt-Jakob disease virus isolations from the Gerstmann-Sträussler syndrome. Brain 104: 559–588

Medori R, Montagna P, Tritschler HJ, LeBlanc A, Cortelli P, Tinuper P, Lugaresi E, Gambetti P (1992a) Fatal familial insomnia: a second kindred with mutation of prion protein gene at codon 178. Neurology 42: 669–670

Medori R, Tritschler H-J, LeBlanc A, Villare F, Manetto V, Chen HY, Xue R, Leal S, Montagna P, Cortelli P, Tinuper P, Avoni P, Mochi M, Baruzzi A, Hauw JJ, Ott J, Lugaresi E, Autilio-Gambetti L,

Gambetti P (1992b) Fatal familial insomnia, a prion disease with a mutation at codon 178 of the prion protein gene. N Engl J Med 326: 444–449

Meggendorfer F (1930) Klinische und genealogische Beobachtungen bei einem Fall von spastischer Pseudosklerose Jakobs. Z Gesamte Neurol Psychiatr 128: 337–341

Monari L, Chen SG, Brown P, Parchi P, Petersen RB, Mikol J, Gray F, Cortelli P, Montagna P, Ghetti B, Goldfarb LG, Gajdusek DC, Lugaresi E, Gambetti P, Autilio-Gambeti L (1994) Fatal familial insomnia and familial Creutzfeldt-Jakob disease: different prion proteins determined by a DNA polymorphism. Proc Natl Acad Sci USA 91: 2839–2842

Mullan M, Houlden H, Windelspecht M, Fidani L, Lombardi C, Diaz P, Rossor M, Crook R, Hardy J, Duff K, Crawford F (1992) A locus for familial early-onset Alzheimer's disease on the long arm of chromosome 14, proximal to the α1-antichymotrypsin gene. Nature Genet 2: 340–342

Oesch B, Teplow DB, Stahl N, Serban D, Hood LE, Prusiner SB (1990) Identification of cellular proteins binding to the scrapie prion protein. Biochemistry 29: 5848–5855

Palmer MS, Dryden AJ, Hughes JT, Collinge J (1991) Homozygous prion protein genotype predisposes to sporadic Creutzfeldt-Jakob disease. Nature 352: 340–342

Pan K-M, Baldwin M, Nguyen J, Gasset M, Serban A, Groth D, Mehlhorn I, Huang Z, Fletterick RJ, Cohen FE, Prusiner SB (1993) Conversion of α-helices into β-sheets features in the formation of the scrapie prion proteins. Proc Natl Acad Sci USA 90: 10962–10966

Parry HB (1962) Scrapie: a transmissible and hereditary disease of sheep. Heredity 17: 75–105

Parry HB (1983) Scrapie disease in sheep. In: Oppenheimer DR (ed) Academic, New York

Pattison IH (1965) Experiments with scrapie with special reference to the nature of the agent and the pathology of the disease. In: Gajdusek DC, Gibbs CJ Jr, Alpers MP (eds) Slow, latent and temperate virus infections. US Government Printing office, Washington, pp 249–257 (NINDB monograph 2)

Pattison IH (1966) The relative susceptibility of sheep, goats and mice to two types of the goat scrapie agent. Res Vet Sci 7: 207–212

Pattison IH, Jones KM (1967) The possible nature of the transmissible agent of scrapie. Vet Rec 80: 1–8

Pattison IH, Millson GC (1961) Scrapie produced experimentally in goats with special reference to the clinical syndrome. J Comp Pathol 71: 101–108

Prusiner SB (1982) Novel proteinaceous infectious particles cause scrapie. Science 216: 136–144

Prusiner SB (1987) The biology of prion transmission and replication. In: Prusiner SB, McKinley MP (eds) Prions–Novel Infectious Pathogens Causing Scrapie and Creutzfeldt-Jakob Disease. Academic Press, Orlando, pp 83–112

Prusiner SB (1989) Scrapie prions. Annu Rev Microbiol 43: 345–374

Prusiner SB (1991) Molecular biology of prion diseases. Science 252: 1515–1522

Prusiner SB (1992) Molecular biology and genetics of neurodegenerative diseases caused by prions. Adv Virus Res 41: 241–280

Prusiner SB (1993) Transgenetics and cell biology of prion diseases: investigations of PrP^{Sc} synthesis and diversity. Brit Med Bull 49: 873–912

Prusiner SB (1994) Inherited prion diseases. Proc Natl Acad Sci USA 91: 4611–4614

Prusiner SB, Hsiao KK (1994) Human prion diseases. Ann Neurol 35: 385–395

Prusiner SB, McKinley MP, Bowman KA, Bolton DC, Bendheim PE, Groth DF, Glenner GG (1983) Scrapie prions aggregate to form amyloid-like birefringent rods. Cell 35: 349–358

Prusiner SB, Scott M, Foster D, Pan K-M, Groth D, Mirenda C, Torchia M, Yang S-L, Serban D, Carlson GA, Hoppe PC, Westaway D, DeArmond SJ (1990) Transgenetic studies implicate interactions between homologous PrP isoforms in scrapie prion replication. Cell 63: 673–686

Prusiner SB, Fuzi M, Scott M, Serban D, Serban H, Taraboulos A, Gabriel J-M, Wells G, Wilesmith J, Bradley R, DeArmond SJ, Kristensson K (1993a) Immunologic and molecular biological studies of prion proteins in bovine spongiform encephalopathy. J Infect Dis 167: 602–613

Prusiner SB, Groth D, Serban A, Koehler R, Foster D, Torchia M, Burton D, Yang S-L, DeArmond SJ (1993b) Ablation of the prion protein (PrP) gene in mice prevents scrapie and facilitates production of anti-PrP antibodies. Proc Natl Acad Sci USA 90: 10608–10612

Prusiner SB, Groth D, Serban A, Stahl N, Gabizon R (1993c) Attempts to restore scrapie prion infectivity after exposure to protein denaturants. Proc Natl Acad Sci USA 90: 2793–2797

Race RE, Graham K, Ernst D, Caughey B, Chesebro B (1990) Analysis of linkage between scrapie incubation period and the prion protein gene in mice. J Gen Virol 71: 493–497

Raeber AJ, Borchelt DR, Scott M, Prusiner SB (1992) Attempts to convert the cellular prion protein into the scrapie isoform in cell-free systems. J Virol 66: 6155–6163

Riesner D, Kellings K, Meyer N, Mirenda C, Prusiner SB (1992) Nucleic acids and scrapie prions. In: Prusiner SB, Collinge J, Powell J, Anderton B (eds) Prion diseases of humans and animals. Horwood, London, pp 341–358

Rogers M, Serban D, Gyuris T, Scott M, Torchia T, Prusiner SB (1991) Epitope mapping of the Syrian hamster prion protein utilizing chimeric and mutant genes in a vaccinia virus expression system. J Immunol 147: 3568–3574

Rogers M, Yehiely F, Scott M, Prusiner SB (1993) Conversion of truncated and elongated prion proteins into the scrapie isoform in cultured cells. Proc Natl Acad Sci USA 90: 3182–3186

Rosen DR, Siddique T, Patterson D, Figlewicz DA, Sapp P, Hentati A, Donaldson D, Goto J, O'Regan JP, Deng H, Rahmani Z, Krizus A, McKenna-Yasek D, Cayabyab A, Gaston SM, Berger R, Tanzi RE, Halperin JJ, Herzfeldt B, Van den Bergh R, Hung W-Y, Bird T, Deng G, Mulder DW, Smyth C, Laing NG, Soriano E, Pericak-Vance MA, Haines J, Rouleau GA, Gusella JS, Horvitz HR, Brown RH Jr (1993) Mutations in Cu/Zn superoxide dismutase gene are associated with familial amyotrophic lateral slerosis. Nature 362: 59–62

Schellenberg GD, Bird TD, Wijsman EM, Orr HT, Anderson L, Nemens E, White JA, Bonnycastle L, Weber JL, Alonso ME, Potter H, Heston LL, Martin GM (1992) Genetic linkage evidence for a familial Alzheimer's disease locus on chromosome 14. Science 258: 668–671

Scott MR, Foster D, Mirenda C, Serban D, Coufal F, Wälchli M, Torchia M, Groth D, Carlson G, DeArmond SJ, Westaway D, Prusiner SB (1989) Transgenic mice expressing hamster prion protein produce species-specific scrapie infectivity and amyloid plaques. Cell 59: 847–857

Scott MR, Köhler D, Foster D, Prusiner SB (1992) Chimeric prion protein expression in cultured cells and transgenic mice. Protein Sci 1: 986–997

Scott MR, Groth D, Foster D, Torchia M, Yang S-L, DeArmond SJ, Prusiner SB (1993) Propagation of prions with artificial properties in transgenic mice expressing chimeric PrP genes. Cell 73: 979–988

Sigurdsson B (1954) Rida, a chronic encephalitis of sheep with general remarks on infections which develop slowly and some of their special characteristics. Br Vet J 110: 341–354

Sparkes RS, Simon M, Cohn VH, Fournier REK, Lem J, Klisak I, Heinzmann C, Blatt C, Lucero M, Mohandas T, DeArmond SJ, Westaway D, Prusiner SB, Weiner LP (1986) Assignment of the human and mouse prion protein genes to homologous chromosomes. Proc Natl Acad Sci USA 83: 7358–7362

St. George-Hyslop P, Haines J, Rogaev E, Mortilla M, Vaula G, Pericak-Vance M, Foncin J-F, Montesi M, Bruni A, Sorbi S, Rainero I, Pinessi L, Pollen D, Polinsky R, Nee L, Kennedy J, Macciardi F, Rogaeva E, Liang Y, Alexandrova N, Lukiw W, Schlumpf K, Tanzi R, Tsuda T, Farrer L, Cantu J-M, Duara R, Amaducci L, Bergamini L, Gusella J, Roses A, McLachlan DC (1992) Genetic evidence for a novel familial Alzheimer's disease locus on chromosome 14. Nature Genet 2: 330–334

Stahl N, Borchelt DR, Hsiao K, Prusiner SB (1987) Scrapie prion protein contains a phosphatidylinositol glycolipid. Cell 51: 229–240

Stahl N, Baldwin MA, Teplow DB, Hood L, Gibson BW, Burlingame AL, Prusiner SB (1993) Structural analysis of the scrapie prion protein using mass spectrometry and amino acid sequencing. Biochemistry 32: 1991–2002

Taraboulos A, Jendroska K, Serban D, Yang S-L, DeArmond SJ, Prusiner SB (1992) Regional mapping of prion proteins in brains. Proc Natl Acad Sci USA 89: 7620–7624

Tateishi J, Kitamoto T (1995) Inherited prion diseases and transmission to rodents. Brain Pathol 5: 53–59

Tateishi J, Sato Y, Ohta M (1983) Creutzfeldt-Jakob disease in humans and laboratory animals. In: Zimmerman HM (eds) Progress in Neuropathology, Vol.5. Raven Press, New York, pp 195–221

Tateishi J, Doh-ura K, Kitamoto T, Tranchant C, Steinmetz G, Warter JM, Boellaard JW (1992) Prion protein gene analysis and transmission studies of Creutzfeldt-Jakob disease. In: Prusiner SB, Collinge J, Powell J, Anderton B (eds) Prion diseases of humans and animals. Horwood, London, pp 129–134

Telling GC, Scott M, Foster D, Yang S-L, Torchia M, Sidle KCL, Collinge J, DeArmond SJ, Prusiner SB (1995) Transmission of Creutzfeldt-Jakob disease from humans to transgenic mice expressing chimeric human-mouse prion protein. Proc Natl Acad Sci USA (in press)

Telling GC, Scott M, Mastrianni J, Gabizon R, Torchia M, Cohen FE, DeArmond SJ, Prusiner SB (In press) Prion propagation in mice expressing human and chimeric PrP transgenes implicates the interaction of cellular PrP with another protein. Cell

Van Broeckhoven C, Haan J, Bakker E, Hardy JA, Van Hul W, Wehnert A, Vegter-Van der Vlis M, Roos RA (1990) Amyloid β protein precursor gene and hereditary cerebral hemorrhage with amyloidosis (Dutch). Science 248: 1120–1122

Van Broeckhoven C, Backhovens H, Cruts M, De Winter G, Bruyland M, Cras P, Martin J-J (1992) Mapping of a gene predisposing to early-onset Alzheimer's disease to chromosome 14q24.3. Nature Genet 2: 335–339

Westaway D, Goodman PA, Mirenda CA, McKinley MP, Carlson GA, Prusiner SB (1987) Distinct prion proteins in short and long scrapie incubation period mice. Cell 51: 651–662

Westaway D, Mirenda CA, Foster D, Zebarjadian Y, Scott M, Torchia M, Yang S-L, Serban H, DeArmond SJ, Ebeling C, Prusiner SB, Carlson GA (1991) Paradoxical shortening of scrapie incubation times by expression of prion protein transgenes derived from long incubation period mice. Neuron 7: 59–68

Westaway D, Cooper C, Turner S, Da Costa M, Carlson GA, Prusiner SB (1994a) Structure and polymorphism of the mouse prion protein gene. Proc Natl Acad Sci USA 91: 6418–6422

Westaway D, DeArmond SJ, Cayetano-Canlas J, Groth D, Foster D, Yang S-L, Torchia M, Carlson GA, Prusiner SB (1994b) Degeneration of skeletal muscle, peripheral nerves, and the central nervous system in transgenic mice overexpressing wild-type prion proteins. Cell 76: 117–129

Wickner RB (1995) Evidence for a prion analog in *S. cerevisiae*: the [URE3] non-Mendelian genetic element as an altered *URE2* protein. Science (in press)

Wilesmith JW, Hoinville LJ, Ryan JBM, Sayers AR (1992a) Bovine spongiform encephalopathy: aspects of the clinical picture and analyses of possible changes 1986–1990. Vet Rec 130: 197–201

Wilesmith JW, Ryan JBM, Hueston WD, Hoinville LJ (1992b) Bovine spongiform encephalopathy: epidemiological features 1985 to 1990. Vet Rec 130: 90–94

Ziegler DR (1993) In: O'Brien SJ (ed) Genetic maps – locus maps of complex genomes, 6th edn. Cold Spring Harbor Laboratory, Cold Spring Harbor pp 4.42–4.45

Poliovirus Biology and Pathogenesis

V.R. Racaniello[1] and R. Ren[2]

1 Introduction

Poliovirus is the causative agent of the acute central nervous system disease known as poliomyelitis. Towards the end of the nineteenth century, epidemics of poliomyelitis began to occur in the United States and in Europe, much to the surprise of the medical community, which had previously viewed the disease as a rarity, seen largely as sporadic cases in infants. Poliovirus was first isolated in 1908 by inoculation of monkeys with a cell-free extract made from the spinal cord of a fatal case of poliomyelitis (LANDSTEINER and POPPER 1908). For the next 40 years, research on the virus provided the necessary information on antigenic types, pathogenesis and immunity required to formulate vaccines that could prevent infection. As a result of this work, two excellent vaccines were developed which have effectively controlled poliomyelitis in much of the world.

The successful introduction of poliovirus vaccines by the early 1960s brought to a halt the investigation of poliovirus pathogenesis. The termination of these

[1] Department of Microbiology, Columbia University, College of Physicians and Surgeons, 701 W. 168th St., New York, NY 10032, USA
[2] Department of Biology, Rosenstiel Basic Medical Sciences Research Center, Brandeis University, 415 South Street, Waltham, MA 02254, USA

studies was in part a consequence of the great expense needed to study pathogenesis, which could only be done in monkeys or chimpanzees. In place of experimental work on pathogenesis, studies on the molecular biology, structure and genetics of polioviruses quickly flourished, and today poliovirus is among the best understood viruses of eukaryotic cells. Unfortunately, many important questions about the pathogenesis of poliomyelitis remain. Recently there has been a resurgence of interest in viral pathogenesis, as new experimental tools make it possible to address old questions. The development of a transgenic mouse model for poliomyelitis (KOIKE et al. 1991; REN et al. 1990) has stimulated new interest in understanding how this virus causes disease. It is therefore an appropriate time to review the pathogenesis of poliomyelitis and to consider the nature of the unsolved problems.

Poliovirus is a member of the Picornaviridae, a large virus family that contains many human and animal pathogens. There are three immunologically defined serotypes of poliovirus, all of which are capable of causing paralytic disease. The viral capsid consists of 60 copies each of four different polypeptides, arranged with icosahedral symmetry. The viral genome is a single-stranded, positive-sense RNA molecule approximately 7500 nucleotides in length. Poliovirus replication begins when the virus attaches to a cell surface receptor, which is a novel member of the immunoglobulin superfamily (MENDELSOHN et al. 1989). The precise events that lead to uncoating of the viral RNA are not known, but ultimately result in delivery of the RNA into the cell cytoplasm, where it is first translated and then replicated. These stages of viral replication have been recently reviewed (RACANIELLO 1995; WIMMER et al. 1993). The complete infectious cycle is relatively short, requiring approximately 6 h from receptor binding to cell lysis.

2 Poliovirus Host Range

Humans are the only known natural hosts for poliovirus. Chimpanzees and certain species of monkeys are susceptible to poliovirus infection by the intracerebral, intraspinal, and oral routes, although the susceptibility of neurons among primates varies. For example, the P1/Mahoney strain causes paralysis when 1–10 tissue culture infectious doses (TCID) are inoculated intracerebrally in cynomolgus monkeys, whereas $10^6–10^8$ TCID are not paralytogenic after intracerebral inoculation of chimpanzees (SABIN et al. 1954). Similar experiments with strains of all three types of poliovirus have established a hierarchy of the sensitivity of primate neurons to infection with poliovirus (SABIN 1956, 1957). The lower spinal cord neurons of the monkey are most susceptible to infection, followed by the brain stem neurons of monkeys, and then the lower neurons of chimpanzees. Since the susceptibility of chimpanzees to oral poliovirus infection is much higher than that observed in human populations, it is believed that human neurons are either as susceptible or less susceptible than those of chimpanzees.

This ranking of neuron susceptibility is the opposite of the susceptibility of the alimentary tract of the primates to poliovirus infection (SABIN 1956). P2/Lansing and P2/MEF1, which are highly neurotropic in monkeys by intracerebral inoculation, do not infect the alimentary tract of monkeys, but infect chimpanzees by the oral route. Attenuated virus strains that have limited infectivity in the alimentary tract of monkeys appear to multiply well in the alimentary tract in chimpanzees and better in humans (SABIN 1956). Certain species of monkeys are not susceptible to poliovirus infection by the oral route (HASHIMOTO et al. 1984).

Most poliovirus strains are host restricted and cause paralysis in primates but not nonprimates, such as the P1/Mahoney strain (LA MONICA et al. 1986). However, by a process of adaptation involving serial passage of viruses in nonprimates, strains of poliovirus, including P2/Lansing, P1/LS$_b$ and a variant of P3/Leon, were adapted to mice and other animal hosts (ARMSTRONG 1939; LI and SCHAEFFER 1953). Some strains of poliovirus are naturally virulent in mice (MOSS and RACANIELLO 1991). Mice inoculated intracerebrally with P2/Lansing develop a disease with clinical, histopathological, and age-dependent features resembling human poliomyelitis (JUBELT et al. 1980a,b). In contrast to the human disease, the virus is not infectious by the oral route, and no extraneural sites of viral replication have been described in mice.

The host restriction of poliovirus P1/Mahoney in mice is controlled at the level of the virus-receptor interaction, because transgenic mice expressing the human poliovirus receptor (PVR) gene can be infected with this strain (KOIKE et al. 1991; REN et al. 1990). The host range of P1/Mahoney can also be expanded to mice by substitution of amino acids 95–104 of capsid protein VP1, the VP1 B-C loop, with the sequence from P2/Lansing (MARTIN et al. 1988; MURRAY et al. 1988). VP1 B-C loop residues may determine host range by modulating contact with the cell receptor. Other mutations, located in the interior of the capsid, that enable P1/Mahoney to utilize a mouse receptor are believed to expand receptor recognition by affecting capsid transitions during cell entry (COLSTON and RACANIELLO 1995; COUDERC et al. 1993, 1994; MOSS and RACANIELLO 1991). The mouse-adapted phenotype of P1/LSa requires amino acid changes in both capsid protein VP1 and proteinase 2A^{pro} (LU et al. 1994). The mechanism by which the proteinase influences neurovirulence in mice is not known.

Poliovirus can infect many cell lines derived from primates. In contrast, nonprimate cell lines are not susceptible to poliovirus infection (HOLLAND and MCLAREN 1959). This host range restriction in cultured cells is determined at the level of the cell receptor. Primate cells contain virus-binding activity while nonprimate cells do not (HOLLAND and MCLAREN 1959). However, one replicative cycle occurs and infectious virus is released when a variety of nonprimate cells are transfected with purified viral RNA (HOLLAND et al. 1959a,b). Furthermore, expression of PVR in mouse L cells results in susceptibility to multicycle viral infection (MENDELSOHN et al. 1989).

Transgenic mice containing the human PVR gene in the germ line are susceptible to infection with neurovirulent strains of all three serotypes (KOIKE et al. 1991; REN et al. 1990). The transgenic mice express PVR transcripts and poliovirus binding sites in a wide range of tissues. Inoculation of PVR transgenic mice with poliovirus leads to development of a fatal paralytic disease that clinically and histopathologically resembles human poliomyelitis. Transgenic PVR mice are therefore a new animal model for investigating poliovirus pathogenesis.

3 Clinical Features of Poliomyelitis

Infection with polioviruses may result in inapparent infection without symptoms, mild (minor) illness, aseptic meningitis, or paralytic poliomyelitis (MELNICK 1990). Most individuals infected with poliovirus experience no symptoms. Approximately 4%–8% experience the minor illness, which is characterized by fever, malaise, drowsiness, headache, nausea, vomiting, constipation, or sore throat and is followed by complete recovery. Only 1%–2% of infected individuals develop paralytic poliomyelitis during an epidemic (BODIAN and HORSTMANN 1965). In spinal poliomyelitis, paralysis is limited to muscles supplied by motor neurons in the spinal cord. The legs are affected more frequently than the arms. In bulbar poliomyelitis, the cranial nerve nuclei or medullary centers are involved. Bulbar poliomyelitis is often fatal due to respiratory or cardiac failure.

4 Course of Poliovirus Infection

4.1 Primary Site of Replication

In a typical infection, virus is ingested and initially multiplies in the oropharyngeal and the intestinal mucosa (BODIAN and HORSTMANN 1965; SABIN 1956). It is not known whether virus multiplies in epithelial or lymphoid cells of the alimentary tract. High levels of PVR RNA accumulate in the human intestinal epithelia, suggesting that these cells are primary sites of poliovirus multiplication (REN 1992). Examination of tissues from infected chimpanzees in the presymptomatic period reveals the presence of virus primarily in tonsillopharyngeal tissue and in the Peyer's patches of the ileum (BODIAN and HORSTMANN 1965). Virus has been isolated from the central nervous system (CNS), tonsillopharyngeal tissue, wall of the ileum, and lymph nodes of humans (SABIN and WARD 1941; WENNER and RABE 1953). However, virus multiplication is as extensive in the throats of the human volunteers without tonsils or adenoids as in those who still had these tissues (SABIN 1956). It is not clear whether virus replicates in these lymphoid tissues or is absorbed into regional lymph nodes after replication in superficial epithelial cells.

Studies on reoviruses show that viruses selectively bind to specialized microfold cells (M cells) overlying Peyer's patches and are transcytosed into lymphoid tissue (BASS et al. 1988). Virus undergoes primary replication in the mononuclear cells of ileal Peyer's patches and in neurons of the adjacent myenteric plexus and spreads directly from the intestinal lumen to the CNS through vagal autonomic nerve fibers (MORRISON et al. 1991; WOLF et al. 1981). Poliovirus has been shown to adhere to and be endocytosed by intestinal epithelial M cells with low efficiency (SICINSKI et al. 1990). The exact mode of initial poliovirus infection requires further investigation.

Unfortunately, Tg PVR mice are probably not suitable for clarifying the early stages of alimentary tract replication. P1/Mahoney fails to multiply in the intestine of PVR transgenic mice made in one laboratory (TgPVR mice; REN and RACANIELLO 1992a). Failure to observe poliovirus replication most likely is due to the absence of PVR RNA transcripts in the intestine of transgenic mice. Replication of P1/Mahoney in the intestine of another line of PVR transgenic mice has been reported, although the efficiency of replication has not been noted and the cell type in which multiplication occurs has not been determined (KOIKE et al. 1993). This transgenic line, called Tg1, was derived by microinjection of a PVR genomic DNA into CD1 mouse zygotes. The LD_{50} of P1/Mahoney in Tg1 mice is about 1×10^2 by intracerebral inoculation and 1×10^7 by peroral inoculation, whereas the LD_{50} of P1/ Mahoney in the TgPVR mice is about 1×10^5 by intracerebral inoculation (REN et al. 1990). It is possible that the differences between these two transgenic mouse lines are due to the different mouse strains (CD1 compared with (C57BL6/J $\times$ CBA/J) F2 mice) used to generate transgenic lines. Alternatively, the P1/Mahoney strain used in these experiments may be different.

Although viral replication in the intestine of TgPVR mice is not detected, these animals developed paralytic disease after oral administration of virus (REN 1992). This phenomenon was also observed in a proportion of monkeys infected orally with poliovirus (SABIN 1956). The paralytic disease may result from infection by olfactory pathways, since viral replication was detected in the nasal mucosa and the olfactory bulb in suckling transgenic mice (REN 1992), or from spread of virus into the body by the trauma of inoculation.

4.2 Viremia

From the primary sites of multiplication, the virus drains into deep cervical and mesenteric lymph nodes and then to the blood, resulting in a transient viremia which carries virus to other susceptible tissues (BODIAN and HORSTMANN 1965). Virus has been detected in the blood of monkeys, chimpanzees, and humans in the early stages of infection (BODIAN 1954a,b; HORSTMANN 1952; HORSTMANN et al. 1954). It is believed that viral replication in extraneural tissues results in maintenance of viremia beyond the first stage, and is required for viral invasion of the CNS (BODIAN and HORSTMANN 1965), but the sites at which this replication occurs in humans is not known. In the experimentally infected chimpanzee, virus is found in very high concentrations in the brown fat of suprasternal, upper axillary, and paravertebral regions (BODIAN 1955). In monkeys infected intramuscularly, large amounts of virus are found in lymph nodes, axillary fat, adrenals, and the inoculated muscle (WENNER and KAMITSUKA 1956, 1957). There is also evidence that replication may occur in cells of the reticuloendothelial system and in the vascular endothelium in monkeys (BLINZINGER et al. 1969; KANAMITSU et al. 1967). In TgPVR mice, poliovirus replication was detected in skeletal muscle, brown adipose tissues, and nasal mucosa (REN and RACANIELLO 1992a).

In most natural infections only a transient viremia occurs. In 1%–2% of infected individuals, the virus enters the CNS, where it replicates primarily in motor neurons within the anterior horn of the spinal cord, the brain stem, and the motor cortex, destroying these cells and producing the characteristic paralysis.

4.3 Entry of Poliovirus into the CNS

The route by which poliovirus enters the CNS is not completely understood. Two possibilities have been suggested which are not mutually exclusive: the virus enters the CNS from blood across the blood-brain barrier (BBB), or it enters a peripheral nerve and is transmitted to the CNS (BLINZINGER and ANZIL 1974; BODIAN 1959; HURST 1936; MELNICK 1985; SABIN 1956). In support of the hypothesis that virus crosses the BBB, it is clear that the viremia preceding paralytic infection is necessary for virus entry into the CNS. Furthermore, the presence of specific antibodies in the blood blocks viral spread in the host and prevents invasion of the CNS (BODIAN and HORSTMANN 1965; MELNICK 1985). Virus antigen has been detected by immunofluorescence in vascular endothelial cells of monkeys infected with poliovirus (BLINZINGER et al. 1969; KANAMITSU et al. 1967), and PVR has been detected in a small percentage of freshly dispersed endothelial cells (COUDERC et al. 1990). These observations suggest that poliovirus may use receptors on endothelial cells to gain access to the CNS from capillaries. However, neither PVR expression nor virus replication is detected in endothelial cells in PVR transgenic mice (REN and RACANIELLO 1992a). Poliovirus spread in transgenic mice is therefore not likely to involve receptors on endothelial cells. The presence of poliovirus RNA in axonal and dendritic processes of neurons suggests that virus may spread through neural pathways.

There is ample evidence in support of the neural spread hypothesis. For example, in monkeys, inoculation of poliovirus into the sciatic nerve results in virus first in the lumbar cord, then in the leg area of the right motor cortex, indicating that poliovirus can spread along nerve fibers in both peripheral nerves and the CNS (HURST 1936). Following intramuscular injection of monkeys with the highly neurotropic poliovirus type 2 MV strain, localization of initial paralysis in the injected limb occurs at high frequency (NATHANSON and BODIAN 1961). Freezing the sciatic nerve blocks spread of this virus from muscle to the CNS. In the Cutter incident, in which children received incompletely inactivated polio vaccine, a high frequency of initial paralysis was observed in the inoculated limb (NATHANSON and LANGMUIR 1963).

   The study of viral spread following intramuscular or intrafootpad injection in TgPVR mice demonstrates that poliovirus enters the CNS through peripheral nerves (REN and RACANIELLO 1992b). This conclusion is based on three experimental observations: First, after intramuscular inoculation, the first limb paralyzed was always the limb that was inoculated. Second, following inoculation of virus into the hindlimb, virus was first detected in the lower spinal cord. Third, sciatic nerve transection blocked poliovirus infection after footpad inoculation. A possible scenario for the spread of poliovirus through peripheral nerves is as follows: after intramuscular or intrafootpad inoculation, virus replicates in muscle cells, binds to poliovirus receptors at the neuromuscular junction and enters the 2° motor axon. Virus then spreads, by retrograde axonal transport, to the 2° motor neuron in the spinal cord. Replication in the neuron cell body results in cell destruction and release of new virus particles, which infect neighboring neurons. When sufficient numbers of neurons are destroyed, paralysis of the innervated limb results. It is not known if PVR is expressed on the surface of the synapses, although it has been reported that human synaptosomes contain poliovirus binding sites (BROWN et al. 1987).

   Based on studies of poliovirus pathogenesis in PVR transgenic mice, combined with observations on the disease in humans, chimpanzee, and monkeys (BODIAN 1955; SABIN 1956), a revised scheme for the pathogenesis of the disease in humans is suggested (Fig. 1). Ingested poliovirus initially replicates in the alimentary tract, possibly in epithelial cells lining the tract. Replication leads to release of virus into the throat and gut lumen and establishment of viremia. Disseminated virus then replicates at an extraneural site—possibly skeletal

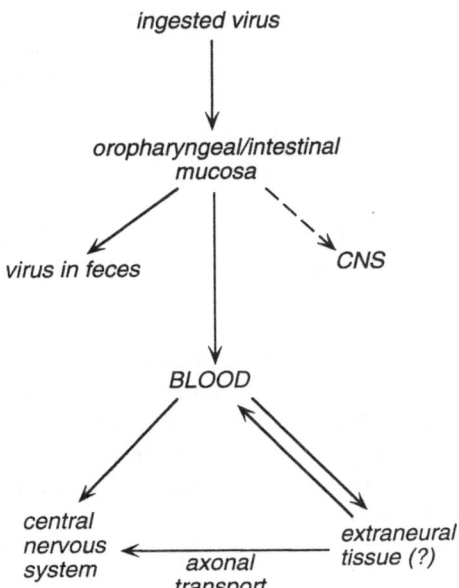

Fig. 1. Hypothetical scheme of poliovirus pathogenesis in humans, based on studies of poliovirus pathogenesis in humans, monkeys, chimpanzees and PVR transgenic mice. *Dotted line,* possible but unproved route of spread by axonal transport from the alimentary tract to the CNS. The extraneural tissue may include skeletal muscle and brown fat

muscle cells—enters peripheral nerves and spreads to the CNS. Virus replication at the extraneural site maintains persisting viremia, which may disseminate infection to multiple sites from which virus may also enter the CNS.

There are other pathways which poliovirus may use to enter the CNS. Reovirus can spread directly from the intestinal lumen to the CNS through vagal autonomic nerve fibers (MORRISON et al. 1991). Because the initial site of poliovirus replication is the alimentary tract, it is possible that poliovirus may spread via a similar pathway in humans (Fig. 1). The observation that 5%–30% of poliovirus infections involve the brain stem is consistent with this route of spread. Polioviruses are found in peripheral ganglia of the alimentary tract prior to invasion of the CNS in monkeys fed poliovirus (FABER 1956; SABIN 1956), and poliovirus replication is detected in neurons in peripheral ganglia in suckling transgenic mice after oral administration of virus (REN 1992). Transmission of virus along nerve fibers from peripheral ganglia might provide an additional route for entry into the CNS. Despite these considerations, in the majority of paralytic infections in humans, virus appears to initially infect the lower motor neurons of the spinal cord, which is consistent with the hypothesis that poliovirus spreads from the muscle to the CNS.

The observation that the persisting viremia which precedes paralytic infection is important for virus spread to the CNS has been used as evidence in support of the hypothesis that virus enters the CNS from the blood. However, persisting viremia may be a result of successful viral replication in skeletal muscle, which leads to release of virus into the blood and virus spreading to the CNS. Factors which increase access of poliovirus to muscle cells and nerve terminals would therefore be expected to increase the incidence of poliomyelitis. Consistent with this idea are the observations that in many cases with initial lumbar or cervical involvement there appears to be a temporal association with recent vigorous exertion or injury of the lower or upper limbs, and bulbar poliomyelitis often develops in patients with previous tonsillectomy (BODIAN and HORSTMANN 1965).

The fact that passive or active immunization against poliovirus terminates viremia and prevents CNS infection has been considered strong evidence that poliovirus enters the CNS through the BBB (BODIAN and HORSTMANN 1965; MELNICK 1985). Studies on the pathogenesis of reovirus serotype 3 in mice indicate that virus spreads by nerves and not by the bloodstream to the CNS, despite the presence of viremia (FLAMAND et al. 1991; TYLER et al. 1986). Anti-viral antibody decreases viremia and prevents appearance of virus in the CNS after inoculation of virus in the hindlimb footpad (TYLER et al. 1989). Recently it was shown that antibody can mediate clearance of alphavirus infection from neurons by restricting viral gene expression (LEVINE et al. 1991). These studies demonstrate that blocking virus entry to the CNS by the BBB is not the only mechanism by which antibody prevents CNS infection. The mechanism by which anti-poliovirus antibody prevents CNS infection clearly requires further study.

4.4 Histopathology of Experimental Poliomyelitis in Primates

The histopathology of experimental poliomyelitis in the primate is well known (HURST 1929). CNS lesions in poliomyelitis consist of neuronal changes and inflammation. Viral replication results in destruction of neurons and the inflammatory process follows as a secondary response (BODIAN and HORSTMANN 1965). There is little evidence for viral replication in other cell types in the CNS.

The characteristic pattern of distribution of poliomyelitis lesions has been shown to be due to two principal factors: (1) the inherent variation of susceptibility of nervous centers to infection, and (2) the restricted movement of virus along certain nerve fiber pathways (BODIAN and HORSTMANN 1965). The motor neurons of the anterior horns of the cervical and lumbar regions are the most sensitive to the virus, followed by neurons in motor nuclei of cranial nerves in the brain stem. In the spinal cord, lesions are largely observed in the anterior horns, although lesions of varying severity are observed in the intermediate, the intermediolateral and the posterior gray columns (BODIAN 1959; BODIAN and HORSTMANN 1965). Lesions may extend to the sensory spinal ganglia. The lesions in the brain are primarily in the brain stem, extending from the spinal cord to the anterior hypothalamus. Lesions in the forebrain are usually mild and restricted to the precentral gyrus (motor cortex) and the neighboring cortex, the thalamus and the globus pallidus. Severe lesions are also often found in the cerebellar vermis and the deep cerebellar nuclei (BODIAN and HORSTMANN 1965).

The sites of poliovirus replication in the CNS of PVR transgenic mice closely parallel those observed in primates (KOIKE et al. 1991; REN et al. 1990; REN and RACANIELLO 1992a). As in primates, the motor neurons of the ventral horns of the cervical and lumbar regions are the most sensitive to the virus, followed by neurons in motor nuclei of cranial nerves in the brain stem. One difference is that poliovirus replicates in the hippocampus of PVR transgenic mice, causing pathological changes in that area.

Poliovirus replication has been detected in neurons in both the ventral and dorsal horns of the spinal cord of TgPVR mice (REN and RACANIELLO 1992a). However, inflammation and neuronal degeneration is localized largely to the ventral horns, suggesting that neurons in the dorsal horn are either infected at a late stage of the disease, or that poliovirus replication in these neurons does not lead to cell destruction. Poliovirus histopathology is observed in the posterior horn (analogous to the dorsal horn of mice) in human and monkey spinal cord, although less frequently than in the anterior horn (BODIAN 1959).

4.5 Tissue Tropism

Viral infections are often localized to specific cells and tissues within the host. This cell and tissue tropism results in distinct disease patterns for different viruses. Because all viruses initiate infection by binding to a specific receptor on the cell surface, the virus-receptor interaction has long been considered the first determinant of tissue tropism. For some viruses, such as human immunodeficiency virus type 1 (MADDON et al. 1986), and Epstein-Barr virus (AHEARN et al. 1988), expression of cell receptors appears to control the pattern of virus infection in the host. However, viral replication in the host may be blocked at steps in the viral life cycle other than receptor binding. For example, the receptor for influenza virus, sialic acid, is ubiquitous, yet viral replication is largely limited to the respiratory tract. Tropism of influenza virus is most likely determined by availability of cellular proteinases required for cleavage of the hemagglutinin (GOTOH et al. 1990; WEBSTER and ROTT 1987).

In the primate host, poliovirus infection is characterized by a restricted tissue tropism despite the presence of virus in many organs during the viremic phase of infection (BODIAN 1955; SABIN 1956). It has long been believed that the cellular receptor is a major determinant of its tissue tropism for the following reasons: (1) Assays for virus binding activity in tissue homogenates revealed a correlation between poliovirus binding and susceptibility to poliovirus infection (HOLLAND 1961). (2) Although poliovirus shows restricted tissue tropism, cells from almost any primate tissue are susceptible to poliovirus infection after cultivation in vitro (ENDERS et al. 1949; HOLLAND and McLAREN 1961; KAPLAN 1955). The acquired susceptibility of cultured cells correlates with appearance of the poliovirus receptor (COUDERC et al. 1990; HOLLAND and HOYER 1962; HOLLAND and McLAREN 1961). However, restriction of viral replication in many tissues may not be due solely to a lack of receptor in these tissues. For example, occasional binding of virus to tissues that are not sites of poliovirus replication has been reported (HOLLAND and McLAREN 1961; KUNIN and JORDAN 1961). Studies on the binding of radiolabeled poliovirus to human regional CNS tissue homogenates showed that the binding activity within the CNS is much more widespread than the restricted distribution of pathologic lesions (BROWN et al. 1987). This observation suggests that factors other than receptor distribution must play a role in determining poliovirus neurotropism.

Whether restricted tissue tropism of poliovirus is due to expression of the PVR might be addressed by studying the tissue distribution of poliovirus receptors. In humans, PVR RNA and protein are expressed in a wide range of human tissues, including those that are not sites of poliovirus infection (FREISTADT et al. 1990; MENDELSOHN et al. 1989). However, little is known about PVR gene expression in individual cell types. It is possible that cells expressing PVR within nonsusceptible tissues are not accessible to poliovirus, or perhaps only a small fraction of cells in these tissues express PVR and as a result virus growth is not detected. Alternatively, poliovirus tissue tropism may not be governed solely by expression of PVR, but may depend on tissue-or cell-specific modification of the PVR, additional factors required for PVR function, or perhaps factors required for subsequent stages in virus replication.

Northern blot analysis shows that PVR transcripts are expressed in a wide range of PVR transgenic mouse tissues (REN et al. 1990). A similar pattern is observed in humans (MENDELSOHN et al. 1989), indicating that the PVR gene used to establish the transgenic lines contains sequence elements necessary for PVR expression. In addition, the multiple spliced forms of PVR mRNAs observed in human tissues (KOIKE et al. 1990) are also present in PVR transgenic mouse tissues (KOIKE et al. 1991; REN and RACANIELLO 1992a).

The results of in situ hybridization of tissues from two different lines of TgPVR mice show that the expression pattern of PVR RNA in the transgenic mouse mimics that in humans (REN and Racaniello 1992a). A significant difference, however, is observed in the expression of PVR RNA in the intestine. Both adult and fetal human intestinal epithelia accumulate high levels of PVR RNA, whereas PVR RNA is present at only low levels in both adult and fetal TgPVR mouse intestine (REN 1992; REN and RACANIELLO 1992a). The low expression of PVR RNA in transgenic mouse intestine may result from the absence of positive cis-acting regulatory element(s) in the PVR gene used to establish transgenic lines, or the presence of tissue-specific negative trans-acting factor(s) for regulatory elements shared between the PVR gene and mouse intestine epithelial cells.

Poliovirus binding sites are also expressed in a wide range of transgenic mouse tissues. This finding correlates with the wide range of expression of PVR proteins in human tissues (FREISTADT et al. 1990), but differs from observations with primate tissues (HOLLAND 1961). The receptor protein in nonsusceptible human tissues may be masked or shielded by its natural ligand, preventing binding of virus; in TgPVR mice the receptor protein might either be expressed in excess or may not interact with an endogenous ligand, permitting poliovirus binding. Alternatively, there may be small numbers of cells in nonsusceptible primate tissues that express active poliovirus binding sites which are difficult to detect. For example, poliovirus binding activity has been irregularly detected in primate liver (HOLLAND 1961), where poliovirus replication is not observed.

Although PVR expression in TgPVR mice is widespread, when transgenic mice are inoculated with poliovirus, viral replication is limited to skeletal muscle, neurons in the CNS, and to a lesser extent in brown adipose tissue, peripheral ganglia, and nasal mucosa (REN and RACANIELLO 1992a). Poliovirus RNA is detected in adult TgPVR mouse skeletal muscle cells of the hamstring after intramuscular inoculation. In addition, poliovirus replicates extensively in skeletal muscle of suckling transgenic mice after oral administration of virus. Replication of poliovirus has also been demonstrated in monkey skeletal muscle after intramuscular inoculation (WENNER and KAMITSUKA 1957). Poliovirus RNA is detected in all neurons of TgPVR mouse spinal cord and in most neurons in the brain stem. In the brain, infected neurons are detected in several areas, including the cerebral cortex, pyramidal layer of the hippocampus, olfactory bulb, thalamus, hypothalamus and deep cerebellar nuclei. Poliovirus replication is not detected in kidney, adrenal gland, thymus or intestine. Therefore, susceptibility of cells to poliovirus appears to correlate with PVR RNA expression in the CNS, muscle, and brown adipose tissues, but not in other tissues.

There are several possible explanations for the failure of poliovirus to replicate in transgenic mouse tissues that express PVR. It is unlikely that the PVR transcripts detected by in situ hybridization are not translated. Organ homogenates from PVR transgenic mice contain poliovirus binding activity, and freshly dispersed transgenic mouse kidney cells express PVR on the cell surface, as shown by their ability to bind poliovirus. Virus may be unable to reach some cells which express PVR RNA, such as developing T lymphocytes of the thymus and epithelial cells of Bowman's capsule. However, tubular epithelial cells in the kidney and endocrine cells in the adrenal cortex should be exposed to circulating viruses. Poliovirus replication was not detected in these cells, indicating that poliovirus tissue tropism is not governed solely by expression of the PVR gene or by accessibility of cells to virus.

Alternative splicing of PVR transcripts might control susceptibility of tissues to poliovirus infection. Human tissues contain both membrane-bound and secreted PVR isoforms, generated by alternative splicing of PVR mRNA (Koike et al. 1990). Expression of secreted PVRs in transgenic mouse tissues might result in neutralization of poliovirus infectivity (Kaplan et al. 1990). RNAs encoding both membrane-bound and secreted PVRs are present in susceptible and nonsusceptible TgPVR tissues (Koike et al. 1991; Ren and Racaniello 1992a). Alternatively, virus binding to nonsusceptible tissues might be blocked in vivo by the natural ligand of the PVR, or poliovirus entry might require factors in addition to the PVR that are lacking in nonsusceptible tissues. For example, expression of human CD4 in rodent cells is not sufficient to render these cells susceptible to HIV-1 infection, due to a block at the level of entry (Maddon et al. 1986).

Poliovirus replication in nonsusceptible tissues might be controlled at stages beyond virus entry, such as translation, replication or assembly. This possibility has been generally discounted in the past, as inoculation of viral RNA intracerebrally into rabbits, chicks, guinea pigs, and hamsters results in one cycle of replication and production of infectious virus (Holland et al. 1959b). However, these experiments indicate only that poliovirus host range restriction (species tropism) is determined at the level of entry in neural cells. Whether or not there is an internal block to poliovirus infection in nonsusceptible tissues remains unresolved.

Although poliovirus infection in primates is restricted, cells from almost any tissue develop susceptibility to infection after cultivation in vitro (Enders et al. 1949; Holland 1961; Kaplan 1955). PVR transgenic mouse kidney cells express poliovirus binding sites but are initially resistant to poliovirus infection. When cultured in vitro, kidney cells develop susceptibility to poliovirus infection after 24 h (Ren and Racaniello 1992a). The basis of the acquired susceptibility to poliovirus infection in these cells is not known, but might involve the induction of factors required for virus entry or replication. Study of the changes that occur in cultured PVR transgenic mouse kidney cells that permit poliovirus infection would provide information on the block to poliovirus infection in these cells and might reveal the mechanism of poliovirus tissue tropism in primates.

An important question is whether studying poliovirus tropism in PVR transgenic mice provides information on tropism of the virus in primates. The pattern of expression of many human genes in transgenic mice is often similar to that observed in humans (reviewed in Palmiter 1986); however, poliovirus binding sites are expressed in all PVR transgenic mouse tissues examined, while binding sites in humans have been detected largely in neural tissues and intestine and occasionally in kidney and liver (HOLLAND 1961; KUNIN and JORDAN 1961). It is possible that the basis of poliovirus tropism in humans differs from that in PVR transgenic mice and is determined by factors that control the ability of the receptor to bind virus. Alternatively, poliovirus binding sites might be expressed in all human tissues, but at low levels or in unstable forms, making their detection difficult in some tissues. Consistent with this idea, poliovirus binding activity can be detected in human fetal liver, which in TgPVR mice expresses PVR RNA (REN 1992). The PVR transgenic mice studied to date contain multiple copies of the PVR gene and express high levels of PVR mRNA, perhaps facilitating detection of binding sites in all organs. Resolution of this question awaits development of more sensitive assays to detect poliovirus binding sites in human tissues.

5 Attenuation of Neurovirulence

Strains of poliovirus may exhibit striking differences in neurovirulence, the ability to replicate in and destroy cells of the CNS. Measurement of neurovirulence is influenced not only by the viral genotype, but also by the animal host employed (e.g., monkeys, chimpanzees, and mice) and the different routes of inoculation used (e.g., intracerebral, intraspinal, intraperitoneal, intramuscular, intravenous, and oral). Viruses with reduced or attenuated neurovirulence may occur naturally or may be isolated by passage of the virus in a different animal host or in various cultured cells. The first attenuated poliovirus strain was isolated by THEILER (1947), who reported that after 50 rapid intracerebral passages in mice, the Lansing strain no longer caused signs of poliomyelitis in rhesus monkeys inoculated intracerebrally. Cultivation of the P1/Brunhilde strain in human nonneural tissue results in a marked reduction of its neurovirulence in monkeys (ENDERS et al. 1949). Serial propagation of wild-type poliovirus strains (P1/Mahoney, P2/YSK, and P3/Leon) in cynomolgus kidney cultures have no effect on virulence for cynomolgus monkeys when low multiplicities of infection are used to initiate the cultures, while rapid passage with large inocula leads to the appearance of avirulent variants, which can be separated from virulent variants by terminal dilution (SABIN et al. 1954).

The Sabin live oral polio vaccine strains were produced by controlled passage of viruses in animals and cultured cells until variants unable to cause paralysis in primates were obtained (reviewed in SABIN and BOULGER 1973). The Sabin vaccine strains cause no paralysis when inoculated into the CNS of animals, yet after oral administration in humans replicate sufficiently in the alimentary tract to induce a protective immune response. The P1/Sabin strain (LSc, 2ab) and P3/Sabin strain (Leon 12a$_1$b) were derived from neurovirulent strains, P1/Mahoney and P3/Leon respectively. The P2/Sabin strain (P712, Ch, 2ab), however, was derived from P712, an isolate from the feces of healthy children of low intraspinal cynomolgous neurovirulence.

The Sabin vaccine strains have been extremely effective in controlling polio-myelitis. Since the Sabin strains were isolated, it has been of great interest to determine the molecular and functional basis for their attenuation phenotypes (reviewed in MINOR et al. 1993). This information has provided ins ght into the biology of poliovirus, enabled better understanding of vaccine-associated disease, and suggests ways to improve the existing vaccines.

To identify determinants of attenuation in each of the three vaccine strains, genomic recombinants have been constructed between the attenuated viruses and closely related neurovirulent strains, and the ability of these recombinants to cause paralysis has been assayed in primates or in mice. Because the poliovirus genome is an RNA molecule, the recombinants have been constructed using cloned cDNAs from which virus can be derived by transfection (RACANIELLO and BALTIMORE 1981). The hosts for these studies include monkeys inoculated intra-cerebrally or intraspinally and either normal or PVR transgenic mice. The suitability of mice for such analyses was shown by studying the mouse neurovirulence of recombinants between the P2/Lansing strain and P3/Sabin, its neurovirulent parent P3/Leon, and P3/119, a type 3 isolate obtained from a vaccine-associated case of poliomyelitis. These experiments demonstrate that the presence of a U at nucleotide 472 in the 5'-noncoding region results in an attenuation phenotype, while the presence of a C at this position results in neurovirulent viruses (LA MONICA et al. 1987). The same mutation attenuates poliovirus neurovirulence in monkeys and in humans (WESTROP et al. 1987). Mutations that attenuate poliovirus neurovirulence in monkeys and in humans have also been shown to attenuate neurovirulence in PVR transgenic mice, demonstrating that this model for polio-myelitis can be used to identify determinants of attenuation (HORIE et al. 1994).

Early studies suggested that attenuation determinants of P1/Sabin were scattered throughout the viral genome, but that a strong determinant was located at nucleotide 480 in the 5'-noncoding region (KAWAMURA et al. 1989; OMATA et al. 1986). By analyzing the neurovirulence of P1/Sabin-P1/Mahoney recombinants and single base mutants of P1/Mahoney in TgPVR mice, attenuating determinants in P1/Sabin have been identified (BOUCHARD and RACANIELLC 1995). Attenu-ating mutations of P1/Sabin are found at base 935 of VP4 (amino acid 4065), base 2438 of VP3 (amino acid 3225), and base 2795 of VP1 (amino acid 1106), while weaker mutations are present at VP1 bases 2749 and 2879 (amino acids 1090 and 1134). A variant of P1/Sabin, selected at elevated temperatures, is neurovirulent in monkeys and contains mutations at bases 2741 and 2795, suggesting that these sites might be involved in attenuation (CHRISTODOULOU et al. 1990). A recombinant of P1/Mahoney containing the 3D$^{po'}$ and 3'-noncoding region of P1/Sabin does not cause paralysis in all of four monkeys inoculated, although the spread values and lesion scores are similar to those of P1/Mahoney, which paralyzed three out of four monkeys inoculated (OMATA et al. 1986). Neurovirulence analysis of revertants of P1/Sabin selected at high temperatures also suggest that base 6203 may be involved in attenuation in the monkey model (CHRISTODOULOU et al. 1990). However, neurovirulence assays of viral recombinants and mutants in a mouse model lead to the conclusion that 3D$^{po'}$ of P1/Sabin is only slightly attenuating (TARDY-PANIT et al. 1993), while results obtained with TgPVR mice indicate that there are no determinants of attenuation in 3D$^{po'}$ of P1/Sabin (BOUCHARD and RACANIELLO 1995). These differing findings suggest that monkeys and PVR transgenic mice differ in their sensitivity to some attenuation determinants.

Three attenuation determinants have been identified in the type 3 vaccine strain, P3/Sabin: a uridine residue at nucleotide 472 in the 5'-noncoding region; a phenylalanine at amino acid 91 of capsid protein VP3, which is also responsible for the temperature sensitive (ts) phenotype of the virus (MINOR et al. 1989; WESTROP et al. 1989); and a threonine at amino acid 6 of capsid protein VP1 (TATEM et al. 1992).

Nucleotide A-481 in the 5' noncoding region and isoleucine at amino acid 143 of capsid protein VP1 have been identified as the major attenuation determinants in P2/Sabin (MACADAM et al. 1991b, 1993; REN et al. 1991). These studies have been carried out both in TgPVR mice and in monkeys, and the results from both experimental systems are in agreement.

It is highly likely that weak determinants of attenuation have not been detected by these experimental approaches. For example, although A-481 and isoleucine at amino acid 143 of VP1 account for most of the attenuation phenotype of P2/Sabin, changes at these positions to the sequences found in the virulent P2/117 strain do not fully restore neurovirulence (REN et al. 1991), suggesting that there are additional determinants of attenuation in P2/Sabin.

All three Sabin vaccine strains contain strong attenuation determinants in the 5' noncoding region of the viral genome. These mutations are believe to reduce viral replication by influencing the translational efficiency of the viral RNA (SVITKIN et al. 1990). Precisely how attenuation determinants in the capsid coding region reduce viral neurovirulence is not clear. These mutations have been proposed to influence the steps of virus binding, and the conformational transitions associated with uncoating of the viral RNA (FILMAN et al. 1989; REN et al. 1991). Other attenuating mutations are believed to reduce the efficiency of viral assembly (MACADAM et al. 1991a). In general, attenuating mutations appear to reduce viral replication sufficiently to prevent invasion of the CNS but do not compromise the immunogenicity of the vaccine. Whether or not the attenuation determinants are cell type-specific and impair viral replication in neural cells to a greater extent than in other cells remains to be determined.

Although the attenuated Sabin strains have an excellent safety record, vaccine-associated paralytic poliomyelitis (VAPP) may occur in recipients at a frequency of one case per 1.2 million first vaccine doses (NKOWANE et al. 1987). This disease is most often associated with types 2 and 3 and rarely with type 1 vaccine strains (ASSAAD and COCKBURN 1982; NKOWANE et al. 1987). Analysis of viruses isolated from cases of VAPP demonstrates that they are revertants of the Sabin strains. One VAPP isolate of P2/Sabin, called P2/117, contains mutations to the parental sequence at nucleotide 481 of the 5'-noncoding region and amino acid 143 of VP1, two positions that have been demonstrated to be important determinants of attenuation (MACADAM et al. 1991b, 1993; POLLARD et al. 1989; REN et al.

1991). VAPP isolates of P3/Sabin contain mutations to the parental sequence at nucleotide 472 of the 5'-noncoding region and amino acid 91 of VP3, or at second sites that suppress the 91 mutation (CANN et al. 1984; MACADAM et al. 1989). The mutations at base 472 and amino acid 91 of VP3 are important determinants of attenuation in P3/Sabin (WESTROP et al. 1989).

The Sabin vaccine viruses undergo extensive mutation during multiplication in the human gut. For example, attenuating mutations in the 5'-noncoding region of the type 2 and type 3 strains rapidly revert during gut replication (EVANS et al. 1985; MINOR and DUNN 1988). Reversion of the type 1 strain occurs in approximately 50% of healthy vaccines. Attenuating determinants elsewhere in the viral genome also have been observed to revert to the neurovirulent sequence. Given that viruses excreted from healthy vaccinees are of increased neurovirulence, the safety record of the live, attenuated poliovirus strains is remarkable. An understanding of why the live vaccine strains are so safe, given their rapid reversion in the gut, is clearly lacking.

6 Concluding Remarks

The establishment of a transgenic mouse model for poliomyelitis has facilitated studies on poliovirus pathogenesis and attenuation. However, it is clear that a number of important problems remain unresolved. For example, the basis for the restricted tropism of poliovirus is not known. While many tissues in TgPVR mice that express PVR are not susceptible to infection in vivo, when placed into culture these cells acquire the ability to support poliovirus replication (REN 1992). Therefore other systems are required to address this important but difficult question.

The nature of the primary site of virus multiplication in the alimentary tract and whether virus may spread from the gut to the CNS must be addressed. Unfortunately, existing PVR transgenic mouse strains do not appear to fully support gut replication, a shortcoming that may be a result of poor PVR expression at that site. Clearly derivation of new transgenic mouse lines with higher PVR expression in the gut is required to address these questions. Orally susceptible PVR transgenic mice would be valuable not only for answering these questions, but for studying the multiplication and reversion of poliovirus in the gut as well as the induction of mucosal immunity.

The availability of PVR transgenic mice susceptible to poliovirus infection has raised the possibility that the monkey neurovirulence test, which has been used for years to assess the safety of the live attenuated poliovirus vaccine, might be replaced. The World Health Organization is currently coordinating a worldwide collaborative study to determine if a neurovirulence assay using PVR transgenic mice can be developed. Advantages of the PVR mice include lower cost and the ability to use greater numbers of test animals, perhaps increasing accuracy of the test. Availability of an orally susceptible PVR transgenic mouse would add a new dimension to the neurovirulence test, which is currently conducted by inoculating virus into the monkey CNS.

The World Health Organization has set a goal of eradicating poliovirus from the globe by the year 2000. Given the current rate of elimination of the disease, this goal may well be met (WARD et al. 1993). Even if the disease is eliminated from the globe, poliovirus will always be present, a consequence of using a live virus vaccine. Hence, it will be necessary to forever vaccinate humans against poliomyelitis. It may therefore be appropriate to consider modifying the currently used vaccine strains so that they are unable to revert to neurovirulence. Furthermore, derivation of more thermostable vaccine strains would assist in distribution of the vaccine in tropical areas. The availability of infectious poliovirus cDNA (RACANIELLO and BALTIMORE 1981) and a transgenic mouse model for poliomyelitis (KOIKE et al. 1991; REN et al. 1990) make such endeavors tantalizingly possible.

References

Ahearn J, Hayward S, Hickey J, Fearon D (1988) Epstein-Barr virus (EBV) infection of murine L cells expressing recombinant human EBV/C3d receptor. Proc Natl Acad Sci USA 85: 9307–9311

Armstrong C (1939) Successful transfer of the Lansing strain of poliomyelitis virus from the cotton rat to the white mouse. Public Health Rep 54: 2302–2305

Assaad F, Cockburn WC (1982) The relation between acute persisting spinal paralysis and poliomyelitis vaccine-results of a ten-year enquiry. Bull World Health Organ 60: 231–242

Bass DM, Trier JS, Dambrauskas R, Wolf JL (1988) Reovirus type 1 infection of small intestinal epithelium jn suckling mice and its effects on M cells. Lab Invest 58: 226–235

Blinzinger K, Anzil AP (1974) Neural route of infection in viral disease of the central nervous system. Lancet ii: 1374–1375

Blinzinger K, Simon J, Magrath D, Boulger L (1969) Poliovirus crystals within the endoplasmic reticulum of endothelial and mononuclear cells in the monkey spinal cord. Science 163: 1336–1337

Bodian D (1954a) Viremia in experimental poliomyelitis. 1. General aspects of infection after intramuscular inoculation with strains of high and of low invasiveness. Am J Hyg 60: 339–357

Bodian D (1954b) Viremia in experimental poliomyelitis. 2. Viremia and the mechanism of the "provoking" effect of injections or trauma. Am J Hyg 60: 359–370

Bodian D (1955) Emerging concept of poliomyelitis infection. Science 12: 105–108

Bodian D (1959) Poliomyelitis: pathogenesis and histopathology. In: Rivers TM, Horsfall FL (eds) Viral and rickettsial infections of man. Lippincott, Philadelphia, pp 479–498

Bodian D, Horstmann DH (1965) Polioviruses. In: Horsfall FL, Tamm I (eds) Viral and rickettsial infections of man. Lippincott, Philadelphia, pp 430–473

Bouchard M, Racaniello VR (1995) Determinants of attenuation and temperature sensitivity in the type 1 poliovirus vaccine strain. J Virol 69: 4972–4978

Brown RH, Johnson D, Ogonowski M, Weiner HL (1987) Type 1 human poliovirus binds to human synaptosomes. Ann Neurol 21: 64–70

Cann AJ, Stanway G, Hughes PJ, Minor PD, Evans DMA, Schild GC, Almond JW (1984) Reversion to neurovirulence of the live-attenuated Sabin type 3 oral poliovirus vaccine. Nucleic Acids Res 12: 7787–7792

Christodoulou C, Colbere-Garapin F, Macadam A, Taffs LF, Marsden S, Minor P, Horaud F (1990) Mapping of mutations associated with neurovirulence in monkeys infected with Sabin 1 poliovirus revertants selected at high temperature. J Virol 64: 4922–4929

Colston E, Racaniello VR (1995) Poliovirus variants selected on mutant receptor-expressing cells identify capsid residues that expand receptor recognition. J Virol 69: 4823–4829

Couderc T, Barzu T, Horaud F, Crainic R (1990) Poliovirus permissivity and specific receptor expression on human endothelial cells. Virology 174: 95–102

Couderc T, Hogle J, Le Blay H, Horaud F, Blondel B (1993) Molecular characterization of mouse-virulent poliovirus type 1 Mahoney mutants: Involvement of residues of polypeptides VP1 and VP2 located on the inner surface of the capsid protein shell. J Virol 67: 3808–3817

Couderc T, Guédo N, Calvez V, Pelletier I, Hogle J, Colbère-Garapin F, Blondel B (1994) Substitutions in the capsids of poliovirus mutants selected in human neuroblastoma cells confer on the Mahoney type 1 strain a neurovirulent phenotype in mice. J Virol 68: 8386–8391

Enders JF, Weller TH, Robbins FC (1949) Cultivation of the Lansing strain of poliomyelitis virus in cultures of various human embryonic tissues. Science 109: 85–87

Evans DMA, Dunn G, Minor PD, Schild GC, Cann AJ, Stanway G, Almond JW, Currey K, Maizel JV (1985) Increased neurovirulence associated with a single nucleotide change in a noncoding region of the Sabin type 3 poliovaccine genome. Nature 314: 548–550

Faber HK (1956) The evolution of poliomyelitic infection. Pediatrics 17: 278–286

Filman DJ, Syed R, Chow M, Macadam AJ, Minor PD, Hogle JM (1989) Structural factors that control conformational transitions and serotype specificity in type 3 poliovirus. EMBO J 8: 1567–1579

Flamand A, Gagner J-P, Morrison LA, Fields BN (1991) Penetration of the nervous system of suckling mice by mammalian reoviruses. J Virol 65: 123–131

Freistadt MF, Kaplan G, Racaniello VR (1990) Heterogeneous expression of poliovirus receptor-related proteins in human cells and tissues. Mol Cell Biol 10: 5700–5706

Gotoh B, Ogasawara T, Toyoda T, Inocencio NM, Hamaguchi M, Nagai Y (1990) An endoprotease homologous to the blood clotting factor X as a determinant of viral tropism in chick embryo. EMBO J 9: 4189–4195

Hashimoto I, Hagiwara A, Komatsu T (1984) Ultrastructural studies on the pathogenesis of poliomyelitis in monkeys infected with poliovirus. Acta Neuropathol (Berl) 64: 53–60

Holland JJ (1961) Receptor affinities as major determinants of enterovirus tissue tropisms in humans. Virology 15: 312–326

Holland JJ, Hoyer BH (1962) Early stages of enterovirus infection. Cold Spring Harb Symp Quant Biol 27: 101–111

Holland JJ, McLaren LC (1959) The mammalian cell-virus relationship II. Adsorption, reception, and eclipse by HeLa cells. J Exp Med 109: 487–504

Holland JJ, McLaren LC (1961) The location and nature of enterovirus receptors in susceptible cells. J Exp Med 114: 161–171

Holland JJ, Mclaren JC, Syverton JT (1959a) The mammalian cell virus relationship III. Production of infectious poliovirus by non-primate cells exposed to poliovirus ribonucleic acid. Proc Soc Exp Biol Med 100: 843–845

Holland JJ, McLaren JC, Syverton JT (1959b) The mammalian cell virus relationship IV. Infection of naturally insusceptible cells with enterovirus ribonucleic acid. J Exp Med 110: 65–80

Horie H, Koike S, Kurata T, Sato-Yoshida Y, Ise I, Ota Y, Abe S, Hioki K, Kato H, Taya C, Nomura T, Hashizume S, Yonekawa H, Nomoto A (1994) Transgenic mice carrying the human poliovirus receptor: new animal model for study of poliovirus neurovirulence. J Virol 68: 681–688

Horstmann DM (1952) Poliomyelitis virus in blood of orally infected monkeys and chimpanzees. Proc Soc Exp Biol Med 79: 417

Horstmann DM, McCollum RW, Mascola AD (1954) Viremia in human poliomyelitis. J Exp Med 99: 355–369

Hurst EW (1929) The histology of experimental poliomyelitis. J Pathol 32: 457–477

Hurst EW (1936) The newer knowledge of virus diseases of the nervous system: a review and an interpretation. Brain 59: 1–34

Jubelt B, Gallez-Hawkins B, Narayan O, Johnson RT (1980a) Pathogenesis of human poliovirus infection in mice. I Clinical and pathological studies. J Neuropathol Exp Neurol 39: 138–148

Jubelt B, Narayan O, Johnson RT (1980b) Pathogenesis of human poliovirus infection in mice. II. Age-dependency of paralysis. J Neuropathol Exp Neurol 39: 149–158

Kanamitsu M, Kasamaki A, Ogawa M, Kasahara S, Imumura M (1967) Immunofluorescent study on the pathogenesis of oral infection of poliovirus in monkey. Jpn J Med Sci Biol 20: 175–191

Kaplan AS (1955) Comparison of susceptible and resistant cells to infection with poliomyelitis virus. Ann NY Acad Sci 61: 830–839

Kaplan G, Freistadt MS, Racaniello VR (1990) Neutralization of poliovirus by cell receptors expressed in insect cells. J Virol 64: 4697–4702

Kawamura N, Kohara M, Abe S, Komatsu T, Tago K, Arita M, Nomoto A (1989) Determinants in the 5' noncoding region of poliovirus Sabin 1 RNA that influence the attenuation phenotype. J Virol 63: 1302–1309

Koike S, Horie H, Dise I, Okitsu H, Yoshida M, Iizuka N, Takeuthi K, Takegami T, Nomoto A (1990) The poliovirus receptor protein is produced both as membrane-bound and secreted forms. EMBO J 9: 3217–3224

Koike S, Taya C, Kurata T, Abe S, Ise I, Yonekawa H, Nomoto A (1991) Transgenic mice susceptible to poliovirus. Proc Natl Acad Sci USA 88: 951–955

Koike S, Horie H, Sato Y, Ise I, Taya C, Nomura T, Yoshioka I, Yonekawa H, Nomoto A (1993) Poliovirus-sensitive transgenic mice as a new animal model. Dev Biol Stand 78: 101–107

Kunin CM, Jordan WS (1961) In vitro adsorption of poliovirus by noncultured tissues. Effect of species, age and malignancy. Am J Hyg 73: 245–257

La Monica N, Meriam C, Racaniello VR (1986) Mapping of sequences required for mouse neuro-virulence of poliovirus type 2 Lansing. J Virol 57: 515–525

La Monica N, Almond JW, Racaniello VR (1987) A mouse model for poliovirus neurovirulence identifies mutations that attenuate the virus for humans. J Virol 61: 2917–2920

Landsteiner K, Popper E (1908) Mikroscopische präparate von einem menschlichen und zwei affentückermarker Wien Klin Wochenschr 21: 1930

Levine B, Hardwick JM, Trapp BD, Crawford TO, Bollinger RC, Griffin DE (1991) Antibody-mediated clearance of alphavirus infection from neurons. Science 254: 856–859

Li CP, Schaeffer M (1953) Adaptation of type 1 poliomyelitis virus to mice. Proc Soc Exp Biol Med 82: 477–481

Lu H-H, Yang C -F, Murdin AD, Klein MH, Harber JJ, Kew OM, Wimmer E (1994) Mouse neurovirulence determinants of poliovirus type 1 strain LS-a map to the coding regions of capsid protein VP1 and proteinase 2A^{pro}. J Virol 68: 7507–7515

Macadam AJ, Arnold C, Howlett J, John A, Marsden S, Taffs F, Reeve P, Hamada N, Wareham K, Almond J, Cammack N, Minor PD (1989) Reversion of the attenuated and temperature-sensitive phenotypes of the Sabin type 3 strain of poliovirus in vaccinees. Virology 172: 408–414

Macadam AJ, Ferguson G, Arnold C, Minor PD (1991a) An assembly defect as a result of an attenuating mutation in the capsid proteins of poliovirus type 3 vaccine strain. J Virol 65: 5225–5231

Macadam AJ, Pollard SJ, Ferguson G, Dunn G, Skuce R, Almond JW, Minor PD (1991b) The 5' noncoding region of the type 2 poliovirus vaccine strain contains determinants of attenuation and temperature sensitivity. Virology 181: 451–458

Macadam AJ, Pollard SR, Ferguson G, Skuce R, Wood D, Almond JW, Minor PD (1993) Genetic basis of attenuation of the Sabin type 2 vaccine strain of poliovirus in primates. Virology 192: 18–26

Maddon PJ, Dalgleish AG, McDougal JS, Clapham PR, Weiss RA, Axel R (1986) The T4 gene encodes the AIDS virus receptor and is expressed in the immune system and the brain. Cell 47: 333–348

Martin A, Wychowski C, Couderc T, Crainic R, Hogle J, Girard M (1988) Engineering a poliovirus type 2 antigenic site on a type 1 capsid results in a chimaeric virus which is neurovirulent for mice. EMBO J 7: 2839–2847

Melnick JL (1985) Enteroviruses: polioviruses, coxsackieviruses, echoviruses and newer entero-viruses. In: Fields BN, Knipe DM, Chanock RM, Melnick JL, Roizman B, Shope RE, (eds) "Virology". Raven, New York, pp 705–738

Melnick JL (1990) Enteroviruses: polioviruses, coxsackieviruses, echoviruses and newer entero-viruses. In: Fields BN, Knipe DM, Chanock, Hirsch MS, Melnick JL, Monath TP, Roizman B (eds) "Virology". Raven, New York, pp 549–605

Mendelsohn C, Wimmer E, Racaniello VR (1989) Cellular receptor for poliovirus: molecular cloning, nucleotide sequence and expression of a new member of the immunoglobulin superfamily. Cell 56: 855–865

Minor PD, Dunn G (1988) The effect of sequences in the 5'-noncoding region on the replication of polioviruses in the human gut. J Gen Virol 69: 1091–1096

Minor PD, Dunn G, Evans DMA, Magrath DI, John A, Howlett J, Phillips A, Westrop G, Wareham K, Almond JW, Hogle JM (1989) The temperature-sensitivity of the Sabin type 3 vaccine strain of poliovirus: Molecular and structural effects of a mutation in the capsid protein VP3. J Gen Virol 70: 1117–1123

Minor PD, Macadam AJ, Stone DM, Almond JW (1993) Genetic basis of attenuation of the Sabin oral poliovirus vaccines. Biologicals 21: 357–364

Morrison LA, Sidman RL, Fields BN (1991) Direct spread of reovirus from the intestinal lumen to the central nervous system through vagal autonomic nerve fibers. Proc Natl Acad Sci USA 88: 3852–3856

Moss EG, Racaniello VR (1991) Host range determinants located on the interior of the poliovirus capsid. EMBO J 5: 1067–1074

Murray MG, Bradley J, Yang XF, Wimmer E, Moss EG, Racaniello VR (1988) Poliovirus host range is determined a short amino acid sequence in neutralization antigenic site I. Science 241: 213–215

Nathanson N, Bodian D (1961) Experimental poliomyelitis following intramuscular virus infection. 1. The effect of neural block on a neurotropic and a pantropic strain. Bull Johns Hopkins Hosp 108: 308–319

Nathanson N, Langmuir A (1963) The Cutter incident: poliomyelitis following formaldehyde-inactivated poliovirus vaccination in the United States during the spring of 1955. III. Comparison of the clinical character of vaccinated and contact cases occurring after use of high rate lots of Cutter vaccine. Am J Hyg 78: 61–81

Nkowane B, Wassilak S, Orenstein W, Bart K, Schonberger L, Hinman A, Kew O (1987) Vaccine-associated paralytic poliomyelitis United States: 1973 through 1984. JAMA 257: 1335–1340

Omata T, Kohara M, Kuge S, Komatsu T, Abe S, Semler BL, Kameda A, Itoh H, Arita M, Wimmer E, Nomoto A (1986) Genetic analysis of the attenuation phenotype of poliovirus type 1. J Virol 58: 348–358

Palmiter RD (1986) Germ-line transformation of mice. Annu Rev Genet 20: 456–499

Pollard SR, Dunn G, Cammack N, Minor PD, Almond JW (1989) Nucleotide sequence of a neurovirulent variant of the type 2 oral poliovirus vaccine. J Virol 63: 4949–4951

Racaniello VR (1995) Early events in infection: receptor binding and cell entry. In: Rotbart HA (ed) "Human enterovirus infections". American Society for Microbiology, Washington DC, pp 73–93

Racaniello VR, Baltimore D (1981) Cloned poliovirus complementary DNA is infectious in mammalian cells. Science 214: 916–919

Ren R (1992) Development and characterization of a transgenic mouse model for poliomyelitis. Columbia University

Ren R, Racaniello V (1992a) Human poliovirus receptor gene expression and poliovirus tissue tropism in transgenic mice. J Virol 66: 296–304

Ren R, Racaniello VR (1992b) Poliovirus spreads from muscle to the central nervous system by neural pathways. J Infect Dis 166: 635–654

Ren R, Costantini FC, Gorgacz EJ, Lee JJ, Racaniello VR (1990) Transgenic mice expressing a human poliovirus receptor: a new model for poliomyelitis. Cell 63: 353–362

Ren R, Moss EG, Racaniello VR (1991) Identification of two determinants that attenuate vaccine-related type 2 poliovirus. J Virol 65: 1377–1382

Sabin AB (1956) Pathogenesis of poliomyelitis: reappraisal in light of new data. Science 123: 1151–1157

Sabin AB (1957) Properties of attenuated polioviruses and their behavior in human beings. In: Rivers TM (ed) "Cellular biology, nucleic acids and viruses". New York Academy of Science, New York, pp 113–133

Sabin AB, Boulger LR (1973) History of Sabin attenuated poliovirus oral live vaccine strains. J Biol Stand 1: 115–118

Sabin AB, Ward R (1941) The natural history of human poliomyelitis. I. Distribution of virus in nervous and non-nervous tissues. J Exp Med 73: 771–793

Sabin AB, Hennessen WA, Winsser J (1954) Studies on variants of poliomyelitis virus: I. Experimental segregation and properties of avirulent variants of three immunologic types. J Exp med 9: 551–576

Sicinski P, Rowinski J, Warchol JB, Jarzabek Z, Gut W, Szczygiel B, Bielecki K, Koch G (1990) Poliovirus type 1 enters the human host through intestinal M cells. Gastroenterology 98: 56–58

Svitkin YV, Cammack N, Minor PD, Almond JW (1990) Translation deficiency of the Sabin type 3 poliovirus genome: Association with an attenuating mutation C472-U. Virology 175: 103–109

Tardy-Panit M, Blondel B, Martin A, Tekaia F, Horaud F, Delpeyroux F (1993) A mutation in the RNA polymerase of poliovirus type 1 contributes to attenuation in mice. J Virol 67: 4630–4638

Tatem JM, Weeks-Levy C, Georgiu A, DiMichele SJ, Gorgacz EJ, Racaniello VR, Cano FR (1992) A mutation present in the amino terminus of Sabin 3 poliovirus VP1 protein is attenuating. J Virol 66: 3194–3197

Theiler M (1941) Studies on poliomyelitis. Medicine 20: 443–462

Tyler KL, McPhee DA, Fields BN (1986) Distinct pathways of viral spread in the host determined by reovirus S1 gene segment. Science 233: 770–774

Tyler KL, Virgin IVth HW, Bassel-Duby R, Fields BN (1989) Antibody inhibits defined stages in the pathogenesis of reovirus serotype 3 infection of the central nervous system. J Exp Med 170: 887–900

Ward NA, Milstein JB, Hull HF, Hull BP, Kim-Farley RJ (1993) The WHO-EPI initiative for the global eradication of poliomyelitis. Biologicals 21: 327–333

Webster RG, Rott R (1987) Influenza virus pathogenicity: the pivotal role of hemagglutinin. Cell 50: 665–666

Wenner HA, Kamitsuka P (1956) Further observations on the widespread distribution of polio-myelitis virus in body tissues following intramuscular inoculation of cynomolgous monkeys. Virology 2: 83–95

Wenner HA, Kamitsuka P (1957) Primary sites of virus multiplication following intramuscular inoculation of poliomyelitis virus in cynomolgous monkeys. Virology 3: 429–443

Wenner HA, Rabe EF (1953) The recovery of virus from regional lymph nodes of fatal human cases of poliomyelitis. Am J Med Sci 222: 292–299

Westrop GD, Evans DMA, Minor PD, Magrath D, Schild GC, Almond JW (1987) Investigation of the molecular basis of attenuation in the Sabin type 3 vaccine using novel recombinant polioviruses constructed from infectious cDNA. In: The molecular biology of the positive strand RNA viruses. Rowlands DJ, Mayo MA, Mahy BWJ (eds) Liss, New York, pp 53–60

Westrop GD, Wareham KA, Evans DMA, Dunn G, Minor PD, Magrath DI, Taffs F, Marsden S, Skinner MA, Schild GC, Almond JW (1989) Genetic basis of attenuation of the Sabin type 3 oral poliovirus vaccine. J. Virol. 63: 1338–1344

Wimmer E, Hellen C, Cao X (1993) Genetics of poliovirus. Annu Rev Genet 27: 353–436

Wolf JC, Rubin DH, Finberg R, Kauffman RS, Sharpe AH, Trier JS, Fields BN (1981) Intestinal M Cells: a pathway for entry of reovirus into the host. Science 2: 471–472

Safety Requirements for Maintenance and Distribution of Transgenic Mice Susceptible to Human Viruses: The Example of Poliovirus-Susceptible Transgenic Mice*

Y. Ghendon and P.-H. Lambert

1 Introduction

During the last few years, considerable progress has been achieved in obtaining transgenic mice containing parts of the human genome and susceptible to human viruses. The establishment of such animal models has proven to be extremely useful. Research leading to the development of improved methods for the prevention and control of serious human viral diseases is often limited by the absence of readily available animal models capable of mimicking human interactions with the viral agent. Examples of viral diseases in which efforts have been restricted by the lack of such animal models include the acquired immunodeficiency syndrome and all forms of viral hepatitis and poliomyelitis. In each case, only certain nonhuman primate species have been shown to be susceptible to infection. The nonhuman primates which are required for such experimental work are expensive to acquire and maintain, often limited in availability, or even an endangered species.

World Health Organization, 20 Avenue Appia, 1211 Geneva 27, Switzerland
*This Chapter summarizes conclusions of an informal consultation organized by the World Health Organization, Geneva, 18–19 November 1993, on maintenance of distribution of transgenic mice susceptible to human viruses.

New technologies have emerged over the past decade that may allow the replacement of these nonhuman primate models of human diseases with novel transgenic animals which express selected human genes. Typically, the foreign gene (transgene) is incorporated into the chromosomal DNA of the transgenic animal. Selective breeding results in the availability of both homozygous and heterozygous animals. Where as common laboratory animals such as mice are naturally nonpermissive for infection with a pathogenic virus because of the lack of an essential human protein (for example, the cellular receptor for a virus), lines of transgenic animals expressing the relevant human gene may be susceptible to infection. This represents a powerful technology which offers great promise of advancing knowledge related to the prevention or treatment of certain infectious diseases.

There is no better example of the power of this technology than that offered by the development, within independent laboratories located in Japan (KOIKE et al. 1991) and the United States (REN et al. 1990), of lines of transgenic mice expressing the human cellular poliovirus receptor (PVR). The data obtained in these two laboratories show that these mice are susceptible to infection with all three serotypes of poliovirus by a variety of inoculation routes. Infections in the mice appear to mimic many features of poliovirus infection in humans. A considerable body of new and potentially useful information concerning the pathogenesis of poliomyelitis has thus emerged from these studies. Equally important, the attenuated and neurovirulent phenotypes of the Sabin vaccine and wild-type strains of poliovirus, respectively, are preserved in the transgenic mice. Although further experience is required, it is not unlikely that these transgenic mice may eventually replace nonhuman primates in the safety testing of attenuated poliovirus vaccines. In addition, the replacement of primates by transgenic mice in biomedical research has important ethical implications. The saving of primates is a most welcome contribution to applied animal protection.

At a theoretical level, however, the development of transgenic mice which are permissive for replication of pathogenic human viruses may pose special public health hazards. In the case of the PVR-positive transgenic mice in particular, the unique opportunities for development of better disease control measures far outweigh any theoretical risks these animal lines may pose. Nonetheless, careful consideration must be given to these potential hazards, and measures must be taken to ensure that they are eliminated to the greatest extent possible.

2 Potential Risks to Public Health Posed by Transgenic Animals

The risks posed by transgenic animals that have been constructed to support the replication of pathogenic human viruses include: (1) the possibility that infected, laboratory-maintained transgenic animals may shed and transmit a pathogenic virus to susceptible humans, e.g., recently established transgenic mice containing a complete genome of human immunodeficiency virus or hepatitis B virus (ABRAMCZUK et al. 1992; JOLICOEUR et al. 1992; ARAKI et al. 1991; CHOO et al. 1991), or (2) the possibility that the transgenic animals might escape from the laboratory and that the transgene might become established within wild animal populations, potentially leading to the creation of a new animal reservoir for a pathogenic human virus, e.g., transgenic mice expressing human cellular PVR and thus susceptible to this virus.

In evaluating these risks, the epidemiological characteristics of a particular virus and the conditions under which naturally infected humans may transmit the virus to others need to be carefully considered. For example, if transgenic mice expressing the PVR were capable of shedding substantial quantities of this virus into the environment, as infected humans do, there would be a significant risk of transmission, as this virus is relatively stable in the environment. However, in making comparisons between the epidemiology of human infections and the potential transmissibility of virus from infected transgenic animals, attention also must be paid to the possibility that expression of the transgene may alter the host cell tropism of the virus and possibly change its transmission patterns. Thus, in order to arrive at an informed estimate of the infectious risks posed by such animals, it is essential to understand the pathobiology of the relevant agent in the transgenic animal. It is also necessary to carry out quantitative studies of susceptibility by different routes of infection and to determine the source and magnitude of any virus released into the environment by infected animals.

Similar considerations hold for the risks posed by the potential establishment of the transgene in the wild animal gene pool. If transgenic animals were capable of efficient transmission of the infectious agent, establishment of the gene for the human PVR in the wild mouse population could have potentially serious consequences. In particular, the presence of a wild animal reservoir for poliovirus could have an adverse impact on current efforts to achieve global eradication of wild-type polioviruses.

3 Transgenic Mice Susceptible to Poliovirus Infection

Studies done in Japan by KOIKE et al. (1991, 1993) show that all lines of transgenic mice expressing the human cellular PVR (TgPVR mice) established in this laboratory develop paralysis after inoculation with virulent polioviruses by the intraspinal, intracerebral, intraperitoneal, intravenous and oral routes. With all routes of inoculation, poliovirus enters the central nervous system (CNS) where it replicates in neurons. Doses of 10^2 PFU of the virulent strain Mahoney of poliovirus type 1 inoculated intracerebrally cause death of TgPVR mice, while a dose of 10^6 PFU of the Sabin vaccine strain of poliovirus type 1 was required to cause paralysis and death of TgPVR mice. Similar observations were obtained when Sabin vaccine poliovirus strains type 2 and 3 were compared with virulent poliovirus strains Lansing type 2 and Leon type 3, respectively. Compared to other routes of inoculation to one TgPVR mouse line which required 10^1-10^4 PFU of virulent type 1 poliovirus strain Mahoney, approximately 10^7 PFU are needed for induction of paralysis by the oral route of infection, whereas about $10^{8.5}$ PFU are needed for induction of paralysis and death by the oral route of infection with the Sabin strain poliovirus type 1. Thus TgPVR mice show different sensitivities to virulent and attenuated strains of poliovirus with oral inoculation and other routes of infection.

To determine whether poliovirus replicates in the gut of TgPVR mice, animals were fed light-sensitive poliovirus type 1 strain Mahoney or strain Sabin. Light resistant virus was identified in the feces, indicating that the virus replicated in the transgenic animals. Some TgPVR mice excreted up to 1000 PFU daily in the feces, until death occurred. Other TgPVR mice excreted virus only during the first 2 days after inoculation. These results suggest that poliovirus replicates at low levels in the gut of TgPVR mice. Surprisingly, continuous excretion of wild-type Mahoney strain poliovirus type 1 was also observed in one nontransgenic control mouse. Further studies are required to determine whether this virus is a mutant selected for its ability to replicate in the mouse alimentary tract.

Oral infection of TgPVR mice with 10^7 PFU of poliovirus type 1 strain Sabin did not elicit neutralizing antibodies. Only a low level of antibodies was elicited after a booster immunization with the same dose of this vaccine strain. In some mice, infection with 3×10^8 PFU induced low levels of antibodies in both TgPVR and nontransgenic mice. This level of antibodies, however, did not protect against challenge with poliovirus type 1 strain Mahoney. These results indicate that poliovirus replication in the alimentary tract of TgPVR mice is not sufficient to induce protection against challenge with virulent poliovirus.

The ability of poliovirus to spread within colonies of TgPVR mice was determined by placing orally infected and uninfected transgenic mice in the same cage. Observation of seven mixed cages suggested that poliovirus is not efficiently transferred from infected to uninfected mice, probably because of the small amount of poliovirus excreted and the low efficiency of oral infection.

In the USA (REN et al. 1990) TgPVR mice were established using PVR genomic DNA from Hela cells. TgPVR mice developed poliomyelitis after inoculation with virulent polioviruses by intracerebral, intraperitoneal, intramuscular and intravenous routes. No paralysis was observed after intracerebral inoculation with high levels of the three Sabin vaccine strains of polio virus. In contrast, between 4×10^3 and 1×10^5 PFU of virulent strains such as poliovirus type 1 strain Mahoney, poliovirus type 2 strain Lansing or poliovirus 3 strain Leon caused paralysis of 50% of mice. Despite widespread expression of PVR mRNA in transgenic mouse tissues, poliovirus replication largely occurred in the brain, spinal cord, and skeletal muscle. These lines of TgPVR mice were also susceptible to infection by the oral route, although development of disease by this route of infection was strongly age-dependent: 1-day old mice were 100% susceptible after infection with 10^8 PFU of virulent poliovirus type 1 strain Mahoney, while 7–14 day old mice were refractory to oral infection.

Despite susceptibility of young TgPVR mice to oral infection, no replication was detected in the alimentary tract after inoculation with large quantities of poliovirus. After oral administration of virus to 1-day old TgPVR mice, infectious virus was detected in the feces for 1–2 days after inoculation, but not thereafter. In addition, viral RNA replication in cells of the intestine was not detected by in situ hybridization. In all TgPVR lines examined, PVR mRNA expression in the intestine was very low. It is therefore likely that these lines of TgPVR mice do not support poliovirus replication in the gut due to low levels of receptor expression. The mechanism by which orally administered poliovirus induces paralysis in 1-day old animals remains to be determined.

In summary, lines of TgPVR mice isolated in two different laboratories are susceptible to poliovirus infection with polioviruses by a variety of inoculation routes. While the resulting disease is clinically and histopathologically similar to human poliomyelitis, the mouse model differs from humans in that alimentary tract replication of orally administered poliovirus is either inefficient or does not occur. TgPVR mice demonstrate profound differences in terms of their susceptibility to neurovirulent (wild-type) and attenuated (vaccine) strains of poliovirus. Detailed information about TgPVR mice susceptible to polioviruses is presented in the chapter written by Dr. V. RACANIELLO (this volume).

4 Potential Applications of Poliovirus-Susceptible Transgenic Mice

4.1 Neurovirulence Testing of Poliovirus Isolates

The diagnosis of poliomyelitis and the identification of the actual etiological agent of neurological disease may require testing of the neurovirulence of individual virus isolates. The approach is particularly important in the virological diagnosis of paralytic cases which are allegedly vaccine associated.

TgPVR mice provide a good model for neurovirulence testing of poliovirus isolates. It was possible to differentiate among a wide range of neurovirulence levels in the TgPVR mice inoculated by either intracerebral or intraperitoneal routes with polioviruses, having high and low levels of neurovirulence in the monkey test (REN et al. 1990; KOIKE et al. 1991, 1993). However, more studies are necessary to establish the sensitivity of TgPVR mice to a large number of poliovirus strains of different origins.

4.2 Oral Poliomyelitis Vaccine—Safety Testing

TgPVR mice should be considered likely to provide a good model for in vivo safety testing of Oral Poliomyelitis Vaccine (OPV). Studies carried out so far are encouraging, demonstrating the capacity of the TgPVR mouse tests to distinguish different levels of neurovirulence over a wide range (LEVENBOOK, personal communication). At present, however, the data are not sufficient to assess the ability of TgPVR mice to differentiate among OPV batches which either passed or failed the monkey neurovirulence test. More studies are thus necessary and this could be accomplished either by genetically modifying the sensitivity of TgPVR mice to poliovirus infection or by seeking optimal experimental conditions (route of inoculation, dose of virus, method for evaluation of the level of neurovirulence).

5 Estimate of the Risk to Public Health Posed by Poliovirus Receptor Transgenic Mice

Although further experiments are needed to better define the natural history of poliovirus infections in TgPVR transgenic mice, it was considered very unlikely that escape of these transgenic mice to the wild would result in the establishment of a new animal reservoir for poliovirus. Nevertheless, given the present state of knowledge, this possibility cannot be completely excluded for the following reasons:

1. During breeding of TgPVR mice, the human cellular PVR is transmitted in a Mendelian dominant mode of inheritance.
2. TgPVR mouse strains may breed readily with other strains of mice.
3. TgPVR mice have no lethal mutations that would prevent their spread in the wild.
4. TgPVR mice are susceptible to oral infection when challenged with high titers of poliovirus.
5. Limited fecal excretion has been observed in some lines of TgPVR mice.
6. Since virus titers in the CNS and muscle tissue of infected TgPVR mice are high, infection might be transmitted to other mice by biting and cannibalism.
7. Wild poliovirus may be isolated from human sewage in some cases at titers upto 10^7, and exposure of mice to human sewage may occur in the wild.
8. Adaptation of poliovirus to more efficient replication in mice is possible withpassage of the virus.
9. Some TgPVR mouse lines do not appear to mount an efficient antibody response to poliovirus following infection, making detection of infected animals difficult.

Nonetheless, several lines of evidence argue strongly against the possibility that PVR-positive transgenic mice could serve as a reservoir for the virus in the wild:

1. The quantities of virus shed in the feces of infected transgenic mice created in Dr. Nomoto's laboratory are 10 000 –1 000 000 times less than required for infection and paralysis in the mice by the oral route (KOIKE et al. 1993). No fecal shedding has been noted from infected transgenic animals created in Dr. Racaniello's laboratory (REN et al. 1990). Further experiments are required

to better define differences in the amount of virus required to establish infection (rather than induce paralysis) in non-transgenic and transgenic mice.

2. In one experiment, a normal nontransgenic mouse was shown to be infected following oral challenge with a high titered poliovirus inoculum. This mouse shed virus for a number of days (NOMOTO, personal communication). Despite this observation, there is no evidence that normal mice act as reservoirs for poliovirus in nature.

3. Limited studies completed thus far indicate that the poliovirus-infected transgenic mice developed in Dr. Nomoto's laboratory do not efficiently transmit virus to uninfected, transgenic cagemates, but in some cases poliovirus transmission occurs from infected to uninfected TgPVR mice. Further studies are required, however, to rule out a low level of transmission.

While the risk posed by escape of TgPVR mice thus appears to be very low, it is essential that practices be adopted which will minimize the possibility of escape of those animals, yet will not interfere with the further development of these very useful animal models for studies of poliovirus pathogenicity, vaccine safety and poliovirus field surveillance.

6 General Recommendations Concerning the Maintenance, Containment and Transport of Transgenic Animals Susceptible to Pathogenic Human Viruses

At the 1993 WHO meeting on the maintenance and distribution of transgenic mice susceptible to human viruses, several recommendations were made:

1. For each line of transgenic animals shown to be uniquely susceptible to a pathogenic human virus, detailed studies should be conducted to determine the natural history of the virus infection in the transgenic animal, including the routes by which the animal is susceptible to infection, the inoculum size required for infection, and the nature and extent of virus shedding by infected animals.

2. For each line of transgenic animals shown to be uniquely susceptible to a pathogenic human virus, a registry should be maintained by the developer of the line which includes detailed information concerning each of the physical locations at which the animals are maintained or to which they have been shipped. At each institution maintaining or receiving such animals for experimental studies, an accounting system should be established which faithfully records the birth or receipt and the final disposition of all transgenic animals.

3. Animals should be maintained in clearly designated locked rooms with well defined, restricted access by laboratory workers and animal handlers. Experiments involving infection of transgenic animals should be carried out in a similar limited-access room which is physically separated from rooms used for breeding or maintenance purposes.

4. Laboratory practices involving a level of biosafety which is generally accepted as appropriate for work with specific pathogenic human viruses and generally accepted standards for the care and use of laboratory animals in biomedical investigation (CIMS 1985; WHO/ICLAS 1992) should be rigorously enforced in laboratories engaged in the development or maintenance of transgenic animals which have unique susceptibility to pathogenic human viruses and in laboratories which utilize these animals for research purposes. Although containment laboratories and closed or barrier systems are primarily developed to keep infectious material either in or out, these systems provide also the most effective methods for containment of transgenic animals. These systems include controlled access for the scientific staff and have been well established in laboratory animal science for many years. Requirements for such high levels of containment should be based upon careful scientific review of the potential risks posed by transgenic animals.

Various laws and regulations, both local and international, specify the general care and management required for the maintenance of mice susceptible to human viruses (Animal Scientific Procedures Act, and others) (Table 1). In some cases, guidelines or regulations concerning Tg mice are also applied (Table 2). It is especially important that Tg mice susceptible to a human virus be prevented from escaping from an animal facility.

5. Where feasible, all laboratory workers and animal handlers who may potentially be exposed to transgenic animals which are susceptible to a human pathogenic virus should receive protective immunization against the specific agent.

6. Transgenic animals must be prevented from escaping during transport, since in some cases they could serve as a new reservoir for the human virus to which

Table 1. Laws and regulations concerning the general care and management of laboratory mice

Animal Scientific Procedures Act (UK)
Animal Welfare Act (USA)
Law Concerning the Protection and Welfare of Animals (Japan)
Tierschutzgesetz (Germany)
The UFAW Handbook on the Care and Management of Laboratory Animals (UFAW, UK)
Guide for the Care and Use of Laboratory Animals (UFAW, UK)
Guide for the Care and Use of Laboratory Animals (ILAR, USA)
Standards Relating to the Care and Management of Experimental Animals (Prime Minister's Office, Japan)
International Guiding Principles for Biomedical Research involving Animals (CIOM)
Laws Regulating Genetic Engineering Issues in the FRG; Society for Laboratory Animal Science: Zur Planung und Structur von
Versuchstierbereichen tierexperimentell tätiger Institutionen 1988
WHO/ICLAS Guidelines for the Establishment of Animal Colonies and Use of Laboratory Animals (WHO, Geneva, in preparation)

Table 2. Guidelines concerning the care and management of transgenic mice

Genetic Manipulation Guidelines on Work with Transgenic Animals (Home Office, UK)
Guidelines for Recombinant DNA Experiments (Ministry of Education, Science and Culture, Japan)
Cornell Directive on the Contained Use of Genetically Modified Microorganisms (USA)
Laws Regulating Genetic Engineering Issues (Germany)

they are susceptible or which they harbor. The general methods for the transport of transgenic mice susceptible to pathogenic viruses are specified in local and international regulations (Table 3).

7. Given that numerous laboratories may wish to have access to transgenic mice susceptible to pathogenic human viruses, careful consideration should be given to the possibility of creating a subline of such transgenic mice which contain a lethal mutation with a phenotype that may be nutritionally or otherwise suppressed under normal laboratory conditions and that would result in rapid death or sterility in the event of escape to the wild. WHO recommend the development of such a modified PVR-positive transgenic mouse line.

Table 4 presents general recommendations concerning methods for maintaining transgenic animals.

7 Specific Recommendations and Principles for Maintaining and Transporting Transgenic Mice Susceptible to Poliomyelitis

1. Due to the potential hazards involved in the maintenance and distribution of TgPVR mice susceptible to poliovirus, it is recommended that the maintenance and use of fertile TgPVR mice be restricted to laboratories which meet all the standards outlined in this report and which are able to provide a level of containment sufficient to exclude escape of any mice.
2. The breeding of TgPVR mice should be restricted to facilities which meet all the standards outlined in this report, insuring that TgPVR mice cannot escape into the environment. Breeding facilities should be completely separate from laboratories in which poliovirus or poliovirus-infected animals are handled. If possible, breeding animals should be maintained in SPF facilities as defined in WHO ICLAS guidelines (Table 3).
3. TgPVR mice to be used in experimental studies or for assessment of vaccine neurovirulence in institutions which do not meet the standards outlined in this report should be castrated before transfer from the breeding laboratory in order to prevent the risk of reproduction of animals inadvertently released

Table 3. Guidelines regarding the transport of transgenic mice

Live Animal Regulations (IATA)
The UFAW Handbook on the Care and Management of Laboratory Animals (UFAW, UK)
Guide for the Care And Use of Laboratory Animals (ILAR, USA)
Standards Relating to the Care and Management of Experimental Animals (Prime Minister's Office, Japan)
International Guiding Principles for Biomedical Research Involving Animals (CIOMS)
WHO/ICLAS Guidelines for the Establishment of Animal Colonies and Use of Laboratory Animals (WHO, Geneva, in preparation)

Table 4. Methods For Maintaining Transgenic Animals

Class	Definition	Minimum Type of Containment
Animals capable of transmitting human pathogens	Transgenic animals containing a complete potentially infectious viral genome	Same level of physical containment required for that pathogens
	Transgenic animals containing a gene conferring susceptibility to a human pathogen	Infected animals should be maintained at a level of containment propriate for that pathogen Levels of containment for uninfected animals should be determined on a case by case basis depending on the epidemiology of the human pathogen and the likelihood that the transgenic mice can contribute to its transmission; level 2 is the minimal recommended level of containment
Other transgenic animals		As required by local regulations[a]

[a] If there are no local regulations, refer to regulations/guidelines noted: WHO/ICLAS Guideline for Maintenance of Animals, WHO, Geneva, 1992 and International Guiding Principles for Biomedical Research Involving Animals, CIOMS, Geneva, 1985.

during this process. Orchidectomy and ovariectomy are easily performed by experienced technicians. Alternative methods of sterilization of TgPVR mice may also be used provided that they guarantee sterilization of TgPVR mice without changing their susceptibility to poliovirus.

4.   Fertile TgPVR mice should not be shipped to any institution unless the receiving institution provides evidence that the animal maintenance facilities fulfill the requirements detailed in this report.

5.   Regarding containment of fertile TgPVR mice, all of the general recommendations in maintenance of transgenic mice susceptible to human pathogens, as detailed above, must be met. In addition, special measures should be taken to prevent escape of fertile TgPVR animals from individual cages and racks. In the event that the mice do escape, there must be a second level of containment so that the mice are not able to escape from the animal room. The following measures are recommended:

5.1. Cages and covers: The cages should be made of hard plastic which are chemical and heat resistant. A recessed cover should be used so that the mice cannot damage the edge of the cage by biting. Fasteners should be attached to the covers.

5.2. Racks: It is recommended that cage racks have sliding doors to prevent mice who have been escaped from the cage from leaving the rack. These racks should be separately ventilated with clean air.

5.3. Rodent barriers: Animals should be maintained in designated locked rooms. There should be a double door entry to the animal room through a screened vestibule. All openings, including drain pipes, should be covered with wire mesh. Movable rodent barriers (minimum 50 cm high) should be provided around the doors of animal rooms. It is recommended that large animal rooms be subdivided by moveable rodent barriers. Mouse traps should be placed in the room and vestibule. All interior walls must be free of cracks and crevices and should be easily cleaned. Ventilation ports should be examined to exclude potential routes of escape. Transport of animals between rooms should be in a sealed box. Isolators are very effective in preventing escape of the animals. Negative pressure isolators are also recommended to prevent the spread of infection.

5.4. Individual TgPVR animals should bear a permanent physical marker.

5.5. A careful animals accounting system must be in place. Daily records concerning individual animals should be maintained. Cage tags should indicate the number and type of transgenic animals and should be in agreement with the laboratory records.

6. Special measures should be taken to prevent escape during transportation of the animals. The containers used for transporting fertile TgPVR mice should be escape-proof. They must be constructed of materials which are microbiologically impermeable. Impact resistant shipping containers for transgenic mice are best set up as a box-in-box system. The outer transport container should be locked (padlock) to prohibit release of the animals by unauthorized persons. It should be clearly indicated on the side of the transport containers that the mice are potentially biohazardous. Animals should not be shipped unless the recipient verifies that the maintenance facilities fulfil the necessary requirements outlined above.

8 Transgenic Mice Containing Parts of or a Complete Genome of Human Viruses

During the last few years a number of transgenic mice containing parts of or a complete genome of human viruses have been developed. These include mice containing *tat* gene (VOGEL et al. 1991) and LTR gene (MORREY et al. 1991; VOGEL et al. 1992) of human immunodeficiency virus 1; tax gene (BILLS et al. 1992) and LTR gene (GONZALEZ-DUNIA et al. 1992) of human Tlymphotropic virus-1, x gene (LEE et al. 1990); genes coding envelope protein (DUNSFORD et al. 1990) of hepatitis B virus; LCR (CHOO et al. 1992) and early gene (TINSLEY et al. 1992) of human papilloma virus; TAg gene of polioma virus (AGUZZI et al. 1990); early region (TAg) of the genome of human papovavirus JE (BEGGS et al. 1990); BNLF-1 oncogene of Epstein-Barr virus (WILSON et al. 1990); early region of the genome of adenovirus type 12 (KOIKE et al. 1990); regulatory genes of the genome of human foamy virus (AGUZZI et al. 1992); early gene 1 of human cytomegalovirus (FURTH et al. 1991); LTR gene of Visna virus (SMALL et al. 1989) and some others.

In some studies, lines of transgenic mice were obtained containing a complete genome of human viruses, e.g., human immunodeficiency virus-1 (ABRAMCZUK et al. 1992; JOLICOEUR et al. 1992); hepatitis B virus (ARAKI et al. 1991; CHOO et al. 1991); and human Tlymphotropic virus-1 (IWAKURA et al. 1991).

In transgenic mice containing a complete genome of human immunodeficiency virus-1, it was found that infectious HIV, indistinguishable from parental virus by immunoblot analysis, was recovered from the spleen, lymph nodes and skin of some of the transgenic mice (LEONARD et al. 1988; ABRAMCZUK et al. 1992) and viral proteins were detected in serum and milk (JOLICOEUR et al. 1992). In sera of transgenic mice containing a complete genome of hepatitis B virus, high titers of HBsAg and HBeAg were found and double and single-stranded DNA of hepatitis B virus were detected in liver (CHOO et al. 1991); 42 nm Dane particles were found in serum which were indistinguishable from particles produced in an infected human liver, and can infect human fetal hepatocytes (ARAKI et al. 1991).

It is clear that more and more transgenic animals containing a complete genome of human viruses will be developed in the future and it was already noted that some of these animals may shed and transmit a pathogenic virus to susceptible humans.

The risks posed by such transgenic animals should be reduced to a minimum. When dealing with such animals, including maintainance and transportation, the same level of physical containment should be used as is required for that pathogen agent which is shed and transmitted by the transgenic animal.

Since humans are the only natural host for several viruses, transgenic animals susceptible to human viruses or containing a complete genome of human viruses can be an excellent tool for studies of human viral diseases. However, one must take into account that such transgenic animals may in principle shed and transmit a pathogenic virus to susceptible humans or establish within wild animal populations a new animal reservoir for a pathogenic human virus. Any manipulation of these animals should be done with special caution to exclude any potentional dangers.

References

Abramczuk JW, Penson DS, Leonard J, Monell-Torrens E, Belcher J, Martin M, Notkins AL (1992) Transgenic mice carrying intact HIV provirus: biological effects and organization of a transgene. J Acquir Innume Defic Syndr 5: 196–203

Aguzzi A, Wagner EF, Williams R, Courtheidge SA (1990) Sympathetic hyperplasia and neuroblastomas in transgenic mice expressing polyoma middle T antigen. New Biol 2: 533–543

Aguzzi A, Bothe K, Anthauser I, Horak I, Rethwilm A, Wagner E (1992) Expression of human foamy virus is differentially regulated during development in transgenic mice. New Biol 4: 225–237

Animals Scientific Procedures Act (UKJ); Animal Welfare Act (USA); Law Concerning the Protection and Control of Animals (Japan); Tier Schutzgestetz (Germany); The UFAW Handbook on the Care and Management of Laboratory Animals (YFAW, UK); Guide for the Care and Use of Laboratory Animals (UFAW, UK); Guide for the Care and Use of Laboratory Animals (ILAR, USA); Standards Relating to the Care and Management of Experimental Animals (Prime Minister's Office, Japan); International Guiding Principles for Biomedical Research Involving Animals (CIOMS); Laws regulating Genetic Engineering Issues in the FRG; Society for Laboratory Animal Science: Zur Planung und Structur von Versuchstierbereichen tierexperimentell tätiger Institutionen 1988; WHO/ICLAS Guidelines for the Establishment of Animal Colonies and Use of Laboratory Animals (WHO, Geneva, in preparation)

Araki K, Nishimura S, Ochiya T, Okybo K, Miyazaki J, Matsubara K, Yamamura K (1991) Production and effect of infectious Dane particles in transgenic mice Jpn J Cancer Res 82: 235–239

Beggs AH, Miner JH, Scangos GA (1990) Cell type-specific expression of JC virus T antigen in primary and established cell lines from transgenic mice. J Gen Virol 71: 151–164

Bills ND, Hinrichs SH, Morgan R, Clifford AJ (1992) Delayed tumor onset in transgenic mice fed a low-folate diet. J Natl Cancer Inst 84: 332–337

Choo KB, Liew LN, Chong KY, Lu RH, Cheng WT (1991) Transgenome transcription and replication in the liver and extrahetatic tissues of a human hepatitis B virus transgenic mouse. Virology 182: 785–792

Choo KB, Chong KY, Liew LN, Hsu HC, Cheng WT (1992) Unregulated and basal transcriptional activities of the regulatory sequence of the type 18 human papillomavirus genome in transgenic mice. Virology 188: 378–383

Dunsford HA, Sell S, Chisari FV (1990) Hepatocarcinogenesis due to chronic liver cell injury in hepatitis B virus transgenic mice. Cancer Res 50: 3400–3407

Genetic Manipulation Guidelines on Work with Transgenic Animals (Home Office, UK); Guidelines for Recombinant DNA Experiments (Ministry of Education, Science and Culture, Japan); Cornell directive on the contained use of genetically modified micro organisms (USA); Laws Regulating Genetic Engineering Issues in the FRG

Gonzalez-Dunia D, Grimber G, Briard P, Brahic M, Ozden S (1992) Tissue expression pattern directed in transgenic mice by the LTR of an HTLV-1 provirus isolated from a case of tropical spastic paraparesis. Virology 187: 705–710

Live Animal Regulations (IATA): The UFAW Handbook on the Care and Management of Laboratory Animals (UFAW, UK): Guide for the Care and Use of Laboratory Animals (ILAR, USA); Standards Relation to the Care and Management of Experimental Animals (Prime Minister's Office, Japan); International Guiding Principles for Biomedical Research Involving Animals (CIOMS); WHO/ICLAS Guidelines for the Establishment of Animal Colonies and Use of Laboratory Animals, Chapter 14 (WHO, Geneva, in preparation)

Furth PA, Hennighausen L, Baker C, Beatty B, Woychick R (1991) The variability in activity of the universally expressed human cytomegalovirus immediate early gene I, enhancer/promoter in transgenic mice. Nucleic Acids Res 19: 6205–6208

International Guiding Principles for Biomedical Research Involving Animals, CIMS, Geneva 1985

Iwakura Y, Tosu M, Yoshida E, Takiguchi M, Sato K, Kitajima I (1991) Induction of inflammatory arthrpathy resembling rheumatoid arthritis in mice transgenic for HTLV-1. Science 253: 1026–1028

Jolicoeur P, Laperriere A, Beaulieu N (1992) Efficient production of human immunodeficiency virus proteins in transgenic mice. J Virol 66: 3904–3908

Koike K, Jay G, Hartley JW, Schrenzel MD, Higgins R, Hinrichs S (1990) Activiation of retrovirus in transgenic mice: association with development of olfactory neuroblastoma. J Virol 64: 3988–3991

Koike S, Taya C, Kurata T, Abe S, Ise I, Yonekawa H, Nomoto A (1991) Transgenic mice susceptible to poliovirus. Proc Natl Acad Sci USA 88: 951–955

Koike S, Horie H (1993) Poliovirus-sensitive transgenic mice as a new animal model. In: Brown F, Lewis BP (eds) Poliovirus Attenuation: Molecular Mechanisms and Practical Aspects. Dev Biol Stand Basel, Karger, 78, pp 101–107

Lee TH, Finegold MJ, Shen R, DeMayo JL, Woo SL, Butel JS (1990) Hepatitis B virus transactivator X protein is not tumorigenic in transgenic mice. J Virol 64: 5939–5947

Leonard JM, Abramczuk JW, Pezen D, Rutledge R, Belcher J, Hakim F (1988) Development of disease and virus recovery in transgenic mice containing proviral DNA. Science 242: 1665–1770

Maintenance and distribution of transgenic mice susceptible to human viruses (1993) WHO Bulletin 71: 497–502

Morrey JD, Bourn SM, Bunch TD, Jackson MK, Sidwell R, Barrows L, Daynes R, Rosen CA (1991) In vivo activation of human immunodeficiency virus type 1 long terminal repeat report by UV type A(UV-A) light plus psoralen and UV-B light in the skin of transgenic mice. J Virol 65: 5045–5051

Ren R, Costantini F, Gorgacz E, Lee JJ, Racaniello VR (1990) Transgenic mice expressing a human poliovirus receptor: a new model for poliomyelitis. Cell 63: 353–362

Small JA, Bieberich C, Chotbi Z, Hess J, Scangos G, Clements JE (1989) The visna virus long terminal repeat directs expression of a reporter gene in activated macrophages, lymphocytes, and the central nervous systems of transgenic mice. Virol 63: 1891–1896

Tinsley JM, Fisher C, Searle PF (1992) Abnormalaties of epidermal differentiation associated with expression of the human papillomavirus type 1 early region in transgenic mice. J Gen Virol 73: 1251–1260

Vogel J, Hinrichs SH, Napolitano L, Ngo L, Jay G (1991) Liver cancer in transgenic mice carrying the human immunodeficiency virus tat gene. Cancer Res 51: 6686–6690
Vogel J, Cepeda M, Tschachler E, Napolitano L, Jay G (1992) UV activation of human immuno-deficiency virus gene expression in transgenic mice. J Virol 66: 1–5
WHO/ICLAS Guidelines for Maintenance of Animals (1992) WHO, Geneva
Wilson JB, Weinbegr W, Jonson R, Yuspa S, Levine A (1990) Expression of the BNLF-1 oncogene of Epstein-Barr virus in the skin of transgenic mice induces hyperplasia and aberrant expression of keratin 6. Cell 61: 1315–1327

Subject Index

Current Topics in Microbiology and Immunology

Volumes published since 1989 (and still available)

Vol. 184: **Dunon, Dominique; Mackay, Charles R.; Imhof, Beat A. (Eds.):** Adhesion in Leukocyte Homing and Differentiation. 1993. 37 figs. IX, 260 pp. ISBN 3-540-56756-9

Vol. 185: **Ramig, Robert F. (Ed.):** Rotaviruses. 1994. 37 figs. X, 380 pp. ISBN 3-540-56761-5

Vol. 186: **zur Hausen, Harald (Ed.):** Human Pathogenic Papillomaviruses. 1994. 37 figs. XIII, 274 pp. ISBN 3-540-57193-0

Vol. 187: **Rupprecht, Charles E.; Dietzschold, Bernhard; Koprowski, Hilary (Eds.):** Lyssaviruses. 1994. 50 figs. IX, 352 pp. ISBN 3-540-57194-9

Vol. 188: **Letvin, Norman L.; Desrosiers, Ronald C. (Eds.):** Simian Immunodeficiency Virus. 1994. 37 figs. X, 240 pp. ISBN 3-540-57274-0

Vol. 189: **Oldstone, Michael B. A. (Ed.):** Cytotoxic T-Lymphocytes in Human Viral and Malaria Infections. 1994. 37 figs. IX, 210 pp. ISBN 3-540-57259-7

Vol. 190: **Koprowski, Hilary; Lipkin, W. Ian (Eds.):** Borna Disease. 1995. 33 figs. IX, 134 pp. ISBN 3-540-57388-7

Vol. 191: **ter Meulen, Volker; Billeter, Martin A. (Eds.):** Measles Virus. 1995. 23 figs. IX, 196 pp. ISBN 3-540-57389-5

Vol. 192: **Dangl, Jeffrey L. (Ed.):** Bacterial Pathogenesis of Plants and Animals. 1994. 41 figs. IX, 343 pp. ISBN 3-540-57391-7

Vol. 193: **Chen, Irvin S. Y.; Koprowski, Hilary; Srinivasan, Alagarsamy; Vogt, Peter K. (Eds.):** Transacting Functions of Human Retroviruses. 1995. 49 figs. IX, 240 pp. ISBN 3-540-57901-X

Vol. 194: **Potter, Michael; Melchers, Fritz (Eds.):** Mechanisms in B-cell Neoplasia. 1995. 152 figs. XXV, 458 pp. ISBN 3-540-58447-1

Vol. 195: **Montecucco, Cesare (Ed.):** Clostridial Neurotoxins. 1995. 28 figs. XI., 278 pp. ISBN 3-540-58452-8

Vol. 196: **Koprowski, Hilary; Maeda, Hiroshi (Eds.):** The Role of Nitric Oxide in Physiology and Pathophysiology. 1995. 21 figs. IX, 90 pp. ISBN 3-540-58214-2

Vol. 197: **Meyer, Peter (Ed.):** Gene Silencing in Higher Plants and Related Phenomena in Other Eukaryotes. 1995. 17 figs. IX, 232 pp. ISBN 3-540-58236-3

Vol. 198: **Griffiths, Gillian M.; Tschopp, Jürg (Eds.):** Pathways for Cytolysis. 1995. 45 figs. IX, 224 pp. ISBN 3-540-58725-X

Vol. 199/I: **Doerfler, Walter; Böhm, Petra (Eds.):** The Molecular Repertoire of Adenoviruses I. 1995. 51 figs. XIII, 280 pp. ISBN 3-540-58828-0

Vol. 199/II: **Doerfler, Walter; Böhm, Petra (Eds.):** The Molecular Repertoire of Adenoviruses II. 1995. 36 figs. XIII, 278 pp. ISBN 3-540-58829-9

Vol. 199/III: **Doerfler, Walter; Böhm, Petra (Eds.):** The Molecular Repertoire of Adenoviruses III. 1995. 51 figs. XIII, 310 pp. ISBN 3-540-58987-2

Vol. 200: **Kroemer, Guido; Martinez-A., Carlos (Eds.):** Apoptosis in Immunology. 1995. 14 figs. XI, 242 pp. ISBN 3-540-58756-X

Vol. 201: **Kosco-Vilbois, Marie H. (Ed.):** An Antigen Depository of the Immune System: Follicular Dendritic Cells. 1995. 39 figs. IX, 209 pp. ISBN 3-540-59013-7

Vol. 202: **Oldstone, Michael B. A.; Vitković, Ljubiša (Eds.):** HIV and Dementia. 1995. 40 figs. XIII, 279 pp. ISBN 3-540-59117-6

Vol. 203: **Sarnow, Peter (Ed.):** Cap-Independent Translation. 1995. 31 figs. XI, 183 pp. ISBN 3-540-59121-4

Vol. 204: **Saedler, Heinz; Gierl, Alfons (Eds.):** Transposable Elements. 1995. 42 figs. IX, 234 pp. ISBN 3-540-59342-X

Vol. 205: **Littman, Dan. R. (Ed.):** The CD4 Molecule. 1995. 29 figs. XIII, 182 pp. ISBN 3-540-59344-6